DEAN SEWELL, PHILIP WATKINS
AND MURRAY GRIFFIN

SPORT AND EXERCISE SCIENCE

AN INTRODUCTION

Second Edition

Routledge
Taylor & Francis Group

LONDON AND NEW YORK

Dean Sewell

The making of the editions of this book has not simply been the coming together and commitment of three authors, but also the hard work and patience of a publishing team, including editors and artists. I would like to thank them for making the book a reality. The knowledge and understanding contained in this book has been gained by scientists over many decades, so it is to the scientific community that we should be grateful. I would also like to thank the excellent physiologists that I have had the privilege to meet and work with (EH being one of them, who passed away in 2011), my academic friends, colleagues and students, past and present for their inspiration and patience with me as a teacher and a researcher. And time with one's family is always compromised in completing these endeavours, so to Victoria, Amy and Jack I am thankful for your patience and support, too.

Philip Watkins

To my parents and to Heather.

Murray Griffin

I would like to thank my colleagues for their forbearance, my students for their comments, and my son Jack who, if he was able to speak, would probably tell me to stop writing and come and play some music.

First published 2005 by Hodder Arnold

Second edition published 2012 by Hodder Education

Published 2013 by Routledge

2 Park Square, Milton Park, Abingdon, Oxfordshire OX14 4RN

711 Third Avenue, New York, NY, 10017, USA

First issued in hardback 2015

Routledge is an imprint of the Taylor & Francis Group, an informa business

British Library Cataloguing in Publication Data

A catalogue record for this book is available from the British Library

ISBN 13: 978-1-138-12898-9 (hbk)
ISBN 13: 978-1-4441-4417-8 (pbk)

Typeset by Integra Software Services Pvt. Ltd., Pondicherry, India

CONTENTS

Contents

Contents

INTRODUCTION

Ever since humans have had a competitive edge, the need to improve performance and athletic ability has been an area of intense study. The last decade in particular has seen significant growth in course and student numbers in the discipline of sport and exercise science.

How can one define sport and exercise science? Sport and exercise science uses the scientific principles from the mainstream sciences, and applies them to the sport and exercise environment in order to, for example, improve performance or health, reduce injury or increase motivation. Sport science finds its ideal home firmly within the world of elite and professional sport. World-class athletes such as Usain Bolt and Paula Radcliffe, as talented as they are, would find it very difficult to achieve their significant levels of sporting excellence without the knowledge and support offered by sport science and sports medicine. Individual training programmes are designed so these athletes can train at the correct intensities, in the right frame of mind, recover appropriately, take in the correct nutrition at the optimum times, and produce their best performances when it matters. Producing top-level performances in modern sport, whether on the track, on the pitch or in the pool, requires an enormous team effort both on and off the field of play. National governing bodies (NGOs) such as the Football Association employ sport scientists (physiology, biomechanics, psychology), physiotherapists, doctors, strength and conditioning coaches and sports nutritionists to support the coaching staff and players in the quest for sporting excellence.

And it is not just in elite level sport where knowledge of sport and exercise science can be beneficial. Many employment opportunities exist in such areas as teaching, lecturing, research, coaching, fitness instructing, sports management and sports development (to name but a few). It is, of course, equally rewarding to study such an exciting and diverse subject area for its own sake at both undergraduate and postgraduate levels.

Historically in the UK, the sports-related courses of the early 1980s were based predominantly at the older and more traditional teacher-training establishments, the roots of which lay firmly in physical education. As a result, these courses were driven by a high practical sports element. Over time, courses became more laboratory-based as sport and exercise science courses were established in science divisions to counter the drop in student numbers on science-orientated courses. The foundations of sport and exercise science lie in the disciplines of physiology, biomechanics and psychology and so the science-based courses had a strong bias towards these subject areas.

However, institutions are now offering specialist pathways, modules and degree programmes in areas like sports injuries, sports therapy, strength and conditioning, sports coaching, nutrition for sport, exercise and health and sports medicine. There has also been a shift from sport to exercise science. This is not surprising in light of successive government drives to promote public health through physical activity, in an attempt to create a healthier nation and, consequently, reduce the burden on the National Health Service. Furthermore, the 2012 Olympic bid was partly built on the legacy of increased participation in sport and a more active population. Whether that is realistic is yet to be seen.

The first edition of **Sport and Exercise Science: An Introduction** was a groundbreaking new text and this second edition continues to fill a significant gap in the market, covering the three key areas of Anatomy and Physiology, Biomechanics and Psychology required for early undergraduate study and beyond. Without a firm grounding in these disciplines at first-year level, progression cannot be made to the more specialist and applied subject areas usually available to students in their Middle and Honours years of their degree programmes. The development in schools during the late 1980s and early 1990s of exam-based A-Level Physical Education also forced some Higher Education establishments to increase the academic content of sports-related courses. The market has became swamped with various advanced level texts for sixth-formers studying physical education and sports-related courses but, until now, there has been little for undergraduates covering all three disciplines at the appropriate level.

In keeping with the above, the text has been divided into three clearly defined sections: Anatomy and Physiology, Biomechanics and Psychology. Each section is divided into separate chapters covering a particular theme or subject. Every chapter employs accessible and easy-to-follow examples, relevant diagrams and photographs from the sport and exercise arena to highlight specific principles that are grounded in academic theory. Time-out features are used to help bridge the gap between the practical and the theoretical, and to introduce the reader to recent experiments and developments within the field. Summaries, further reading suggestions and review questions provide the student with an excellent opportunity to extend their knowledge and promote critical thinking. Accompanying this textbook is a fully interactive, comprehensive and free companion website http://cw.tandf.co.uk/sport/ that complements and enhances student learning.

Sport and exercise science will continue to grow as a highly popular subject area to study at university. As these courses become more specialized and more prolific, and course titles change to satisfy the demands of employers or government policy (such as widening participation), **Sport and Exercise Science: An Introduction** will remain an excellent introductory core text for any student studying sport and exercise in Higher Education.

Dean Sewell
Philip Watkins
Murray Griffin

Guide to companion website

To support students in their learning, there is a companion website for this second edition of Sport & *Exercise Science*, providing free access to a range of invaluable resources such as:

- interactive multiple-choice questions to test students' progress
- revision questions and answers that reinforce students' understanding
- audio files that summarise key topics
- animations that demonstrate complex processes of exercise physiology and biomechanics
- glossary that explains key sport and exercise terminology clearly.

Terms in bold within the book can be found within the online glossary. To access the glossary and many other resources, go to http://cw.tandf.co.uk/sport/.

I Anatomy and Physiology

1 Sport and Exercise in the Life Sciences

Chapter Objectives

In this chapter you will learn about:

- The genomic era, 60 years after the structure of DNA was revealed.
- The feats of scientific endeavour and human performance that have advanced considerably and are set to continue to do so.
- The basic structure of the cell and its function in relation to exercise and health.
- The basic features of DNA structure and information transfer.
- The role of reactions and enzymes in metabolism and the basic function of molecules and macromolecules.

Introduction

Exercise scientists joke (although there is a hint of truth in everything we say) that to succeed in sport you are well advised to choose your parents carefully. Each individual is significantly influenced by their own genetic make-up (genotype), as well as by environmental factors such as living conditions during growth and development – genotype plus environmental factors produces our phenotype.

In this chapter we will first consider where sciences such as human anatomy and physiology of exercise fit in the bigger picture of life, the universe and everything. The biology of exercise resides in the domain of the life sciences, which encompasses a diverse range of scientific fields such as psychology, medicine, marine biology and environmental biology. How might our pursuit of an increasing knowledge and practice base in human anatomy and the physiology of exercise relate to other branches of the life sciences?

We can celebrate two important events at the forefront of endeavour in sport and exercise, and in science that occurred around 60 years ago. These were the feats of reaching the highest point on earth (Everest) for the first time and the discovery of the double helical structure of DNA (Figure 1.1) and, thus, the first precisely defined amino acid sequence of a protein.

Reaching the top of Sagarmartha (the Nepalese name for Mount Everest) cannot be achieved without considerable physiological, psychological and biomechanical endurance. Whilst not being technically difficult to climb, it requires the support of a highly skilled team, including members trained in the life sciences. Multi-tasking and multidisciplinarity are fundamental for any expeditionary and scientific team. The environment is extreme, and respect for this and for the people, flora and fauna and the health economy of the Himalaya are important factors. The support of the indigenous people is also essential. They are genetically adapted, both physiologically and psychologically, to survival at higher altitudes. This is exemplified by the new record set in 2004 by the 26-year-old Sherpa, Pemba Dorjee, who reached the summit of Everest, from base camp, in just over eight hours, a journey that is normally scheduled over four days, with overnight recovery periods.

Modern day teams of Western climbers need to include medical, paramedical, physiological, nutritional and psychological support in order to achieve success. The physiologist on the 1953 Everest Expedition was Griffith Pugh, who had accompanied Eric Shipton to the Himalaya in 1952 and wrote a report from which a number of useful lessons could be learnt. In his account of the ascent, Hunt (1953) notes 'This climbing party was further enlarged by the attachment of two others . . . . The first was Griffith Pugh, a physiologist employed in the Division of Human Physiology of the M.R.C., who had a long experience of what may be termed mountain physiology.' There was considerable discussion about including 'members whose objectives are different from those of the rest of the team. But there was no denying the contribution made by a study of physiology to the problem of Everest in the past'. One of Pugh's roles was to look after food and diet for the expedition. The sport and exercise scientist can

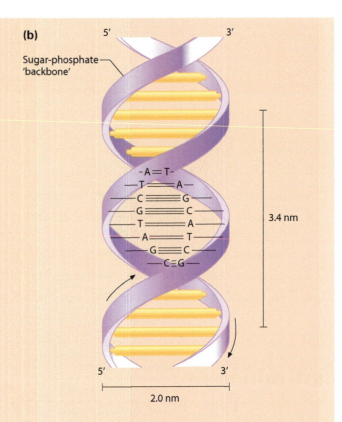

Sugar-phosphate 'backbone'

5'   3'

3.4 nm

-A═T-
—T═══A—
—C═══G—
—G═══C—
—T═══A—
—A═══T—
—G═══C—
—C≡G—

5'   3'

2.0 nm

△ **Fig. 1.1** Two important events of 1953: (a) the ascent of Everest and (b) the discovery of the double-helical structure of DNA.

perform a key role in such a team and must be able to participate fully and understand where everyone 'is coming from'. The ultimate team is greater than the sum of its component parts and the team effect is to add value, to provide a platform for individuals to excel. At the same time it must be remembered that such individuals are impotent without the support of the other team members.

The ascent of Everest can be compared with other sport and exercise pinnacles, such as becoming the fastest person in the world on the track or in the pool, or in a dinghy, kayak or rowing boat. Again, nearly 60 years ago (in 1954), the mile was run by Roger Bannister in less than 4 minutes for the first time. In 2012 the record stands at around 3 minutes 43 seconds (Hicham El Guerrouj, in 1999). In 1952, the world record for the marathon of 2 hours and 25 minutes (Yun Bok Suh, Korea, in the 1947 Boston Marathon, having been knocked to the ground by a dog, suffering a serious gash and broken shoelace) was broken by James Peters in 2 hours and 20 minutes, despite him being hit by a car 8 miles into the race! Women were not even 'allowed' to run such a distance competitively. Now the marathon has been run in less than 2 hours and 4 minutes (Kenyan, Patrick Macau Musyoki in Berlin, 2011) at an average speed just over 20 km/hour, and Paula Radcliffe holds the women's world record time of 2:15:25 set in 2003. An interesting account of the fight to establish the women's Olympic marathon race can be found in Lovett (1997). People now speculate *when*, not *if*, the marathon will be run in less than 2 hours – requiring an average speed of just over 21 km/hour. A current estimate is that this will be achieved by 2028.

These are achievements and challenges of a sporting nature, but equally important challenges and other dimensions for exercise biology in the twenty-first century can be found in the field of medicine. The epidemics of modern diseases, such as cardiovascular disease, diabetes, obesity and cancer, are a threat. Knowledge of these medical conditions, prevalent in industrialized countries are an opportunity for exercise biologists to shape the future of human morbidity and mortality through a greater dependence on primary prevention. The relative importance of this was described as the need to wage war on modern chronic diseases (Booth *et al.*, 2000).

Previous U.K. government health policies called National Service Frameworks (see the section 'Health' in Chapter 6) set out targets for the prevention and control of such diseases. Related to this is the opportunity to relate and apply functional exercise biology to the genomic potential (and susceptibility) of an individual through new-found knowledge of the human genome.

This is the context for this chapter, briefly introducing the key features of molecular biology and the chemistry of life that sport and exercise scientists might contemplate in order to take a holistic approach to the subject as it sits in the life sciences. So let us begin with the exciting human genome revelations of our time.

Plunging into the gene pool

Sixty years ago the first protein amino acid sequence was defined and the double helical structure of DNA was explained; both were exciting landmarks in science. In 1996 the first ever DNA sequence of a eukaryotic organism (a yeast) was completed, and in 1990 an international project – the Human Genome Project (HGP) – began to map and sequence the whole of the **human genome**. The working draft sequence of the human genome was published in 2001 and in the subsequent years final touches were made to the sequences and map of the human genome. This is quite simply what it is – a map. Maps are of little use unless they can be interpreted and used to their maximum potential. Enter the navigator – the physiologist!

The cost of DNA sequencing has fallen dramatically, such that whilst the HGP took 13 years and cost $3 billion, now the human genome can be sequenced in a day for £4,000. Brandler (2011) cites the view of a Chief Executive of a biotech firm suggesting that all babies born from around 2020 will have their genetic code mapped at birth. Such knowledge is a double-edged sword – knowing that you carry variants that put you at an increased risk of developing type 2 diabetes could make you rush out, hire a personal trainer, start exercising more often, and get your BMI the right side of 25. Making appropriate lifestyle choices to minimise the risks of disease is one benefit of knowing the risks, but not everyone will make such choices, and lifestyle habits are notoriously resistant to change (discussed more in Chapter 6), and difficult to sustain. For other medical conditions, there might be little one can do to prevent or delay onset, thereby causing undue stress. Furthermore, genetic risk assessment could lead to genetic hypochondria (Pääbo, 2001) causing many to spend their lives waiting for a disease that may never arrive.

So what is the human genome? It is the complete genetic make-up (total of all the genes in the cell) of an individual. In humans, the genome consists of 23 pairs of **chromosomes** contained in the nucleus of the somatic cell (i.e. any cells other than an ovum or sperm). These chromosomes are made up of 22 autosomes and one sex chromosome. The 22 autosomes are present as homologous pairs, having, for example, the same length and centromere position. The sex chromosomes of the female are also homologous – there are two XX chromosomes. Males, however, have one X and one Y chromosome and only small parts of the X and Y chromosomes are homologous.

Each chromosome contains the double-helical **deoxyribonucleic acid (DNA)** molecule, as shown in Figure 1.1(b). The human genome contains 30 000 different genes, a gene being a sequence of DNA that codes for one polypeptide. Peptides are sequences of amino acids – the building blocks of proteins – and will be discussed in a little more detail under the heading 'Molecules and macromolecules', on page 15.

DNA is the genetic building block of what we are, and the stuff of life that can be boiled down to a sticky residue in the bottom of a test tube and played around with to modify organisms, including humans. There are two major plans of cellular organization – prokaryotic (microorganisms such as bacteria and archaea) and more complex eukaryotic. The simpler prokaryotes have DNA that is not separated from the rest of the cell in a nucleus, neither do they have the typical organelles in the cell, like those of a eukaryote.

The **prokaryotic** world is characterized by its ubiquitous distribution, its rapid growth and short generation time, its tremendous biochemical versatility and genetic flexibility, and its consequent usefulness to experimental biologists, who in recent years have exploited these properties to great advantage (in the science of molecular biology).

The **eukaryotes** are plants, animals, fungi and protozoa and have cells with distinct membrane-enclosed compartments (organelles, e.g. mitochondria), including a nucleus that contains DNA. An exception to this rule is the red blood cell (erythrocyte), which lacks a nucleus and mitochondria. Plant and animal cells are differentiated into numerous forms, each specialized for the different functions that they perform in the organism. In the mammalian body, we find many different cell types; for example, the striated muscle cell, which is multinucleate and full of mitochondria to produce the energy necessary for contraction, and the motor neuron, which has long extensions of cytoplasm that conduct electrical potential (Figure 1.2).

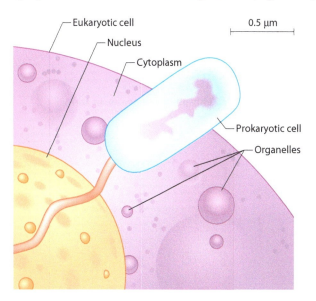

Eukaryotic cell
Nucleus
Cytoplasm
0.5 µm
Prokaryotic cell
Organelles

△ **Fig 1.2** Typical organization of eukaryotic and prokaryotic cells. This illustration shows a bacteria (prokaryotic) invading a human (eukaryotic) cell. The bacterial cell is much simpler and lacks the organelles found in the eukaryotic cell. From Campbell & Reece (2002).

Each chain of the DNA molecule is made up of four types of chemical building blocks called **nucleotides**. A nucleotide is formed from the combination of a molecule type called a base with a sugar (to form a nucleoside) and the addition of a phosphate. Deoxyribose sugars linked by phosphodiester bonds form the backbone of DNA (Figure 1.3).

In the DNA helix, the backbone on the outside consists of alternating sugar and phosphate molecules. The **base molecules** (attached to the sugars) form the 'rungs' of the helix 'ladder', and adenine (A) forms a link only with thymine (T), and guanine (G) only with cytosine (C). The bases A and T pair via two hydrogen bonds whereas G and C have three (Figure 1.1 (b)). The pairing of the bases AT/GC explains one of the principal findings on the road to the discovery of the structure of DNA – that the amount of A in DNA always equals the amount of T, while there are always the same amounts

of G and C. James Watson and Francis Crick (1953) reported their molecular model of the double helical structure of DNA, its structure suggesting the basic mechanism for DNA replication, and in 1962 were awarded a Nobel Prize (along with Maurice Wilkins).

In the molecular biology laboratory, a method that amplifies DNA in the test tube – the **polymerase chain reaction** (PCR) – has been developed. This technique now underpins much of modern biology and medicine. It is the basis of genetic testing, of obtaining DNA 'fingerprints' from crime scenes, or extracting DNA from fossils. The standard PCR method involves a number of cycles of heating and cooling but this may be improved by a helicase method that uncoils double-stranded DNA in cells and allows all of the amplification process to be done at a single, lower temperature.

A very small amount of DNA is found in mitochondria (the 'mitochondrial genome'). Although the vast majority

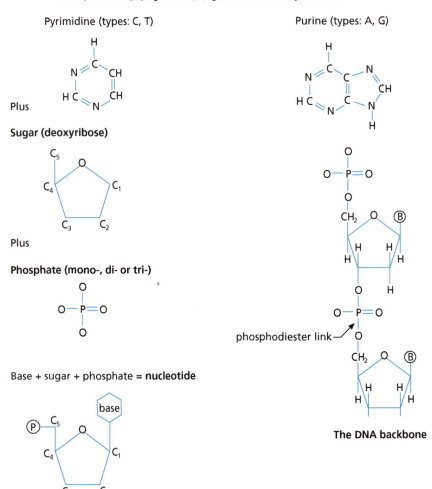

△ **Fig 1.3** DNA is made up of four bases – adenine (A), cytosine (C), guanine (G) and thymine (T) – linked to sugar and phosphate molecules in the form of a chain.

of proteins are encoded by nuclear genes, mitochondrial DNA codes for a small number of proteins (about 13).

Sequences of DNA are described by writing the sequence of bases in one strand. The strands of DNA in a human cell are estimated to be around 2 m in length, so how is it all packed into a small nucleus? Basically, DNA is packaged into a nucleosome that consists of proteins known as **histones**. Coiling of DNA around histone proteins enables the long strands to be tightly packed into a nucleoprotein complex known as **chromatin**.

When a cell divides, chromatin becomes more condensed and can be seen as chromosomes under light microscopy. Chromosomes have been numbered and genes that are responsible for (encode) particular proteins can be identified as being localized to the regions, bands and sub-bands of an arm of that chromosome (Figure 1.4).

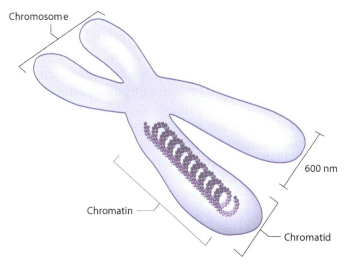

△ **Fig 1.4** The chromosome.

The structure of DNA enables a process of nucleic acid replication to occur and produce the information required to build proteins. During DNA replication the two chains break away from each other and each becomes a template for the formation of two complementary DNA strands. The flow of genetic information (**expression**) continues when information is retrieved from one of the available strands (the template strand). The segment of DNA containing a gene is used to produce a single strand of messenger ribonucleic acid (mRNA) in a process called **transcription**.

RNA is required to read the DNA code and translate it into proteins. RNA is structurally different from DNA in a number of ways: it is usually single stranded, its sugar is ribose instead of deoxyribose, and uracil replaces thymine as one of its bases. The mRNA strand formed in transcription diffuses out of the nucleus into the cytoplasm where the genetic message is read by ribosomes to

produce a polypeptide. The process of **translation** – the formation of a polypeptide under the control of mRNA in the cytoplasm – has three stages: initiation, elongation and termination. This requires a further type of RNA found in cytoplasm, transfer RNA (tRNA). In fact, RNA exists in three forms – mRNA, rRNA and tRNA (see **Box 1.1**), and there are also microRNAs.

Box 1.1

Forms and roles of RNA

- **Messenger RNA** (mRNA) is formed in the transcription of a segment of DNA that encodes a protein sequence. Transcription is the DNA-directed synthesis of RNA.

- **Ribosomal RNA** (rRNA) facilitates the interaction of mRNA and tRNA. This results in the translation of mRNA into protein.
  Translation is the RNA-directed synthesis of a polypeptide.

- **Transfer RNA** (tRNA) carries amino acids to ribosomes. Ribosomes are the organelles in the cell that carry out protein synthesis.

- **microRNAs** (miRNA) are small RNA molecules, typically 20–25 nucleotides in length, that do not encode proteins but instead regulate gene expression.

Genes carry the code for a sequence of amino acids to be formed (amino acids are the building blocks of proteins and will be discussed later in the chapter) and translation results in amino acids being linked together to form an amino acid chain or polypeptide. The genetic code for the 20 amino acids is the same in all organisms, and each is encoded by a triplet of nucleotide bases. 'Codes' of three bases specify a particular amino acid, and as there are 64 possible combinations of three bases, there are plenty to specify all 20 amino acids. The codon AUG, for example, codes for the amino acid methionine and also acts as a 'start' signal for ribosomes to begin translating the mRNA at that point. There are three codons that function as 'stop' signals (UAA, UAG and UGA; Figure 1.5).

In the genetic code there is redundancy; for example codons CAC and CAU both specify histidine, but there is no ambiguity (neither codon specifies another amino acid). Some amino acids have four or more codons. This genetic code was deciphered during the 1960s and Marshall Nirenberg received the Nobel Prize for Medicine in 1968 for his contribution to the work.

The proteins synthesized through this process create what is termed the **proteome** – the complete protein make-up of an individual – and the analysis by measurement of proteins in terms of their presence and relative

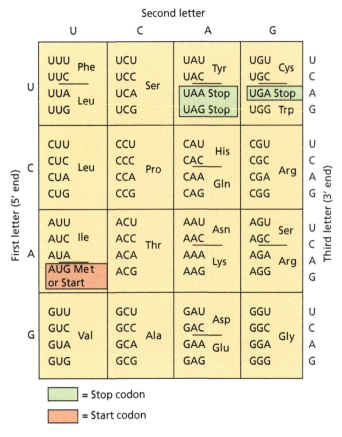

Second letter

△ **Fig 1.5** The genetic code and codons. Sixty-one of the 64 possible codons specify amino acids. There are also start and stop codons that signal the ribosome to start or stop translation.

abundance is termed **proteomics**. There are around 50 000 proteins in the human body. The emerging field of proteomics allows us to better understand protein expression and formation.

A recent exciting scientific development in molecular biology has been the discovery of microRNAs. These are small RNA molecules that do not encode proteins but instead regulate gene expression. MicroRNAs (also written as μRNAs and miRNAs) are non-protein coding small RNAs, approximately 20–25 nucleotides in length that regulate gene expression at the post-transcriptional level. They act by negatively regulating gene expression, interacting with target mRNAs to inhibit translation or induce cleavage of the message. The cellular functions of human miRNAs are little known, however, since the discovery of the first miRNA gene in *C. elegans* the roles of these molecules in biological processes are surfacing. It is conceivable that the dysregulation of microRNAs may participate in the pathogenesis of prevalent human diseases such as cancer and diabetes and in ageing. It has been suggested that microRNA genes (estimated to be about 1 per cent of all human genes) regulate protein production for 10 per cent or more of all human

genes. Micro-RNAs feed into the same pathway as short-interfering RNAs (siRNAs), double-stranded RNAs (dsRNAs) that mediate gene silencing by RNA interference (RNAi). The biogenesis of microRNAs are derived from two major processing events, driven by sequential cleavages by the RNAse-III enzymes Drosha and Dicer (for reviews see He and Hannon, 2004; Kim 2005). MicroRNAs are transcribed, producing primary-microRNAs (pri-miRNAs) which are then subjected to processing, resulting in the excision of a pre-micro-RNA. These pre-microRNAs are then recognized and transported from the nucleus to the cytoplasm and further processed to ultimately be incorporated into the RNA-induced silencing complex (RISC) where they act. There are emerging methodologies for the expression profiling of microRNAs, and the implication of roles and involvement in developmental and disease processes emphasises the importance of developing such methodologies to obtain detailed expression profiles in human tissues.

Since the year 2000, a group of workers have been building a human gene map for physical performance and health-related fitness phenotypes. In 2001, it was possible to describe 73 chromosomal loci where there was evidence of association or linkage with a performance or fitness phenotype in sedentary or active people, in adaptation to acute exercise or for training-induced changes. The 2002 map included 90 gene entries and quantitative trait loci (QTL), plus two on the X chromosome (Perusse *et al.*, 2003). The work is concentrating on physical performance phenotypes that include cardiorespiratory endurance, elite endurance athlete status, muscle strength and exercise intolerance, and with health-related fitness phenotypes such as exercise heart rate, blood pressure, heart size and shape, body composition, and metabolic factors.

By reviewing papers published up to the end of 2007, the latest human gene map for physical performance and health-related phenotypes includes 214 autosomal gene entries and QTL, and seven X chromosome gene entries. Moreover there are 18 mitochondrial genes in which sequence variants have been shown to influence fitness and performance phenotypes (Bray *et al.*, 2009).

Evolution, diversity and classification

Form reflects function at different structural levels within and between species – the existence of a backbone, for example, represents the evolutionary transition from a relatively sedentary lifestyle to a more active one. Those animals that have a backbone are known as vertebrates, one of the subphyla of the phylum chordata. Life science has a hierarchical structure or taxonomy for classifying species into groups, then placing groups into bigger groups. There are numerous levels; for example closely related species are first placed in the same genus, and genera (plural for genus) are grouped into families. Using humans (species Homo sapiens) as an example, Figure 1.6 illustrates the scheme.

In the past, the highest level of classification was said to be the kingdom, of which there at least five, possibly more, based on DNA comparison technology. However, in recent years a further, higher level of classification has been introduced that of **domains**. The three domains of life are Bacteria, Archaea and Eukarya, based on the distinction between the two fundamental or principal cell types. The domains Bacteria and Archaea contain organisms that are mostly unicellular and are prokaryotic cells. The domain Eukarya contains the organisms with eukaryotic cells, consisting of organisms belonging to the four kingdoms of Protista, Plantae, Fungi and Animalia.

If species have a likeness due to shared ancestry, they are considered to be homologous, however, not all likeness can be considered as homology. Some species from different branches of the evolutionary tree may have features resembling one another if they have similar ecological or functional roles and natural selection has shaped analogous adaptations. This is known as convergent evolution. Birds and mammals, for example, both have four-chambered hearts (see Figure 3.13 in Chapter 3) but are descended from different reptilian ancestors and evolved independently – this similarity due to convergence is called analogy, not homology.

Many species have athletic abilities; those of some species have been honed (like the human sprinter) or instilled domestically (like the thoroughbred horse or greyhound dog) or are simply still necessary to escape

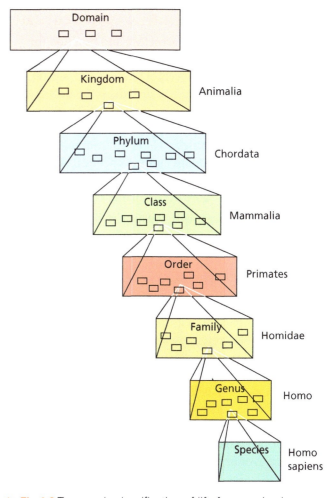

△ **Fig 1.6** Taxonomic classification of life forms using humans as an example. From Campbell & Reece (2002).

predation (like the pronghorn antelope). The basic athletic capabilities of these animals are compared in Table 1.1. It is difficult to place speeds under every category; for example, whilst husky dogs are renowned for their sled-pulling capabilities, it is difficult to estimate an average speed for such a laden race. Nevertheless, it gives us some idea of our relative capabilities compared with other animal species.

	Sprint	Middle distance	Long distance
Human	10 m/s (100 m)	6 m/s (10 km)	5.6 m/s (42 km)
Dog	16 m/s (500 m)	No races	No unladen races
Horse	20 m/s (400 m)	No flat races	4.6 m/s (100 km; with safety stops)
Camel	No races	9 m/s (10 km)	No races
Antelope	22 m/s (1000 m)[a]	18 m/s (11 km)[b]	No record

[a] Data from Burton and Burton (1969).
[b] Data from McNeill Alexander (1991).

△ **Table 1.1** Some known approximate maximal performance speed characteristics of athletic species (metres per second; m/s) and the respective distances.

Levels of biological organization

The cell is the basic unit of structure and function for the human body and will be discussed in more detail here. In the hierarchical levels of biological organization, the **organelle** (such as the cell nucleus) is the next level down from the cell. Organelles carry out different functions within the cell, requiring the use of many molecules (such as proteins), themselves built up from many atoms (such as the most abundant atom or chemical element in the body by mass, oxygen). Atoms and molecules will also be discussed in more detail further on in this chapter.

As a multicellular organism, the human body requires cells to be organized into **tissues** (such as muscle tissue), the next level up from the cell in biological organization. Tissues – groups of similar cells – form functional units, and this level of human organization will be discussed in detail in Chapter 2 (see the section 'Tissue types'). Tissues form into an organ (such as the heart), which may have a variety of different tissues within it (contractile and excitatory tissues, tissues forming valves), with specific functions. Organs form part of a physiological system, and in Chapter 3 we will look at a number of different physiological systems. These physiological systems come together to form an organism, the human body, which is a member of a biological community that includes many other species of organisms.

As a species, we have evolved alongside an enormous number of other species, including plants and animals. In the introduction to this chapter, a challenge facing human populations – the emergence of modern chronic diseases – was mentioned. Many animal species face equally challenging times, often as a result of the actions of humans who, for example, hunt and/or destroy the normal habitats and food sources of species such as the giant panda, rhinoceros and the Bengal tiger. It might be considered that this is part of the evolutionary process that has brought us thus far, and has brought us a diversity that boasts about 1.5 million species, including about 50 000 vertebrate animal species and about six times as many plant species.

Let us consider in more detail the basic unit of structure and function for the human body – the cell.

The cell

The cell is the simplest collection of material that can be considered as living. Single cell organisms exist, but more complex organisms such as plants and animals exist as multicellular structures containing many varieties of specialized cells. There are two principal types of cell that can be distinguished by their structural organization – prokaryotic and eukaryotic cells. You might recall from earlier in this chapter that prokaryotes (microorganisms such as bacteria) have DNA that is not separated from the rest of the cell in a nucleus; nor do they have the membrane-enclosed compartments known as organelles. Eukaryotic cells (e.g. of plants, animals, fungi) are more complex – they have various types of organelles, including a nucleus that contains DNA. Organelles perform particular functions such as lipid production or protein synthesis.

Cell organelles

By definition, eukaryotes have a nucleus, which contains most of the cell's DNA, enclosed by a double layer of membrane. The DNA is kept compartmentalized from the rest of the cell contents, the cytoplasm, where most of the cell's metabolic reactions occur. It is within this cytoplasm that many distinct organelles can be found (Figure 1.7).

Two types of organelle are involved in the genetic control of the cell – the nucleus and ribosomes. As the name suggests, the **nucleus** (3–10 μm in size) is the central or most crucial cell organelle. It is separated from the cytoplasm of the cell by a nuclear envelope consisting of two membranes. It is the site of storage and replication of most of the cell's hereditary material. All the chromosomal DNA is held in the nucleus, packed into chromatin fibres, and when a cell prepares to divide, it is these fibres that coil up and become visible as chromosomes.

Each eukaryotic species has a characteristic number of **chromosomes**, a typical human cell has 46 chromosomes in its nucleus, apart from the human egg and sperm cells which have 23 chromosomes. DNA in the nucleus directs protein production in the cytoplasm by dictating the sequence of bases in messenger RNA, which travels to the cytoplasm and binds to ribosomes. As the ribosome moves along the mRNA the genetic message is translated into polypeptides of a specific amino acid sequence.

Ribosomes can either be bound to the rough endoplasmic reticulum or free in the cytoplasm, and as described earlier they are the organelles that carry out protein synthesis. They make proteins that are destined for insertion into membranes, for packaging into other organelles or for export from the cell. In pancreatic cells, for example, the bound ribosomes make a number of secretory proteins including the hormone insulin (endocrine role) and digestive enzymes (exocrine role), whereas the free ribosomes mainly make proteins that remain dissolved in the cytoplasm.

There are many different membranes in eukaryotic cells, forming the **endomembrane system**. These membranes are related either by direct physical contact or through

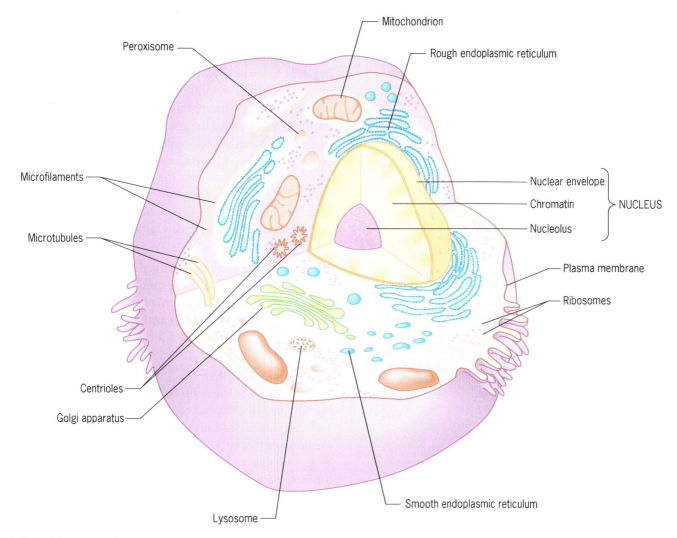

Peroxisome

Mitochondrion

Rough endoplasmic reticulum

Microfilaments

Microtubules

Nuclear envelope

Chromatin — NUCLEUS

Nucleolus

Plasma membrane

Ribosomes

Centrioles

Golgi apparatus

Lysosome

Smooth endoplasmic reticulum

△ **Fig 1.7** Drawing of a generalized animal cell and the common structures found inside.

the transfer of membrane segments (vesicles). They vary in structure and function including metabolic behaviour as described below.

There are two distinct regions of the **endoplasmic reticulum** (ER) – smooth and rough. Rough ER is so called because it has ribosomes attached to its membrane; smooth ER lacks ribosomes. In liver cells, smooth ER has an important role in carbohydrate metabolism and in the synthesis of lipids and detoxification of drugs and poisons. Muscle cells also demonstrate a specialized function of smooth ER: the ER membrane actively pumps calcium ions from the cytoplasm into the cisternal space (tubules and sacs). When the muscle cell is stimulated, calcium rushes back across the ER membrane into the cytoplasm and triggers contraction of the muscle cell (this is discussed in more detail in Chapter 4). The main functions of the rough ER are synthesis of secretory proteins and membrane production. Most secretory proteins are glycoproteins, and once these are formed they depart the

ER wrapped in the membrane of vesicles, which transport the protein from one part of the cell to another. After leaving the ER the vesicles travel to the Golgi bodies.

Golgi bodies (sometimes called the Golgi apparatus) are a system of stacked, membrane-bound, flattened sacs involved in modifying, sorting and packaging macromolecules for secretion or for delivery to other organelles. It can be considered the final packaging location for proteins and lipids. You can think of Golgi bodies as mail rooms; just as a letter is taken to the mail room to be sent to another location, proteins and lipids are transported to the Golgi body for shipment to other locations. Within the Golgi body, these proteins and lipids are labelled with a sequence of molecules (an address) which tell the body where these products should be delivered. Secretory vesicles pinch off from the Golgi apparatus and take cell secretions packaged at the Golgi apparatus, such as hormones or neurotransmitters, to the cell surface for release.

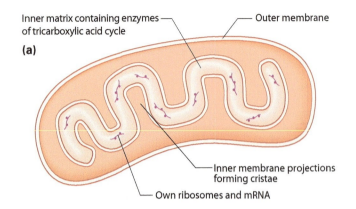

(a)

Inner matrix containing enzymes of tricarboxylic acid cycle

Outer membrane

Inner membrane projections forming cristae

Own ribosomes and mRNA

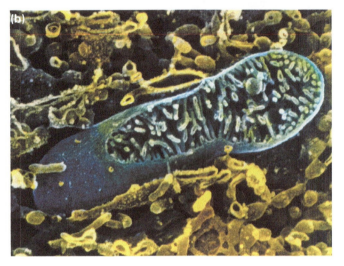

(b)

△ **Fig 1.8** Mitochondrial structure: (a) schematic representation (b) electron microscope image.

Mitochondria (0.5 μm diameter) carry out the oxidative degradation of nutrient molecules in almost all eukaryotic cells (Figure 1.8). They can be thought of as the powerhouses of the cell. They have a smooth outer membrane and a highly folded inner membrane, thereby forming an inner and outer compartment. Mitochondria are often found close to the cellular machinery they supply with ATP. In muscle, for example, they are neatly sandwiched between the contractile elements of the muscle cell. Skeletal muscle mitochondria will be discussed further in Chapter 4.

Lysosomes are membrane-bound sacs of enzymes that the cell uses to digest macromolecules. It is important to emphasize the need for the intracellular compartmentalization as lysosomes provide a space where the cell can digest macromolecules safely, without allowing the enzymes to roam freely in the cell. All major classes of macromolecules, e.g. proteins, fats and nucleic acids, can be hydrolysed by lysosomal enzymes. Lysosomes also use their enzymes to recycle the cell's own material, dismantling digested material and recycling useful parts. A human liver cell, for example, recycles half of its macromolecules each week (Campbell and Reece, 2008).

Whilst lysosomes bud from the endomembrane system, **peroxisomes** do not, instead growing by incorporating lipids and proteins made in the cytoplasm, and increasing in number by dividing into two when they become too large. Peroxisomes are specialized compartments bound by a single membrane and they contain oxidative enzymes that protect the cell from its own production of hydrogen peroxide (H_2O_2). Peroxisomes can also break down fatty acids into smaller molecules (beta-oxidation), ready for transportation to the mitochondria. In peroxisomes, long and very long fatty acids (i.e. more than 20 carbon atoms in length) are broken down through beta-oxidation (shorter chain fatty acids are broken down in the mitochondria).

Organs that are actively involved in detoxification such as the kidney and liver have many peroxisomes in their cells. Peroxisomes are also involved in bile acid and cholesterol synthesis and in amino acid and purine metabolism. In recent years it has started to emerge that numerous genetic disorders are associated with defects in the peroxisome.

Proteasomes are the main engines of non-lysosomal protein degradation and recycling in the cell – misfolded and short-lived proteins are their primary target. They are not membrane bound, instead being like barrels or rings with enzymatic regions on the inner surface. In test tube experiments, 'large' proteasomes of eukaryotic cells sediment in an ultracentrifuge with a coefficient of 26S, capped with regulatory protein complexes sedimenting with 19S.

The chemistry of life

Chemical elements and compounds

Before the molecule there is the atom of an element, and at the beginning of all life are these chemical elements, the four most common elements in living systems being carbon (C), hydrogen (H), oxygen (O) and nitrogen (N). Around 60 per cent of the body mass of humans is accounted for by oxygen, around 19 per cent by carbon, and around 10 per cent by hydrogen. Only about 30 elements are thought to be important in biological systems.

A reminder of the structure of the typical atom is needed here in order to understand some of the concepts and methods used in the study of sport, exercise and health. These concepts include the use of chemical elements (isotopes) to study metabolism or the food and other substances we might ingest in an attempt to improve health or exercise performance.

Atoms, ions and molecules

The atom of an element consists of a nucleus and orbiting electrons. Contained in the nucleus are protons that are positively charged, and neutrons that have no charge. Circulating electrons have a negative charge (Figure 1.9).

The **atomic mass** of an element is based on the number of protons and neutrons in the nucleus, whilst the **atomic number** of an element is equal to the number of protons it has. Carbon, for example, has six protons and six neutrons in the nucleus, plus six circulating electrons (^{12}C). The equal number of protons and electrons means that the atom has a neutral charge. Atoms can change their charge – an ion is an atom that has lost or gained one or more electrons and therefore lost or gained charge. By gaining an electron the atom becomes more negatively charged and is known as an **anion**. By losing an electron, the atom becomes more positively charged and is known as a **cation**.

Electrons circulate round the atom in energy shells, each of which has a maximum electron capacity; the first shell has a maximum capacity of two, the next energy shell can have a maximum of eight.

The number of neutrons in the nucleus in an atom can change (and thereby the mass), giving rise to different isotopes of some elements. Isotopes might be radioactive,

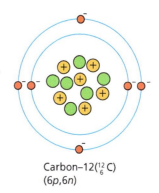

Carbon–12($^{12}_{6}$C)
(6p,6n)

△ **Fig 1.9** Schematic diagram of a carbon atom. The carbon atom nucleus consists of six protons (+) and six neutrons. Six electrons circulate rapidly around the nucleus, one inner shell with two electrons, and an outer shell with four electrons.

such as ^{14}C, or 'stable', such as ^{13}C. A radioactive atomic nucleus spontaneously disintegrates to form an atom of another element, thereby emitting radiation. The radioactive isotope ^{14}C disintegrates to become ^{14}N, emitting beta (electron) rays. The atoms ^{14}C and ^{14}N have the same mass but are chemically distinct (C has an atomic/proton number of 6, N has an atomic number of 7).

Isotopes are used as biological tools to 'trace' molecules in biological systems. Stable isotopes have the advantage of not decaying and not emitting radiation, and thereby present no danger when used as biological tools. Stable

---

TIME OUT

STABLE ISOTOPES AND THEIR USE IN HEALTH AND EXERCISE RESEARCH

Stable isotopes do not spontaneously disintegrate and are non-radioactive. The most abundant isotope of carbon is ^{12}C (Figure 1.9). The stable isotope ^{13}C has 6 protons and 7 neutrons and naturally contributes about 1.1 per cent of the total carbon pool. By enriching a molecule with atoms of a stable isotope, the molecule can be used as a 'tracer' in a physiological system.

Water can be labelled with stable isotopes in order to measure energy expenditure over a period of several days in humans. The doubly labelled water (DLW) method requires a participant to drink a small amount of water containing ^{2}H and ^{18}O. The two isotopes mix in the water contained in the body and are slowly metabolized. The ^{2}H mixes with the body water pool only, whilst the ^{18}O mixes with the water and the bicarbonate (HCO_3) pool. The difference in the rates of disappearance of the two isotopes (measured in samples of urine collected from one to four weeks) can be used to calculate carbon dioxide production. This information is then used to estimate energy expenditure over the chosen time period using calorimetric equations. Methods of estimating energy balance are discussed further in Chapter 5.

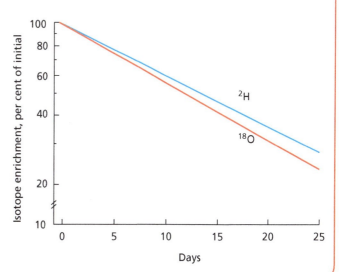

isotopes are increasingly used in nutrition and metabolism research (see **Time Out** *Stable Isotopes and their use in health and exercise research*), for example to estimate human energy expenditure in free-living conditions and to measure whole-body substrate oxidation *in vivo* in humans during exercise (e.g. van Loon, 2001).

Reactions and enzymes in metabolism

Chemical bonds and reactions

To understand energy metabolism and the metabolic processes that occur when we exercise, we need to understand the basic principles of some chemical bonds and reactions. The electrons of an atom are responsible for its chemical behaviour, and electrons can be given up or shared between atoms, for example to form molecules. A molecule is made up of two or more atoms joined by ionic or covalent bonds – the two principal forms of chemical bonding. The sharing of electrons is known as bonding, and whilst not all chemical bonds are formed in the same way, they all involve the electrons of atoms. An **ionic bond** is one which is formed as a result of the attraction of the opposite charges, such as the sodium ions (Na^1) and chloride ions (Cl^-) of sodium chloride (NaCl), the salt we add to food. Sodium tends to give up its lone outer electron to fill up the outer shell of chlorine (Figure 1.10(a)). This process of giving up electrons becomes less favourable from an energetic viewpoint if more than three electrons might be given up.

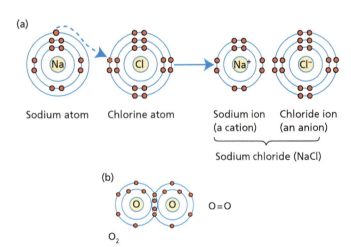

(a)

Sodium atom   Chlorine atom   Sodium ion   Chloride ion
(a cation)   (an anion)

Sodium chloride (NaCl)

(b)

O=O

O_2

△ **Fig 1.10** Ionic and covalent bonding examples. (a) NaCl is an example of an ionic bond (giving up/accepting electrons). A valence electron is transferred from sodium to chlorine, completing a full valence shell for both atoms. (b) Two oxygen atoms form a molecule, O_2, by sharing two pairs of valence electrons to complete the outer valence shell. The atoms are joined by a double covalent bond.

Oxygen, which has six electrons in its outer shell, gains stability by sharing electrons. An example of this is the molecule that it forms with itself (O_2; see Figure 1.10(b)). This is known as **covalent bonding** – the mutual attraction of two atoms to share electrons. When one electron pair is shared between two atoms, a single covalent bond is formed (represented by a single line between atoms in bond line formulae; H–H), whereas double covalent bonds are formed when two atoms share four electrons, two from each atom. Carbon dioxide has two double covalent bonds (O=C=O). Covalent bonds are stronger than hydrogen bonds. Hydrogen bonds arise from the attraction between the charge on the hydrogen atom (slightly positive) and the charge on the nearby oxygen or nitrogen atom (slightly negative). They are important to the structures of nucleic acids, proteins, and other macromolecules.

Water in biology

Water has an important role in biology. From a global perspective, around 75 per cent of the earth's surface is covered in water, and life began in water, carrying on for about 3 billion years before developing on land. In nature water exists in three states – gas, liquid and solid. From a human physiological perspective, the body is composed of approximately 60 per cent water (see also 'Body composition measurement' in Chapter 5), and whilst we can survive for a long time without food, water is more essential for survival. Cells are surrounded by water and it is the major cellular constituent (about 80 per cent).

Chemically, water appears to be a simple molecule, consisting of two hydrogen atoms and an oxygen atom linked by single covalent bonds. However, an unusual feature is that it is a **polar** molecule – it has opposite charges at opposite ends of the molecule. This means that the molecules tend to stick together as a result of electrical attraction. Oxygen tends to attract the shared electrons more strongly than the hydrogen atoms, because the oxygen atom has a slight excess of charge and the hydrogen a slight excess of negative charge. Hydrogen bonds form as a result of the slightly positively charged hydrogen of one molecule being attracted to the slightly negatively charged oxygen of another.

Acid–base concepts

In Chapter 3 we will discuss in a little more detail the concept of **homeostasis** – the condition in which the body's internal environment remains relatively constant, within physiological limits. The maintenance of a stable extracellular fluid is important, and a particular parameter that needs to be regulated is the pH of that fluid. This creates a challenge to the body. It is continually producing by-products as a result of metabolism, in particular carbon dioxide as a result of respiration, the metabolism of amino acids and the generation of lactic acid during muscle action. These processes result in the production of hydrogen ions (H^+) (see **Box 1.2**).

Box 1.2

Lactic acid and lactate

When lactic acid molecules are produced at 'physiological' levels of acidity in the cells of the body, they dissociate to form lactate (an anion) plus hydrogen ions

Lactic acid ($C_3H_6O_3$)
dissociation
Lactate (anion: $C_3H_6O_3^-$)   plus Hydrogen ion
                                                          (cation: H^+)

See also Figure 4.16 in Chapter 4.

The definition of an acid and a base (Bronsted–Lowry) can be summed up in the equation:

$$Acid \rightleftharpoons Base + Proton$$

A molecule that liberates H^+ (i.e. a proton donor) and forms anions is called an acid, whilst a substance that accepts H^+ (i.e. a proton acceptor) and forms cations is called a base.

The number of 'free' hydrogen ions (H^+) in a solution give it its acidic properties and it is conventional to describe this concentration in negative logarithm units rather than in moles of H^+. Thus:

$$pH = \log_{10} - \frac{1}{[H^+]}$$

For example, the pH of a solution containing 1×10^{-6} mole [H^+] per litre is 6. A change in concentration (concentration is conventionally indicated in physiology by square brackets [ ]) by a factor of 10 becomes a change in pH of 1 unit, whereas a two-fold change corresponds to a pH change of about 0.3. Increasing [H^+] results in a decrease in pH.

Molecules and macromolecules

Earlier in this chapter we discussed the scientific discovery of the structure of DNA. DNA is a nucleic acid, and along with carbohydrates, proteins and lipids, is one of the group of **macromolecules** – giant molecules – that are our skin and bone, our fuel stores and our insulation. Macromolecules have different levels of structure – the primary structure is the multiple, repeating units of molecules that give the macromolecule its backbone; the secondary structure is the way the backbone is shaped or folded, using internal bonds, often maximizing the number of hydrogen bonds; and the tertiary structure is the way the chains link together to form a three-dimensional shape using weaker bonds and ionic interactions.

Monosaccharides such as the blood sugar glucose (Figure 1.11), and also galactose and fructose, can combine using a glycosidic linkage to form molecules of two units, **disaccharides**, such as maltose (glucose–glucose), lactose (glucose–galactose) or sucrose, the common table sugar (glucose–fructose).

This process of linkage polymerisation can continue to make many units in a row; when there are 3 to 9 it is termed an **oligosaccharide**, and when there are more than 10 it is termed a **polysaccharide**. The polysaccharides are found in plants and animals, and the major polysaccharides of interest to sport and exercise are the storage form of carbohydrate in plant foods, starch, and the storage form of carbohydrate in the animal/human body, glycogen. Starch is found in the storage organs of plants such as the fruits and can be broken down to glucose as required by the plant, and when ingested and digested by animals. Whether a disaccharide or a polysaccharide, dietary carbohydrate has to be broken down to its component monosaccharides before absorption can occur. We will discuss the digestive system in Chapter 3.

TIME OUT

BLOOD AND MUSCLE PH AT REST AND AS A RESULT OF HIGH-INTENSITY EXERCISE

The pH of the blood ranges from around 7.35 to 7.45 and is regulated by a number of mechanisms. The pH of muscle is lower than this (around 7.1) because it is the source of more by-products of metabolism that generate hydrogen ions. When we exercise at a high intensity, we generate large amounts of lactic acid that dissociates into lactate and free hydrogen ions. The fall in muscle pH from 7.1 to around 6.5 corresponds to a small increase in the free [H^+] compared with the lactic acid generated. This is because one of the homeostatic mechanisms of the body is to limit the rise in [H^+] by 'quenching' H^+. This is done by a variety of buffering systems. Nevertheless, the relatively small disturbance of muscle acid–base status is generally considered to be one of the causes of fatigue during high-intensity exercise. Attempts to limit the rise in muscle [H^+] in order to improve performance have included 'bicarbonate loading', whereby athletes (human and equine) have ingested a large quantity of sodium bicarbonate prior to a competitive sporting event (see Time Out *Bicarbonate loading in sport and exercise*).

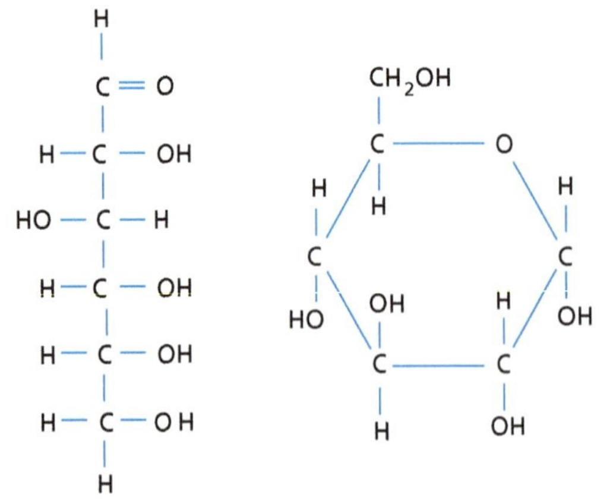

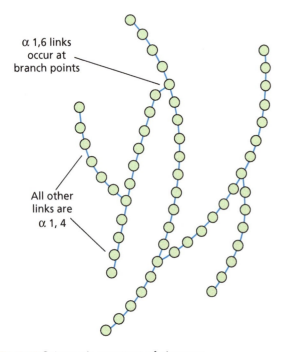

△ **Fig 1.11** The chemical structure of glucose ($C_6H_{12}O_6$). Molecules such as glucose can be represented as a straight/open chain of carbon atoms or as a ring form.

Animals store polysaccharide, as glycogen, in the liver and in muscle. The glycogen content of liver and muscle can be manipulated by diet and exercise to aid competitive endurance exercise performance. Glycogen, from the Greek *glykr*, meaning 'sweet', plus *gen* from the verb *gennaein*, 'to produce', has a similar structure to starch but the glucose molecules are primarily linked by alpha-1, 4-glycosidic bonds (links between the first and fourth carbon atoms in the adjoining molecules), with alpha-1, 6-glycosidic bonds forming branches about every 10 glucose molecules. A knowledge of the structure of glycogen (Figure 1.12) is important for sport and exercise scientists when considering, for example, the steps in the synthesis and degradation of glycogen in skeletal muscle.

Other plant polysaccharides that are important to humans include the non-starch polysaccharides (NSPs), also known as dietary fibre (e.g. cellulose), which are largely undigested by humans. The constituent monosaccharides of these do not become available as an energy source.

Protein (from the Greek for 'first things') was the name suggested by the early nineteenth-century chemist, Jöns Jacob Berzelius and used, in 1838, by Gerardus Johannes Mulder for a group of nitrogen-containing materials that he had found in animal and plant materials (Rose, 1979). You may remember from earlier in the chapter that ribosomes are the organelles that carry out protein synthesis. They make proteins for that particular cell or for transport to other cells and in the information pathway ribosomal RNA (rRNA) facilitates the interaction of mRNA and tRNA, resulting in the translation of mRNA into protein. So, the life of the cell is very much centred around proteins.

Proteins are made up of **amino acids**, and we talked of these when describing the way that genes carry the code for a sequence of amino acids to be formed by the cell – in other words, the amino acid sequences of proteins are genetically determined. The genetic code for the 20 amino acids is the same in all organisms, and each is encoded by a triplet of nucleotide bases. All amino acids have the same basic chemical structure (**Box 1.3**) and they are joined together by covalent linkages (peptide bonds) that combine the amino group of one amino acid with the carboxyl group of the next. The simplest amino acid structure is that of glycine.

α 1,6 links occur at branch points

All other links are α 1, 4

△ **Fig 1.12** Schematic structure of glycogen

Box 1.3

Structure of amino acids

Amino acids have the same basic structure – an amino group ($-NH_2$) and a carboxylic acid group ($-COOH$) attached to the same carbon atom. A further group is attached to this same carbon atom and this differs between each of the 20 amino acids.

Glycine is a small hydrophobic amino acid with a hydrogen atom (H) as its side chain, whilst **arginine** has a more complex, basic side chain. Both glycine and arginine are substrates for creatine synthesis, just one of their biochemical functions.

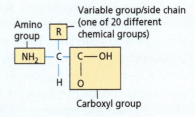

Amino group

Variable group/side chain (one of 20 different chemical groups)

Carboxyl group

△ **Fig 1.B** Structure of amino acids

Two amino acids joined together form a molecule termed a **peptide**. A peptide bond has the same function as a glycosidic bond in the structure of saccharide molecules (e.g. disaccharides, polysaccharides). In fact, two amino acids joined together are called a **dipeptide**. There are some physiologically interesting dipeptides (carnosine for example in human skeletal muscle), present in tissues of animals that are adapted to long periods of hypoxia (e.g. deep diving under water) and to high-speed running (see **Time Out** *Histidine-containing dipeptides in skeletal muscle and exercise capability*, and for a recent review of the possible roles of carnosine, see Hipkiss, 2009)

The dipeptide formed by the joining of two amino acids leaves free the amino group of one amino acid and the carboxyl group of another, and so they can continue the chain formation to form tripeptides, and beyond this polypeptides, known as proteins. Our main concern is with the proteins in the body that contain several hundred amino acids and have distinct and complex physiological functions, such as the blood protein haemoglobin (**Box 1.4** *Haemoglobin – its importance in health and disease*), and insulin, a small protein of less than 60 amino acids and a molecular weight of 12 000 produced by cells of the pancreas. Insulin was the first protein to be sequenced, a landmark in science in 1953, published in the scientific literature in 1956 and earning Frederick Sanger and team a Nobel Prize in 1958 – showing for the first time that a protein has a precisely defined amino acid sequence.

Box 1.4

Haemoglobin – its importance in health and disease

Haemoglobin is the oxygen transporter in red blood cells and consists of four polypeptide chains. Oxygen binds to haemoglobin (see Figure 3.20 in Chapter 3).

In disease, the defined amino acid sequence that is the normal protein might only differ by one amino acid alteration or displacement. This is the case, for example, in **sickle cell anaemia**, which occurs as a result of a single amino acid change (valine instead of glutamate) in the beta chain of the haemoglobin protein. It results in sickle-shaped red cells that can become trapped in small blood vessels, impairing the circulation and damaging various organs. It is a genetically transmitted disorder, with an incidence of around four per thousand in the black population. In those who inherit the abnormal gene from both parents (are homozygous), about 50 per cent of the red blood cells in the circulation are sickled. In people who are heterozygous for the abnormal gene, only about 1 per cent of red cells in the circulation are sickled, and although these people do not normally exhibit symptoms, hypoxia (a reduction in the partial pressure of oxygen in the air, e.g. at altitude) and high-intensity exercise, particularly a combination of the two, can be problematic.

TIME OUT

HISTIDINE-CONTAINING DIPEPTIDES IN SKELETAL MUSCLE AND EXERCISE CAPABILITY

Carnosine, an imidazole dipeptide consisting of the amino acids β-alanine and L-histidine, is present in human skeletal muscle at a content of around 20 mmol/kg dry muscle and around 125 mmol/kg dry muscle in the thoroughbred horse (Sewell *et al.*, 1992). Although it is accepted that carnosine and its analogues (e.g. anserine and ophidine) should play some physiological or metabolic role, no unified hypothesis exists that can satisfactorily explain their role. One potential role is as an intracellular buffer to counter the effect of hydrogen ion accumulation in association with lactic acid production. Intense exercise results in the production of large amounts of lactic acid – this accounts for about 94 per cent of the H^+ released within muscle during exhaustive exercise (see also Time Out *Blood and muscle pH at rest and as a result of high-intensity exercise*). The ability of carnosine to buffer H^+ is because of the effectiveness of the imidazole ring of histidine to bind hydrogen ions and, in addition to other systems, limit the fall in pH in muscle.

A further potential role that has been proposed for carnosine is as an endogenous antioxidant, with potential to reduce free radical damage to tissues.

Whatever the physiological role, the interesting comparative physiological perspective is that animal species adapted to high-speed running and to prolonged periods of hypoxia (muscle acidifying type activities) have a higher muscle content of histidine dipeptides than other species.

The sequence of amino acids is regarded as the primary structure of the protein, whilst folding of the primary chain (to achieve the largest number of hydrogen bonds between different parts of the chain) is called the secondary structure. Further features of the shape of a protein form its tertiary and quaternary structure. Insulin consists of two chains of amino acids, bound together with disulphide bridges – relatively strong covalent bonds that aid the tertiary structure. Disulphide bridges are cysteine to cysteine connections, cysteine being one of the sulphur-containing amino acids. The sulphydryl groups (–SH) in each cysteine molecule unite to form the disulphide bridge.

Proteins can be divided into those that are straight chained (fibrous proteins; e.g. keratin in hair and nails or collagen in skin) and those that are coiled together into irregular shapes (globular proteins; most hormones and enzymes). The tertiary structure of globular proteins, created by some relatively weak bonds (such as hydrogen and ionic bonds), is exposed at non-physiological temperatures and pH. Egg proteins denature, for example, when they are heated and precipitate to form the familiar texture of scrambled egg. Once denatured, it is very difficult for proteins to be renatured, and this is another example of the important roles of homeostasis – maintaining the integrity of the body's proteins by ensuring pH and temperature are kept relatively constant.

Nucleic acids are the next class of macromolecules that we will consider, albeit briefly. The material originally isolated was named nuclein because it came from the nucleus of cells, but we now know that nucleic acids are more widely distributed in tissues. The role of nucleic acids in the replication of genetic information and the synthesis of protein does, however, give an appropriate significance to the name. DNA is the deoxyribose-containing nucleic acid that is found in the nucleus of the cell (but there is also a small amount of mitochondrial DNA), the other class being ribose-containing (RNA), which as we have heard earlier in the chapter, is distributed more widely.

The DNA and RNA occur in combination with protein, and the protein found attached to DNA in the cell nucleus contains large amounts of arginine, one of the hydrophilic amino acids. These proteins are known as **histones** and, as described earlier, they enable the long strands of DNA in the cell to be tightly packed into chromatin by acting as a core around which the DNA coils. Nucleic acids are relatively more robust than proteins and, as a result, they can be extracted from cellular material using acids, alkalis and heat treatments that degrade protein but not nucleic acid.

The double-stranded structure of DNA was described earlier and you might like to refer back to page 6. Nucleic acids contain purine and pyrimidine bases, sugars (either deoxyribose or ribose) and phosphoric acid. The chemical building blocks of nucleic acids are called nucleotides. A nucleotide is formed from the combination of a molecule called a base, with a sugar (to form a nucleoside) and the addition of a phosphate. In RNA, which is usually single stranded, its sugar is ribose instead of deoxyribose, and uracil replaces thymine as one of its bases. You might also recall that RNA exists in three forms (mRNA, rRNA and tRNA). There are also microRNAs.

Lipids are the final class of macromolecule that we need to consider. Most people think of this class of macromolecules as 'fats', but fats are just one type of lipid of great importance to us, along with oils, phospholipids and steroids. Lipids are quite a diverse group of macromolecules. Fats and oils differ with respect to whether they are solid or liquid at normal temperatures. Saturated fats have no double bonds between their carbon atoms and tend to be solid at room temperature, whereas unsaturated fats (one or more double bonds) are liquid at room temperature and are therefore considered to be oils. We will be looking at dietary aspects of fat intake in Chapter 5.

A 'typical' fat, if there could be such a thing, is a straight-chained fatty acid, say 18 carbon atoms in length, combined with glycerol to form a monoglyceride. If three fatty acids are attached to the glycerol the resulting molecule is called a triglyceride (sometimes referred to as a triacylglycerol) (Figure 1.13). Some fatty acids have common names derived from the source from which they were first isolated. For example, oleic acid, a C18 monounsaturated fatty acid, was originally isolated from olive oil.

Inside the body, the primary function of body fats for humans is as a store of energy. Nature originally intended this store of fat to fluctuate depending on the availability of food, but in the developed world of the twenty-first century we have the problem of overnutrition, which manifests itself as overweight and obesity in large numbers of people in some countries. The health implications of this are discussed in Chapter 6.

Phospholipids are important to us as components of cell membranes, and because of their chemical structure (hydrophobic at one end, hydrophilic at the other) they give membranes their distinctive structural and functional properties. Phospholipids contain glycerol linked to two fatty acids and a phosphate group. One of the fatty acids is polyunsaturated and attached to carbon-2 of glycerol (Figure 1.13(b)). The phosphate group of a phospholipid is attached to the glycerol at a point where a third fatty acid would be in a triglyceride, and this is linked to one of a variety of compounds (labelled 'R' in Figure 1.13(b)).

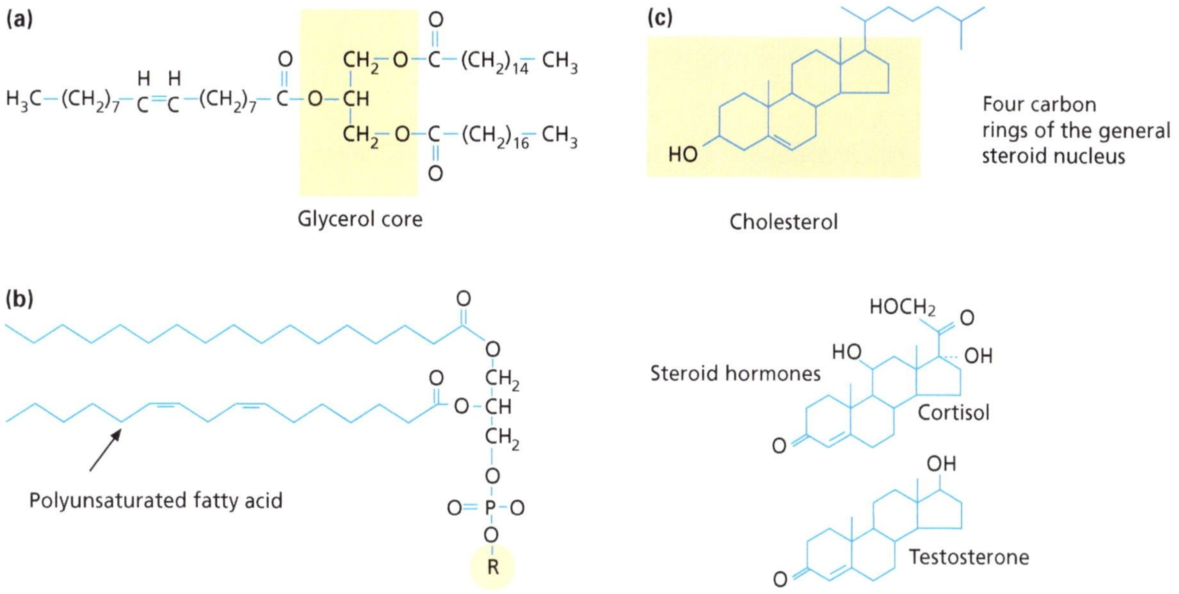

△ **Fig 1.13** (a) Basic structure of a triglyceride (also called triacylglycerol). (b) Basic structure of a phospholipid. (c) Basic structure of cholesterol and steroid hormones.

Steroids and **sterols** are types of lipid that are important in maintaining homeostasis through their role as precursors of hormones (e.g. testosterone and aldosterone) and as constituents of cell membranes. The general steroid nucleus structure is that of four carbon rings joined together (Figure 1.13(c)), cholesterol having an alcoholic group attached to the steroid nucleus. Cholesterol is the parent compound (or precursor) of steroids in the body, for example of vitamin D and the bile acids of digestion, as well as a number of hormones.

Enzymes and enzyme activity

Enzymes are proteins that increase the rate of (catalyse) specific chemical reactions and control the rate of metabolism. All enzymes are proteins but not all proteins are enzymes. The molecular structure of an enzyme allows it to join with specific substrates and then remove itself when the transformation has been completed. Sometimes the action of an enzyme is reversible – it can convert substrate A to B or convert B to A, i.e. drive the reaction in either direction. Enzymes are very specific to the substrate. For example, in the small intestine the enzyme sucrase breaks down the disaccharide sucrose (common table sugar) into one molecule each of the monosaccharides glucose and fructose. Enzyme molecules achieve this specificity by having an active site that binds to the substrate molecule.

Enzymes are sometimes conjugated, that is they contain non-protein components called co-factors that are bound tightly to the enzyme and are essential for the enzyme to function properly. Co-factors can be organic (e.g. co-enzymes) or inorganic (e.g. magnesium or zinc) and because they are not used up in a reaction they can be recycled. Co-enzymes are similar but not attached as tightly. In most cases they are changed during the reaction, an example being NAD^+, which is reduced to NADH.

The physiological environment of the body affects the way in which enzymes work. For example, they are affected by pH, temperature, substrate concentration and inhibitors. Each enzyme has an optimal pH, normally between pH 5 and 9, and a suboptimal pH will cause the hydrogen bonding holding the protein (enzyme) in its active shape to break down. You might recall that hydrogen bonds are relatively weak bonds that are used to form the secondary and tertiary structure of proteins. Some enzymes have to be able to cope with acidic environments, such as those found in the stomach.

Temperature is another important condition for enzyme activity, the optimal temperature for human enzymes being around 37°C. In lower than optimal temperatures enzyme activity is slowed, while at temperatures higher than optimal activity will increase to a point, whereafter the hydrogen bonds may be broken and the enzyme denatured, rendering the enzyme permanently unable to function.

If there is insufficient substrate for an enzyme to act on (substrate depletion) then activity will diminish, the reaction rate being a function of the enzyme concentration, the substrate concentration and the Michaelis constant.

The Michaelis constant (K_m) for an enzyme is the substrate concentration at which the velocity of the reaction is half of its maximum (V_{max}). The Michaelis constant is a measure of the affinity of an enzyme for its substrate, the lower the K_m, the higher the affinity. A further factor affecting enzyme activity is inhibition, either competitive – when enzyme action is prevented as active sites are occupied, or non-competitive – when inhibitors bind to the enzyme at other sites, not occupying the active sites but altering the conformation of the protein. Competitive inhibition increases K_m whereas non-competitive inhibitors tend to leave K_m unchanged.

Enzymes can exist in more than one form, known as isoforms. Different isoforms will catalyse the same reaction but might be found in different tissues and have different kinetics, specificities or catalytic capabilities. Lactate dehydrogenase (LDH) is the enzyme that catalyses the reversible interconversion of pyruvate to lactate and is made up of four subunits, the subunit existing in two forms M and H. Therefore, as you can probably imagine, there are five possible combinations of four subunits. The M form is found predominantly in skeletal muscle, whereas the H form predominates in cardiac muscle. The M form predominates in tissues with a relatively higher glycolytic capacity, whereas the H form is associated with tissues that have a relatively high oxidative capacity. Some of these terms might seem difficult to understand at this point, but will be explained when we discuss skeletal muscle metabolism in Chapter 4.

It is useful to illustrate some types or classifications of enzymes using examples in biochemical reactions that occur in the body. There are six classes of enzymes: (1) oxidoreductases, (2) transferases, (3) hydrolases, (4) lyases, (5) isomerases and (6) ligases. In the international classification system, each enzyme is given a unique four-digit number. The first number indicates the class of enzyme, the second and third numbers refer to sub- and sub-subclasses, and the fourth number indicates the specific enzyme reaction.

Lactate dehydrogenase, for example, which is an **oxidoreductase**, is given the number 1.1.1.27. Dehydrogenases are a subclass of oxidoreductase enzymes that add or remove hydrogen atoms (H). Another example is succinate dehydrogenase (1.3.99.1), which converts succinate to fumarate in the tricarboxylic acid (TCA) cycle by the removal of H (metabolic cycles are discussed in Chapter 4).

Transferase enzymes transfer a chemical group from one substrate to another. In glycolysis (another metabolic cycle), hexokinase (2.7.1.1) transfers/adds a phosphate molecule to glucose to form glucose 6-phosphate.

Hydrolases form two products from a single substrate by hydrolysis, chemically breaking down water to do so. For example, AMP deaminase (3.5.4.6) breaks down AMP to form inosine monophosphate and ammonia, which can result in exercise hyperammonaemia during very high-intensity exercise (see Chapter 4).

Lyases break down carbon–carbon bonds, adding a group across the bond. An example is enoyl-CoA hydratase (4.2.1.17), which reacts with water to convert acyl-CoA to hydroxyacyl-CoA. This is a step in beta-oxidation – the breakdown of fatty acids to produce acetyl-CoA which can then enter the TCA cycle.

Isomerases catalyse the isomerization of a range of substrates. Triose-phophate isomerase (5.3.1.1), for example, is found in glycolysis, where it converts dihydroxyacetone phosphate to glyceraldehyde 3-phosphate.

Finally, **ligases** form new bonds, consuming energy (ATP). An example is acetyl-CoA carboxylase (6.4.1.1), an enzyme of fatty acid metabolism.

Chemical energy from ATP

The energy liberated from dietary nutrients through reactions occurring in metabolism results in both heat production and the conservation of energy within the compound adenosine triphosphate (ATP). ATP consists of a nucleoside (adenosine) and three phosphate molecules (Figure 1.14) and it is the universal energy currency in all biological systems, acting like a chemical battery. ATP was first isolated from muscle and chemically characterized by the Danish scientist Einar Lundsgaard, in 1929.

The role of ATP as a highly reactive nucleoside triphosphate is to increase the chemical reactivity of the substrates that it phosphorylates. Its two functional parts – the nucleoside and the triphosphate arm – allow it to recognize and bind to specific enzymes, and to confer high chemical reactivity, respectively. The metabolically

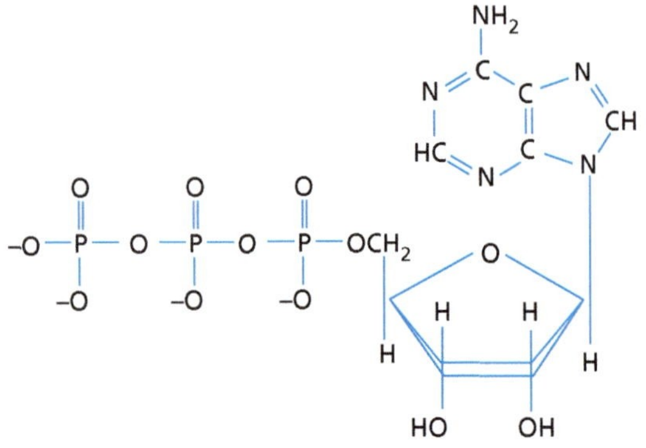

△ **Fig 1.14** The structure of adenosine triphosphate (ATP). ATP is a nucleotide consisting of the purine base adenine, a five-carbon sugar D-ribose and a triphosphate unit.

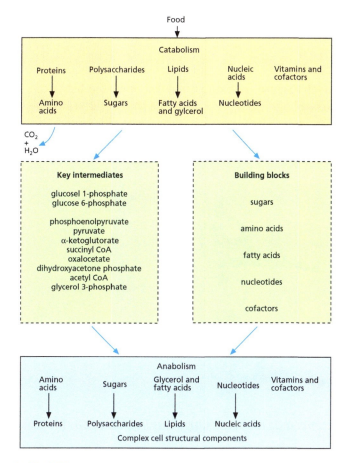

Food

Catabolism

| Proteins | Polysaccharides | Lipids | Nucleic acids | Vitamins and cofactors |

| Amino acids | Sugars | Fatty acids and glycerol | Nucleotides |

$CO_2 + H_2O$

Key intermediates

glucosel 1-phosphate
glucose 6-phosphate

phosphoenolpyruvate
pyruvate
α-ketoglutorate
succinyl CoA
oxalocetate
dihydroxyacetone phosphate
acetyl CoA
glycerol 3-phosphate

Building blocks

sugars

amino acids

fatty acids

nucleotides

cofactors

Anabolism

| Amino acids | Sugars | Glycerol and fatty acids | Nucleotides | Vitamins and cofactors |

| Proteins | Polysaccharides | Lipids | Nucleic acids |

Complex cell structural components

△ **Fig 1.15** Some important intermediate compounds generated by catabolism and used in anabolism.

active form of ATP is its magnesium salt, $MgATP^{2-}$. When ATP is broken down to adenosine diphosphate (ADP), as happens for example in muscle during exercise, the battery is topped up through synchronous rephosphorylation of ADP in a number of metabolic pathways, the most important, quantitatively, being the mitochondrial electron transport chain.

The term **metabolism** comes from a Greek word (*metabole*) meaning 'to change, convert or transform'. Metabolism can be subdivided into **anabolism**, the biosynthesis of more complex molecules from simpler ones, and **catabolism**, involving the network of chemical pathways in which complex molecules are broken down into smaller molecules (Figure 1.15).

The breakdown and resynthesis of ATP in the context of skeletal muscle energetics, metabolism and pathways of energy metabolism are discussed in detail in Chapter 4.

Structure and function

The idea of how the form of the body fits its function can be viewed from a number of structural levels, from the molecule to the organism. At the level of the organism,

the human body and its movement might be viewed in a biomechanical way, like an engineer views a machine. By analysing the human body we gain clues about what it does and how it does it, and this can be applied to everyday activities such as walking, running, jumping, lifting and throwing, types of movement that underpin many sporting activities. We have skeletons of rigid bone that require joints so we can move. What kind of joints do we need and how many? Why do we have arms that are in two parts – would the arm be more functional if we had two elbows? (McNeill Alexander, 1992).

The biomechanics of sport and exercise is presented in Part 2 of this book, but at a level below that of the organism, we might consider the form and function of the organ and physiological system that is responsible for a key component of the mammalian body – the circulation of blood.

The flow of blood is driven by the heart, an organ which has innate rhythmicity for continuous work and strong muscular walls to pump blood with enough force to deliver it to the lungs and to the body. It is divided into chambers to allow collection and delivery, without mixing, to two circulations (systemic and pulmonary). Blood flows away from the heart in vessels that have relatively thick walls (arteries), which are able to withstand higher pressures than the thinner walled vessels (veins) that return the blood. Like the heart, veins have valves, which allow blood to flow only towards the heart. The realization that blood flows in only one direction in each blood vessel is attributed to the 1628 publication of William Harvey, a physician and scientist whose experiments included restricting blood flow in the upper arm (with a ligature around the soft tissue just above the elbow), preferably of a lean person who had been exercising.

The whole cardiovascular system operates at a pressure that is optimal to maintain blood flow through all of the vital organs, and has the ability to shift blood flow according to different demands (such as the response to exercise) and with the plasticity to adapt to longer term, repeated bouts of exercise (such as the increase in cardiac volume with endurance training).

Scientific study

The human body is a very complex organism that is difficult to understand fully if always treated as a whole. There are numerous ways to study the body scientifically in order to gain a greater understanding of human and exercise physiology, not all of which involve experimentation in humans. Whole body human and exercise physiology experimentation has its limitations, guided by ethical considerations and committees established for

---

TIME OUT

CLASSICAL HARVEY EXPERIMENT

If during hand-grip exercise blood flow is occluded with a cuff such that arterial blood can flow but venous blood cannot escape, the hand and lower arm become swollen. The small swellings that appear along the lines of the vessels are valves, and if two fingers are used to squeeze blood from a section of vein between two valves, when the upper finger (nearer to the elbow) is lifted, the small section of vein will remain empty. If the lower finger (nearer the wrist) is lifted after emptying, the empty section will refill. This demonstrates that blood can flow past the valve(s) in the vein(s) towards the heart but not away from the heart.

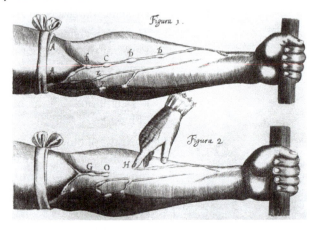

△ Typical anatomical drawing from Harvey's *Exercito anatomica de motus et sanguinus in animalibus* (1628).

---

scientists to be accountable to. Nevertheless, it is possible to judiciously use tissue sampling, arterial, venous and tissue cannulation and catheterization and many other techniques under medical supervision. An example of tissue sampling in exercise biology is that of skeletal muscle. Much use of the needle biopsy technique has been made in the last 40 years for muscle and exercise physiology research (see Figure 2.2).

Comparative physiology allows us to use samples from the animal kingdom, an example in exercise biology being the study of domesticated or wild athletic species, such as the racehorse or greyhound dog. If we wish to use highly invasive techniques on systems, tissues or organs to gain understanding, it might necessitate the use of animals. Progress in surgery has been possible, for example, by experimenting on other animals, including mammals, before being attempted on humans. Animal experimentation in biomedical research is tightly controlled in the UK by the Home Office, through laws that are designed to minimize suffering of animals and to limit the amount of experimentation. Most of the animals that are used in medical research are rodents (e.g. rats and mice). Caution has to be exercised not to assume that animals have the same anatomy, physiology and metabolism as humans.

Rats, for example, have a liver with five lobes, whereas the human liver has four lobes. And in a genetically modified mouse model lacking muscle glycogen synthase, it has been suggested that muscle glycogen is not essential for exercise (Pederson *et al.*, 2005) – far from the case in humans.

A further aspect of the scientific process is that we find it difficult to understand the multisystem whole body of an organism, and instead tend to break it down into smaller parts. However, a system broken down into its constituent parts does not necessarily represent the original sum of the parts. This concept – reductionism (reducing complex systems into smaller, more manageable components for studying) – can be interpreted at different levels. Molecular biologists can view the cell as complex, whilst cell biologists see the organ/system as complex. Human physiologists, interested in the processes and function of living things, like to consider the intact organism, or at least samples taken from an intact living, functioning body. The idealistic goal is for scientists at each of these levels to have an open mind as to the merits of the whole body through to the reductionist approach and to combine knowledge, through integrative scientific projects, that truly take us from 'molecule to man'.

Summary

- Genotype and phenotype influence our ability in sport and exercise, and our health.
- The human genome has now been mapped, and is being used in scientific research to understand human physiology, health and disease.
- A human gene map for physical performance and health-related fitness is developing.
- The basic unit of structure and function for the body is the cell, and the cell has highly specialized organelles, such as the nucleus and mitochondria.
- The four most common chemical elements in living systems are carbon, hydrogen, oxygen and nitrogen, and atoms have a nucleus containing protons and neutrons, with electrons circulating.
- Molecules such as sugars, fatty acids and amino acids are the building blocks of the macromolecules – nucleic acids, proteins, carbohydrates and lipids – that have numerous roles in the body.
- Enzymes catalyse specific chemical reactions such as the action of lactate dehydrogenase on pyruvate, and control the rate of metabolism.
- ATP is the universal energy currency of the cell and its continuous high rate of production is imperative for skeletal muscle energetics.
- There are a number of scientific approaches that contribute to our knowledge of the body, however, our interest is primarily in the whole, intact functioning human body.

Review Questions

1. The first ever DNA sequence of a eukaryotic organism was completed in 1996. What was the organism? What further whole-organism DNA sequences have since been completed?
2. What is the difference between genotype and phenotype?
3. Humans share the common feature of life with other organisms. What are the essential characteristics of life?
4. Explain the synthesis of protein, from the DNA code to the exit of a protein from the cell. Where in the cell do (a) transcription and (b) translation take place?
5. What are the five most abundant chemical elements in the body? Macromolecules are made up of the four most abundant plus the sixth, phosphorus – what are the main roles of each class of macromolecule?
6. What are the major components found in human cells and what are their functions?

References

Booth FW, Gordon SE, Carlson CJ, Hamilton MT (2000) Waging war on modern chronic diseases: primary prevention through exercise biology. *Journal of Applied Physiology* 88: 774–787.

Brandler W (2011) The genetic revolution: Implications for our health, children, and gym attendance. *The Biochemist* 33 (4) 42–44.

Bray MS, Hagberg JM, Perusse L, Rankinen T, Roth SM, Wolfarth B, Bouchard C (2009) The human gene map for performance and health-related fitness phenotypes: the 2006–2007 update. *Medicine and Science in Sports and Exercise* 41: 35–73.

Burton M, Burton R (1969) *The International Wildlife Encyclopedia*, Vol 14. New York: BPC Publishing.

Campbell NA, Reece JB (2002) *Biology*, 6th edn. San Francisco: Pearson Education.

He L, Hannon GJ. (2004). MicroRNAs: Small RNAs with a big role in gene regulation. *Nature Reviews Genetics* 5, 522–531.

Hipkiss AR (2009) Carnosine and its possible roles in nutrition and health. *Advances in Food and Nutrition Research* 57: 87–154.

Hunt J (1953) *The Ascent of Everest*. London: Hodder and Stoughton.

Kim VN. (2005). MicroRNA biogenesis: Coordinated cropping and dicing. *Nature Reviews Molecular and Cell Biology* 6, 376–385.

McNeill Alexander R (1991) It may be better to be a wimp. *Nature* 353: 696.

McNeill Alexander R (1992) *The Human Machine*. London: Natural History Museum Publications.

Pääbo S (2001) Genomics and society: the human genome and our view of ourselves. *Science* 291, 1219–1220.

Pederson BA, Cope CR, Schroeder JM, Smith MW, Irimia JM, Thurberg BL, DePaoli-Roach AA, Roach PJ (2005) Exercise capacity of mice genetically lacking muscle glycogen synthase. *The Journal of Biological Chemistry* 280: 17260–17265.

Perusse L, Rankinen T, Rauramaa R, Rivera MA, Wolfarth B, Bouchard C (2003) The human gene map for performance and health-related fitness phenotypes: the 2002 update. *Medicine and Science in Sports and Exercise* 35: 1248–1264.

Rose S (1979) *The Chemistry of Life*, 2nd edn. Harmondsworth: Penguin Books.

Sewell DA, Harris RC, Marlin DJ *et al.* (1992). Estimation of the carnosine content of different fibre types in the middle gluteal muscle of the thoroughbred horse. *Journal of Physiology* 455: 447–453.

van Loon LJC (2001) The effects of exercise and nutrition on muscle fuel selection. PhD thesis, Maastricht University, The Netherlands.

Watson JD, Crick FHC (1953) Molecular structure of nucleic acids: a structure for deoxynucleic acids. *Nature* 171: 738.

Further reading

Harvey, W (1628) *Exercitatio anatomica de motus et sanguinus in animalibus.*

Lovett C (1997) *Olympic Marathon.* Westport, CT: Greenwood Publishing.

2 Human Anatomy and Physiology

Chapter Objectives

In this chapter you will learn about:

- The anatomical and physiological systems of the human body.
- The different tissue types, including epithelial, connective, muscle and nervous tissues.
- The functions of the skeleton, including the bones and joints with a major influence on human movement.
- The identification of the muscles of different parts of the body and their relevance to human movement in exercise and sport.

Introduction

In this chapter we look at the anatomical and physiological systems that form the basis of protection (the integumentary system), support (the skeletal system) and movement (the muscular system). The slant will be towards anatomy, rather than physiology. In a later chapter we will look at systems from more of a physiological perspective (Chapter 3) and further into the section we will take an even closer look at aspects of the muscular system, namely skeletal muscle structure and function (Chapter 4).

What do we mean by 'anatomy' and 'physiology'? Human anatomy is the branch of biology that investigates the structure of our body, for example the shape and size of bones, muscles and organs. Anatomy also explores the relationship between structure and function. Human physiology, on the other hand, is the scientific study of the process, or function of the living body. We will consider the levels of the study of the living body in Chapter 3. Exercise physiology is the study of the functional short-term changes (responses) that occur when we exercise, as well as the longer term changes (adaptations) that occur when we regularly and repeatedly exercise.

Anatomy

Human anatomy has a very long and distinguished history. The first to study this area were the Greeks, and the 'Father of Medicine', Hippocrates, is regarded as being one of the founders of the study of anatomy. The Greek physician Herophilus (circa 300 BC) became known as the 'Father of Anatomy', being the first to dissect the human body. In the Roman period, Galen wrote extensively about the

inner workings of the human body – albeit with several fundamental errors – and in later centuries, Mondino d'Luzzi (1276–1326), Leonardo de Vinci (1452–1519) and Vaselius (1514–1564) all made very important contributions to anatomical study. William Harvey (1578–1657; see **Time Out** *Classical Harvey experiment* in Chapter 1) and William Hunter (1718–1783) were important English anatomists, and it was Hunter who introduced the technique of embalming still used today. Perhaps the most influential anatomist of all was Scotland's Henry Gray (1827–1861), who wrote the bestselling medical text *Gray's Anatomy*, but whose life was cut short by smallpox when he was just 34.

Anatomy might be considered at a variety of levels. Developmental anatomy, the study of the changes that occur from conception to adulthood is not of primary concern in this introductory text. We are, however, interested in cytology (the study of cells) and of the tissues made up of those cells. In this chapter, after thinking about the basic features of tissue types, the integumentary system and the skeletal system, we will take a regional anatomical approach, studying areas of the body. In Chapter 3, we will take a systemic physiological approach, studying groups of structures that have a common function.

Studying and visualizing the human body

In your study of human anatomy, you will, we hope, see the body in a variety of different ways. You may use yourself and your fellow students ('living' anatomy) to identify the anatomical features that are used by sport and exercise scientists. For example, to make anthropometric measurements it is important to know how to find anatomical landmarks such as an acromial process or the iliac crest ('surface' anatomy). This type of study can be supported by the use of anatomical models that can be taken apart and explored.

Histological anatomy is the study of tissues of the body and this is another important way of understanding the form and function of the human body. In medicine, anatomy has traditionally been studied using human cadavers (cadaveric anatomy). This was made compulsory for all medical students at the University of Edinburgh in 1826. In 1828 a notorious scandal took place in Scotland involving 'body-snatchers' Burke and Hare, who were found guilty of murdering 16 people in order to maintain a steady supply of cadavers for dissection by medical students. In order to foil any attempts to steal the bodies of the recently dead, many people made arrangements to have their remains interred within the relatively secure crypt of a local church.

It might be possible for your tutors to arrange a visit to a medical school anatomy teaching laboratory, to study aspects of the human body in this way, but increasingly, anatomical/medical imaging is producing new and exciting ways of visualizing the living body without the need for cadavers. These are of great importance to the sports medicine physician. Of the older techniques, you will be familiar with X-rays (radiographic anatomy) (Figure 2.1 (a)), and some will also have seen ultrasound images, particularly of a foetus in the uterus (Figure 2.1(b)). Computerized tomography (CT) images are computer-analysed X-ray images. Tomography is a technique that makes a picture of sections or slices of an object. The use of radio waves in an electromagnetic field enables magnetic resonance images

(MRI) to be produced and magnetic resonance spectroscopy (MRS) has been used in exercise science research, for example to estimate the extent of liver glycogen resynthesis (Casey et al., 2000).

More direct in vivo measures of human liver glycogen content have been made, but with considerably greater difficulty, using a biopsy technique. This is because liver biopsies are only considered appropriate for diagnostic but not research purposes and are generally excluded in healthy subjects. A further technique that can identify the metabolic state of tissues is positron emission tomography (PET), which is particularly useful in brain imaging.

The imaging techniques mentioned here are very expensive and sophisticated clinical tools that are available mainly for disease and injury diagnosis and management. Some medical imaging techniques can be used in our understanding and estimation of body composition, primarily in patient groups rather than in sport and exercise subjects. We will consider such possibilities in Chapter 5.

Sometimes we need to look more closely at the tissues we wish to study, in which case invasive methods are used. A mildly invasive technique that is used judiciously as a 'window on metabolism' in exercise science research is the muscle biopsy. This technique, which samples a piece of human muscle and allows for its analysis in the laboratory, has informed much of what we know of muscle and exercise metabolism.

The word biopsy or *Bios* and *Opsis* means 'life vision'. Muscle samples for the study of metabolism were traditionally obtained by open biopsy, a technique still used

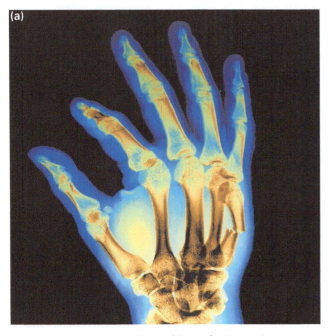

△ **Fig. 2.1** (a) A sports injury – X-ray of a broken bone.

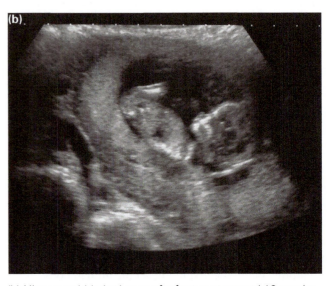

(b) Ultrasound 'dating' scan of a foetus at around 12 weeks.

(a)

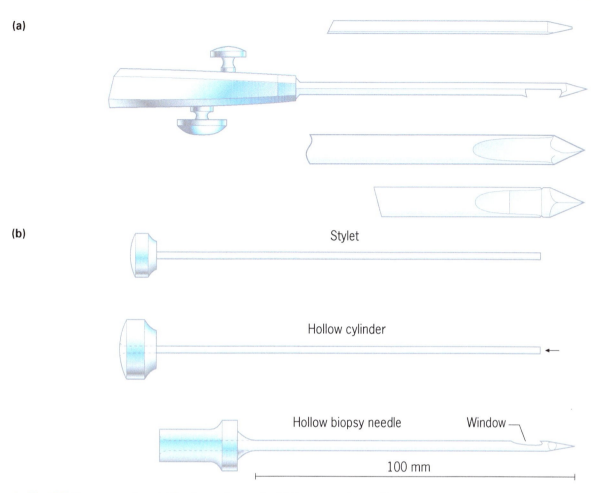

(b)

Stylet

Hollow cylinder

Hollow biopsy needle   Window

100 mm

△ **Fig. 2.2** Biopsy needles: (a) Duchenne's needle (b) Bergstrom's needle.

when large tissue volumes are required. Considerably simpler, more rapid and less traumatic, however, is the technique of needle biopsy. A use of a biopsy needle was first described by the French neurologist Guillaume Benjamin Amand Duchenne in 1865 and was used to study the muscle of patients suffering from muscular dystrophy. It was modified and re-introduced by Jonas Bergstrom in Sweden, nearly 100 years later, to study the electrolyte status of human skeletal muscle (Bergstrom, 1962). The application of the technique to the study of changes in muscle glycogen by Eric Hultman (Bergstrom and Hultman, 1966; Hultman, 1967) marked the first use of the technique in human exercise physiology, and by the 1970s the technique started to be used in equine exercise physiology (Lindholm and Piehl, 1974).

The needle biopsy work of Bergstrom and others in humans was carried out in the lateral portion of the *m. quadriceps femoris* muscle group (namely the *m. vastus lateralis*; see Figure 2.16 on page 43), which was chosen because of its relative lack of major blood vessels and nerves (Harris *et al.*, 1992) (Figure 2.2).

Anatomical terms

The human body is described by anatomists in an eloquent and specific way, with much of the language being derived from Latin and Greek. The 'anatomical position' is probably the best place to start. This is a position that you should be able to imagine with this simple description – standing upright, arms down, with the face, feet and palms facing forward (Figure 2.3).

Descriptive terms are used to describe body positions when we are lying down (**recumbent**), for example the **supine position** is with face directed upwards, the **prone position** is with the face downwards, and a **semi-recumbent position** is one between lying down and sitting (i.e. face directed upwards and supported, for example, with cushions behind our back and shoulders).

Directional terms and planes

Anatomical terms are used to explain the location of a body part, particularly in relation to another. For example, the thigh is **superior** to (i.e. above) the knee, whereas

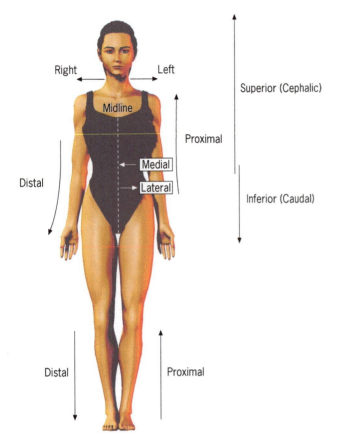

△ **Fig. 2.3** The anatomical position.

the ankle is **inferior** to (i.e. below) the knee. These terms are used in pairs, and there are others:

- **Proximal** means nearer to the trunk, whereas **distal** means further from. An example of the way in which these terms can be used is to say that the elbow is proximal to the wrist, and the wrist is distal to the forearm.
- **Anterior** means towards the front of the body, whilst **posterior** means towards the back of the body. The nose is anterior to the ears, and the ears are posterior to the eyes.
- **Medial** refers to being nearer to the imaginary midline of the body or a part of the body, **lateral** means further away from the midline. The nose is medial to the eyes, and the ears are lateral to the nose.
- **Superficial** means nearer to the surface of the body, whilst **deep** refers to being further from the body surface. Note that these terms are not the same as **internal** (inside the body) and **external** (outside the body).
- **Ipsilateral** means the same side, whilst **contralateral** means the opposite side.

Peripheral is a term that is sometimes used to describe structures such as nerves that are directed away from the centre of the body – peripheral nerves lead away from the brain and spinal cord for example. Hands and feet have

a **dorsal** surface (back of the hand, upper surface of the foot) whilst the hand has a **palmar** surface (palm) and the foot has a **plantar** surface (sole).

In describing muscles, the **origin** is the end of a muscle that is relatively fixed to the bone during contraction, whilst the point of **insertion** is the end of the muscle that moves during its contraction. The **belly** of a muscle is its widest, fleshy, contractile part that then narrows and becomes a **tendon** – a fibrous, non-contractile, cord-like part that attaches to bone.

The body consists of an **axial** region (head to crutch excluding the arms) and an **appendicular** region (arms and legs) and is made up of, from the inside of the body outwards, the thoracic and abdominal cavities containing internal organs of the body (viscera), and bones forming the supporting framework of the body. Muscles are attached to the bones via tendons, and blood vessels, nerves and lymphatic vessels form neurovascular bundles that take a course between the muscles and along fascial planes. The whole of the body is covered by a loose connective tissue (deep fascia then superficial fascia) and by skin (dermis and epidermis).

Some of these terms are incorporated into Figure 2.3, which also shows the anatomical position. You might take the opportunity of taking a class colleague and describing to them the parts of your body and how the position of one part is positioned relative to another part of your body, using anatomical terminology.

Planes are a useful way of imagining and describing the body (Figure 2.4). They are imaginary flat surfaces, so, for example, the **frontal (coronal) plane** is one that describes

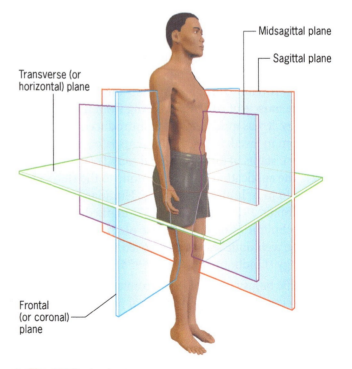

△ **Fig. 2.4** Body planes.

the body as if you are facing the side of the body and making a lengthwise cut, dividing the body into asymmetrical anterior (front) and posterior (back) parts. If we look at someone face to face, we see the **mid-sagittal plane** dividing the left and the right sides of the body lengthwise along the midline into symmetrical parts. Still facing a person, if we take a line off-centre and divide the body into asymmetrical left and right sections, we have the **sagittal plane**. A **transverse (horizontal) plane** is one that runs parallel to the ground, dividing the body into superior and inferior parts. The system of planes is not only used for the whole body but for parts of the body, including internal organs and tissues.

Tissue types

In the previous chapter we considered the cell. Cells form tissues – a collection of cells of a similar type. There are four principal tissue types: epithelial, connective, muscle and nervous tissues. Histology is the study of tissues and in your course you will hopefully be given the opportunity to look at tissue sections under a microscope. Tissues such as muscle and epithelial tissue are sometimes obtained for research and diagnostic purposes by taking a biopsy sample, as described earlier in this chapter with particular reference to skeletal muscle sampling. Autopsy tissue is that taken from the organs of a dead body, often to determine a cause of death, and the study of abnormal tissues is termed histopathology. In histology and histopathology, tissue is embedded (often in wax) to provide support, then very thin sections of tissue are made using a sharp blade mounted in a microtome. The sections are put onto a microscope slide, treated with chemicals and stains, and then a coverslip is put over the section to preserve the tissue slice.

Epithelial tissue

Along with connective tissue, epithelial tissues are the most diverse in form. Epithelial tissue has little extracellular material between the cells, and covers external and internal surfaces of the body such as the lungs, and cavities such as the mouth. It has a basement membrane, beneath which are blood vessels that do not reach the epithelium but gases and substrates diffuse to the blood vessels. Epithelial tissue has specialized cell contacts such as gap junctions and desmosomes, a 'free' surface not attached to other cells, and can divide by mitosis to replace old with new epithelial cells.

The functions of epithelial tissues are to protect structures, form barriers to particles, allow substances to pass through, secrete substances and absorb substances. Epithelial tissue types include **simple, stratified** and **pseudostratified**, and additionally, the cells of these tissues can be of different shapes, such as **squamous** (flat, scale-like), **columnar** or **cuboidal**. Epithelial tissue in the lung bronchioles or lining the stomach and gastrointestinal tract, for example, is termed simple columnar epithelium. It can have cilia (in the bronchioles of the lungs) or villi (in the small intestine) and this tissue type lends itself to the movement of particles and absorption. In contrast, the epithelial tissue that is found in sweat glands and salivary gland ducts is of the stratified cuboidal type and it lends itself particularly to secretion (Figure 2.5).

Connective tissue

The function of connective tissue is to enclose and separate organs and tissues or to connect one tissue to another. For example, the capsule that surrounds the liver and the tissues that connect muscles to bone are both made up of connective tissue. Like bone, connective tissue supports and moves the body, physically protecting underlying structures, and semi-rigid tissue (such as cartilage) supports structures such as the nose and the surfaces of joints. As a storage tissue, connective tissue can store fat (adipose tissue) and minerals (in bone), and adipose tissue also provides protection and insulation.

(a)

(b)

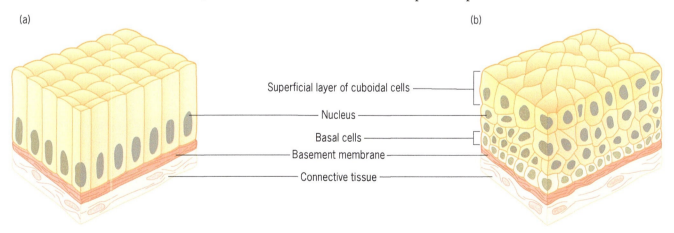

Superficial layer of cuboidal cells

Nucleus

Basal cells

Basement membrane

Connective tissue

△ **Fig. 2.5** Epithelial tissue examples: (a) simple columnar epithelium (b) stratified cuboidal epithelium.

Connective tissue diversity is such that blood and immune system cells are conveniently classified as connective tissues. Blood provides for the transport of substances such as gases and nutrients, and cells and substances of the immune system provide protection from external agents and tissue injury.

In the adult body, connective tissues can be classified into about 11 different fibre or cell types (**Box 2.1**).

Box 2.1

Some fibre and cell types of connective tissues

Extracellular components	Collagenous fibres
	Reticular fibres
	Elastic fibres
	Ground substance
Fixed cells	Fibroblasts
	Adipose tissue cells (adipocytes)
	Macrophages
	Mast cells
	Reticular cells
Mobile cells	Leukocytes
	Plasma cells

Extracellular components

The extracellular components of connective tissue make up the matrix and include ground substance and protein fibres. **Ground substance** is the component between the protein fibres and cells, providing support and binding or 'glue'. It contains water and various molecules such as proteins and polysaccharides (e.g. hyaluronic acid, chondroitin sulphate, dermatan sulphate). Sometimes known as glycosaminoglycan (GAG), these polysaccharides are associated with proteins known as **proteoglycans**.

Proteoglycans have a protein core with covalently bonded sulphate side chains. Another protein present in the matrix is fibronectin, an adhesion protein, which binds to collagen fibres and ground substance, linking the two together.

The three types of **protein fibre** – collagen, reticular and elastic – are embedded in the matrix and between cells (Figure 2.6). Their function is to strengthen and support connective tissues. **Collagenous fibres** are white in colour and made up mainly of the protein collagen. They are essentially non-elastic and are the most common fibres, found in all types of connective tissue. When supporting organs, collagenous fibres are arranged in a loose, pliable way, in contrast to when found in tendons, where they are tightly packed and inelastic. When collagen is boiled it turns into a soft gelatin, which is why stewing meat makes it more tender. In contrast, tanning of leather by the addition of tannic acid toughens leather by converting collagen into a firm, much less soluble material.

Reticular fibres consist of collagen arranged in bundles with a coating of glycoprotein. They are found in areolar tissue, adipose tissue and smooth muscle, and are produced by fibroblasts. The connective tissue that covers (the stroma) or supports several soft organs such as the spleen has an abundance of reticular fibres. These fibres also help to form the basement membrane.

Elastic fibres are smaller in diameter than collagen fibres and, not surprisingly, are the most elastic or 'rubbery' of the protein fibres. They are found in tissues that need to be capable of being stretched but then returning to shape after the stretching force has been removed. Tissues of the lungs, blood vessels and skin have an abundance of elastic fibres, consisting of the protein elastin and surrounded by the glycoprotein fibrillin. A defect in the gene coding for fibrillin results in abnormal development of the elastic fibres, known as Marfan syndrome.

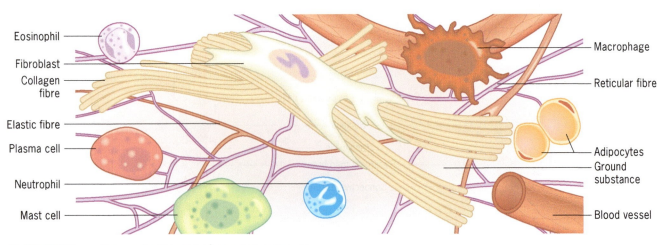

△ **Fig. 2.6** Some fibre and cell types of connective tissues.

TIME OUT

GLUCOSAMINE AND CHONDROITIN SUPPLEMENTATION

The supplement industry has tried to take advantage of the extracellular constituents and substrates of connective tissue – global sales of glucosamine supplements reached almost £1.3 billion in 2008, an increase of about 60 per cent compared with 2003. Though little evidence exists as to whether ingested supplements of glucosamine and chondroitin (e.g. sold as sulphate salts) raise serum concentrations, or reach their intended destination (e.g. connective tissue, bone), it is claimed that they can alleviate the symptoms of osteoarthritis. Most clinical trial data, however, is not consistent and research methodology is flawed. The proposed mechanism of action of glucosamine and chondroitin supplementation is not understood but glucosamine is a substrate for articular cartilage glycosaminoglycan (GAG) synthesis in the joint space. GAG traps water, making ground substance more jelly-like.

If there is the potential for chondroitin or glucosamine to alleviate the significant disability and impaired quality of life brought about by osteoarthritis then this can only be a good thing, possibly reducing the dependency on non-steroidal anti-inflammatory drugs and joint replacement surgery. This can only be realized through large, well-designed clinical trials and in 2002 the UK National Health Service Health Technology Assessment Programme commissioned work to examine the role of glucosamine supplementation in osteoarthritis. However, in 2010, the outcome of a network meta-analysis (Wandel *et al.* 2010) of patients with osteoarthritis of the hip or knee concluded that glucosamine and chondroitin do not reduce joint pain or have an impact on narrowing of joint space. Therefore, they advised that health authorities and health insurers should not cover the costs of these supplements.

Fixed cells

Fibroblasts exist in loose and dense connective tissue and are capable of moving small distances – for example, to a wound from adjacent connective tissue. They assist in the 'glueing' together of cells, that is, the secretion of extracellular matrix (formation of fibrous connective tissue such as the three protein fibre types – collagen, reticular and elastic). It is this matrix and the cells within it that are often referred to as **connective tissue**, as this conglomeration connects cells together into tissues and tissues into organs. The term '-blast' means to bud or sprout, so in cartilage the equivalent cell is known as a chondroblast, and in bone, an osteoblast. In cartilage and bone, the cells that they mature into are known as chondrocytes and osteocytes respectively.

Adipose tissue cells are specialized cells of connective tissue that store fat. The consensus is that the number of adipocytes we have remains relatively constant, but the capacity to increase our adipose tissue/fat stores is brought about by the ability of adipose cells to greatly increase in size. Adipose tissue is found beneath the skin in the hypodermis (subcutaneous fat; see Figure 2.8) and is also found around organs such as the heart and kidneys in order to protect and cushion (visceral fat). Fat tends to accumulate in different parts of the body depending on gender; in women it accumulates around the thighs and hips, and in men mainly in the abdomen, between the shoulder blades and around the waist.

Macrophages are derived from a type of white blood cell (leukocyte) called monocytes and they have the capacity to engulf and destroy foreign particles and organisms through a process called phagocytosis (see Chapter 3). **Mast cells** are rarely mobile, and are found in connective tissues at potential points of entry to the body such as the airways. When activated, mast cells release histamine, increasing the permeability of blood vessels and contracting smooth muscle. They also produce heparin that reduces blood clotting. **Reticular cells** are flat, star-shaped cells also involved with the immune response and associated with reticular fibre formation and the formation of other cells of connective tissue.

Mobile cells

Mobile cells or cells of the blood are classified as connective tissues not because they connect anything, but because they have the same embryonic origin (mesenchyme) and because blood has cells, fibres and ground substance – the three components of any connective tissue. **Leukocytes** (or white blood cells, for example eosinophils and neutrophils) are produced in the bone marrow and lymphatic tissue and there are several types, with roles such as phagocytosis. Leukocytes and the immune system are discussed in more detail in Chapter 3, as are plasma cells. **Plasma cells** are a specific type of leukocyte derived from B lymphocytes. They produce antibodies (known as the primary immune response) and have a role in defence against infection.

What might be considered as 'normal' connective tissues can be divided into loose (or areolar) connective tissue and dense (or collagenous) connective tissue, based on the arrangement and density of their extracellular fibres. **Areolar connective tissue** contains several of the cell types and protein fibres we have described. **Dense connective tissue** contains fewer cells and more numerous, thicker (e.g. collagenous) fibres than loose connective tissue.

Later in this chapter we will consider the bones and joints of the body. Cartilage and bone tissue are two types of specialized connective tissue, and are the principal components of these structures, so they need to be discussed in the context of musculoskeletal anatomy and physiology. **Cartilage** is a specialized connective tissue that supports and aids the movement of joints. The cells of cartilage are known as chondrocytes, and cartilage consists of a dense network of collagen and elastic fibres embedded in chondroitin sulphate, forming a relatively rigid matrix, second only to bone in its rigidity of the body tissues.

Cartilage itself can be of one of three types – hyaline cartilage, fibrocartilage and elastic cartilage. **Hyaline cartilage** is the most abundant in the body, the weakest of the three types and surrounded by a perichondrium (a layer of dense, irregular connective tissue). **Fibro-cartilage** lacks a perichondrium, is the strongest type of cartilage and is the type that makes up the intervertebral discs of the spine. **Elastic cartilage** is made up of collagen, proteoglycans and elastic fibres, the elastic fibres being dispersed throughout the elastic cartilage matrix. Rugby front row forwards will be aware, for example, of the reasonably rigid but also elastic properties of the external ears, sometimes becoming 'cauliflower' ears.

Bone, our final topic for brief discussion under the heading of connective tissue types, is the hardest of the connective tissue types and consists of living cells (osteocytes) and a mineralized matrix. Osteocytes are located within holes in the matrix called lacunae. There are two types of bone: **cancellous** (or 'spongy') bone and **compact** bone, based on how the matrix and cells are organized.

The basic unit of compact bone is known as the osteon or Haversian system. Osteons are cylinders made up of four parts:

- the **lamellae**, rings of matrix containing calcium and phosphates;
- **lacunae**, small cavities that house the osteocytes;
- **canaliculi** that radiate like spokes from the lacunae;
- a **central canal** that contains blood vessels, nerves and lymphatic vessels.

Cancellous bone does not have osteons, instead it consists of columns of bone that contain trabeculae, an irregular lattice of thin columns of bone containing osteocytes, lacunae, canaliculi and lamellae.

A number of factors adversely affect peak bone mass and bone loss, including physical inactivity, low calcium intake, underweight, smoking and alcoholism. Almost half the content of bone is made up of calcium and phosphorus and these dietary minerals, and magnesium, are of obvious importance. Many young adults, especially women, fail to achieve a recommended intake of calcium. Vitamin D is involved in the ossification process and a deficiency can slow bone growth rate, and vitamin C deficiency can inhibit bone growth.

Ageing can result in a loss of calcium from bone and/or its utilization, and a decreased ability to generate materials for the bone matrix. Bone marrow decreases and the metabolic disorder **osteoporosis** can develop, especially in post-menopausal women, increasing susceptibility to fracture as a result of a loss of skeletal mass and density. Bone mineral density can be measured in people at risk of osteoporosis using dual-energy X-ray absorptiometry (DEXA; see **Time Out** *Bone mineral density measurement and bone health*).

Osteocytes are the main cells of fully developed bone, but there are other types of bone cell capable of changing their roles in the growing and adult skeleton. Osteogenic cells are unspecialized stem cells, which, like all connective tissues, are derived from the mesenchyme. **Osteogenic** cells are the only cells to undergo cell division (by mitosis) and can be transformed into bone building cells (osteoblasts) during healing. **Osteoblasts** – bone-forming/building cells – synthesize and secrete unmineralized ground substance and collagen fibres, and initiate calcification. **Osteoblasts** are normally found in the growing portions of bones and they become osteocytes by surrounding themselves in matrix and becoming trapped in their secretions. **Osteoclasts** are very large cells, formed from the fusion of monocytes. They digest protein and minerals from the bone matrix in a natural breakdown process known as resorption.

Muscle tissue

There are three principal types of muscle – smooth, cardiac and skeletal – and whilst they might be classified differently and have different characteristics, they are all specialized for contraction in their own way, meaning that they contract/shorten and are responsible for movement. In fact, they all have four major functional properties:

- **Contractility** – they attempt to shorten in order to produce tension
- **Excitability** – they are stimulated to act by neural and chemical signals

TIME OUT

BONE MINERAL DENSITY MEASUREMENT AND BONE HEALTH

The mass of bone mineral can be estimated by a clinical scanning technique known as **dual-energy X-ray absorpti-ometry (DEXA)**. The technique can be used in the scanning of the whole body or a part of the body depending on the size of the scanner and uses a beam of photons at two known energies. With a two-energy photon source, the attenuation of both beams as they pass through tissues can be observed and information can be obtained about both the bone mineral mass and the soft tissue mass, and of the lean-to-fat ratio in the soft tissue. Although the measurement involves radiation, the dose is very low. Computer software can use the data obtained to estimate bone mineral content and density. DEXA measurement is mentioned again in Chapter 5 with reference to body composition and its use in models of body composition estimation.

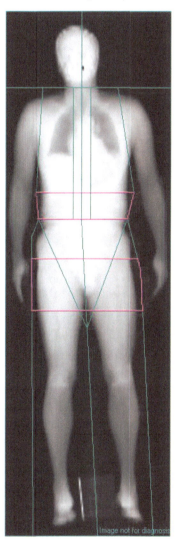

(a)   (b)

(a) DEXA whole body image of soft tissue in an adult male, in which lungs and empty stomach are discernible.
(b) DEXA whole body image of adult male skeleton.

- **Extensibility** – they can be stretched
- **Elasticity** – they return to their original length after stretching.

Cardiac muscle and skeletal muscle are both striated, but smooth muscle is not. On the other hand, while both smooth and cardiac muscle are under involuntary control, skeletal muscle is under voluntary control. Striations (or the organization of the contractile apparatus) and control are broad categories under which muscle types can be classified, however, let us look more closely at how each type is uniquely distinguishable.

The general functional properties of muscle tissues outlined above are as a consequence of the characteristics of the cell that make up that particular type of muscle.

Skeletal muscle (striated, under voluntary control) is attached to the skeleton by way of a tendon, and the cells are often long and arranged in parallel, both anatomically and mechanically. Skeletal muscle cells function independently, and the force generated by skeletal muscle *in vivo* is the sum of the force generated by a number of cells. Skeletal muscle is normally relaxed and is recruited to generate force/tension, and therefore movement.

Cardiac muscle (striated, under involuntary control) has structural and functional characteristics that are intermediate between those of skeletal and smooth muscle. It is located only in the heart, whereas **smooth muscle** is widespread throughout the body, in hollow organs and tissues such as the gastrointestinal tract, blood vessels and respiratory airways. The cells of smooth muscle cannot function as independent cells and are arranged as a sheet, connected in series and in parallel.

Muscle cells and structures are sometimes described using slightly different terms that you should be aware of. Muscle cells are sometimes referred to as **myocytes** (and fat cells as adipocytes, liver cells as hepatocytes). The term 'sarco' is also used with reference to skeletal muscle (e.g. sarcopenia) and components of muscle cells, such as **sarcoplasm** (cytoplasm) and **sarcoplasmic reticulum** (endoplasmic reticulum – described in Chapter 1 as one of the components of the eukaryotic cell).

The cells of the three types of muscle have different features, such as shape, nucleation and cell-to-cell attachment. Such features will be described and explained here in more detail for each muscle type in turn. Typical images of the types of muscle are illustrated in Figure 2.7 but it is also interesting to take a look at these features under the microscope yourselves, in histology tissue samples.

Skeletal muscle

Skeletal muscles are attached to bone via tendons and are composed of fasciculi contained within the muscle fascia (sometimes known as the epimysium). Each fascicle contains numerous muscle cells, termed fibres, between which there are dotted muscle spindles – collections of specialized muscle fibres (intrafusal fibres) that detect changes in the length and rate of change of length of fibres surrounding them.

The 'normal' muscle cells or fibres contained within a fascicle are termed **extrafusal fibres**. Skeletal muscle fibres can be subdivided into two major types based on their characteristics (discussed in detail in Chapter 4). They contain **myofibrils** (specialized intracellular

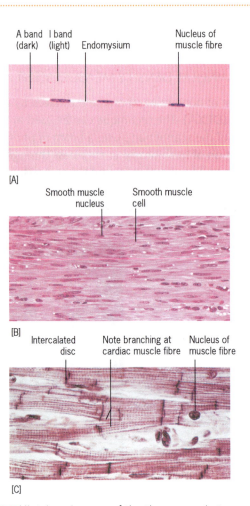

△ **Fig. 2.7** Histology images of the three muscle types (longitudinal sections): (a) skeletal (b) smooth (c) cardiac.

structures) and **myofilaments** (cytoskeletal elements), formed of the contractile proteins **actin** (thin filaments) and **myosin** (thick filaments). Skeletal muscle cells or fibres have the most developed sarcoplasmic reticulum of all the muscle cell types, and several nuclei that are peripherally located. The mitochondrial density varies, depending on the fibre subtype. Transverse tubules (T-tubules) are only found in striated muscle types and run in a direction perpendicular to the surface of the cell membrane into the central portions of the cell/fibre, enabling action potentials to be rapidly transmitted from the surface of the fibre to its central portion.

Skeletal muscle is stimulated (or 'innervated') via a motor neuron axon and its terminal branches which form neuromuscular junctions with the muscle fibre. The series of events linking muscle excitation (presence of an action potential in the fibre) to muscle contraction (cross-bridge formation to enable filaments to slide together) is described in detail in Chapter 4.

Actin filaments are anchored to a structure called the Z-line, whereas in smooth muscle they are anchored to dense bodies. To create tension in skeletal muscle, actin filaments slide between myosin filaments to shorten the distance between two Z-lines, known as the **sarcomere**.

The key features of skeletal muscle are summarized in **Box 2.2**.

Box 2.2

Key features of skeletal muscle

- Skeletal muscle exists in a variety of anatomical and morphological forms.
- The cells are bound together by connective tissue and are contained in bundles (fasiculi) that are themselves bound together in a fascia.
- Cells, known as fibres, are multinucleate and contain contractile and structural proteins forming an orderly filamentous structure.
- There are two major types of skeletal muscle fibre, and fibre contraction is controlled by large motor nerves.
- Skeletal muscle contains highly specialized stretch receptors known as muscle spindles that aid in proprioception (the sense of the body's position in space).

Smooth muscle

Smooth muscle, sometimes referred to as visceral muscle, is found as the major component of hollow organs and structures such as the urinary bladder, blood vessels and the gastrointestinal tract. It does not attach to the skeleton but develops force to provide motility (e.g. to move food down the gastroinstestinal tract) or alter the shape of an organ (contract the bladder to expel urine). It therefore has structural and functional diversity quite unlike the other muscle types in order to sustain contractions and maintain organ and tissue dimensions. The absence of an attachment to the skeleton means that coordinated cellular responses require complex control for cell-to-cell communication (neural and hormonal control achieve this). The fibrillar contractile apparatus and therefore force transmission must also be able to operate in a non-linear configuration. As muscle cells are often continuously active, ATP utilization needs to be minimized and to be efficient.

The diversity of smooth muscle makes it difficult to categorize. Some smooth muscle is normally relaxed and responds to stimulation intermittently or phasically, such as the urinary bladder, while in other areas it is normally contracted/has a degree of tension, such as the smooth muscle of the airways. Multi-unit smooth muscle has many discrete units that act independently of each other and is often stimulated by nerves to contract (i.e. is **neurogenic**). The walls of large blood vessels and the base of hair follicles (to produce 'goose bumps') are examples. Single-unit smooth muscle on the other hand is mainly self-excitable (i.e. is **myogenic**) and acts as a single unit; the uterus and the bladder are like this. Smooth muscle also forms sphincters, structures that regulate the movement of solids and liquids along passageways (the bladder, oesophogeal, ileocaecal and pyloric sphincters are examples).

Under the microscope, smooth muscle cells can be seen to be elongated and interwoven, with one, centrally located nucleus. The cells are joined by a variety of junctions, including **gap junctions** (communicating junctions) and **desmosomes** (adhering junctions), the numbers and types of junction varying widely between subtypes of smooth muscle. As in other types of muscle, the intracellular proteins responsible for contraction are **myosin** and **actin**, but there is relatively more actin (forming thin filaments) than myosin (forming thick filaments) in smooth muscle than in skeletal muscle. Because the actin filaments are anchored to dense bodies and the thin and thick filaments are not arranged to form myofibrils or the sarcomere pattern seen in other muscle types, smooth muscle is not striated. A third type of filament – intermediate – is also found in smooth muscle. This is composed of a protein called **desmin**. In general, smooth muscle contains much less protein than skeletal muscle and the myosin content is much lower. While the amounts of actin and tropomyosin are the same in both types of muscle, smooth muscle does not contain the protein **troponin**. Instead of it there are two other thin filament proteins called **caldesmon** and **calponin**.

As in all muscle types, Ca^{2+} has a central role in the excitation–contraction process, however, smooth muscle cells have no T-tubules and a poorly developed sarcoplasmic reticulum. Excitation of smooth muscle results in a rise in the cytosolic Ca^{2+}, mainly from the extracellular fluid, but also from the sparse sarcoplasmic reticulum stores. The role of **calmodulin** (calcium-modulated protein) is to form a complex with Ca^{2+}, which then activates myosin kinase in order for the enzyme to catalyse the phosphorylation of myosin. Phosphorylated myosin can then bind with actin to form cross-bridges, resulting in contraction.

Smooth muscle has several characteristic features related to its ability to contract and maintain tension. The speed of contraction is slow but prolonged, and whilst ATP energy utilization is relatively inefficient in order to phosphorylate myosin, the low rate of cross-bridge cycling required for the maintenance of tension (for comparatively less ATP consumption) results in relative efficiency, suited to sustained tension with little energy consumption and fatigue resistance. Length and tension are less closely linked in smooth muscle than in striated muscle, which means that smooth muscle can exert tension at both extremes of its length and similar tensions at most lengths.

The key features of smooth muscle are summarized in **Box 2.3**.

Box 2.3

Key features of smooth muscle

- Comparatively small cells and linked by a variety of junctions serving both mechanical and communication roles.
- Either multi-unit (neurogenic; requiring stimulation by its autonomic nerve supply to contract) or single-unit (myogenic; able to initiate its own contraction).
- The cellular proteins include actin and myosin and tropomyosin, but not troponin. Another protein, calmodulin, has a key role in contraction by forming a complex with Ca^{2+}.
- The length–tension characteristics are quite different from striated muscle types.
- Single-unit smooth muscle is able to maintain tension at a variety of lengths, making it particularly suitable to form the walls of distensible, hollow organs.

Cardiac muscle

Cardiac muscle is only found in the heart. We shall consider this type of muscle only briefly here, but because of the particular interest to sport and exercise science of the heart and of skeletal muscle we shall look in more detail at both in later chapters. Like smooth muscle cells, cardiac muscle cells have a single nucleus, centrally located. They are rich in mitochondria because they are dependent on producing energy via oxidative metabolism. Cells are joined via numerous desmosomes and gap junctions at areas of intertwined membranes forming specialized structures called **intercalated discs**. Both intercalated discs and the striations formed by thin and thick filaments, arranged in an orderly fashion, can be seen under the microscope in longitudinal sections of muscle.

The key features of cardiac muscle are summarized in **Box 2.4**.

Box 2.4

Key features of cardiac muscle

- Cardiac muscle has many structural and functional characteristics intermediate between smooth and skeletal muscle.
- Contains delicate connective tissue, supported by a rich capillary network and an abundance of mitochondria to meet the high oxidative metabolic demands of continuous activity.
- Cardiac fibres are joined together in a branching network and the ends of adjacent cardiac muscle cells have specialized intercellular junctions called intercalated discs.
- The contractile proteins are arranged in a similar way to skeletal muscle but striations are more difficult to see because of the irregular branching shape of the cells.
- Cardiac thin filaments contain troponin and tropomyosin to create the site of Ca^{2+} action for cross-bridge formation.

Nervous tissue

Nervous tissue is found in the brain, spinal cord and nerves and is capable of transmitting electrical signals in the form of action potentials, carried by **neurons** (conducting cells) and supported by neuroglia cells. Neurons have a **cell body** which contains the nucleus, **dendrites** which are extensions of the cell body, and an **axon** which is much longer than dendrites and conducts action potentials away from the cell body. Neurons with only one dendrite and an axon are called **bipolar neurons**, whilst those with several dendrites (there can be about 10, of more than 1 mm in length in neurons of the somatic subdivision of the peripheral nervous system) are called **multipolar neurons**. Neuroglia are cells of the central nervous system (brain and spinal cord) and the peripheral nervous system that protect, nourish and insulate neurons. We discuss the nervous system in greater detail in Chapter 3, and nerve and muscle connections in Chapter 4.

The integumentary system

The term integumentary refers to those structures that cover the outside of the body such as skin, hair and nails. Hence, we can see a lot more of the integumentary system than of other tissues and organs of the body. The integumentary system protects, senses, regulates temperature and even provides for vitamin synthesis (vitamin D when skin is exposed to sunlight).

The outermost structure, the **skin**, consists of three principal layers – the epidermis (outermost layer), the dermis and the hypodermis. Penetrating the layers of the skin are structures such as hairs and their follicles, sebaceous glands, smooth muscle, nerves, arteries and veins, and sweat glands (Figure 2.8). All in all, the skin is a very complex organ, metabolically active and whilst protective, it is also vulnerable to injury and disease. This becomes relevant to sport and exercise science, when for example, considering the risk of skin cancer in those who take part in sports or exercise in the outdoors without adequate protection against the sun.

From the inside, the deepest layer, the **hypodermis** anchors the skin and connects to muscle and bone, supplies nerves and blood vessels and contains loose connective tissue. It can contain a very large proportion of the total adipose tissue fat store of the body. When we take skinfold measures (see Chapter 5), we attempt to estimate the fat content of the body based upon the adipose tissue layer in the skin.

The outer layer or **epidermis** consists of a number of layers, the outer ones of which are made up of cells that are dead and flattened (stratum corneum), whereas the cells of inner layers are alive, cuboidal and rapidly dividing. There is little blood supply to the epidermis; nutrients are supplied through diffusion from the vascular network of the underlying dermis. The newly formed cells of the inner layers of the epidermis 'push' the older cells to the top and allow for a loss of dead cells at the skin surface. Cells of the epidermis are joined by desmosomes and contain an irregular network of filaments rich in the protein keratin. This forms the keratinized layer, which provides strength

and an airtight, relatively waterproof and impervious protective layer. The keratinized layer is thickest in areas of the body that are subjected to the most stress, such as the soles of the feet. A breakdown in the keratinization process results in a variety of skin problems, such as psoriasis, in which 'scales' of skin flake away, sometimes causing bleeding from small blood vessels at the top of the dermis.

Underneath the epidermal layers is the **dermis**, a connective tissue layer made up of elastin fibres for flexibility, collagen fibres for strength, blood vessels and nerve endings. The blood vessels supply the dermis and epidermis with nutrients and also have a role in temperature regulation. The diameter of these vessels and hence the blood flow can change in order to restrict (vasoconstriction) or enhance (vasodilation) blood flow depending on the need to retain or lose heat respectively. The exocrine glands of the skin – sweat and sebaceous glands – penetrate the epidermis and dermis, along with hair follicles. Hair follicles consist of a shaft, which protrudes above the skin surface, and a root, which is lined with keratin-producing cells that secrete keratin and other proteins to form the hair shaft. Cells of the sebaceous glands produce an oily secretion rich in lipids known as sebum. Most sebaceous glands are connected by a duct to an adjacent hair follicle to enable release of sebum which oils the hair and skin surface, helping to waterproof the skin and to prevent drying and cracking.

Finger- and toenails are a specialized form of keratinized product derived from living epidermal structures. They protect the ends of the digits and help in grasping small objects. Nails have a body (the portion visible) and a root (covered by skin). The nail bed is visible through the clear nail and appears pink because of the small blood vessels in the dermis.

Sweat glands are located over most of the body and release a solution containing a wide variety of organic and inorganic solutes through small openings or pores onto the surface of the body. This is one of the mechanisms the body uses to regulate temperature through evaporation of sweat as it spreads over the skin surface. Individuals vary hugely in their capacity to sweat, and some people have far fewer sweat glands than others. Sweat loss in sport and exercise can be considerable and can result in dehydration if it is not countered by increased fluid intake. During a marathon, a runner losing 2 litres per hour could end the run with a 7 per cent loss in body mass (assuming a 70 kg runner and a 2 hour 30 minute marathon time). Exercise performance is impaired at dehydration losses of more than about 1 per cent of body mass so you can see that it is vital to replace fluid to maintain exercise performance (see **Time Out** *Exercise performance is compromised by a loss in body mass caused by sweating*).

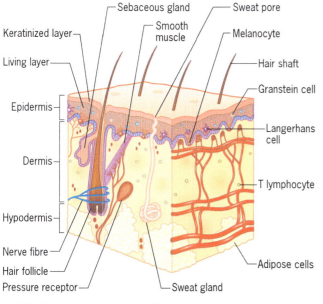

△ **Fig. 2.8** Anatomy of the skin.

Sweat loss rate depends on many factors, including exercise intensity and the ambient temperature and humidity. Sweat that drips from the skin without evaporating (e.g. when saturation/humidity is high, or when sweat rate is high) reduces the effectiveness of sweating as a means of cooling. As a result of a decrease in blood volume through sweat loss during exercise in a hot and/or humid environment it can be difficult for the body to maintain blood flow to the skin, therefore reducing heat loss and causing body temperature to rise further.

Skin colour is determined by pigments in the skin, circulating blood and the thickness of the outer layer, the stratum corneum. Specialized cells called melanocytes produce the group of pigments termed **melanin** that are also responsible for hair and eye colour. The amount of each type of melanin (e.g. black, brown, yellow or red) pigmentation is responsible for the different skin colours of the races of the world and is under genetic control. In the production of melanin the amino acid tyrosine is converted to dopaquinone by the enzyme tyrosinase. The Golgi apparatus of melanocytes package melanin into vesicles known as melanosomes, which then move into the cell processes of the melanocytes. Keratinocytes then phagocytose the melanocyte processes, taking up the melanin in melanosomes. On exposure to sunlight, melanin that is already present and the stimulation of further melanin production combines to 'tan' the skin as a protective measure against exposure to harmful ultra-violet light rays.

The skeleton and movement

Cartilage, joints and bones together make up the skeletal system. There are 206 bones in the human body, and whilst it is possible to apportion the bones to different parts of the body and build up a list of all of them, it is not the intention for you to be able to do so here. The emphasis in this section of the chapter is to introduce you to the key features of the skeleton and its major bones (Figure 2.9) and joints, such that you will have an image that can be built upon when we identify and discuss the actions of the muscles of particular parts of the body in the next section.

Activity 2.1

Independent learning task: Find out about vitamin D

1. What are the functions of vitamin D?

2. Is the synthesis of vitamin D by the skin on exposure to light our only source of this vitamin?

3. Can you remember the function of vitamin D mentioned earlier in the chapter in relation to bone health?

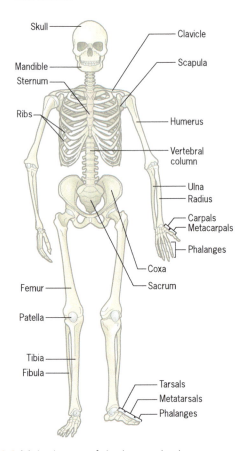

Axial skeleton

Skull — Clavicle — Scapula — Mandible — Sternum — Ribs — Humerus — Vertebral column — Ulna — Radius — Carpals — Metacarpals — Phalanges — Coxa — Sacrum — Femur — Patella — Tibia — Fibula — Tarsals — Metatarsals — Phalanges

△ **Fig. 2.9** Major bones of the human body.

The **axial skeleton** forms the upright axis of the body and comprises the skull and the trunk, including the sternum, ribs, vertebral column and the sacrum. The **appendicular skeleton** on the other hand comprises the bones of the limbs and girdles ('belts') – the pectoral (or shoulder) girdle and pelvic (or hip) girdle.

Describing movement

Much of our anatomical interest in sport and exercise is based on movement of joints. When discussing movement involving joints and muscles, it is important to speak a common language. This helps the sport and exercise scientist to be able to communicate effectively with, for example, sport and exercise physicians and physiotherapists. In the later part of the chapter dealing with major muscles of the body, we discuss these muscles in relation to the major (synovial) joints of the body.

Joints may be flexed, extended, abducted, adducted, or externally/internally rotated. These terms are defined in **Box 2.5**

Box 2.5

Terms used to describe joint movements

- **Flexed** (forward movement) – Distal part bent forwards in the upper limb or backwards in the lower limb.
- **Extended** (backward movement) – Distal part bent backwards in the upper limb or forwards in the lower limb.
- **Abducted** (out away from the body) – Distal part moved away from the midline.
- **Adducted** (in towards the body) – Distal part moved towards the midline.
- **Externally (laterally) rotated** (rotation out away from the body) – Distal part rotated away from the midline.
- **Internally (medially) rotated** (rotation in towards the body) – Distal part rotated towards the midline.

Types of joint

Three basic kinds of joint – fibrous, cartilaginous and synovial – are classified based on their structure. Fibrous and cartilaginous joints have these specific connective tissue types binding the bones together, whilst synovial joints have a fluid-filled joint capsule. Joints can also be classified according to the degree of motion possible, but this is less common and we will restrict our thoughts primarily to the structural classification and, in particular, to synovial joints.

Synovial joints are arguably of greatest interest in sport and exercise, allowing movements of one rigid structure (usually a long bone) with another adjacent rigid beam, usually another long bone, or, for example, the pelvic girdle. Bones and joints and their role as levers are discussed in detail in Section 2, 'Biomechanics'. Synovial joints normally have two joint surfaces, a synovial cavity and surrounding muscles to move the joint. There are three main types of synovial joint:

- **Ball (and socket) joint** – this allows movement in all three planes/axes. Examples include the hip joint and the gleno-humeral joint at the shoulder.
- **Hinge joint** – this allows movement in one plane/axis only. Examples include the elbow and ankle joints.
- **Complex joints** – these may allow movement like a hinge or a ball, but have additional movement. An example is the knee joint, which is a little like a hinge but is capable of additional rotational movements.

Synovial joints are bathed and nourished by the synovial fluid, allowing considerable movement between articulating bones. Movements of the joint encourage both lubrication of the joint and nourishment of the articular cartilage.

The range of movement (ROM) of a joint depends upon a number of factors, including the type of joint, the age of a person (younger people tend to have a greater ROM), genetic factors and past injury or deformity. Agonist muscles cause movement in a joint, whilst antagonist muscles also act, but weakly in comparison with the agonist(s).

Before discussing the major joints of the limbs of the body further, we should briefly review the skeletal anatomy of the two girdles.

The appendicular skeleton girdles

The **pectoral girdle** attaches the upper limbs to the axial skeleton and therefore contains two pairs of bones attaching the arms to the body: the **scapula**, commonly known as the shoulder blade, and the clavicle, commonly known as the collar bone. The scapula is a flat, almost triangular bone (Figure 2.10) with an acromial process (or acromion) that can be felt at the tip of the shoulder – try to locate each of yours. The distance between the tips of the acromial processes is known as the **biacromial distance**.

The **pelvic girdle** provides the place of attachment for the lower limbs, support for the weight of the body and protection for internal organs. It is a bony ring formed by the two hip bones (right and left coxae) and the sacrum (Figure 2.11).

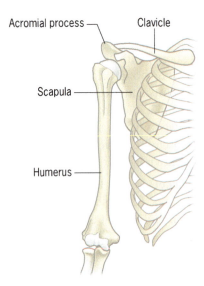

△ **Fig. 2.10** Schematic view of the right scapula and clavicle (collar bone).

The femur articulates with the socket where the parts of the hip bone converge – the acetabulum of the hip bone. The sacrum is a sequence of five vertebrae, fused together into a wedge shape, and at the end of the spinal column are four terminal vertebrae known as the coccyx. Each coxa is a large concave plate made up of the ilium, pubis and ischium. The two coxae join together anteriorly at the symphysis pubis, and at the posterior join with the sacrum.

In the female body, the pelvic girdle protects a developing foetus and forms a passageway through which the foetus moves during delivery. The female pelvis is usually smaller but broader than the male pelvis, with a larger

and more rounded inlet and outlet. The distance between the most lateral points of the iliac crest of the coxae is known as the **bicristal distance.**

Major muscles and joints of the body

In the following sections it is intended to introduce you to the major muscles of particular parts of the body that are of importance to sport and exercise, using as a focus the limbs and joints that they are most closely related to. It is by no means a comprehensive survey of the muscles of the human body, but it is reasonable for a sport and exercise science student to be familiar with, and perhaps be able to name and locate, about a hundred of the 400 or so muscles of the body. This might seem quite a tall order, but in your future career, whether as a teacher, a fitness instructor or working alongside sports medicine specialists, your student, clients or colleagues will expect you to have such knowledge and command of your specialist subject area.

As an overview, Figure 2.12 shows schematic figures of the superficial muscles of the human body from the anterior view (a) and posterior view (b). These figures feature about 45 muscles as a starter!

Muscle anatomy of the leg
The hip

The hip is a ball and socket joint capable of multi-directional movements, the joint capsule being strong at the front and back but quite weak underneath. The

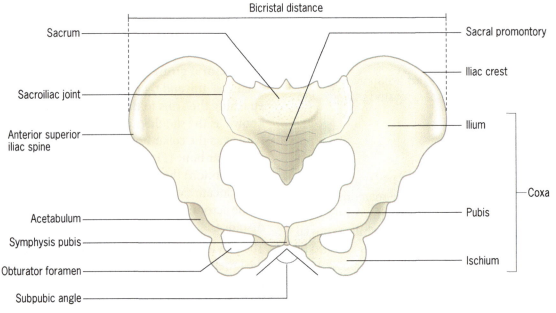

△ **Fig. 2.11** The pelvic girdle, iliac crest and bicristal distance.

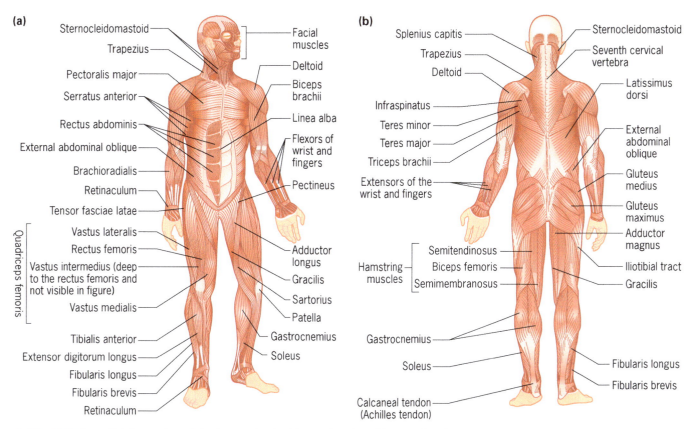

(a)

- Sternocleidomastoid
- Trapezius
- Pectoralis major
- Serratus anterior
- Rectus abdominis
- External abdominal oblique
- Brachioradialis
- Retinaculum
- Tensor fasciae latae
- Vastus lateralis
- Rectus femoris
- Vastus intermedius (deep to the rectus femoris and not visible in figure)
- Vastus medialis
- Tibialis anterior
- Extensor digitorum longus
- Fibularis longus
- Fibularis brevis
- Retinaculum

Quadriceps femoris

- Facial muscles
- Deltoid
- Biceps brachii
- Linea alba
- Flexors of wrist and fingers
- Pectineus
- Adductor longus
- Gracilis
- Sartorius
- Patella
- Gastrocnemius
- Soleus

(b)

- Splenius capitis
- Trapezius
- Deltoid
- Infraspinatus
- Teres minor
- Teres major
- Triceps brachii
- Extensors of the wrist and fingers
- Semitendinosus
- Biceps femoris
- Semimembranosus

Hamstring muscles

- Gastrocnemius
- Soleus
- Calcaneal tendon (Achilles tendon)

- Sternocleidomastoid
- Seventh cervical vertebra
- Latissimus dorsi
- External abdominal oblique
- Gluteus medius
- Gluteus maximus
- Adductor magnus
- Iliotibial tract
- Gracilis
- Fibularis longus
- Fibularis brevis

△ **Fig. 2.12** Superficial muscles of the human body: (a) anterior view (b) posterior view.

acetabulum of the pelvis creates the socket and the ball is made up from the head of the femur bone. The rim of the socket is very shallow, allowing flexion/extension, adduction/abduction, and medial/lateral rotation. Most of the muscles that move the hip have their origin in the pelvis and the muscles that pass over the hip from the pelvis have actions on more than one joint, such as the rectus femoris (**Box 2.6**).

The muscles of the buttocks stabilize the hip joint when we walk or run, and provide for a backward drive of the leg, particularly when walking or running uphill. The muscles of the groin swing the leg towards the midline of the body (are adductor muscles) and originate from the pubic bone. The posterior and lateral muscles (gluteus maximus, medius, minimus and tensor fascia latae) are the external rotators of the hip.

The gluteus maximus or large buttock muscle straightens and adducts the hip, rotates the thigh outwards and aids in straightening the knee, whilst the gluteus medius is an abductor and rotator of the hip. The smallest buttock muscle, the gluteus minimus has the same function as the gluteus medius. Both stabilize the pelvis and are active when we walk.

Box 2.6

Major muscles of the hip

Buttock muscles

- Gluteus maximus (large)
- Gluteus medius (intermediate)
- Gluteus minimus (small)

Groin muscles

- Gracilis
- Adductor longus
- Adductor magnus
- Adductor brevis

Hip flexors

- Iliacus and psoas major (they have different origins but the same insertion, and are known under the collective name of iliopsoas)
- Sartorius (this, the longest muscle in the body, has many different functions)

The adductor muscles pass across the hip but have their origin on the pubic bone/groin area. The adductor longus (long), magnus (large) and brevis (short) all adduct the

Activity 2.2

Laboratory investigation: Range of movement of the hip

Measure the range of movement of the hip of a class member using the goniometers in your laboratory. Fill in the table below with your observed measurements.

	Expected range of movement (degrees)	Observed range of movement (degrees)
Flexion	115	
Extension	30	
Abduction	50	
Adduction	30	
External rotation	50	
Internal rotation	35	

hip. Additionally, the adductor brevis can rotate the hip outwards, the adductor magnus can rotate it inwards. Also in the groin, the gracilis (slender thigh muscle; which passes across the hip and knee joints with its origin on the pubic bone and insertion on the tibia) adducts the hip and bends the knee, rotating it inwards.

The muscles that flex the hip, the iliacus and psoas major, are attached to the lesser trochanter (point of insertion). They bend the hip, rotate the leg outwards and can also bend the vertebral column sidewards. A simple stretching exercise for the iliopsoas muscles is shown in Figure 2.13.

The sartorius (tailor's) muscle is the longest in the body and has many functions, including bending, abducting and rotating the hip outwards. It also bends the knee and rotates it inwards. It is termed the tailor's muscle because it is used in crossing the legs, a pose traditionally used by tailors when hand sewing.

Push right hip forward to stretch right iliopsoas. Keep low back flat and maintain upright posture.

△ **Fig. 2.13** Stretching exercise for the iliopsoas. Push the right hip forward to stretch the right iliopsoas. Maintain an upright posture and keep the lower back flat.

The knee

The thigh has only one bone, the femur, and its distal end has medial and lateral condyles that articulate with the tibia. Proximal to the condyles are medial and lateral **epi**condyles which can be felt on either side of the knee as they are normally covered by only a thin layer of skin (Figure 2.14). The distance between the medial and lateral epicondyles of the femur is known as the **knee distance**.

The knee is a modified hinged synovial joint with a knee cap (patella) lying on the top, located within the tendon of the quadriceps femoris muscle group. There can be flexion/extension movement of the joint. Slight rotation can occur but this is not a true movement. The knee has four principal ligaments:

- ACL (anterior cruciate ligament)
- PCL (posterior cruciate ligament)
- MCL (medial collateral ligament)
- LCL (lateral collateral ligament).

These act as stabilizers of the joint and prevent the joint moving in an unwanted range. A lack of or damage to knee ligaments therefore reduces the stability of the knee, making full recovery of knee injuries of this nature particularly important in contact sports to reduce the likelihood of recurrence.

Between the ends of the femur and tibia there are two half-moon shaped pieces of cartilage called the menisci, lying on the inside (medial meniscus) and the outside (lateral meniscus). These act as a shock absorber for the joint as well as providing nutrition to the joint via synovial fluid (Figure 2.15).

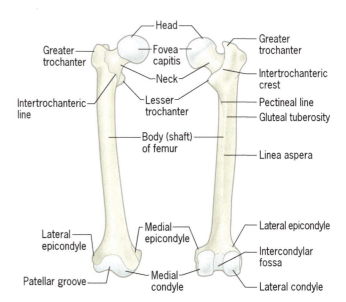

△ **Fig. 2.14** The left and right femur – joining the hip with the knee.

Activity 2.3

Laboratory investigation: Range of movement of the knee

Measure the range of movement of the knee of a class member using the goniometers in your laboratory. Fill in the table below with your observed measurements. Compare the left with the right knee.

	Expected range of movement (degrees)	Observed range of movement (degrees)
Flexion	135	Left:
		Right:
Extension	−10 (hyper-extension)	Left:
		Right:

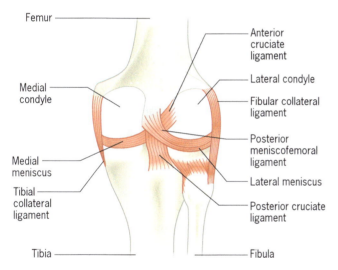

△ **Fig. 2.15** Posterior deep view of the right knee showing cruciate ligaments and menisci.

The knee extensors (muscles that straighten the knee when working concentrically, i.e. shortening) are the rectus femoris and the three vasti muscles. These vasti muscles are the external vastus (lateralis), the central vastus (intermedius) and the internal vastus (medialis). Together with the rectus femoris, the vasti form what is known as the four-headed thigh muscle group – **quadriceps femoris** (Figure 2.16).

The muscles that act as antagonists to the knee extensors are those of the back of the thigh. The collective name of this muscle group is the **hamstrings** and they act across both the hip and knee joints:

- The **biceps femoris** (two-headed thigh muscle) – straightens (extends) the hip, inserting onto the head of the fibula (lower leg).

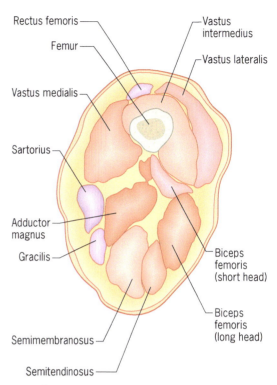

△ **Fig. 2.16** Schematic diagram of cross-section of the right mid-thigh (viewed from above) to show location and relative size of muscles at this point. Based on Engstrom *et al.* (1991).

- The **semi-tendinosus** and **semi-membranosus** – insert onto the tibia and they both straighten the hip and bend the knee and rotate it inwards.

Other muscles of the back of the thigh that join this group but are not strictly speaking the hamstrings are the muscles of the groin (gracilis, sartorius, adductor brevis, longus and magnus). Some of these can be seen on the schematic section through the mid-thigh shown in Figure 2.16.

Some simple stretching exercises for the quadriceps and hamstring muscle groups are shown in Figure 2.17.

The lower leg, ankle and foot

The lower leg is the part of the lower limb between the knee and the ankle. It consists of two bones: the **tibia**, commonly known as the shin, and the **fibula**. The ankle is a hinge joint made up of the tibia, fibula and the talus. The tibia and fibula run down from the knee and make a cup-like tunnel. Sitting in this tunnel is the box-shaped bone known as the **talus**. The talus moves backwards and forwards, in and out of the tunnel, creating the movements of dorsiflexion (toe towards the knee) and plantarflexion (pointed toes). Below the talus is the **calcaneus** (heel bone). The movement of the talus on the calcaneus along with the navicular bone create the action of adduction/abduction (in and out).

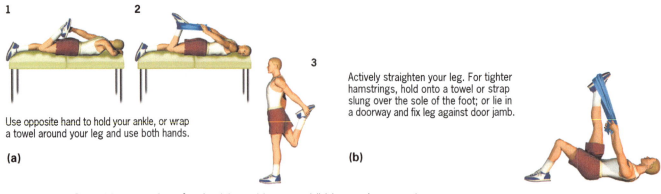

1

2

Use opposite hand to hold your ankle, or wrap a towel around your leg and use both hands.

(a)

3

Actively straighten your leg. For tighter hamstrings, hold onto a towel or strap slung over the sole of the foot; or lie in a doorway and fix leg against door jamb.

(b)

△ **Fig. 2.17** Stretching exercises for the (a) quadriceps and (b) hamstring muscle groups.

Activity 2.4

Laboratory investigation: Range of movement of the ankle

Measure the range of movement of the ankle of a class member using the goniometers in your laboratory. Fill in the table below with your observed measurements. Compare the left with the right ankle.

	Expected range of movement (degrees)	Observed range of movement (degrees)
Flexion	50 (pronation)	Left:
		Right:
Extension	20 (dorsiflexion)	Left:
		Right:

The muscle group at the back of the lower leg is known as the calf muscle, or the three-headed **triceps surae**. This consists of the two-headed **gastrocnemius** (one head from each of the condyles of the femur) and the flat, **soleus muscle**. These muscles form the heel tendon known as the Achilles tendon, which inserts into the heel bone. The gastrocnemius and soleus muscles allow flexion of the ankle such as the ankle position adopted when swimming the crawl (plantarflexion). In the movement of the foot, the most important flexors are the calf muscles, but movements are enhanced by muscles at the front of the leg, behind the shin. These include the tibialis anterior, extensor hallucis longus, extensor digitorum longus and the peroneus (or fibularis) longus and brevis. The extensor hallucis longus and extensor digitorum longus are responsible for dorsiflexion and pronation of the ankle (Figure 2.18).

Some simple stretching exercises for the gastrocnemius and soleus muscles are shown in Figure 2.19.

Muscle anatomy of the trunk

Muscles of the trunk can be categorized into those around the vertebral column, those of the thorax and those of the abdominal wall. The erector spinae group has three subgroups: the iliocostalis, the longissimus and the spinalis (Figure 2.20). Some simple stretching exercises for the erector spinae group are shown in Figure 2.21.

In the thorax, the muscles are mainly involved in the process of breathing (e.g. internal and external intercostals muscles and the diaphragm; Figure 2.22).

The abdominal wall muscles flex and rotate the vertebral column (Figure 2.23).

Muscle anatomy of the upper limbs

In the shoulder, the **supraspinatus** adducts and rotates the arm outwards, the **teres major** adducts the arm and turns it inwards, the **teres minor** and infraspinatus rotate the arm outwards, and the subscapularis rotates the arm inwards. They all have their origin on the shoulder blade (scapula) and are inserted into the upper arm (Figure 2.24).

The **levator scapulae, rhomboids** (minor and major), **serratus anterior** and **trapezius** have their origin on the trunk and are inserted on the shoulder blade. The levator scapulae raises the shoulder blade ('levate' meaning raise) and the rhomboids adduct and rotate the shoulder blade inwards. The trapezius adducts and rotates the shoulder blade outwards, turns the head and bends the neck backwards. The serratus anterior stabilizes the shoulder blade when the (outstretched) hand presses against an object.

The **pectoralis major** adducts the arm and rotates it inwards, the deltoid acts in all movements of the upper arm and latissimus dorsi swings the arm backwards and rotates it inwards. They have their origin on the trunk and their insertion on the arm.

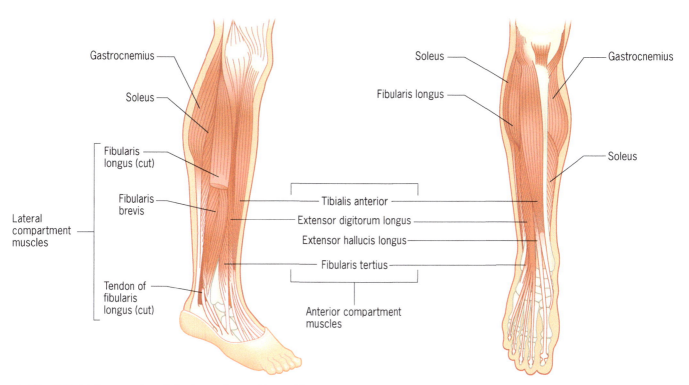

Gastrocnemius
Soleus
Fibularis longus (cut)
Fibularis brevis
Lateral compartment muscles
Tendon of fibularis longus (cut)

Soleus
Fibularis longus
Gastrocnemius
Soleus

Tibialis anterior
Extensor digitorum longus
Extensor hallucis longus
Fibularis tertius
Anterior compartment muscles

△ **Fig. 2.18** Major muscles of the lower leg, ankle and foot.

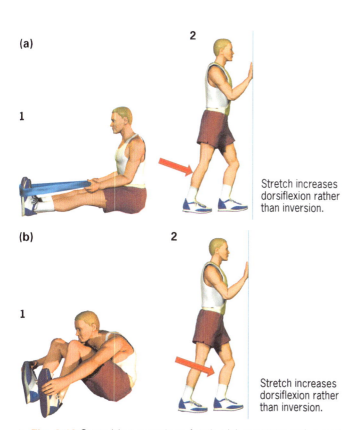

(a)
1
2
Stretch increases dorsiflexion rather than inversion.

(b)
1
2
Stretch increases dorsiflexion rather than inversion.

△ **Fig. 2.19** Stretching exercises for the (a) gastrocnemius and (b) soleus muscles.

The elbow

The elbow is an example of a hinge joint, the ulna and the humerus being the bones either side of the joint. The humerus is the single bone of the upper arm, running from the shoulder to the elbow and at its distal end it articulates with the two forearm bones – the radius and the ulna. The medial and lateral epicondyles function as the

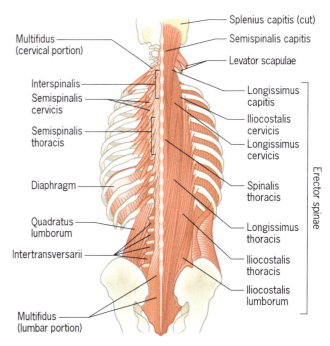

Multifidus (cervical portion)
Interspinalis
Semispinalis cervicis
Semispinalis thoracis
Diaphragm
Quadratus lumborum
Intertransversarii
Multifidus (lumbar portion)

Splenius capitis (cut)
Semispinalis capitis
Levator scapulae
Longissimus capitis
Iliocostalis cervicis
Longissimus cervicis
Spinalis thoracis
Longissimus thoracis
Iliocostalis thoracis
Iliocostalis lumborum

Erector spinae

△ **Fig. 2.20** Major muscles of the trunk – vertebral column.

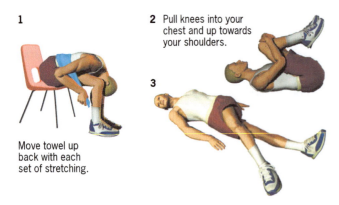

1

2 Pull knees into your chest and up towards your shoulders.

3

Move towel up back with each set of stretching.

△ **Fig. 2.21** Stretching exercises for the erector spinae group.

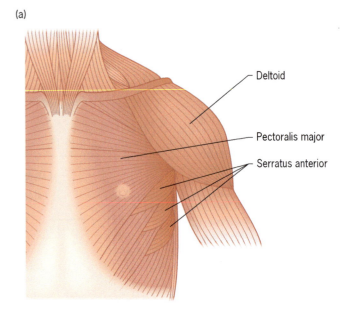

(a)

Deltoid

Pectoralis major

Serratus anterior

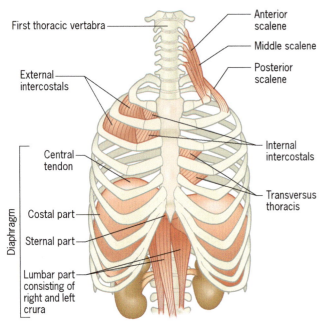

First thoracic vertabra

External intercostals

Diaphragm

Central tendon

Costal part

Sternal part

Lumbar part consisting of right and left crura

Anterior scalene

Middle scalene

Posterior scalene

Internal intercostals

Transversus thoracis

△ **Fig. 2.22** Major muscles of the trunk – thoracic muscles.

(b)

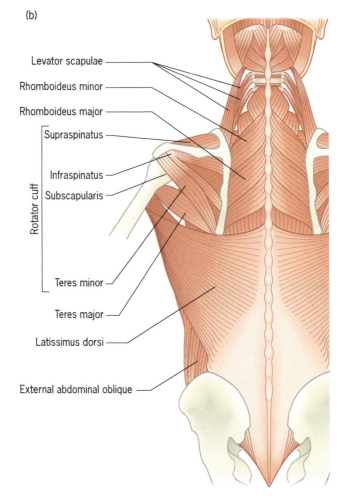

Levator scapulae

Rhomboideus minor

Rhomboideus major

Rotator cuff

Supraspinatus

Infraspinatus

Subscapularis

Teres minor

Teres major

Latissimus dorsi

External abdominal oblique

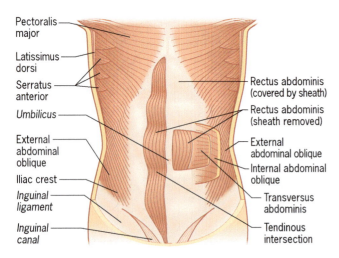

Pectoralis major

Latissimus dorsi

Serratus anterior

Umbilicus

External abdominal oblique

Iliac crest

Inguinal ligament

Inguinal canal

Rectus abdominis (covered by sheath)

Rectus abdominis (sheath removed)

External abdominal oblique

Internal abdominal oblique

Transversus abdominis

Tendinous intersection

△ **Fig. 2.23** Major muscles of the trunk – abdominal wall muscles.

△ **Fig. 2.24** Major muscles of the upper limb (a) anterior view (b) posterior view.

Laboratory investigation: Range of movement of the shoulder

Measure the range of movement of the shoulder of a class member using the goniometers in your laboratory. Fill in the table below with your observed measurements. Compare the left with the right shoulder.

	Expected range of movement (degrees)	Observed range of movement (degrees)
Flexion	170	Left:
		Right:
Extension	50	Left:
		Right:
Abduction	170	Left:
		Right:
Adduction	50	Left:
		Right:
External rotation	70	Left:
		Right:
Internal rotation	70	Left:
		Right:

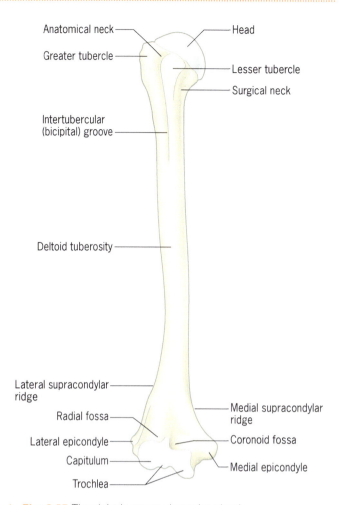

△ **Fig. 2.25** The right humerus (anterior view).

points of attachment for the muscles of the forearm and the elbow distance is that distance between the medial and lateral epicondyles of the humerus (Figure 2.25).

Flexion is possible through action of the **biceps** muscles, whilst extension is possible as a result of **triceps** action. The important flexors of the elbow are the biceps brachii (biceps 5 two heads, brachii 5 upper arm), which bends and supinates the elbow and swings the shoulder joint forwards, and the brachialis and brachoradialis which both bend the elbow.

The muscle responsible for extension/stretching of the elbow is the three-headed **triceps brachii**. One head is attached to the shoulder blade, the other two to the upper arm. Like the biceps brachii, it passes across both the shoulder and the elbow joints (Figure 2.26).

The wrist

There are several major muscles that affect both the forearm and the wrist (Figure 2.27), including: **extensor digitorum** (or finger extensor), to straighten the fingers, the wrist and the elbow; **extensor carpi radialis longus** and **brevis**, to extend and abduct the wrist joint; **extensor carpi ulnaris**, to extend and adduct the wrist; **flexor digitorum superficialis**, to flex the fingers and the wrist; **flexor carpi radialis**, to flex

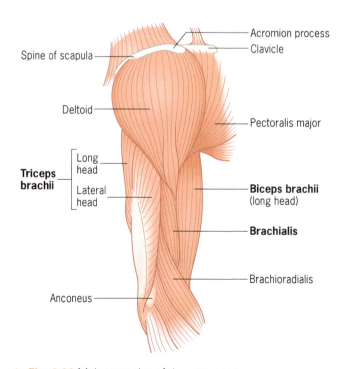

△ **Fig. 2.26** Major muscles of the upper arm.

Activity 2.6

Laboratory investigation: Range of movement of the elbow

Measure the range of movement of the elbow of a class member using the goniometers in your laboratory. Fill in the table below with your observed measurements. Compare the left with the right elbow.

	Expected range of movement (degrees)	Observed range of movement (degrees)
Flexion	150	Left:
		Right:
Extension	0	Left:
		Right:
External rotation	85 (supination)	Left:
		Right:
Internal rotation	70 (pronation)	Left:
		Right:

and abduct the wrist; and **flexor carpi ulnaris**, to flex and adduct the wrist.

This concludes this part on muscular and skeletal anatomy, and, indeed, the chapter. We have only looked at major muscles that you may come across as a sport and exercise practitioner, and there are many more that we have not described, such as the muscles of the scalp, face, mouth, and those of the hand. Obviously these remain very important

muscles, but one of the objectives of this text is as an introduction to the sport and exercise sciences, and if you need to study muscle anatomy at a deeper level you might consult the texts listed under Further reading. The role of joints and their attachments as levers in human movement are considered in Section 2, the Biomechanics section of this book.

Summary

- Human anatomy has a long and distinguished history, Hippocrates being considered one of the founders of the study of anatomy.
- The human body can be studied and viewed in a variety of different ways – from the surface, using cadavers, through the microscope and as a result of high-technology imaging.
- In anatomy and medicine, specific terminology and planes are used to describe the body.
- The body is composed of a variety of tissue types, including epithelial, connective and muscle tissues.
- The integumentary system covers the body and protects, regulates and senses.
- The skeletal system comprises cartilage, joints and bones and provides support and points of attachment for the skeletal muscles of the body.
- Synovial joints are of great interest to sport and exercise science, the three main types being ball, hinge and complex.
- The major muscles of the body act around joints such as the hip, knee, ankle, shoulder, elbow and wrist.
- There are about four hundred muscles in the body, about one hundred of these are of considerable interest to the sport and exercise scientist.

Review Questions

1. What are proteoglycans? In what tissues might they be found and why are they an important structural component?

2. Name the three main types of synovial joint and give an example of each. Choose one of these types of joint that is of particular interest to you in sport and exercise and describe its movement.

3. Explain what subtypes of joint the knee and elbow are. Do the same for the hip and shoulder.

4. Draw a schematic diagram of a section through the mid-thigh. Label and name the muscles that you will see in such a section. What are the common terms that describe the two major muscle groups?

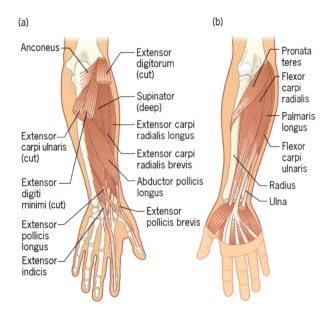

(a)

Anconeus
Extensor digitorum (cut)
Supinator (deep)
Extensor carpi radialis longus
Extensor carpi ulnaris (cut)
Extensor carpi radialis brevis
Extensor digiti minimi (cut)
Abductor pollicis longus
Extensor pollicis longus
Extensor pollicis brevis
Extensor indicis

(b)

Pronata teres
Flexor carpi radialis
Palmaris longus
Flexor carpi ulnaris
Radius
Ulna

△ **Fig. 2.27** Major muscles of the lower arm and wrist (a) posterior, deep view (b) anterior, superficial view.

References

Bergstrom J (1962) Muscle electrolytes in man. Determined by neutron activation analysis on needle biopsy specimens. A study on normal subjects, kidney patients and patients with chronic diarrhoea. *Scandinavian Journal of Clinical and Laboratory Investigation* 14 (Suppl 68): 1–110.

Bergstrom J, Hultman E (1966) Muscle glycogen synthesis after exercise. An enhancing factor localised to the muscle cells in man. *Nature* 210: 309–310.

Casey A, Mann R, Banister K, Fox J, Morris PG, Macdonald IA, Greenhaff PL (2000) Effect of carbohydrate ingestion on glycogen resynthesis in human liver and skeletal muscle, measured by ^{13}C MRS. *American Journal of Physiology, Endocrinology and Metabolism* 278: E65–E75.

Engstrom CM, Loeb GE, Reic JG, Forrest WJ, Avruch L (1991) Morphometry of the human thigh muscles. A comparison between anatomical sections and computer tomographic and magnetic resonance images. *Journal of Anatomy* 176, 139–156.

Harris RC, Soderlund K, Hultman E (1992) Elevation of creatine in resting and exercised muscle of normal subjects by creatine supplementation. *Clinical Science* 83: 367–374.

Hultman E (1967) Muscle glycogen in man determined in needle biopsy specimens. *Scandinavian Journal of Clinical and Laboratory Investigation* 19: 209–217.

Lindholm A, Piehl K (1974) Fibre composition, enzyme activity and concentrations of metabolites and electrolytes in muscles of Standardbred horses. *Acta Veterinaria Scandinavica* 15: 287–309.

Wandel S, Juni P, Tendal B, Nuesch E, Villiger PM, Welton NJ, Reichenbach S, Trelle S (2010) Effects of glucosamine, chondroitin, or placebo in patients with osteoarthritis of hip of knee; network meta-analysis. *British Medical Journal* 341: c4675.

Further reading

Jarmey C (2003) *The Concise Book of Muscles*. Chichester: Lotus Publishing.

Seeley RR, Stephens TD, Tate, P (2008) *Anatomy and Physiology*, 8th edn. New York: McGraw-Hill.

Williams PL ed. (2009) *Gray's Anatomy: the anatomical basis of medicine and surgery*, 40th edn. Edinburgh: Churchill Livingstone.

Wirhed R (1997) *Athletic Ability and the Anatomy of Motion*, 2nd edn. New York: Mosby.

3 Human Physiological Systems

If you learn to see clearly and to move freely in the half-lights of human physiology, you will see better and blunder less when you have to act in the quarter-lights of exercise physiology.

with apologies to Augustus D Waller, MD 1888
St Mary's Hospital Medical School

Chapter Objectives

In this chapter you will learn about:

- The major physiological systems important to sport and exercise science.
- The structure and function of key organs related to these physiological systems.
- The nervous system and the control of movement, including locating and understanding the role of the major endocrine glands, the general characteristics of hormone action, and the site and stimulus of release, and principal actions of a number of hormones.
- The structure of the cardiovascular system and of cardiac muscle and how it relates to the function of the heart.
- Lung structure and function, and the mechanism of respiratory exchange.
- The components of both the lymphatic and immune systems and how the components of both systems protect us from the threat of infection.
- The gastrointestinal, hepatic and renal systems, their structure and modes of action.

Introduction

In this chapter we will look at a range of physiological systems that serve particular types of function. The human body can be described as comprised of a number of discrete organ/physiological systems. Some of these have been discussed in other chapters of this section of the book (**Box 3.1**).

Box 3.1

Organ/physiological systems of the human body and where we discuss them

Chapter 2
- Integumentary system – Structures that cover
- Skeletal and muscular systems – Support and movement – Anatomical aspects

Chapter 3
- Nervous, endocrine systems – Control and communication
- Cardiovascular system and cardiac muscle – Transport
- Respiratory system – Exchange
- Lymphatic and immune system – Protection and defence
- Digestive, hepatic and renal systems – Regulation

Chapter 4
- Skeletal muscle (muscular system) – Physiological, metabolic, cellular and molecular aspects

The reproductive system of both males and females, whilst of obvious importance in biology, are of less importance to exercise biology at an introductory level and are not considered in this book.

As mentioned in the introduction to Chapter 2, human physiology is the scientific study of the process or function of the living body. As a postgraduate student of muscle and exercise metabolism, this author was struck by the writing of an eminent physiologist in a classical physiology text. Joseph Barcroft (1934), discussing the principle of maximal activity wrote 'It seems that the most uniform measure of an organ is in terms of the maximum it can accomplish. And why not? The purpose of the organ is to do work and not merely to vegetate.' As students of sport and exercise science, this is a philosophy that might pervade both your academic and personal endeavours.

In this chapter, our primary concern is the consideration of the living body at the organ/systemic level. In the previous chapter, we considered the skeletal and muscular systems as part of our anatomical consideration of protection, support and movement, and in this chapter we will consider the heart and cardiac muscle as part of the cardiovascular system. Due to the importance of skeletal muscle to sport and exercise, we will take a closer look at this tissue at the functional, metabolic, cellular and molecular levels in Chapter 4.

Homeostasis

Throughout life, the body is faced with a variety of challenges. These might be internal, such as exercise, illness and disease, or external, such as injury or environmental stress and they may be acute (of short duration) or chronic (of long duration). Our ability to survive depends on maintaining a relatively constant environment within the body; the existence and maintenance within tolerable limits of this constant environment within the body is known as **homeostasis**. The 'constancy' required is in the form of temperature control, maintenance of body water and control of the composition and nature of the extracellular fluid (ECF), including pH and osmotic pressure.

Homeostasis is of interest to us because challenges to this state can come from exercise or from additional environmental stress during sports-related activities, such as a low partial pressure of oxygen in the atmosphere in high altitude mountaineering. An acute bout of exercise, if it is of very high intensity, can be seen as a challenge, for example, to the maintenance of intracellular pH (pH inside the cells), whilst a prolonged bout of moderate exercise in the heat can be seen as a challenge to the maintenance of a desirable core temperature and blood volume status. Physiological changes that occur as a result of these disturbances or challenges to homeostasis involve the coordination and integration of a wide range of physiological systems, most of which will be described in this chapter.

When we repeat exercise challenges to homeostasis on a regular basis and over a period of time (i.e. exercise training), the body demonstrates its 'plasticity' or the ability to change to suit its needs.

The response to challenges to homeostasis are driven by negative-feedback mechanisms (**Box 3.2**). This mechanism of regulating homeostatic deviation is based on making the deviation smaller or resisting it. The minimum components of the homeostatic mechanism are normally a **receptor**, specialized to detect change, an **effector**, for example the cardiovascular system, and **coordinating** and **integrating mechanisms** that provide communication between the receptor and the effector. The nervous system and endocrine system provide communication.

> ## Box 3.2
>
> **Homeostasis – an example of negative feedback**
> 1. Blood pressure rises as a result of the action of the heart – detection is by baroreceptors in blood vessels (in the carotid artery).
> 2. This sends a signal to the brain.
> 3. The brain's control centre responds to the increase in blood pressure.
> 4. The brain sends a signal for heart to slow rate and force.
> 5. Blood pressure is reduced.

In evolutionary terms, when considering physical activity in relation to whole-body homeostasis, the normal state is a cycle of exercise/exercise recovery/exercise, etc., and when considering diet, the normal state is feast/fast/feast, etc., thereby ensuring, through exercise and diet, a metabolism that involves a constant turnover of tissue glycogen and intramuscular triglyceride stores. At the level of the whole body, it is therefore more relevant to consider lack of exercise, or physical inactivity as the 'disturbance' of homeostasis. As Barcroft (1934) points out 'The condition of exercise is not a mere variant of the condition of rest, it is the essence of the machine'. A gene-based or evolutionary approach to scientific enquiry in exercise physiology and metabolism is therefore to consider activity not as the intervention or experimental condition, but as the control condition, and sedentary individuals as the treatment condition.

The nervous and endocrine systems

Before discussing the communication between tissues provided by the nervous and endocrine systems, we need to briefly review the features of cellular physiology and communication that will help us to better understand the whole of physiological communication. In Chapter 1 we looked at the cell and its components. Of the total body water, about 55 per cent is contained inside the cell (intracellular fluid: ICF), and about 45 per cent forms the extracellular fluid (ECF) that surrounds the cell. The ECF is a medium through which exchanges between cells and their external environment occurs and it has a high content of sodium (Na^+) and chloride (Cl^-) ions. The ICF is high in potassium (K^+) and organic anions.

The plasma membrane separating the ICF from the ECF is composed of globular proteins embedded in a lipid bilayer, enabling lipid soluble solutes to permeate the plasma membrane by dissolving and diffusing in the lipid bilayer. Additionally, permeation can be through temporary combination with a specific membrane component. Ions and other molecules may cross the cell plasma membrane through channels (**membrane channels**) or with carriers (**carrier-mediated transport**). Carrier-mediated transport can be divided into **facilitated diffusion**, a passive process of transport whereby movement is driven by the electrochemical potential gradient, and **active transport**. Active transport is dependent on metabolic energy (ATP; metabolic energy production in the context of skeletal muscle metabolism is discussed in Chapter 4). Active transport is important, for example, for maintaining normal ionic concentrations such as K^+, Ca^{2+} and H^+ that are essential for many intracellular activities, for maintaining the ionic gradients that are the basis of resting membrane potentials and **action potentials** (used for signalling in nerves and for the activation of muscle), and for regulating cell volume (by transporting Na^+ out of the cell).

Whilst information exchange between cells is brought about by the nervous and endocrine systems, the final step in cellular communication is brought about by chemical signals (**neurotransmitters** in the nervous system, **hormones** in the endocrine system) that either diffuse between cells or reach cells via the blood. These are known as **first messengers**. Neurotransmitters include acetylcholine, which we will discuss in relation to the neuromuscular junction in Chapter 4, and the catecholamines dopamine and serotonin. A cell can also communicate with an adjacent cell through junctions such as gap junctions which transmit both chemical and electrical signals.

First messengers bind to the plasma membrane of a cell or to intracellular receptors to influence the behaviour of that cell, opening membrane channels or activating **second messenger systems.** Second messengers such as cyclic AMP (cAMP), Ca^{2+} and nitric oxide link first messengers (transmitters and hormones) to intracellular

responses, for example ion channel actions, gene expression and enzyme action. The cellular response is brought about after binding of the chemical signal (ligand) with a receptor at the cell membrane surface and in some second messenger systems this is brought about firstly by activation of G proteins, regulatory proteins that bind guanosine triphosphate (GTP).

The nervous system

The nervous system is a system of communication and control. One might also add the general role of integration to this. Specifically:

- it receives sensory information from receptor organs such as the skin, the ears and the eyes;
- it processes this information and generates responses;
- it regulates and coordinates so as to maintain homeostasis;
- it is responsible for higher functions such as emotion, memory and thinking; and
- it also controls muscular movement.

Afferent nerve fibres link receptors to coordinating systems in the central nervous system (CNS) and **efferent** nerve fibres carry information from the CNS to the effector organs such as muscle. Figure 3.1 illustrates the organization of the nervous system and its components and the way in which signals are transmitted and how they are integrated.

The CNS consists of the brain and the spinal cord; the peripheral nervous system (PNS) is made up of the nerves and structures that connect the CNS to the sensory organs to detect sensory information such as touch, sound and sight. There are 12 pairs of cranial nerves that emanate from the brain, and 31 pairs of spinal nerves that stem from the spinal cord.

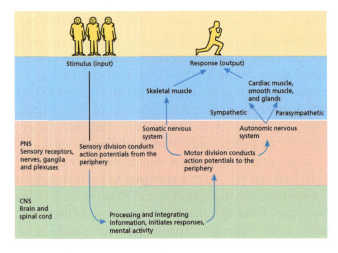

△ **Fig 3.1** The components of the nervous system.

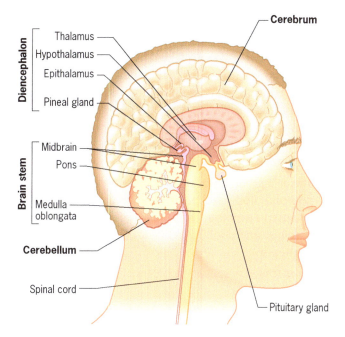

Diencephalon
- Thalamus
- Hypothalamus
- Epithalamus
- Pineal gland

Brain stem
- Midbrain
- Pons
- Medulla oblongata

Cerebrum

Cerebellum

Spinal cord

Pituitary gland

△ **Fig 3.2** The brain (sagittal section, medial view) and its four major subdivisions.

The brain consists of four major subdivisions – the **brainstem**, the **cerebellum**, the **diencephalon** and the **cerebrum** (Figure 3.2).

The brainstem links the cerebrum with the spinal cord and is composed of the **midbrain**, the smallest part of the brainstem that is responsible for conducting nerve impulses from the cerebrum to the pons, medulla and spinal cord, the **pons** (bridge), which contains bundles of axons connecting to different parts of the brain including the left and right sides of the cerebellum, and the **medulla oblongata**, which is a continuation of the spinal cord and has roles in the control of the cardiovascular and respiratory systems.

The cerebellum, situated posterior to the medulla and pons, contains about half of the neurons in the brain yet only makes up about 10 per cent of the brain mass. The function of the cerebellum is to evaluate and integrate movement as a result of muscular action. It does not initiate any movement, though it does participate in each movement through connections to and from the cerebral cortex. Sensory input from organs responsible for balance (see the vestibular apparatus later) and movement (muscles, tendons) is continuously monitored. In muscles and tendons, muscle spindles and Golgi tendon organs (described in Chapter 4) are responsible for the proprioceptive information that is fed back to the cerebellum.

The importance of the role of the cerebellum in ongoing motor activities such as skilled muscular activities (e.g. running, catching, throwing and speaking) is exemplified when the correction of errors, the smoothing of

movement and the coordination of activity breaks down. A cerebellar lesion caused by a stroke or tumour results in tremors and jerky movement, a condition known as ataxia, from the Greek 'lack of order'. Such damage to the cerebellum, however, does not affect sensation or muscle strength.

The diencephalon and cerebrum are sometimes considered together as the third major subdivision of the brain but can also be seen as separate entities. The diencephalon, nearest to the cerebrum is covered over by the intact brain and contains four parts. The **thalamus** is an intermediate relay and processing station for most sensory inputs that reach the cerebral cortex from the spinal cord, cerebellum and brainstem. The **hypothalamus** lies under the thalamus and has important functions such as integration within the autonomic nervous system (a subdivision of the peripheral nervous system), and regulation of temperature, water and electrolyte balance and food intake. The **epithalamus**, containing the pineal gland/body which secretes melatonin, and the **subthalamus**, which contains tracts and nuclei that connect to motor areas of the cerebrum, are the further parts of the diencephalon.

The cerebrum is the largest part of the brain and is divided into two halves known as the cerebral hemispheres, left and right. These consist of major divisions or lobes, namely the frontal, parietal, temporal, occipital, insula and limbic. The outer rim of grey matter is known as the cerebral cortex and is shaped by ridges (gyri), grooves (sulci) and depressions (fissures). The cerebrum controls voluntary movement and coordinates mental activity using its sensory areas (those that receive and interpret sensory impulses), motor areas that control muscle movement, and association areas that function in emotional and intellectual processes. Parts of the cerebrum and diencephalon form the limbic system that functions in emotional aspects of behaviour, and survival such as memory, reproduction and nutrition.

The size of the brain is related to the size of an individual, weighing about 1.4 kg in a typical 70-kg male, and is not related to intelligence. It needs to be protected and this is achieved by the connective tissue membranes that surround it, the cerebrospinal fluid and the skull. The membranes surrounding the brain are called the **meninges**; the most superficial is the **dura mater**, which is the thickest and is the one that holds the brain in place within the skull, preventing it from moving too freely. Between the dura mater and the next, thin meningal membrane, the **arachnoid mater**, there is a gap called the **subdural space**. The **subarachnoid space** is the gap between the arachnoid mater and the **pia mater**, the membrane that is bound tightly to the surface of the brain. The subarachnoid space is filled with **cerebrospinal fluid** (CSF), a clear, colourless liquid similar to serum. The brain therefore floats, and is protected against sudden movements and blows to the head by CSF, as well as being maintained in the chemical environment it needs.

The brain receives blood through the internal carotid arteries and the vertebral arteries. The arteries in the brain are located in the subarachnoid space and they quickly divide into capillaries lined by endothelial cells, constituting the blood–cerebrospinal fluid barrier (blood–brain barrier).

There are 12 pairs of **cranial nerves**, each of which has one or more functions. Part of the peripheral nervous system, they are numbered (by Roman numerals) and are also known by a name. The basic or principal functions are listed in **Box 3.3**.

The significance of particular cranial nerves can be seen in many clinical conditions. For example, Bell's palsy is a form of facial paralysis affecting the VIIth cranial (facial) nerve. It can cause one side of the face to droop, the eye to remain open and can affect taste, though few people suffer the symptoms for long, most completely recovering within 3–9 months. The cause of the swelling of the nerve and hence the paralysis is still not known, but it may be inflammatory rather than viral in origin. In the Scottish Bells Palsy trial (Sullivan *et al.*, 2007), early treatment with prednisolone (a corticosteroid/anti-inflammatory drug) significantly improved the chances of facial function recovery, whilst there was no evidence of benefit of acyclovir (an anti-viral drug) given alone or an additional benefit of acyclovir in combination with prednisolone.

The **spinal cord** is also protected by three meninges that are continuous with the cranial meninges – the dura mater, arachnoid mater and the innermost pia mater. The spinal cord propagates nerve impulses (white matter) and receives and integrates information (grey matter), illustrated in a typical transverse section of the thoracic spinal cord (Figure 3.3(a)).

A **reflex** is a predictable, involuntary and rapid response to a stimulus, and a **reflex arc** (Figure 3.3 (b)) is activated in response to events such as standing on a sharp object in the swimming pool. All reflex arcs consist of the same five elements:

- **Receptors**, e.g. sensory receptors in muscles, skin or sensory organs
- **Afferents** – sensory neuron axons/afferent nerve fibres of the receptors
- **Central neurons/integration** – excitatory and inhibitory inputs to these neurons
- **Efferents**, e.g. motor neuron axons
- **Effectors**, e.g. skeletal muscles.

A reflex that you might want to try out in the physiology laboratory is the knee jerk or patellar reflex (Figure 3.4). This is an example of a simple stretch reflex evoked by tapping a tendon, often used in medical and sports medical practice to check for neurological responsiveness in illness or injury. The brief muscle stretch caused by the tapping

Box 3.3

The cranial nerves

I Olfactory nerve	Sensory function	Olfaction (smell)
II Optic nerve	Sensory function	Vision
III Oculomotor nerve	Motor function[a]	Light accommodation
IV Trochlear nerve	Motor function	Visual proprioception/eye movement
V Trigeminal nerve	Mixed sensory/motor function	Touch sensation/mastication
VI Abducens nerve	Motor function	Visual proprioception/eye movement
VII Facial nerve	Mixed sensory/motor function[a]	Taste, facial expression
VIII Vestibulocochlear nerve	Sensory function	Hearing and balance
IX Glossopharyngeal nerve	Mixed sensory/motor function[a]	Taste, swallowing
X Vagus nerve	Mixed sensory/motor function[a]	Circulation, digestion
XI Accessory nerve	Motor function	Neck and shoulder movement
XII Hypoglossal nerve	Motor function	Speech, swallowing

[a]Have a parasympathetic component.

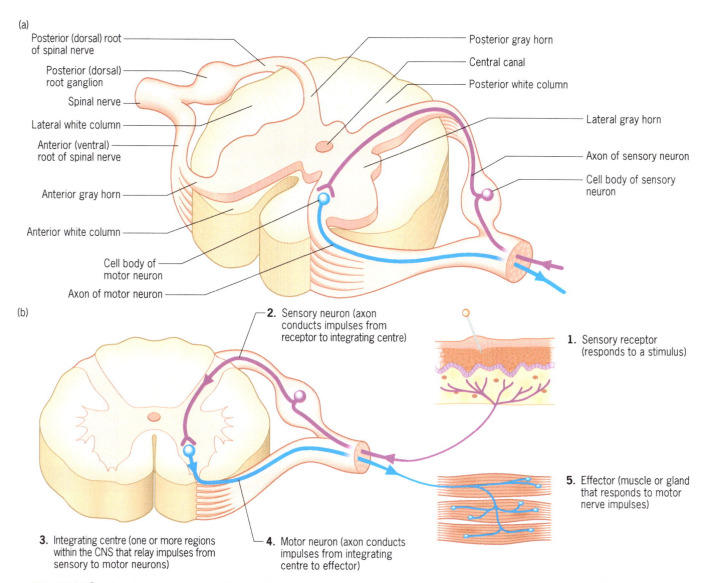

(a)

Posterior (dorsal) root of spinal nerve

Posterior (dorsal) root ganglion

Spinal nerve

Lateral white column

Anterior (ventral) root of spinal nerve

Anterior gray horn

Anterior white column

Cell body of motor neuron

Axon of motor neuron

Posterior gray horn

Central canal

Posterior white column

Lateral gray horn

Axon of sensory neuron

Cell body of sensory neuron

(b)

2. Sensory neuron (axon conducts impulses from receptor to integrating centre)

1. Sensory receptor (responds to a stimulus)

5. Effector (muscle or gland that responds to motor nerve impulses)

3. Integrating centre (one or more regions within the CNS that relay impulses from sensory to motor neurons)

4. Motor neuron (axon conducts impulses from integrating centre to effector)

△ **Fig 3.3** (a) Section through the thoracic spinal cord showing the organization of the white and grey matter and (b) the principal components of the reflex arc.

of the tendon excites the sensitive muscle spindle afferents (muscle spindles are described in Chapter 4).

The 31 pairs of **spinal nerves** emerge at regular intervals from the spinal/vertebral column and are identified based on their location relative to the vertebral column regions (Figure 3.5). There are eight pairs of cervical nerves (C1–C8), 12 pairs of thoracic nerves (T1–T12), five pairs of lumbar nerves (L1–L5), five pairs of sacral nerves (S1–S5), and one pair of coccygeal nerves (Co1). Spinal nerves provide the communication between the spinal cord and the nerves that innervate specific regions of the body.

In the knee/patellar tendon jerk spoken of above and illustrated in Figure 3.4, the root nerve responsible is the L4 (lumbar nerve) acting through the femoral (peripheral) nerve (Figure 3.6). Furthermore, in human movement of the lower limb for example, knee extension is

caused by the quadriceps muscle group, which is supplied by the femoral nerve which has its root in the lumbar nerves L3 and L4. The root and peripheral nerve supply of some important muscle groups can be seen in Figure 3.6.

The PNS can be subdivided into the **afferent**, or sensory division, responsible for signals (action potentials) from the sensory receptors to the CNS, and the **efferent**, or motor division, responsible for action potentials from the CNS to effector organs such as glands and muscles. Furthermore, the efferent or motor division can be further subdivided into the **somatic nervous system**, responsible for transmitting signals from the CNS to skeletal muscles (under voluntary control), where the effectors are skeletal muscles, and the **autonomic nervous system**, responsible for transmitting signals from the CNS to effectors such as

55

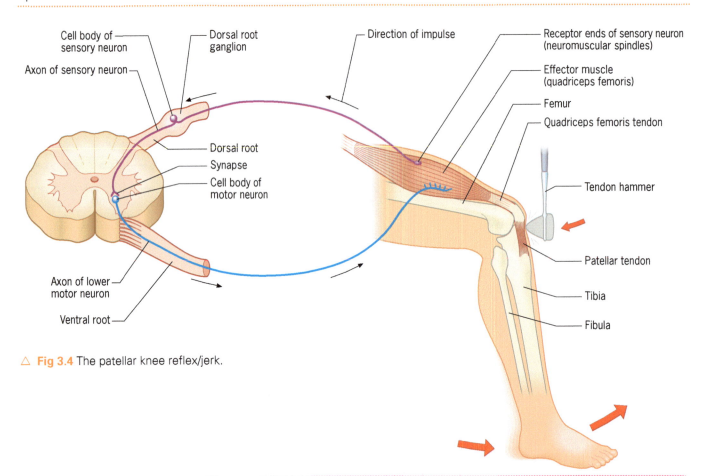

Cell body of sensory neuron
Dorsal root ganglion
Direction of impulse
Receptor ends of sensory neuron (neuromuscular spindles)
Axon of sensory neuron
Effector muscle (quadriceps femoris)
Femur
Quadriceps femoris tendon
Dorsal root
Synapse
Cell body of motor neuron
Tendon hammer
Patellar tendon
Axon of lower motor neuron
Tibia
Ventral root
Fibula

△ **Fig 3.4** The patellar knee reflex/jerk.

Box 3.4

Motor function: the nerve supply of some important muscle groups

Movement	Muscle group	Root nerve(s)	Peripheral nerve(s)
Upper limb:			
Shoulder abduction	Deltoid	C5	Axillary
Shoulder adduction	Latissimus dorsi Pectoralis	C7	
Elbow flexion	Biceps brachioradialis	C5-6	Radial
Elbow extension	Triceps	C7	Radial
Wrist extension	Extensor muscles of forearm	C7	Radial
Wrist flexion	Flexor muscles of the forearm	C7 (with some C8 contribution)	Median & ulna
Lower limb:			
Hip flexion	Iliopsoas	L2 (with some contribution from L3)	
Hip adduction	Adductors	L2,3	Obturator
Knee extension	Quadriceps	L3,4	Femoral
Knee flexion	Hamstrings	S1 (with some contribution from L5)	Sciatic
Ankle inversion	Tibialis anterior Tibialis posterior	L4,5	Sciatic (peroneal) Sciatic (tibial)
Ankle dorsiflexion	Tibialis anterior (& other anterior compartment muscles)	L4 (with some contribution from L5)	Sciatic (peroneal)
Ankle plantarflexion	Gastrocnemius	S1,2	Sciatic (tibial)

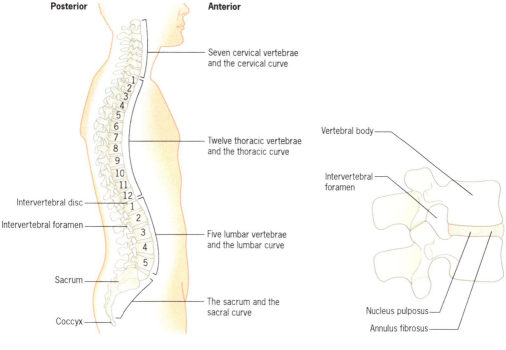

Posterior   Anterior

Seven cervical vertebrae
and the cervical curve

Twelve thoracic vertebrae
and the thoracic curve

Intervertebral disc

Intervertebral foramen

Five lumbar vertebrae
and the lumbar curve

Sacrum

The sacrum and the
sacral curve

Coccyx

Vertebral body

Intervertebral
foramen

Nucleus pulposus

Annulus fibrosus

△ **Fig 3.5** The vertebral column typically consists of a total of 26 vertebrae: 7 cervical, 12 thoracic, 5 lumbar, 1 sacrum and 1 coccyx. Spinal nerves emerge from intervertebral foramina.

cardiac and smooth muscle, and some glands (involuntarily controlled) (Figure 3.7).

The connection from the CNS in the somatic nervous system extends from the cell bodies of somatic motor neurons all the way to skeletal muscle as a single neuron pathway. The connection between nerve and muscle and the role of the somatic nervous system leading to the contraction of skeletal muscle is discussed further in Chapter 4.

The autonomic nervous system obtains sensory input primarily from interoceptors (e.g. those located in blood vessels that provide information about the internal environment) and is responsible for involuntary control from the limbic system, hypothalamus, brainstem and spinal cord. It operates on a two-neuron pathway system whereby **preganglionic** neurons extend from the spinal cord to autonomic ganglia and

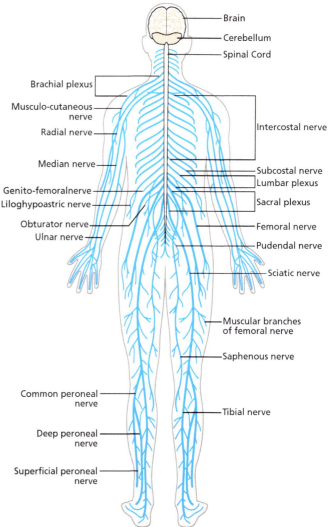

Brain
Cerebellum
Spinal Cord

Brachial plexus
Musculo-cutaneous nerve
Radial nerve

Intercostal nerve

Median nerve

Subcostal nerve
Lumbar plexus

Genito-femoralnerve
Liloghypoastric nerve

Sacral plexus

Obturator nerve
Ulnar nerve

Femoral nerve

Pudendal nerve

Sciatic nerve

Muscular branches
of femoral nerve

Saphenous nerve

Common peroneal nerve

Tibial nerve

Deep peroneal nerve

Superficial peroneal nerve

△ **Fig 3.6** Major nerves of the body.

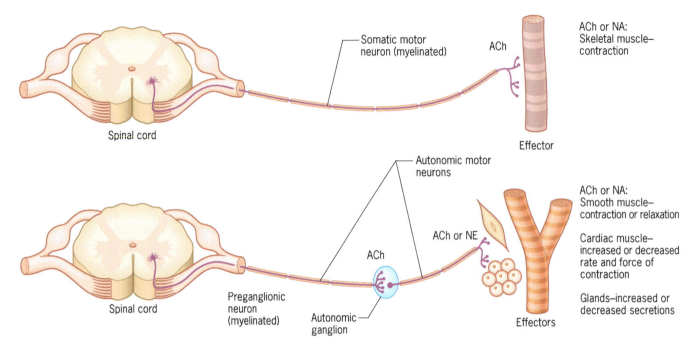

△ **Fig 3.7** Motor neurons in the somatic nervous system and the autonomic nervous system (ACh – acetylcholine; NA – noradrenaline).

postganglionic neurons extend from autonomic ganglia to effectors. Preganglionic neurons have their cell body in the brain or spinal cord and these first neurons of the series extend to autonomic ganglia located outside the CNS. The cell bodies of **postganglionic** neurons located in the ganglia then transmit signals to the effector organs and tissues.

The autonomic nervous system can be subdivided into the **sympathetic nervous system**, responsible for the physiological consequences of, for example, being chased by an opponent, and the **parasympathetic nervous system**, responsible for the physiological consequences of, for example, eating a large meal.

Nerve endings of neurons secrete neurotransmitters, which are principally either acetylcholine (**cholinergic** neurons) or noradrenaline (**adrenergic** neurons). A number of other neurotransmitter or neuromodulator substances have also been found in autonomic nervous system neurons, the specific functions of which are being investigated.

Neuron axons send messages electrochemically, meaning that the movement of chemical signals in the form of ions with electrical charge (important ones for the nervous system being sodium and potassium) cause an electrical signal to be transmitted. When at rest, a neuron is not sending a signal and the inside of the cell has a negative charge relative to the outside. This amounts to a difference in the voltage between the inside and the outside of the cell of around –70 mV and is known as the **resting membrane potential**.

An action potential or flurry of electrical activity sends an impulse down an axon, away from the nerve cell body. This is as a result of a depolarizing current that makes the inside of the cell electrically more positive than when at rest. A critical change of polarity has to occur, known as a threshold (at around –55 mV) before the neuron sends an action potential. The action potential is the same size regardless of the size of neuron, and as long as the threshold level is reached, a signal is sent – this is known as the 'all or nothing principle'.

The axons of most of our neurons are covered by layers of protein and lipid called the **myelin sheath** and are therefore known as myelinated neurons (Figure 3.8). Compared with unmyelinated fibres, nerve impulse conduction in myelinated fibres is increased because the myelin sheath electrically insulates the neuron. In order for the impulse to travel along the neuron it must leap or jump across unmyelinated gaps called nodes of Ranvier. This type of nerve conduction is called **saltatory conduction**, from the Latin *saltare* – to leap/jump. In a disease such as multiple sclerosis (MS), an autoimmune disease of the nervous system, myelin sheaths are destroyed. About 100,000 people in the UK have MS, it is most commonly diagnosed between the age of 20 and 40 years, with twice as many women having MS than men. The prevalence in Scotland is higher than in any other country in the world.

A further factor in the speed of conduction of nerve impulses is the diameter of the fibre – large-diameter fibres conduct impulses faster than smaller ones.

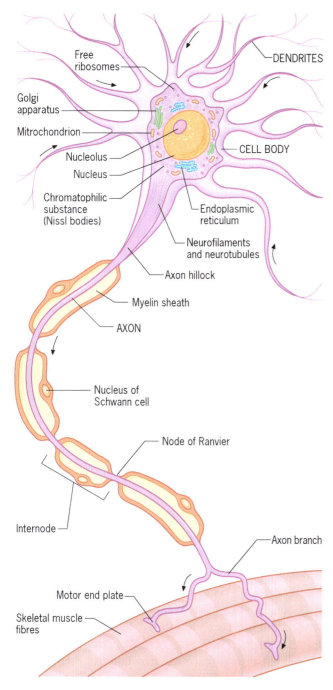

△ **Fig 3.8** Structure of a typical neuron showing dendrites, the cell body, an axon and axon terminals.

The senses

Special senses are localized to specific parts of the body, such as the eye, and whilst they are important to sport and exercise science, only a brief description will follow. These may well be topics that you need to cover in detail at a later time, depending on the specific nature of your course.

Olfaction

Our sense of smell is thought to be much more sensitive than our sense of taste. Most adults can detect about 10 000 different odours, although there are some substances, for example, the poisonous gas carbon monoxide, that are not detectable. Olfactory receptor cells located in a specialized region of the roof of the nasal cavity are connected via their axons directly to olfactory bulbs, specialized structures of grey matter in the olfactory region of the brain.

Olfactory cells are the only nerve cells in mammals capable of continual degeneration and replacement. In contrast, taste receptor cells are replaced less frequently as we age, explaining why our sense of taste may diminish with age.

Taste

Several thousand taste buds are located mainly in the tongue surface, with a few on the epiglottis, soft palate and pharynx. They contain several cell types – taste cells, basal cells and supporting cells – and open onto the surface of the tongue into taste pores and hairs. Taste cells are chemoreceptors, chemicals enter the taste pores and react with the receptor cells. They contain Na^+, K^+ and Ca^{2+} channels which, in response to chemical stimulation, lead to the release of neurotransmitters that generate action potentials in the afferent fibres of the nerves innervating the taste buds.

Taste (gustation) and smell (olfaction) are clearly interrelated sensations, as when we are unable to smell, food tastes become less distinctive.

Vision

Vision is a further specialized and dominant sense in which the brain uses the information from light-sensitive receptors in the retina to create a representation of the space around us. In doing so, a large part of the brain is used, and it is therefore of great importance in perception, and hence in sport and exercise psychophysiology.

The eye is the receptor organ for vision. It has three peripheral layers, an outer fibrous layer containing the sclera and cornea, a middle layer containing the vascular choroid, muscular ciliary body, lens and iris, and an inner neural layer, the retina containing rods and cones. Six muscles are responsible for eyeball movement. The optic nerves contain about one-third of all the afferent nerve fibres carrying information to the CNS; these leave the eye at the optic disc.

Hearing and balance (equilibrium)

Both hearing and balance are sensed by the inner ear, which is one of the three distinct anatomical compartments of the ear: the external ear, the middle ear containing the tympanic membrane (eardrum), the tympanic cavity, auditory tube and ossicles, and the inner ear containing the vestibule, semicircular ducts, cochlea and spiral organ.

The ear has two functional units known as the **auditory apparatus**, used for hearing, and the **vestibular apparatus**, used for posture and balance. The auditory apparatus is innervated by the cochlear nerve and the vestibular apparatus by the vestibular nerve, branches of the VIIIth cranial (vestibulocochlear) nerve (see Box 3.3).

The vestibular system provides information on spatial orientation and movement of the head (dynamic equilibrium) as well as regulating movement of the body and limbs and maintaining body posture. Static equilibrium refers to the position of the head with respect to gravity. For further understanding and learning of these major concepts and their importance in human movement you should consult texts about sensory physiology and applied human movement.

The endocrine system

Although referred to as a 'system', the endocrine system is, in fact, composed of ductless glands scattered all over the body (Figure 3.9) that affect the function of target cells of tissues. The endocrine system functions cohe-

sively by secreting substances or hormones either into the local tissue (paracrine and autocrine actions) or into the blood, where a hormone regulates or directs a more distant target. **Paracrine** actions are those that occur when cell secretions travel a short distance via the extracellular space to act on nearby cells, whilst **autocrine** actions are those that occur in the cells that produced the secretion.

Endocrinology is the study of how homeostasis is maintained by the action of hormones, how growth and development are regulated, and the interaction with the nervous system in control of stress responses. Several hormones influence exercise and energy metabolism; for example, during exercise, hormones such as insulin, glucagon, cortisol and the catecholamines (adrenaline and noradrenaline) are mostly responsible for substrate availability and utilization.

A hormone is a secretion of an endocrine gland which usually interacts with other hormones in a self-regulatory feedback system. They can be classified in different ways: by chemical structure (e.g. peptide, glycoprotein, catecholamine, steroid), by site of action (e.g. cell membrane, cell nucleus, cytoplasm), or by the nature of the action (e.g. opens ion channels, interacts with a membrane receptor to initiate a second messenger cascade, binds to nuclear or cytoplasmic receptors). A number of hormones, including cortisol and testosterone, are derived from the cholesterol molecule, which has a steroid nucleus with an alcohol group attached (see Figure 1.13 in Chapter 1). Remember we said earlier that second messengers such as cyclic AMP, nitric oxide, serve to link 'first messengers' (transmitters and hormones) to intracellular responses.

All hormones have some functions and features in common. They have a type of action that might for example be metabolic, kinetic or behavioural. They have an origin, are secreted and most are transported. For most this means synthesis, packing in vesicles (except steroid hormones), intracellular storage (except extracellular storage of thyroid hormones), and they are often bound to carrier proteins in the blood for transportation. The titre or concentration in the blood is kept very low (circa nano/picomole concentrations), which has the advantage that when the hormone is needed, small absolute changes cause large relative changes. This is brought about by release cascades, providing considerable amplification; for example, 0.1 µg of corticotrophin-releasing hormone (CRH; in the hypothalamus) releases 1.0 µg of adrenocorticotrophic hormone (ACTH: corticotrophin), which causes the release of 50 µg of corticosteroid (also termed glucocorticoid). The release of cortisol, the major glucocorticoid, is the end point of this endocrine axis, in which a hormone released from the hypothalamus stimulates a gland (anterior pituitary) to

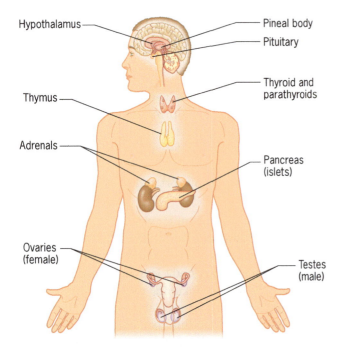

Hypothalamus

Pineal body

Pituitary

Thyroid and parathyroids

Thymus

Adrenals

Pancreas (islets)

Ovaries (female)

Testes (male)

△ **Fig 3.9** The major endocrine glands of the body.

release a hormone to stimulate a gland (adrenal cortex) to release a hormone (cortisol). This example is the hypothalamic–pituitary–adrenal (HPA) axis.

The pituitary gland and hypothalamus

Whilst these glands are anatomically distinct, functionally they are closely interlinked and dependent on each other, increasing the complexity of order and the process of hormone production. The pituitary gland is located directly below the hypothalamus (Figure 3.9). There are two anatomically and functionally distinct lobes – the anterior and posterior. The anterior lobe is also known as the adenohypophysis, and the posterior lobe as the neurohypophysis. The hypothalamus releases hormones that control the secretions of the anterior pituitary lobe and hormones released by the posterior pituitary are synthesized and regulated by neuronal centres in the hypothalamus.

The hypothalamus and anterior pituitary are linked by an extensive system of blood vessels called the hypothalamic-hypophyseal portal, and hormones produced in the hypothalamus are transported via the portal vessels to the anterior pituitary gland, where they either stimulate or inhibit the release of pituitary hormones – for example thyrotrophin-releasing hormone (TRH) stimulates the release of thyroid-stimulating hormone (TSH), and growth hormone-releasing hormone (GHRH) and growth hormone-inhibiting hormone (GHIH) stimulate or inhibit growth hormone (GH). Growth hormone contributes to the growth of bone and muscle.

The hypothalamus controls the release of anterior pituitary hormones. It excretes releasing and inhibiting hormones and controls blood and local levels of specific substances (e.g. TRH, CRH, gonadotrophin-releasing hormone (GnRH), GHRH, GHIH). The anterior pituitary produces TSH, which stimulates the thyroid hormones thyroxine and triiodothyronine via thyroid follicular cells. Production of these hormones increases metabolic rate, the sensitivity of the cardiovascular system to sympathetic nervous system activity, and the homeostasis of skeletal muscle. The anterior pituitary also produces ACTH, which stimulates the adrenal cortex to produce and secrete adrenal cortex steroids (e.g. cortisol).

The posterior pituitary has only its location in common with the anterior pituitary. Hormones released by the posterior pituitary (antidiuretic hormone (ADH), also known as vasopressin, and oxytocin) are synthesized and regulated by neuronal centres in the hypothalamus – hormones are transported as vesicles along nerve fibres from the hypothalamus, and when stimulated by nerve impulses are released. ADH increases water reabsorption via kidney tubules, which results in an increase in blood pressure.

The thyroid and the parathyroid

The thyroid gland (Figure 3.10) is responsible for the production of thyroxine (T_4) and triiodothyronine (T_3). The hypothalamus releases TRH which controls the production of TSH via the anterior pituitary, which in

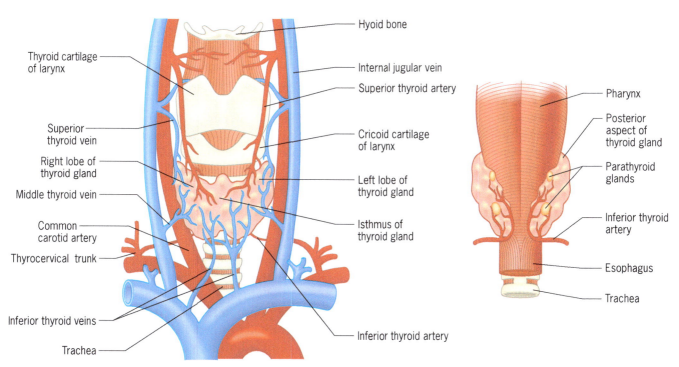

Thyroid cartilage of larynx

Superior thyroid vein

Right lobe of thyroid gland

Middle thyroid vein

Common carotid artery

Thyrocervical trunk

Inferior thyroid veins

Trachea

Hyoid bone

Internal jugular vein

Superior thyroid artery

Cricoid cartilage of larynx

Left lobe of thyroid gland

Isthmus of thyroid gland

Inferior thyroid artery

Pharynx

Posterior aspect of thyroid gland

Parathyroid glands

Inferior thyroid artery

Esophagus

Trachea

△ **Fig 3.10** The thyroid gland.

turn controls the thyroid gland production of T_4 and T_3. As mentioned earlier, these hormones increase metabolic rate and control homeostasis of skeletal muscle. This gland is located in the neck, in front of the trachea (Figure 3.10). Two types of cells make up the thyroid gland – the follicular cells which synthesize and secrete T_3 and T_4, and the parafollicular cells which synthesize and secrete calcitonin, decreasing plasma calcium and phosphate concentrations by inhibiting bone resorption. In direct opposition, the parathyroid hormone (PTH), secreted by the parathyroid glands, increases plasma calcium. An imbalance of plasma calcium and phosphate ions can cause nerve conduction changes, bone tissue degradation and growth, and muscle tetany.

The adrenal glands

The adrenal (ad-renal – 'upon the kidneys') glands are situated on the superior tip of each kidney. The two parts of each adrenal gland are the medulla – the inner portion, and the cortex – the outer portion, surrounding the medulla. The medulla and cortex are functionally distinct, producing different hormones which affect different target organs and tissues. The major hormones secreted by the medulla are adrenaline and noraderenaline (sometimes known as epinephrine and norepinephrine). The secretory cells of the medulla receive direct innervation from the sympathetic division of the autonomic nervous system (see Figure 3.1) and secrete these hormones, which act via sympathetic receptor sites around the body. Adrenaline and noradrenaline reinforce the sympathetic nervous system responses to stress and blood pressure regulation, for example they dilate coronary blood vessels, increase heart rate, and decrease the rate of fatigue in skeletal muscle.

The bulk of the adrenal cortex is formed by the zona fasciculata, one of its three distinct zones. This zone primarily secretes glucocorticoids (from the cortex (corticoid), influencing glucose metabolism (gluco)). Cortisol, corticosterone and cortisone are the three major types of glucocorticoid, cortisol accounting for most of the activity. One of the effects of glucorticoids is to stimulate gluconeogenesis (see Chapter 4). They also increase protein and lipid breakdown, have an anti-inflammatory effect, and depress immune responses. Heavy, prolonged exercise is associated with an increased level of 'stress' hormones such as cortisol and adrenaline, which can itself therefore inhibit some aspects of immune function.

The zona glomerulosa of the adrenal cortex secretes aldosterone, a mineralocorticoid, increasing sodium retention in kidney tubules (see the section 'Renal and urinary system' later in this chapter). The zona reticularis secretes small quantities of sex steroids (gonadocorticoids).

The pancreas

The pancreas is both an endocrine and an exocrine organ (see the section 'The digestive, hepatic and renal system' later in this chapter). In its endocrine role, the two major hormones it produces are insulin (released in the fed state) and glucagon (released in the fasted state). The pancreas also produces somatostatin (sometimes known as growth hormone inhibitory hormone) and is an example of a hormone that is secreted by more than one endocrine gland (the other being the hypothalamus).

The ingestion of carbohydrates results ultimately in the absorption of monosaccharides, the major one being glucose. In response to an increase in blood glucose concentration, insulin is released by exocytosis from the beta (β) cells of the islets of Langerhans of the pancreas (the β cells form about 60 per cent of the islet cells). Insulin is secreted into the hepatic portal vein (see Figure 3.29) and thus reaches the liver directly. Although the half-life of insulin in the blood is only a few minutes, it lowers blood glucose concentration by facilitating glucose uptake in muscle and adipose tissue, and by inhibiting hepatic glucose output.

Insulin is a peptide with two chains joined by disulphide bridges, and the peptides bind to membrane receptors to exert their effects via second messengers. The amino acid sequence of insulin was determined by Sanger in 1953, a landmark in biochemistry because it was the first time a protein was shown to have a defined amino acid structure. After a meal, insulin concentrations rise and this stimulates the increased uptake and utilization of glucose in almost all cells of the body, thereby reducing glycaemia (the level of glucose in the blood). The homeostatic 'set-point' for blood glucose concentration is between 4 and 5 mmol/l; a concentration of above 5 mmol/l is termed hyperglycaemia and a level below 4 mmol/l is termed hypoglycaemia. As one might expect, in response to exercise we need muscle to be utilizing glucose, not storing it in the liver and in muscle, so insulin concentration goes down when exercise begins (Figure 3.11).

Glucagon is produced by the alpha (α) cells, which constitute approximately 25 per cent of the islet cells. The basic action of glucagon is antagonistic to insulin and promotes the breakdown of glycogen (glycogenolysis) in the liver, hence increasing glycaemia. Blood glucagon levels either remain the same or rise in response to exercise, depending on whether we are trained or untrained. Like insulin, glucagon is also a peptide hormone, comprising 28 amino acids. Somatostatin is a short-chain peptide produced by the remaining delta (δ) cells of the pancreatic islets.

DIABETES MELLITUS

The most prevalent disorder of glucose metabolism is diabetes mellitus. This is the result of an absolute or a relative lack of the hormone insulin. The pancreas is no longer able to produce enough or any insulin and/or tissues become insensitive to the insulin that is circulating in the blood.

There are two types of diabetes mellitus (DM): type 1 or insulin-dependent DM (IDDM) and type 2 non-insulin-dependent DM (NIDDM). The latter is more prevalent (approximately 90 per cent of all diabetic cases) and the number of cases worldwide, primarily in the developed world, is rising dramatically. In 1985 there were about 30 million people with diabetes, but by 2010 there were around 285 million. In China there are believed to be around 92 million diabetics, and in the USA around 25 million. By 2030 it is estimated that there will be around 438 million people with diabetes (about 7.8% of the adult population).

Type 2 diabetes used to be termed 'late-onset' diabetes (i.e. later in life), but this is no longer an appropriate term as we now have an increasing number of cases of type 2 diabetes in (mainly overweight) children and adolescents. The causes of type 2 diabetes include physical inactivity, ageing, high-fat diets, and overweight and obesity.

Exercise therefore has a role in diabetes prevention (at best), delay (at worst) and also in its management. This presents a challenge and an opportunity to those trained in the sport and exercise sciences – to convince people of the merits of altering their physical activity and dietary lifestyles for the benefit of their long-term health.

△ **Fig 3.11** A typical insulin response to exercise (based on Gyntelberg *et al.*, 1977). Reproduced with permission.

Other sources of hormones

The **pineal body** is located on the roof of the diencephalon. Several chemicals that have hormonal activity, including melatonin, which is derived from serotonin, have been isolated from it but their functions are still unclear, though it is known that melatonin affects the modulation of sleep/wake patterns. Other glands and organs such as the kidneys, thymus, heart, gastrointestinal tract and placenta also have hormonal-producing functions. These, along with the reproductive organs and the hormones they produce are not major introductory topics

for the sport and exercise sciences. Although the reproductive hormones are not discussed here, skeletal muscle is, for example, influenced by testosterone and other hormones and this will be considered briefly in Chapter 4.

The cardiovascular system

In looking at the cardiovascular system, we will initially consider the heart in the context of the whole cardiovascular system, only briefly considering basic heart structure and function in the first instance. Then, because of the importance of the heart to exercise biology, a more detailed knowledge of the anatomy and physiology of cardiac muscle and the heart will be developed in the later part of this section.

The cardiovascular system is responsible for circulating blood and maintaining blood pressure throughout the body. In circulating the blood it ensures that gases and substrates are delivered to tissues, and that gases and metabolites are carried away from tissues. The **heart**, obviously a key component of the system, generates blood pressure, directs one-way blood flow and regulates blood supply. The basic structure of the human heart, like that of all mammals, has four chambers (two atria and two completely separated ventricles) and a double circulation (Figure 3.12), enabling oxygen-rich blood on the left side of the heart to be handled separately from oxygen-poor blood received and pumped on the right side. With no mixing of the blood between the different sides of the

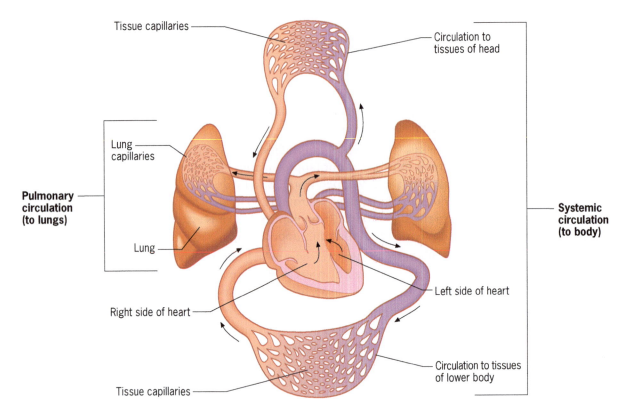

Tissue capillaries

Circulation to tissues of head

Lung capillaries

Pulmonary circulation (to lungs)

Lung

Systemic circulation (to body)

Left side of heart

Right side of heart

Circulation to tissues of lower body

Tissue capillaries

△ **Fig 3.12** The heart and the pulmonary and systemic circulation of the blood.

heart, and with a double circulation that restores pressure after blood has passed through the blood vessels of the lungs, delivery of oxygen to all parts of the body for oxidative metabolism is enhanced. At each heart beat the atria contract to fill the ventricles and the ventricles contract to eject blood from the heart, either to the lungs (through the pulmonary arteries) or to the body (through the ascending and descending aorta).

To prevent blood from flowing the wrong way in the heart, valves are needed. The **valves** between the atria and the ventricles open when the atria contract but shut when the ventricles contract. These heart valves are constructed with two flaps (the bicuspid valve between the left atrium and left ventricle) or three flaps (the tricuspid valve between the right atrium and the right ventricle).

Your study of the structure of the heart (Figure 3.13) can be enhanced in a number of ways. In your course you may have access to anatomical models of the heart. It may also be possible for your tutors to obtain some pig hearts (e.g. from the local abattoir). The pig heart is of particular interest because it is of similar size to the human heart. With some simple dissecting instruments (i.e. scalpel and scissors) you should be able to dissect the heart and identify and describe its key features, including its chambers and major vessels. For some students, it may also be possible for your tutors to arrange a visit to a Medical School anatomy-teaching laboratory where embalmed hearts can be looked at.

When presented with any of these opportunities you should try to do the following:

- Identify the structures passed by blood flowing through the heart
- Relate the anatomy of the heart to its pumping function
- Identify the structures found in each of the four chambers and consider their significance
- Compare and contrast the anatomical characteristics of the left and right sides of the heart
- Identify the arterial supply and venous drainage of the heart.

Marfan syndrome, a defect in the gene coding for fibrillin resulting in abnormal development of elastic fibres, was mentioned in Chapter 2 when we were discussing connective tissue. The most severe complication of Marfan syndrome is the weakening of the walls of the aorta, which can suddenly burst. Look carefully for the aorta, the main artery emerging from the heart, when you look at a heart model or when dissecting the heart.

Before we move on further, we should think a little about blood itself. You may be familiar with the simple ways in which blood samples are obtained – for example by pricking of soft fleshy areas such as the finger tips, ear lobe or heel. This, however, produces only a small amount of blood that is also not necessarily representative of the general venous circulation. You may well have had a sample taken from a

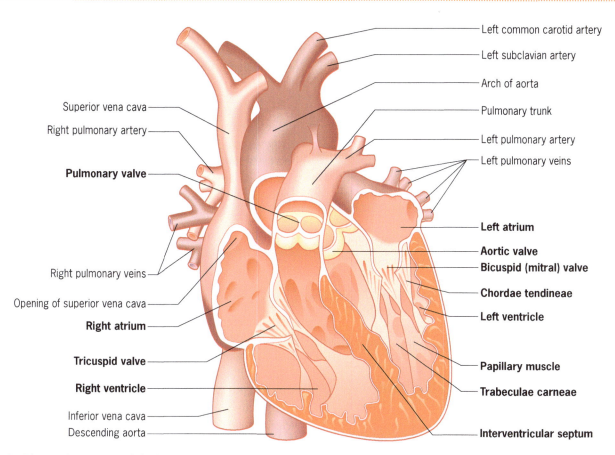

Left common carotid artery

Left subclavian artery

Arch of aorta

Pulmonary trunk

Left pulmonary artery

Left pulmonary veins

Left atrium

Aortic valve

Bicuspid (mitral) valve

Chordae tendineae

Left ventricle

Papillary muscle

Trabeculae carneae

Interventricular septum

Superior vena cava

Right pulmonary artery

Pulmonary valve

Right pulmonary veins

Opening of superior vena cava

Right atrium

Tricuspid valve

Right ventricle

Inferior vena cava

Descending aorta

△ **Fig 3.13** Internal structure of the heart.

TIME OUT

ERYTHROPOIETIN IN SPORT

Erythropoietin (EPO) stimulates the production of red blood cells, and drugs that are used to treat severe anaemia (a lower than normal number of red blood cells) have recently been abused in sport. Epoietin is a drug that is identical to naturally occurring EPO because it is manufactured using recombinant DNA technology. Two forms of epoietin are manufactured, alpha and beta, and a second generation hyperglycosylated derivative with a longer half-life (darbepoietin) is also manufactured. A third generation recombinant human erthyropoeitin – continuous erythropoietin receptor activator (CERA) – became available more recently, and in 2007 was considered to be undetectable; however, testing and re-testing of 2008 Tour de France samples revealed some positive CERA tests.

The use of epoietin (and the practice of blood doping) is dangerous in healthy individuals because it increases blood pressure, and the increased number of red blood cells raises the viscosity of the blood, increasing the resistance to blood flow. These factors therefore increase the workload of the heart.

EPO has been abused in endurance sports (e.g. cycling, cross-country skiing) because of the potential to increase the concentration of circulating red blood cells and as a result increase endurance exercise performance. Several athletes, particularly cyclists, have died of heart attacks and strokes linked to the suspected use of epoietin over the past 20 years. Needless to say, its use is banned in sport and exercise, however, it is difficult to detect due to its similarity to naturally occurring EPO.

vein in the antecubital fossa of the forearm. This method of obtaining a sample with the use of a hypodermic needle and syringe, or vacuum container, is known as venepuncture.

Blood has two major components, **plasma**, containing dissolved substances including proteins, electrolytes, nutrients, gases, waste products, and **formed elements**, containing cells and cell fragments. When blood is allowed to stand in a tube, it will begin to coagulate or stick together. When this process is over, the clot of red blood cells (RBCs) can be separated from the clear liquid called **serum**. Serum therefore does not contain clotting factors. If instead, the tube that the blood is collected in contains an anticoagulant such as heparin or EDTA, the cellular components can be separated from the clear liquid which is called plasma. Plasma therefore still contains deactivated clotting factors. When centrifuged, a blood sample containing anticoagulant separates into two major fractions – the heavier RBCs at the bottom making up about 45 per cent of the volume, with the remaining 55 per cent volume plasma. In between these two components is a thin coat of white blood cells and platelets.

There is a lot more to think about where blood is concerned, including the way in which blood cells are formed, their life cycles, how blood clots when we bleed, and blood groups and types; however, the introductory nature of this text makes these topics something that you might consult in more specialized texts. Issues that relate to blood that are of interest to sport and exercise include blood 'doping', the practice of infusing blood cells prior to competition, increasing the number of circulating red blood cells in an attempt to increase the oxygen-carrying capacity of the blood, and also the use of erythropoietin (see **Time Out Erythropoietin**).

Cardiac muscle and the heart

So far we have looked at the heart in the context of the whole cardiovascular system, stopping only briefly to consider heart structure and its basic function, but due to the importance of striated muscle to homeostasis and to sport and exercise it is important to establish a more detailed knowledge of the anatomy and physiology of cardiac muscle.

The heart is a pump and, along with the rest of the cardiovascular system, is responsible for circulating blood and maintaining blood pressure throughout the body. In circulating the blood it ensures that gases and substrates are delivered to tissues, and that gases and metabolites are carried away from tissues. In circulating oxygen and carbon dioxide, it has a key link with the respiratory system. Cardio-respiratory physiology will be considered later in this chapter, but first, how does the heart, in contrast to skeletal muscle, contract in a coordinated manner, involuntarily (thankfully!), minute after minute, day after day, year after year? Where do the signals to contract come from, and how do the cells transmit the wave of contraction?

Electrical properties of the heart

It has been known since the experiments of Galvani and Volta in the early 1800s that electrical phenomena are involved in the spontaneous contractions of the heart.

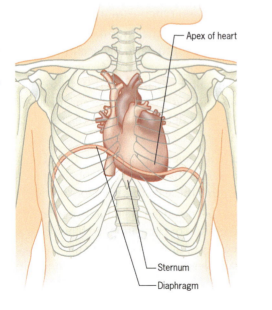

TIME OUT

HEART SIZE AND ENDURANCE TRAINING

The heart can be visualized using medical imaging techniques. It is often noted that the heart of some individuals is larger than others, despite being of similar body size. In women, the heart volume is usually around 0.5–0.6 litres, whereas in a well-trained endurance female athlete or swimmer the heart volume can be around 0.8–0.9 litres. On average men have larger heart volumes that might reach 1.7 litres or so (for example in a large, elite, trained cyclist).

Later, in the nineteenth century, Kollicker and Moller (1856, cited in Johansson, 2001) observed the twitching of a skeletal muscle preparation placed in contact with the surface of the heart – in time with each cardiac contraction. The first accurate recording of the electrocardiogram (ECG) was made in 1895 and, to put these physiological events into a more recent perspective, the first heart transplant was in 1967, the patient surviving for 18 days. Considering that heart transplantation is now relatively common and very successful, it can be seen that there has been an exponential advance in the knowledge and practice of cardiac physiology and medicine in the past 200 years. For an interesting history of electrical properties of the heart and related topics, you might take a look at **www.ecglibrary.com**.

Electrical events that normally take place in the heart initiate its contraction. These events reflect the ordered, electrical behaviour of cardiac muscle cells, conducting an action potential through the tissue (the myocardium). The summation of these electrical events can be detected by electrodes attached to the surface of the body, measuring electrical potential differences. Basic structural characteristics of cardiac muscle cells were discussed in Chapter 2 when we looked at muscle cell types. Interwoven myocardial cells are joined at their boundaries by intercalated discs, often visible in longitudinal sections of tissue when observed under the light microscope. The majority of myocardial cells form the contractile myocardium, whilst a smaller number constitute the specific excitation–contraction (EC) system. The intercalated discs contain gap junctions, permitting cell-to-cell conduction of excitation, and the myocardial cells are electrically coupled, forming what is known as a functional syncytium.

There are two main types of action potential in cardiac muscle fibres, those that have a fast response (normal fibres, Purkinje fibres) and those that have a slow response (fibres of the sinoatrial (SA) and atrioventricular (AV) nodes). The resting potential of slow fibres is less negative than that of the fast fibres. The average resting potential of the cardiac cells is around –70 mV to –80 mV (the inside of the cell having a negative charge relative to the outside), with Na^+ being the predominant extracellular ion and K^+ being the predominant intracellular ion. Excitation–contraction is ultimately brought about by Ca^{2+}, as it is in skeletal muscle.

Cardiac cell action potentials are characterized by depolarization, then a partial, early repolarization, followed by a plateau phase, before the final repolarization phase. As shown in Figure 3.14, the action potential consists of:

● **Depolarization** – reversal of the resting potential is brought about by a change in the permeability of the

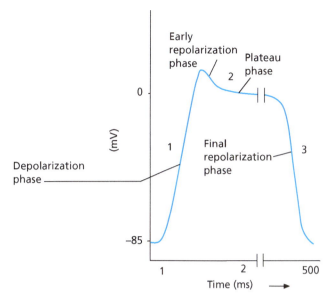

△ **Fig 3.14** Phases of the action potential in cardiac muscle.

membrane. There is a rapid diffusion of Na^+ into the cell and the inside becomes positively charged relative to the outside. After a short, partial early repolarization, there is a plateau phase, during which there is little overall change in charge; Ca^{2+} and Na^+ are moving inwards, K^+ moving out. Electromechanical/excitation–contraction coupling is brought about by Ca^{2+}.

● **Repolarization** – this re-establishes the resting potential (the negative charge on the inside of the cell membrane surface is restored). The principal ionic events are the closing of the Ca^{2+} channels (i.e. Ca^{2+} ions stop diffusing in) and the K^+ channels open (i.e. K^+ ions diffuse out).

The duration of the whole action potential is from 200 ms (in 'fast' fibres) to 500 ms (in 'slow' fibres), much slower than in skeletal muscle (see Chapter 4). During the action potential plateau the heart is inexcitable (which is known as the absolute refractory period) and during repolarization excitability gradually returns to normal (known as the relative refractory period).

The contractile myocardium is formed by the contractile myocardial cells of the atria and ventricles and these carry out the mechanical pumping work of the heart. Excitation–contraction system cells generate and rapidly propagate the excitation to the contractile myocardium. These EC fibres have fewer fibrils and mitochondria, and more cytoplasm than the contractile fibres. Bundle-branch and Purkinje fibres have larger diameters and a higher conduction velocity. The basic heart 'wiring' consists of the major EC system components known as the sinoatrial node, the atrioventricular node and the bundle of His with its left and right branches (Figure 3.15).

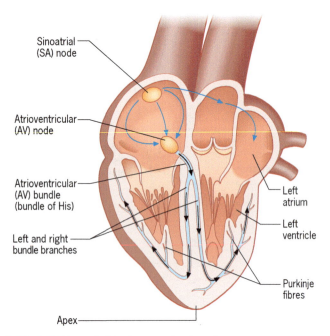

△ **Fig 3.15** Heart 'wiring' or conduction diagram.

AV bundle (also known as the bundle of His) into the tissue between the two ventricles – the interventricular septum. Here the action potentials travel down the left and right bundle branches to the apex of each ventricle before being carried by the Purkinje fibres to the ventricular walls.

The normal resting heart rate of adult humans is between 60 and 80 beats per minute (bpm), though lower resting heart rates are commonly found in people who exercise regularly and have done for some time, particularly if the exercise is of an endurance nature. This bradycardia (heart rate less than 60 bpm) can be perfectly normal. A resting heart rate of less than 35 bpm might still be normal but worthy of clinical enquiry if the athlete is concerned. Tachycardia (a heart rate in excess of 100 bpm) is a normal response to exercise, but resting tachycardia is not normal. Maximal exercising heart rate is around 200 bpm in young adults and this has a tendency to decline with age.

The heart has its own natural excitation properties of automaticity and rhythmicity that are intrinsic to cardiac tissue, and although influenced by the nervous system, its function does not always require intact nervous pathways. A heart can continue to beat outside of the body if kept in the correct physiological conditions. The region that generates impulses at the greatest frequency is the SA node, the 'pacemaker', and from the SA node, action potentials that originate there travel across the wall of the atrium to the AV node. The action potentials pass through the AV node and along the

The conduction of action potentials through the myocardium produces potential differences that can be measured at the surface of the body. The recording obtained by measuring these signals is known as the electrocardiogram (ECG). The ECG signals obtained are different depending upon the placement or configuration of surface electrodes. The most familiar ECG waveform is that seen as if the heart is being 'viewed' from the apex of the heart. There are four principal ECG events – depolarization and repolarization of the atria, and depolarization and repolarization of the ventricles.

TIME OUT

ELECTROCARDIOGRAMS

It is not necessary here to describe the measurement of ECG, suffice to say that you may well look at this in more detail in your course, as a means of understanding the structure and function of the heart – but not as a means of interpreting the ECG which is a task of medically qualified practitioners. Given the 'viewpoint' termed aVR, which is the 'view' from the right shoulder, almost diametrically opposite from the 'view' from the apex, and considering the principal ECG events and the ECG signal direction 'rules' described in the text, you might be able to predict the shape of the ECG waveform that is obtained from the aVR lead when measured. You might also work out, or at least find out, the shape of the other waveforms obtained from augmented and bipolar ECG leads. Refer also to Figure 3.15, which illustrates the directions of the action potentials that cause contraction of the chambers of the heart.

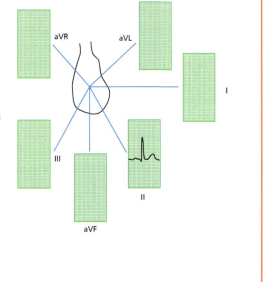

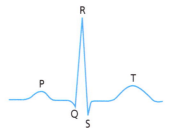

△ **Fig 3.16** The basic shape of the normal ECG 'viewed' from the apex with conventional labels, chosen arbitrarily in the early days of ECG history.

The ECG signal directions can be interpreted as follows:

- Depolarization moving towards the (positive) electrode (or 'viewpoint') results in an upward deflection of the ECG.
- Repolarization moving towards the electrode results in a downward deflection of the ECG.

In this way, knowing which way the heart is being viewed, for example from the apex, we can understand the ECG waveform and explain the waveform in terms of the ECG events (Figure 3.16). The first upward deflection (P wave) represents atrial depolarization, whilst atrial repolarization is masked by the relatively larger signal resulting from ventricular depolarization as the signal moves from the AV node, down the AV bundle and along the bundle branches (the QRS complex). The further upward deflection (T wave) is the result of the ventricular repolarization and is also an upward deflection because the summation of action potentials is moving away from the electrode.

In summary, action potentials travel from the SA node towards the AV node, with excitation spreading to both atria. Transmission of the action potential is relatively slow from the atria to the ventricles because the fibres are thin, but this allows time for atrial contraction to propel blood into the ventricles before they contract. The highly specialized conducting bundles increase the conduction velocity through the ventricles and the arrangement of the conducting system results in ventricular contraction beginning near the apex and progressing upwards through the ventricles, making pumping more efficient.

Mechanical and contractile properties of the heart

The heart is a pump, in fact two pumps working together, repetitively contracting (a phase known as **systole**), and relaxing (**diastole**). In this way, the heart moves blood through the circulatory system from areas of higher pressure to lower pressure. The phases combine to form the cardiac cycle which takes about 1 second to complete

Activity 3.1

Cardiac muscle

1. What speed of responses do cardiac cells exhibit?
2. Are cardiac muscle cell action potentials slower or faster than those of skeletal muscle?
3. What chemical element brings about excitation–contraction?
4. What is the natural pacemaker of the heart?
5. From which node does the action potential originate and is also where impulses are generated at the greatest frequency?
6. What does the P wave of the ECG represent?
7. Which branch of the autonomic nervous system has greatest influence on heart rate at rest?

if the heart is at rest. Systole, which lasts for about 0.27 seconds is characterized first by an isovolumic increase in the tension of the ventricular walls during contraction. The ventricles continue to contract, and during the period of ejection, the semilunar valves (of the aorta and pulmonary trunk) are pushed open and blood flows from the ventricles to the pulmonary trunk and the aorta. At the beginning of diastole, the ventricles relax and blood begins to flow back, causing the semilunar valves to close. When these semilunar valves are closed, all heart valves are closed and there is a period of isovolumic relaxation when no blood flows into the ventricles (end-diastolic volume).

When at rest, the atria aid filling of the ventricles in a relatively passive way as a result of their greater pressure of blood compared with the relaxed ventricles; when we exercise, however, atrial contraction has a more active role in filling the ventricles because less time is available for passive filling when heart rate is increased.

The volume of blood ejected at each contraction by any one ventricle/side of the heart is known as the **stroke volume**, which is the difference between ventricular filling and ventricular emptying. When the ventricles are filled, the volume is known as the **end-diastolic volume**, and the volume of blood remaining in the heart after contraction is known as the **end-systolic volume**. The quantity of blood pumped by the heart per unit of time is termed the **cardiac output**. This is defined as the quantity of blood ejected from the heart during each beat (stroke volume), multiplied by the heart rate (f_H). A typical cardiac output at rest in an adult is around 5000 ml/min, or 5 litres per minute. This, for example, is the product of the pumping frequency (f_H) of around 70 beats per minute and a stroke volume of around 72 ml per beat.

Fick (1829–1901) originally devised a principle as a technique for measuring cardiac output (CO). Fick postulated that the blood flow to an organ (he defined the human body as an organ in the first instance) can be calculated using a marker substance (e.g. oxygen). So if three things are known or can be measured:

Oxygen consumption (VO_2)

The oxygen content of venous (deoxygenated blood; C_v)

The oxygen content of arterial (oxygenated blood; C_a)

Then $VO_2 = (CO \times C_a) - (CO \times C_v)$

Rearranged, this can be:

$$\text{Cardiac output} = \frac{VO_2}{C_a - C_v}$$

Note that $C_a - C_v$ is also known as the arteriovenous oxygen difference (a-vO_2 difference: see also the section 'Oxygen transport in blood', later in this chapter).

Thus, Cardiac output =

$$\frac{\text{Oxygen consumption}}{\text{Arteriovenous oxygen difference}} \times 100$$

Blood pressure in healthy young adults has a systolic peak of around 120 mmHg, and a diastolic trough of around 80 mmHg and the total resistance to flow in the systemic circuit of the body at rest is made up of that of the aorta and arteries (about 25 per cent), the arterioles (about 40 per cent), the capillaries (about 20 per cent), and the venous system (about 15 per cent). The total resistance to blood flow is known as the **total peripheral resistance** (TPR), and as pressure is related to volume flow, and resistance, mean arterial blood pressure (MABP) can be stated as follows:

$$\text{MABP} = \text{Cardiac output} \times \text{TPR}$$

As described earlier, cardiac output is the product of rate and volume, and TPR is regulated quite automatically, and by sympathetic nerves and adrenaline, so we have no direct control of MABP – it is determined only by cardiac output and TPR.

One response to exercise of the cardiovascular system is the increase in cardiac output from around 5 litres at rest to between 20 and 30 litres during maximal exercise. The response is due to an increase in stroke volume in the rest-to-exercise transition, and an increase in heart rate. Heart rate can reach 200 bpm or more in some individuals. Maximal cardiac output differs between people primarily due to differences in body size and the extent to which they might be endurance trained. An improvement in cardiac performance brought about by endurance

training occurs as a result of changes in stroke volume (increased), heart rate (decreased for a set workload), ventricular mass and volume (increased).

Cardiac performance is determined by factors such as **preload**, **afterload** and heart rate. Preload is related to the end-diastolic volume when there is passive tension in the ventricles. The Frank–Starling law of the heart refers to the principle that the more the heart fills during diastole, the greater the force of contraction during systole, so the higher the end-diastolic volume (EDV), the more forceful the next contraction is. Afterload is related to the tension required for the ventricles to develop sufficient isometric (*iso* – same, *metric* – length) tension to open the aortic valve and hence the aortic pressure is regarded as equivalent to its afterload. Thereafter, the ventricles start to eject blood, shortening against the afterload by contracting isotonically (iso – same, *tonic* – tension).

As you develop knowledge of the responses and adaptations to exercise, you will learn to understand that people who have lower resting heart rates (e.g. some endurance-trained athletes) have an increased resting stroke volume because filling time is prolonged and preload is larger. For now we will move on from the cardiovascular system and consider how the oxygen and carbon dioxide that the blood circulates enters the body and is delivered to tissues.

The respiratory system

The air that we breathe contains approximately 21 per cent oxygen, a trace of carbon dioxide and the balance (around 79 per cent) nitrogen. We utilize oxygen in metabolic processes such as oxidative phosphorylation, resulting in expired air containing only around 16 per cent oxygen, and we generate carbon dioxide in processes such as the TCA cycle; around 5 per cent of expired air is carbon dioxide. Although the movement of oxygen and carbon dioxide between air and blood is by simple diffusion, unlike some simple organisms the human body is too large to survive without a specialized respiratory system – the lung. The primary function of the lung is to exchange gas but it has other roles, such as the metabolism of some compounds, the filtering of toxic materials from the circulation, and a reservoir for blood. The blood–gas barrier is very thin and gas is brought to one side of the barrier by airways and blood to the other side by blood vessels.

Air enters the body through the nose and nasal cavity (Figure 3.17), and the mouth, and travels through the **pharynx** which acts as both an air passage during breathing and a food passage from the mouth to the oesophagus. The **larynx** (sometimes called the voice box) is the passageway for air between the pharynx and the airways. The airways consist of a series of branching tubes which

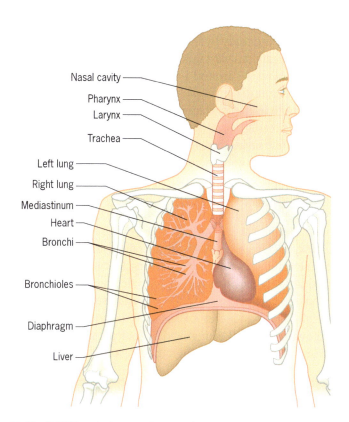

△ **Fig 3.17** Respiratory system anatomy.

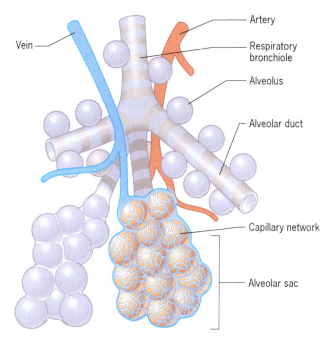

△ **Fig 3.18** Structure of the terminal airways and alveoli.

become narrower, shorter and more numerous as they penetrate deeper into the lung.

The **conducting airways** consist of the trachea, the left and right main bronchi, the lobar and segmental bronchi and terminal bronchioles, serving the function of leading inspired air to the gas exchanging regions of the lung. The **respiratory zone**, where gas exchange takes place, consists of the respiratory bronchioles (with occasional alveoli budding from their walls), the alveolar ducts, completely lined with alveoli (the site of gas exchange) and covered by a network of capillaries (Figure 3.18). As blood moves across the network of capillaries it gains oxygen and loses carbon dioxide to the gas in the alveolus.

During inspiration, the volume of the thoracic cavity increases and air is drawn into the lung. The increase in thoracic volume and the resulting intrapulmonary pressure change is brought about partly by the contraction of the diaphragm and partly by the action of the external intercostal muscles, which raises the ribs. Gas movement is by bulk flow to the level of the respiratory bronchioles and by diffusion beyond them. During exercise, inspiratory movements are assisted by accessory inspiratory muscles, allowing for large increases in tidal volume. We can measure tidal volume and other so-called 'static' lung volumes using a spirometer and in the physiology laboratory there might be a wedge bellows type, or an electronic

spirometer, or an on-line gas analysis system, capable of performing lung function tests. Some of the lung volumes and capacities that can be measured using different types of spirometry are illustrated in Figure 3.19.

Lung function tests (how well we can breathe in and out), peak flow and spirometry tests (how well the lungs are achieving their role) and blood gas tests (how well the lungs are getting oxygen into the blood) are carried out in clinical medicine in the diagnosis and management of conditions such as asthma, chronic obstructive pulmonary disease and lung cancer. Asthma is the most prevalent allergic disease in the UK, affecting around 1.1 million children and 4.1 million adults. In Scotland, the only deaths recorded for allergic disease (for example 100 in 2005) were for asthma. Asthma medication in elite sport and exercise can be controversial, for example in a

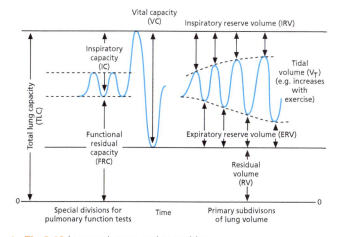

△ **Fig 3.19** Lung volumes and capacities.

doping control context, with athletes needing to declare what asthma medication they are taking, and for all drugs except salbutamol (a bronchodilator) have in place a Therapeutic Use Exemption (TUE). This is because of the perceived advantage that might be gained by non-asthmatics using asthma medications, and the potential anabolic effects of using such medications, generally in larger doses than is used for therapeutic use. The National Anti-Doping Agency in the UK is UK Anti-Doping (UKAD; see www.ukad.org.uk), which since 2010 has been a body independent of UK Sport, who previously had the doping control role.

Generally lung volumes and static capacities are related to body size, more so than aerobic capacity, and little difference is seen in pulmonary function (e.g. forced vital capacity, total lung capacity, forced expiratory volume) and in resting ventilation (e.g. minute ventilation, breathing rate, tidal volume) between good marathon runners and healthy control subjects (Mahler et al., 1982).

Despite the relatively high values of minute ventilation (V_E) reported in the study by Mahler et al. (1982), at rest, typically an adult might have a tidal volume (V_T) of 500 ml and we breathe at a rate of around 12 breaths per minute, resulting in a V_E of around 6 litres per minute. Breathing frequency and V_T both increase when we exercise, frequency up to about 50 breaths per minute and V_T up to about 3.5 litres per minute. The increase in V_E during exercise is accomplished initially by an increase in V_T, but after V_T reaches around 60 per cent of the vital capacity, further increases in V_E are accomplished primarily by increasing breathing frequency. The high values of breathing frequency and V_T mentioned above would result in a V_E of about 175 litres per minute. In large, male, endurance-trained athletes, V_E can reach even higher values, of 200 litres per minute or more. In addition to the increased rate and depth of breathing, the increase in gas exchange during exercise is also accomplished by an increased rate of diffusion of oxygen and carbon dioxide in the lungs and an increase in pulmonary perfusion (blood flow through the lung capillaries).

TIME OUT

LUNG VOLUMES AND STATIC CAPACITIES ARE RELATED TO BODY SIZE

As shown by the following table from the study by Mahler et al. (1982), simple lung function parameters such as pulmonary function and resting ventilation are unable to distinguish athletes from the general healthy population. No statistically significant differences between these values were found.

Measure	Marathon runners (n = 20)	Controls (n = 20)
Anthropometric		
Age (years)	27.8	27.4
Stature (m)	1.75	1.76
Surface area (m²)	1.82	1.89
Pulmonary function		
FVC (litres)	5.13	5.34
TLC (litres)	6.91	7.13
FEV_1 (litres)	4.32	4.47
FEV_1/FVC (%)	84.3	83.8
MVV (l/min)	180	176
Resting ventilation		
VE (l/min)	11.9	11.9
Breathing rate (breaths/min)	11	11
Tidal volume (litres)	1.16	1.06
From Mahler et al. (1982). Reproduced with permission.		

The alterations in ventilation and pulmonary perfusion enable oxygen uptake to increase to meet the energy demands of tissues. At rest, when we have a cardiac output of around 5 litres per minute, about 50 ml of oxygen is extracted per litre of blood. This equates to an oxygen uptake of about 250 ml per minute. Oxygen extraction is the difference between the oxygen content of arterial and venous blood (a-vO_2 difference). At rest, a typical arterial oxygen concentration might be about 200 ml/l and venous oxygen content might be about 150 ml/l. This example of the difference represents the amount of oxygen the **cardio-respiratory** system removes as blood passes around the body (50 ml of oxygen extracted).

Oxygen transport in blood

Transporting oxygen from the lungs to body tissues is one of the respiratory functions of the blood, and arterializing, or oxygenating mixed venous blood is one of the respiratory functions of the lungs. The amount of oxygen transported by the blood is a function of the cardiac output and the amount of oxygen carried by each millilitre of blood (oxygen content of the blood) – most of the oxygen is carried by combining chemically with haemoglobin, and only a small amount is dissolved in the plasma. The amount of oxygen available for the body at rest is therefore around 5000 ml (cardiac output) multiplied by 0.2 ml oxygen/ml blood (arterial oxygen content) which is around 1000 ml/min. As estimated above (oxygen uptake at rest), we use only around 250 ml/min of this available oxygen, leaving around 75 per cent of the unextracted oxygen in the mixed venous blood (75 per cent saturated) which is an 'oxygen reserve'.

Haemoglobin enables oxygen to be carried in the blood and each molecule of haemoglobin can transport four oxygen molecules. The capacity to carry oxygen varies with the concentration of haemoglobin in the blood, and if haemoglobin concentration in the blood is around 150 g per litre of blood (a typical adult male value), and when saturated each gram of haemoglobin can transport around 1.34 ml of oxygen, then 201 ml of oxygen is combined with haemoglobin per litre of blood. The further factor that determines the amount of oxygen bound to haemoglobin is the partial pressure of oxygen in the blood and the relationship between these two factors is known as the **oxyhaemoglobin association/dissociation curve**. The normal curve can be seen in Figure 3.20.

The shape of the curve shown in Figure 3.20 is for normal adult blood at 37°C and at pH 7.40, the normal pH of the blood at rest. In the higher part of the curve (blood partial pressure of oxygen, P_{O_2} = 90–100 mmHg), when the arterial partial pressure of oxygen (Pa_{O_2}) is 100 mmHg

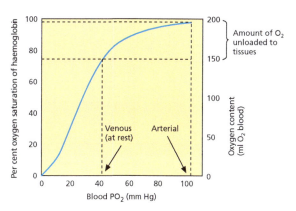

△ **Fig 3.20** The normal oxyhaemoglobin association/ dissociation curve.

(13.3 kPa), haemoglobin is 97.5 per cent saturated. Increasing Pa_{O_2} will therefore not add much more oxygen to the haemoglobin. Conversely, a reduction in Pa_{O_2}, for example, when ascending to altitude, does not result in a large reduction in the loading of oxygen to haemoglobin, enabling us to cope quite well in such circumstances. The lower portion of the curve concerns the release of oxygen, and a typical mixed venous partial pressure of oxygen (Pv_{O_2}) is 40 mmHg, a point where haemoglobin is still 75 per cent saturated. The steep part of the curve below this point means that the tissues are protected because they are able to withdraw large amounts of oxygen from the blood for relatively small decreases in P_{O_2}.

There are factors that cause a shift in the oxyhaemoglobin dissociation curve such as changes in **temperature** and **hydrogen ion concentration** (pH). Temperature affects affinity for oxygen – a fall in temperature increases the affinity for oxygen, and thereby reduces the release of oxygen to the tissues – this is seen as a shift to the left. An increase of body temperature (hyperthermia) increases the release of oxygen (bottom part of curve), seen as a shift to the right, but reduces oxygenation (top part of curve). Hydrogen ion concentration partly affects the rate of reaction between haemoglobin and oxygen. Alkalaemia (increasing pH) shifts the curve to the left (Figure 3.21) whilst decreasing pH (acidaemia) results in a shift to the right. The right shift of the curve as a result of a higher hydrogen ion concentration during exercise (i.e. the decrease in oxygen affinity of haemoglobin when pH decreases or concentration of carbon dioxide increases, for example) is sometimes known as the **Bohr effect** after Christian Bohr (1855–1911) (Figure 3.21). Exercise often results in a combination of both increasing body temperature (e.g. to 39°C) and metabolic acidosis, both factors contributing to a rightward shift of the curve.

Acid–base status in the whole body will depend on the response and integration of a number of systems. Metabolic

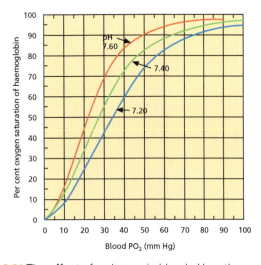

△ **Fig 3.21** The effect of a change in blood pH on the oxygen saturation of haemoglobin.

acidosis, for example, causes an increase in ventilation, stimulated by the rise in arterial hydrogen ion concentration. We shall consider briefly the contribution of extra-cellular buffering mechanisms to acid–base homeostasis. Carbon dioxide produced from the degradation of energy-rich substrates is transported by the circulation to be excreted fully by the lungs. No other excretory route exists and CO_2 balance is achieved when the amount excreted matches the amount produced by cellular metabolism. The body contains a large reservoir of extractable CO_2, a large portion of which is present in chemical combination. Approximately 1 per cent of the extractable CO_2 is dissolved as such in body water, and a negligible fraction is present in the gaseous state within the lungs. The CO_2 produced in tissues is transported by the blood to the pulmonary capillaries for excretion. Most of the CO_2 entering the circulation is converted to non-gaseous form for transport. The $[H^+]$ of the blood at any moment is determined by the ratio of the partial pressure of CO_2 (P_{CO_2}) and the concentration of bicarbonate $[HCO_3^-]$. Therefore the regulation of the pH of the blood plasma is accomplished by controlling the P_{CO_2} and the concentration of HCO_3^-, the two components of the major buffer in plasma. Carbon dioxide is hydrated to carbonic acid, catalysed by carbonic anhydrase, and dissociates to form a hydrogen ion and a bicarbonate ion.

$$CO_2 + H_2O \Rightarrow H_2CO_3 \Rightarrow H^+ + HCO_3^-$$

The CO_2–HCO_3 system minimizes the change in $[H^+]$ but permits some change to occur by allowing kidney function to fix the concentration of base, and lung function to fix the concentration of dissolved CO_2. Both P_{CO_2} and $[H^+]$ are involved in the control of respiration. The kidneys can contribute to the regulation of pH by varying the amount of $[H^+]$ excreted in the urine and the amount of HCO_3^- reabsorbed from the glomerular filtrate. The capacity of the kidneys to alter blood pH is much less than that of the lungs, but when given enough time to act (about three days) they can correct almost completely for deviations of blood pH from around 7.40.

The blood provides approximately 30 per cent of the whole body buffer capacity and two-thirds of this is contributed by the carbon dioxide–bicarbonate buffer system that involves the plasma and red blood cells. Haemoglobin provides about 8 per cent of whole body buffering capacity. The plasma proteins make a small contribution (about 2 per cent). The proportion of buffering provided by haemoglobin occurs through its combination with CO_2 to form carbaminohaemoglobin. This capacity is greatest when haemoglobin is completely free of oxygen. The blood from resting tissues is normally 70 per cent saturated with oxygen and allows for some CO_2 to be carried as carbaminohaemoglobin. This adds to a portion being carried in the form of carbamino compounds with the plasma proteins. More CO_2 can be carried away from active tissues because the haemoglobin has given up more oxygen. Further aspects of acid–base balance, and buffering capacity in skeletal muscle are discussed in Chapter 4.

The lymphatic and immune systems

The lymphatic system is generally considered an integral part of the vascular system (blood vessels), however, in the context of a broader perspective of human physiology, that is, an introduction to sport and exercise science, we will consider the lymphatic system alongside the immune system because one of its major roles is to defend the body against disease. The other important functions of the lymphatic system are the return of excess filtered fluid, the transport of absorbed fat, and the return of filtered protein. First, though, let us look at what the system is made up of.

The lymphatic system consists of organs, glands and tissues (Figure 3.22), including the thymus, the spleen, the tonsils, lymph nodes and a network of thin-walled, one-way vessels – **lymph capillaries** – lying in the interstitial space near blood capillaries. The walls of lymph and blood capillaries are quite similar, but lymph capillaries have larger spaces between endothelial cells, making them permeable to fluid and protein. These small capillaries converge to form larger **lymph vessels** that pass through **lymph nodes** found in the larger lymph vessels and eventually empty via the **lymphatic duct** and the **thoracic duct** to the thorax.

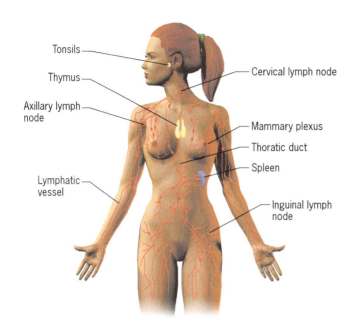

Tonsils

Thymus

Axillary lymph node

Lymphatic vessel

Cervical lymph node

Mammary plexus

Thoratic duct

Spleen

Inguinal lymph node

△ **Fig 3.22** Components of the protection and defence system include skin, tonsils, lymph vessels, lymph nodes, thymus, spleen and bone marrow.

Lymph is the interstitial fluid forced out of blood capillaries into spaces between the cells under the pressure of the blood. When it collects in the lymph capillaries it is known as lymph (from the Latin 'clear water'). The rate of lymph flow ranges from about 2 to 4 litres per day, obviously far less compared with the approximate 7000 litres of blood circulated through the circulatory system. The cells of the lymph are white blood cells (leukocytes), particularly of the class of white blood cell called lymphocytes, many of which are stored in lymph nodes. These white blood cells perform the same protective functions as the cells circulating in the bloodstream and these will be described under the topic of the immune system.

The functions that the lymphatic system performs include the return of the excess filtered fluid, relatively insignificant in rate but significant in terms of the 5 litres of blood volume of the body. The defence role includes mopping up bacteria in the interstitial fluid that are engulfed by the phagocytic cells located in the lymph nodes. The lymphatic system also packages dietary fats that have been digested into particles that are too big to gain access to blood capillaries but can enter the lymph capillaries, and, the plasma proteins that leak out of the blood capillaries can also gain easy access to the lymph capillaries and be returned to the general circulation, preventing a build-up of fluid in the interstitial fluid (and loss of plasma volume) as a result of protein-osmotic drive.

The **tonsils** and other lymphoid tissues such as the lymph nodes, the **appendix**, and tissue associated with

the gastrointestinal tract (sometimes known as Peyer's patches), exchange lymphocytes with the lymph, and produce antibodies and sensitized T lymphocytes from the lymphocytes residing there, and macrophages present in these tissues remove microbes. The **spleen**, an organ that we can sometimes do without if clinical needs result in removal, helps in the exchange of lymphocytes with the blood, produces antibodies and sensitized T lymphocytes from the lymphocytes residing there, and its macrophages remove microbes. It also acts as a small storage deposit for red blood cells in humans. In the horse, the splenic red blood cell reserve is much more impressive, as evidenced when a horse exercises maximally – its packed cell volume (also known as haematocrit) typically increases from around 40 per cent, to 60 per cent, making the increase in oxygen transport capacity associated with red blood cell release an important factor in the large aerobic capacity of the racehorse.

The packed cell volume in humans does not change significantly when we exercise, however, the effect of the dangerous (and not-permitted) practice of **blood doping** aims to raise the packed cell volume by around the same amount as that seen naturally in the trained, exercising horse.

The **thymus** is the site of maturational processing for T lymphocytes and produces the hormone thymosin. Removal of the thymus from adult humans seems not to have any particular effect, but it is more important during growth and development. It becomes less important as we age, when it atrophies and thymosin production decreases.

The lymphatic system is one of the defence methods the body has, and in the previous chapter we also learnt that one of the roles of the skin (integumentary system) is to act as the first line of defence to reduce the entry of external agents such as disease causing microorganisms (e.g. viruses and bacteria). To use a soccer or field hockey analogy, consider the two strikers in a 4–4–2 formation as the first line of defence (as long as they are behind the ball when the attack begins!). They are not specialist defenders, but they have a defensive role nevertheless. There are, however, a variety of other, more specialist ways that we can become susceptible to foreign agents, and the immune system provides us with further protection to combat biological and chemical threat. First, let us put into perspective the variety of features and parts of the body's defences (**Box 3.5**).

The second line of defence of the body, after the skin, is provided by semi-specific (innate) mechanisms provided by the inflammatory response, phagocytic cells and some proteins. This is a general response, if the forwards have been beaten, then we hope we have the midfield to cope with

Box 3.5

Overview of the body's defences

Non-specific defence

- First line of defence
- Skin and its secretions (e.g. sweat)
- Mucous membranes and its secretions (e.g. tears)

Semi-specific defence

- Second line of defence (innate immunity)
- Chemical, e.g. pyrogens, histamine
- Cellular, e.g. phagocytic leukocytes
- Biochemical, e.g. protein and lipid molecules
- Inflammation

Specific defence

- Third line of defence (adaptive immunity)
- Lymphocytes
- Antibodies

(a) Semi-specific (innate) defence

Neutrophil — Macrophage — NK

Acute phase proteins — Complement

Inflammation — Opsinization — Phagocytosis

Cell lysis — Phagocytosis — Cell lysis

(b) Specific (adaptive) defence

T Cells — B Cells

Helper — Cytotoxic — Antibodies (immuno-globulins)

Cytokines (interferon)

---

the ongoing attack. The third line of defence is specific or adaptive immunity, responding to specific disease-causing agents. Importantly, this response improves each time we are exposed to the foreign substance – rather as our muscle strength improves as we exercise regularly. If this third line of defence (our specialized defenders), fails to snuff out the threat, we only have the goalkeeper to save us!

The immune system has a key role in defence but it does seem to be a remarkably complex system. Why is this so? The types of foreign substance we have to defend against are numerous and diverse, ranging from environmental pollutants to the more obvious biological threats, such as viruses, bacteria, fungi and parasites (both protozoan and metazoan). Such infectious agents differ in their size, habitat, mode of attack and rate of multiplication. Not only that, but the diseases and conditions they inflict upon us have different stages – invasion/infection, multiplication, spread and disease – requiring different combative strategies. Such are the roles and requirements of the system. Why is knowledge of the immune system important for sport and exercise science? (see **Box 3.6**).

The different stages in the response to infection include the body's awareness (often before we perceive it), an immediate response, a delayed response, the destruction,

Box 3.6

The potential effect of physical activity on the immune system

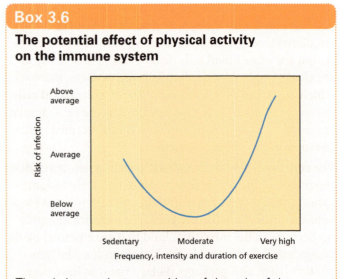

There is increasing recognition of the role of the immune system in overtraining. Exercise is considered to be of benefit to the immune system but when pushing human sport and exercise endeavours to their limit, there is a danger of training too hard and too much, pushing the body past a point of optimal (immunological) benefit. This hypothesis was put forward by Nieman (1990).

White blood cells
Granulocytes
Neutrophil

Basophil

Eosinophil

Agranulocytes
Lymphocyte

Monocyte

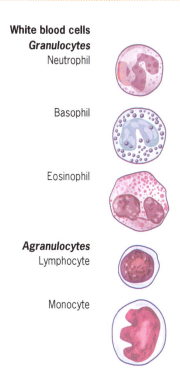

△ **Fig 3.23** Types of leukocyte (white blood cells).

elimination or neutralization of the agent and its products, and finally immunity. There are a variety of cells and molecules that constitute the innate immune response. **Interferons** are small proteins stimulated by cells when infected with a virus. Alpha-interferon is produced by leukocytes, beta-interferon by fibroblasts and gamma-interferon by T cells. The **complement** system is a group of normally inactive proteins that when stimulated set about attacking the plasma membranes of foreign cells. Of the semi-specific cellular responders, **natural killer cells** (NK cells), found mainly in the circulation, are like large granular lymphocytes, but they are not particularly specific and they destroy virus-infected cells and cancer cells by lysing their membrane, inducing apoptosis (programmed cell death) of the infected cell. In order to protect adjacent cells, NK cells secrete an interferon. **Leukocytes** or white blood cells (Figure 3.23) are produced in the bone marrow and lymphatic tissue.

Of the leukocytes that are involved in semi-specific/innate immunity, **neutrophils** are amongst the first cells to leave the blood and enter infected or inflamed tissues, phagocytosing foreign agents and releasing lysosomal enzymes that kill bacteria and viruses and cause inflammation. **Phagocytosis** is the process by which foreign particles are enveloped (endocytosis) and destroyed by phagocytic cells (**Box 3.7**).

Monocytes, another type of leukocyte, become **macrophages** when they leave the blood and enter tissues,

becoming much larger and having the capacity to phagocytose much larger particles than neutrophils can. As they enlarge, macrophages also increase their number of lysosomes and mitochondria. **Basophils** and **eosinophils** are also mobilized, leaving the bloodstream and entering tissues, and **mast cells**, non-motile cells found in connective tissue at potential points of entry to the body such as the airways, are activated, releasing histamine. Eosinophils release enzymes that help to break down chemicals released by basophils and mast cells, so, at the same time as proinflammatory mediators are being released, mechanisms are activated to control and regulate the inflammatory response.

Exercise causes an increase in the number of circulating leukocytes. This is known as exercise **leukocytosis** and is dependent upon the intensity and duration of the bout of exercise (Figure 3.24). There leukocytes are from the population of cells (the 'marginated' pool) that are normally attached to the vascular endothelium and are encouraged to join the circulating pool as a result of factors such as increased blood flow and increased hormone concentrations (e.g. adrenaline, cortisol).

We have so far spoken freely of the process of inflammation without describing it and why it is an essential part of the protective process. The inflammatory response is one that is designed to protect us, both when the body is infected by a pathogen, and for example, when we receive an acute physical insult, such as a sports injury. We

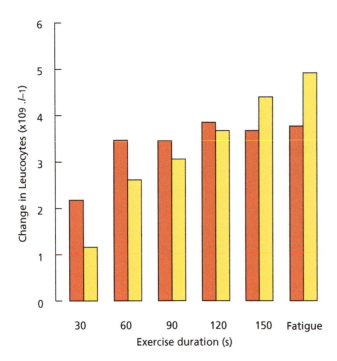

△ **Fig 3.24** Increases in blood leukocyte count (leukocytosis) following different durations of running. Red bars represent mean data from samples obtained immediately after exercise, yellow bars are values 5 minutes after exercise. Adapted from Gleeson *et al.* (1995). Reproduced with permission.

are all no doubt aware of the clinical characteristics of a traumatic blow to the body – by a cricket or hockey ball, or by an accidental blow from the boot of a rugby player – these are pain, swelling, redness, heat and loss of function. Whether we take heed or not, this is nature's way of trying to stop us from immediately using the injured area. The main physiological events occurring are the widening of blood vessels (vasodilation), activation of endothelial cells and mediators, an increase in the permeability of the blood vessels to cells (vascular permeability), and chemotaxis (chemical release to attract phagocytes to the scene). Blood proteins move into the tissue and water follows the proteins by osmosis, causing swelling (oedema).

One of the aims of applying ice to an inflamed area is to reduce swelling and reduce the pressure in the tissue which can also stimulate neurons, causing the sensation of pain. Locally, the response is for macrophages to recognize the problem and cytokines and other inflammatory mediators to be released. Monocytes and neutrophils are recruited to the site and plasma and proteins begin to accumulate. If mast cells are present in the damaged tissue then these may be activated to release histamine to amplify the response.

Specific or adaptive immune defence is dependent upon a cell type known as **lymphocytes** and consists of two types of response – those that are antibody mediated (or

humoral; involving body fluids or humors) and those that are cell mediated. The adaptive immune system brings specificity and memory to the aid of the defence of the body. There are two types of lymphocyte with different antigen receptors and surface molecules. Antibody production is brought about by **B lymphocytes** that mature in the bone marrow. Cell-mediated immunity involves the production of activated **T lymphocytes** that directly attack unwelcome foreign cells. T lymphocytes originate in the bone marrow but migrate from there to mature and differentiate in the thymus (Figure 3.25).

In early cell development, genes code for antibodies to create antigen-specific receptors. T lymphocyte antigen receptors are generated in each T cell, helping B lymphocyte (antibody) responses. When B lymphocytes come into contact with an antigen for the first time they divide and develop into 'plasma' cells, which produce antibodies that circulate in the blood. This is known as the primary response. Helper T lymphocytes amplify the process of binding of the antigen to the B lymphocyte's surface receptor by secreting **cytokines**. Cytokines are a family of mediators that includes the interleukins, and these provide cell-to-cell communication, important for the production of lymphocyte plasma cells. Not all B lymphocytes develop immediately into antibody-producing (plasma) cells. Instead, some remain inactive and retain their smaller size, circulating in the blood for a short time, but remaining in lymphoid tissue until the next exposure to the same antigen that produced the primary response. These B lymphocyte memory cells, having remembered the previous exposure, enable us to respond faster and to release more antibody than in a primary response.

The basic unit of antibody chemical structure consists of four polypeptide chains – two identical pairs of light and heavy chains. Antibodies bind to 'foreign' antigen molecules and belong to one of five major classes: IgM, IgG, IgA, IgE and IgD. They are represented as 'Ig' because

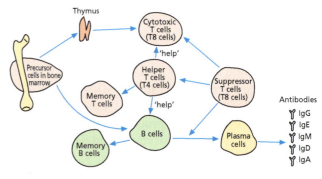

△ **Fig 3.25** Origin and differentiation sites of T and B lymphocyte cells.

they are sometimes called immunoglobulins (globulin proteins involved in immunity), or even gamma globulins. IgA is present in serum and secretions, helping out close to the first line of defence. IgG is the most abundant antibody in serum and IgM is the earliest antibody to be produced after first contact with an antigen. IgE is the mediator or central player of the allergic response (i.e. in allergic diseases such as asthma, eczema, hayfever).

As you might now appreciate, the protection and defence of the body is not a particularly simple system to get to grips with. Sometimes even the body fails to get to grips with the challenges that infectious agents bring and the defence system can experience faults of various kinds. Immunodeficiency might be considered a failure of the system to function properly, whilst autoimmune disease represents confusion in the system – the failure to differentiate between self antigens and foreign antigens. Autoimmune diseases affect around 3 per cent of the population and women appear more likely to develop autoimmune disease than men.

Another fault in the system is that of over-reaction (hypersensitivity/allergy), where the normal immune responses are stimulated but bring with them some unwanted and undesirable side effects. Allergies, with symptoms varying from mild to severe, affect about 17 per cent of the population.

This description of the structure and function of the major aspects of the protection and defence afforded by the lymphatic and immune systems should stand you in good stead for any future development of your interest and knowledge in this area. For sport and exercise it is a growing area of research activity and one that will no doubt result in advances that will most likely benefit the health and well-being of the elite performer who is pushing their body to its limits, and in doing so running the risk of overtraining and its consequences.

The digestive, hepatic and renal systems

The digestive system

As a result of not being able to store adequate phosphagens (ATP and phosphocreatine) to sustain even the most modest of everyday tasks, never mind perform exercise in the sporting context, we need to obtain our energy from food and ultimately convert it into the universal energy currency of ATP. The gastrointestinal tract or digestive system allows us to take in food from the outside (endogenous energy) and either utilize it immediately, or to store it in a variety of ways. As will be discussed in Chapter 5, the result of taking

in more energy than we require results in an addition to our body fuel stores and a gain in body mass. Conversely, the result of not taking in sufficient energy to match our expenditure results in utilization of our body energy stores and a net loss in body mass.

The digestive process involves what is termed the **alimentary tract**. The gastrointestinal (GI) tract is a term used to describe the stomach and intestines. In essence, the alimentary tract can be considered as a long tube (about 9 m in length) running from the mouth to the anus (Figure 3.26). It serves to break down large molecules of food into small soluble molecules that can be absorbed, circulated and used by cells.

Chemical digestion is aided by secretions and mechanical digestion is aided by processes including chewing. The breakdown begins in the mouth with chewing and enzymatic action (e.g. salivary amylase). Salivary glands include the three largest pairs – the parotid, submandibular and sublingual glands – producing around a litre of saliva a day, serving to moisten the walls of the mouth and lubricate food for chewing. They also provide amylase which begins the digestion of carbohydrates.

Swallowing of food has a voluntary phase that occurs in the mouth and an involuntary phase that involves the pharynx and oesophagus. The **pharynx** acts as both an air passage during breathing and a food passage during swallowing, whereas the **oesophagus** transports food from the pharynx to the stomach, passing through the diaphragm supported by the oesophageal hiatus on one side and the phreno-oesophageal ligament on the other side. Food then enters the body or corpus of the stomach through the cardiac sphincter (Figure 3.27). Before leaving the stomach into the duodenal bulb, food passes through the pylorus, containing the pyloric sphincter.

The proximal region of the stomach acts as a reservoir for food, and the distal stomach region, aided by retropulsion (a shearing effect that fragments larger food particles) mixes and breaks down food before controlling its release through the pyloric sphincter into the duodenum. The third major function of the stomach is to secrete gastric juices and enzymes, including hydrochloric acid, pepsinogen and gastric lipase. These secretions initiate the digestive process and kill most of the foreign agents that enter the stomach with food. The contractile activity of the stomach is regulated by myogenic, neural and hormonal mechanisms. By regulating motility, gastric emptying is controlled precisely, thereby presenting food to the duodenum at a rate designed to match the digestive and absorptive capacities of the small intestine.

Some alcohol is metabolized in the stomach by alcohol dehydrogenase before absorption. It is believed that there

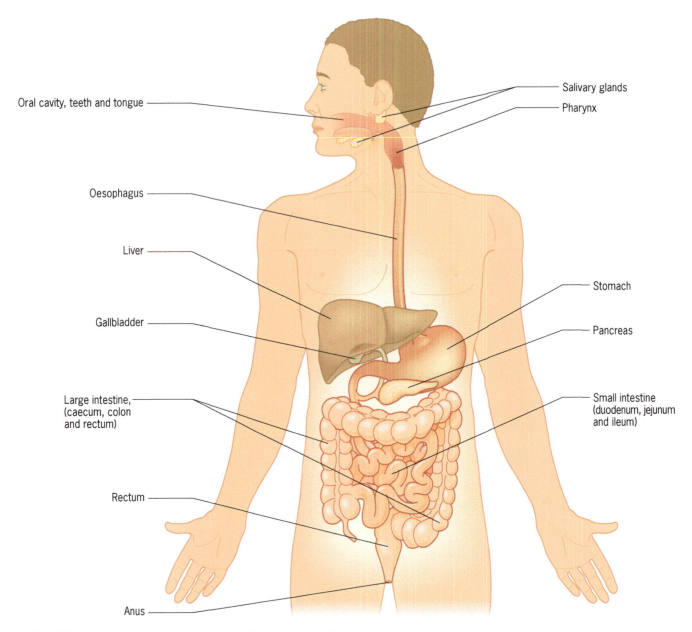

Oral cavity, teeth and tongue

Salivary glands

Pharynx

Oesophagus

Liver

Stomach

Gallbladder

Pancreas

Large intestine,
(caecum, colon
and rectum)

Small intestine
(duodenum, jejunum
and ileum)

Rectum

Anus

△ **Fig 3.26** Overview of the gastrointestinal/alimentary tract. Major parts from mouth to arms.

are marked variations in gastric alcohol dehydrogenase activity between men and women and ethnic groups, and a high fat intake prior to consumption of an alcoholic drink slows absorption.

As food (or chyme as it is by this stage) passes into the duodenum, it is added to by secretory products of the liver, the gallbladder and the pancreas (acting in its exocrine role), through ducts that merge and join the duodenum and are controlled by the sphincter of Oddi, a little way after the duodenal bulb. The **pancreas** is a gland about 150 mm in length, lying across the abdominal wall, behind the stomach. Earlier in this chapter, under the heading of the endocrine system, the part of the pancreas that consists of the endocrine cells, the islets of Langerhans, was discussed. These only make up about 1 per cent of the pancreas, the remaining 99 per cent of cells being exocrine cells, forming groups called acini. Acini secrete about 1.5 litre of fluid per day into the small intestine, consisting of water, enzymes, salts and sodium bicarbonate. **Sodium bicarbonate**, an alkali, is important in buffering the acidic chyme, inhibiting pepsin and providing an optimum pH for intestinal enzymes to operate. The digestive enzymes of pancreatic secretions hydrolyse the major groups of nutrients and include pancreatic lipase (to break down triglycerides), pancreatic amylase (to break down polysaccharides

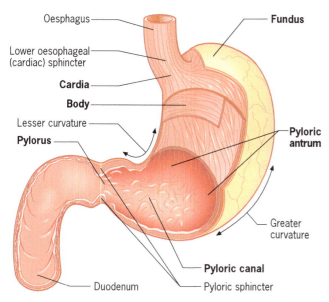

Labels: Oesphagus, Lower oesophageal (cardiac) sphincter, **Cardia**, **Body**, Lesser curvature, **Pylorus**, Duodenum, **Fundus**, **Pyloric antrum**, Greater curvature, **Pyloric canal**, Pyloric sphincter

△ **Fig 3.27** The stomach and its regions.

into dissacharides), and proteolytic enzymes such as trypsinogen, chymotrypsinogen and procarboxypeptidase (to break down proteins).

The **gallbladder** accumulates bile that is produced in the liver. The production of bile by the liver will be discussed later, but bile is ejected by contraction of the gallbladder and relaxation of the sphincter of Oddi. Secretion is only required following a meal in order to digest fats, and between meals the contraction of the sphincter of Oddi causes bile to accumulate. Bile salts are important in the digestion and absorption of fats and the absorption of fat-soluble vitamins (A, D, E, K), because of their emulsifying action. Bicarbonate

is again important to help neutralize the acid chyme entering the duodenum from the stomach.

The start of the small intestine is the **duodenum**. After trypsinogen (from the pancreatic secretions) is secreted into the duodenum, it is activated by the enzyme enterokinase and becomes trypsin, which converts chymotrypsinogen and procarboxypeptidase into their active forms – chymotrypsin and carboxypeptidase, respectively. The small intestine sections (duodenum, jejunum and ileum) are responsible for mixing chyme from the stomach with pancreatic and gallbladder secretions, exposing the contents to the small intestine wall for the final stages of digestion, and for absorption to occur. They also propel, through segmentation, pendular contractions and peristalsis, the intestinal contents towards the large intestine.

The **jejunum** and **ileum** are the major sites of nutrient absorption, aided, as in the duodenum by villi which increase the surface area available. The major enzymes of the small intestine, bound to the membranes of the microvilli (extensions on the surface of the villi), are the **disaccharidases** maltase, lactase and sucrase, which break down maltose (glucose–glucose), lactose (glucose–galactose) and sucrose (glucose–fructose) into their constituent monosaccharides, and **peptidases**, which break down peptide bonds between amino acids.

When food is completely broken down into basic molecules, nutrients move through epithelial cells of the villi by a number of mechanisms, for example glucose, galactose and amino acids by active transport with sodium, fructose by facilitated diffusion, and short-chain fatty acids by simple diffusion.

TIME OUT

SOME FEATURES OF GASTRIC EMPTYING IN RELATION TO HYDRATION AND EXERCISE

You will become aware of the need to be fully hydrated in order to exercise safely and to perform maximally. In certain sport and exercise situations we need to drink before, during and after exercise but there are particular features of gastric (stomach) emptying that are important to consider, including:

- Ingested drinks are retained in the stomach for variable periods of time and therefore the longer they spend in the stomach, the less effective they will be at replacing carbohydrate and/or electrolyte and/or water.
- A higher volume of fluid leads to a higher emptying rate.
- Exercise up to around 70 per cent of maximum oxygen uptake has little or no effect on emptying, but short-term, high-intensity exercise will slow or even stop gastric emptying.
- Increasing the energy content of a drink will slow emptying at rest and during exercise. Emptying is also slowed by solutions with high osmolality relative to that of body fluids.
- There is a large variation among individuals in the rate of emptying, for example as a result of 'practising' drinking during exercise – it is important to do so before competition.
- Factors such as environmental/weather conditions may influence choice of drink and athletes should experiment during training to establish need and tolerance.

The basic functions of the large intestine are to remove water, compact undigestible material and to prepare waste for elimination from the body. The large intestine starts from where the small intestine finishes, a point known as the ileocaecal junction. The caecum is a small, blind sac which extends past the ileocaecal junction. The colon can be divided into four sections – ascending, transverse, descending and sigmoid sections. Mucus is secreted from the mucosal cells of the large intestine to protect the lining and to lubricate and aid movement of material. The reabsorption of most of the remaining water and salts and the accumulation of undigested waste is carried out by the colon, which, unlike the small intestine, does not contain villi and thus has a surface area that is only 3 per cent of the size of the absorptive surface of the small intestine. The colon extends into the pelvis and ends at the rectum, a straight tube that ends at the anus.

Assimilation of fuels

The components of our diet are the macronutrients (carbohydrate, lipids and proteins (and alcohol)) and the micronutrients (vitamins, minerals and trace elements). The 'fuels' or macronutrients are absorbed at different rates – carbohydrates and proteins being absorbed more quickly than lipids. A normal mixed nutrient meal will result in carbohydrates being absorbed over a 2–4 hour period. Carbohydrate is stored in tissues such as the liver and in muscle, aided by the action of insulin, and excess carbohydrate can be converted to free fatty acids and stored as triglyceride. Undigestible carbohydrate, the fraction remaining after treatment with a variety of digestive enzymes, is sometimes referred to as fibre, but not all components are fibrous, some are soluble. Undigestible plant carbohydrates include cellulose, lignin and pectin. Some analytical methods allow for analysis of specific polysaccharides and the sum of such undigestible carbohydrates is then referred to as non-starch polysaccharides (NSP).

The digestion of lipids is slower than that of carbohydrate and protein because lipids delay gastric emptying. Some lipids are absorbed through the lymphatic channels as triglygeride-containing particles called chylomicrons. If a blood sample is taken after a high fat meal it can result in a cloudy plasma or serum sample because of the presence of these chylomicrons. Chylomicron triglycerides are either broken down to glycerol and free fatty acids (by lipoprotein lipase), which then diffuse or are transported by fatty acid-binding proteins into tissues such as the liver or muscle to be oxidized to provide energy, or are stored as triglyceride, for example in adipose tissue.

Due to the importance of carbohydrate and lipids as fuel for exercise, the way in which these macronutrients are metabolized (catabolized) as fuel for tissues will be covered in detail in Chapter 4, using skeletal muscle metabolism as an example. We shall also consider some of the health-related problems associated with the balance in consumption of carbohydrate and lipids in the diet in Chapter 5. Protein, however, is not normally used as a fuel for exercise, nor is it currently considered to be particularly over- (or under-) consumed. After a meal, amino acid concentrations in the blood rise at a similar rate to glucose and are transported in the plasma and by red blood cells to the liver and to muscle.

With regard to dietary protein intake as a whole, requirements are normally easily met by a balanced and healthy nutrient intake – that is, obtaining protein from a variety of normal food sources (e.g. meat, fish, pulses, nuts and dairy products). Protein requirements of exercise are usually met using the same principles, and aided by an increase in energy intake that matches the higher energy expenditure. Providing the diet is balanced (e.g. 10–15 per cent of energy obtained from protein), any apparent 'increased' requirement for exercise above the recommended 0.75–0.9 g protein per kg body mass for adults can be obtained through normal food intake. There is little need to obtain extra protein from amino acid mixtures or protein shakes, despite the unregulated and largely unsubstantiated claims of some supplement producers.

A small minority of highly active or specifically trained individuals may benefit from supplementing a balanced diet with extra protein, but it is more appropriate to approach nutrient and energy balance needs and requirements from an evidence base (on an individual basis) and to address optimization through normal food intake. We shall look at ways of measuring nutrient and energy intake in Chapter 5. Only in extreme circumstances, primarily in disease states, are macronutrient (and micronutrient) supplements a necessity.

There is a dietary need for a whole range of vitamins, minerals and trace elements, collectively known as **micronutrients**. You might wish to refer to a nutrition text to find out more about these individually. Most healthy and active individuals in the Western world with access to a range of foodstuffs have little or no need for supplementation if consuming a healthy balanced diet. One exception to this however is folic acid, if planning a pregnancy. Daily supplementation of 400 µg of folic acid (i.e. in addition the 200 µg consumed in the diet normally) significantly reduces the risk of neural tube defects (e.g. spina bifida) in the newborn. Supplementation should ideally be for at least 3 months before and 3 months after conception. The consumption of some micronutrients as supplements (for example those containing the fat-soluble vitamins A, D, E and K) can sometimes result in overconsumption, leading to toxicity of tissues and metabolic disturbances.

Fluid intake

Although macronutrients can be stored and we can survive without food for many days, we do not store excess amounts of fluid, and in order to maintain fluid balance (i.e. replace losses with intake), we need to drink fluids regularly. We can only live a few days without drinking. The total body water of an average adult male (70 kg) might be 42 litres (60 per cent of body mass) and about 2.6 litres of this is turned over daily. If leading a sedentary lifestyle, our fluid intake is in the form of water in solid food (about 1.1 litres), as drinks (about 1.2 litres), and as the water produced by metabolism (about 300 ml). Our loss of fluid is as urine (about 1.3 litres), through breathing and sweating (about 1.2 litres) and in faeces (about 100 ml). Deviations from this pattern are caused mainly by activity and environmental conditions, and as much as 10 litres per day might need to be consumed during exercise training in a hot climate.

Sweat contains sodium chloride but its concentration (normally about half that of internal body fluids) can be less as a result of exercise training or adaptation to hot

climates. It is for this reason that as a result of heavy sweat losses some electrolyte replacement is useful in addition to water replacement. This does not necessarily need to be as part of a rehydration drink, as we normally obtain plenty of salt in the diet, particularly in processed foods. Electrolytes in a rehydration drink can, however, enhance gastric emptying and intestinal absorption. The amount and formulation of drinks depends on individual preferences and circumstances. Some general advice for drinking in sport and exercise is outlined in **Box 3.8**.

Box 3.8

General advice for drinking in sport and exercise

- Start exercise well hydrated
- Weigh yourself before and after training sessions as an indication of fluid loss
- Drink during exercise if possible
- Consider a drink with some sodium
- Know and observe your own body (e.g. typical fluid losses, urine colour)
- Drink plenty during meals

The liver and biliary system

The liver, often considered part of the digestive system, is an organ but can also be considered a gland; the liver is therefore the largest gland in the body. A gland is a secretory organ from which secretions (not necessarily hormones) may be released into the blood, a cavity or onto a surface. A far as the liver is concerned, its production of up to 1 litre of bile each day makes it part of the digestive system, but it has several other important functions, including the processing and/or storage of nutrients (e.g. glycogen, vitamins), the synthesis of new molecules (e.g. cholesterol, carnitine, plasma proteins), the metabolism of ingested compounds (e.g. alcohol, drugs) and the detoxification of harmful chemicals.

The liver is wedge-shaped and lies under the diaphragm in the upper right region of the abdominal cavity. It consists of two major lobes and two minor lobes (Figure 3.28).

The porta hepatis (gate of the liver) lies on the inferior surface and is where the hepatic portal vein, the hepatic artery and the hepatic duct, other ducts and nerves enter and exit the liver. Arterial branches from the abdominal aorta that supply the stomach, intestine, pancreas and spleen are drained by the hepatic portal vein into the liver. About 75 per cent of the blood received by the liver is from this source, the remainder coming from the hepatic artery. Blood from the liver drains via hepatic

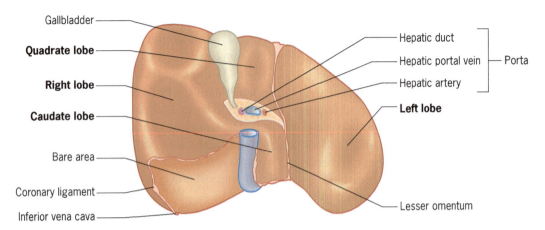

△ **Fig 3.28** The lobes of the liver and related anatomical features.

veins into the inferior vena cava. The splanchnic circulation (digestive system and related organs) receives around 25 per cent of the cardiac output at rest, but during exercise, when cardiac output rises from around 5 litres at rest to 25 litres, blood flow is distributed to where it is most needed – to skeletal muscle (around 85 per cent of blood flow), at the expense of blood flow to regions and organs such as the splanchnic region which only receives about 1 per cent of the cardiac output.

As well as the major blood-carrying vessels of the liver, bile ducts (bile canaliculi) are formed by the bile capillaries that come together after collecting bile from the liver cells. Canaliculi lie between cells within hepatic cords that radiate out from the central vein of each lobule, and these hepatic cords are made up of the functional cells of the liver – the **hepatocytes**. The major functions of the **hepatocytes** are:

- Production of bile
- Storage of nutrients
- Interconversion of nutrients
- Detoxification of harmful substances
- Phagocytosis of cells, organisms and debris
- Synthesis of proteins and transporter molecules.

Bile contains three components that are important for digestion – bile salts, bile pigments and bicarbonate (HCO_3^-). The contents of bile aid the digestion and absorption of lipids by emulsification, they aid in the absorption of fat-soluble vitamins, and provide an excretory route for pigments, cholesterol, steroids, heavy metals and drugs. Bile salts are derivatives of cholesterol, secreted into the bile and they eventually pass into the duodenum with the other biliary secretions. Bile pigments are derived from contents of cells such as haemoglobin from red blood cells, giving bile its greenish-yellow colour. If the bile pigment bilirubin is allowed to enter the extracellular

fluids, then the skin takes a yellowish colour and we call the condition **jaundice**. Bicarbonate production by the duct cells is increased by the action of secretin and assists in neutralizing the gastric acid that enters the duodenum.

The storage role of liver includes storage of fat-soluble vitamins, copper and glycogen. The liver normally contains around 240 mmol glucosyl units per kg of its weight (normally around 1.4 kg), but after a short fast (e.g. overnight), most of the glycogen has been broken down for use as circulating blood glucose (Figure 3.29). Unless we eat, the liver will remain depleted of glycogen, only a small amount (about 40 mmol glucosyl units per kg of liver) will be present due to gluconeogenesis. If we continue to fast for a day or so or starve for longer (say a week), and then eat a high-carbohydrate diet, the amount of liver glycogen will immediately increase to much higher than normal levels (around 500 mmol glucosyl units per kg of liver). This is known as a **supercompensatory effect**.

The metabolism of alcohol, as described earlier, can begin in the stomach. Once absorbed, alcohol (ethanol)

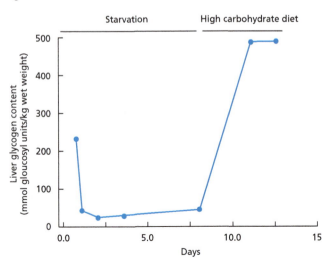

△ **Fig 3.29** Liver glycogen replenishment is affected by diet.

spreads quickly throughout the body water. Blood alcohol is cleared as a result of metabolism by the liver, using a number of pathways, principally via the enzyme alcohol dehydrogenase, but also via the microsomal ethanol-oxidizing system. This enzyme system is not normally detected in liver cells but is permanently induced when alcohol consumption is more or less continuous. A further metabolic route for alcohol is indirectly via catalase when hydrogen peroxide breakdown is catalysed by catalase, resulting in the oxidation of alcohol if present. The first product of the metabolism of alcohol, acetaldehyde, is further metabolized to acetate and then to acetyl-CoA. Acetaldehyde dehydrogenase is an important enzyme in alcohol metabolism and detoxification because of the toxicity of acetaldehyde.

Whilst the liver is able to metabolize moderate amounts of alcohol, it can be adversely affected by heavy drinking. Acetaldehyde and peroxides are a possible cause of cellular damage, and the development of liver disease. Disease develops in stages: fatty liver and fibrosis may occur in the early stages without symptoms. When alcoholic hepatitis occurs, cells die and liver function begins to deteriorate, leading to cirrhosis. Some of the effects of alcohol on the central nervous system and health issues related to alcohol consumption are discussed in Chapter 6. The liver is a major defence against by-products of metabolism, some of which, like acetaldehyde and ammonia, a by-product of amino acid metabolism, can be toxic. Transient increases in plasma ammonia as a result of high-intensity exercise (see Chapter 4 Exercise hyperamonaemia) can be tolerated, but blood ammonia levels must be kept low. As it is not readily removed by the kidneys, hepatocytes convert ammonia to urea, which is then excreted in urine.

The **biliary system** consists of the gallbladder, the left and right hepatic ducts coming together as the common hepatic duct, the cystic duct, and the common bile duct. The common bile duct is the post-mergence of the common hepatic duct and the cystic duct. The gallbladder has a storage function, capable of storing a small amount of the bile being continuously produced and secreted by the liver, and it is stimulated to contract by cholecystokinin, a hormone produced by the duodenum. If, however, the gallbladder causes a medical problem it can be removed (cholecystectomy) without too much detriment to the digestive process. Cholesterol secreted by the liver can precipitate in the gallbladder to produce **gallstones**, occasionally passing out and into the cystic duct, causing a blockage. Gallstones, normally removed by surgery if they do cause a blockage, can result from drastic weight-loss programmes, particularly in obese people. This is one reason why it is important to only reduce weight gradually (no more than 1 kg per week), preferably through combining a reduction in energy intake with an increase in energy expenditure (discussed in Chapter 5).

Renal and urinary system

The renal system consists of the two kidneys, which receive blood from the cardiovascular system in order to perform their regulatory and excretory roles. The kidneys are responsible for the regulation of fluid volume and osmolality, electrolyte balance and acid–base balance. They also deal with the excretion of metabolic and toxic products, and the production and secretion of hormones (renin, prostaglandins, kinins, erythropoietin and 1,25-dihydroxyvitamin D_3).

The pair of kidneys, often referred to as kidney-bean shaped, are positioned above the waist, around the level of the last thoracic and the first few lumbar vertebrae. The right kidney sits lower than the left as a result of the space occupied by the liver on the right side. Each kidney (Figure 3.30) is about 110 mm long, 60 mm wide and 30 mm thick, and is surrounded by the layers known as the **renal capsule** (deep layer), the **adipose capsule** and the **renal fascia** (superficial). These layers protect the kidneys from trauma and hold them firmly in the abdominal cavity.

The kidneys receive around 20 per cent of the cardiac output at rest, but during exercise, when cardiac output can rise from around 5 litres at rest to 25 litres, blood is distributed to where it is most needed – to skeletal muscle, at the expense of blood flow to regions and organs such as the kidneys which only receive about 1 per cent of the cardiac output.

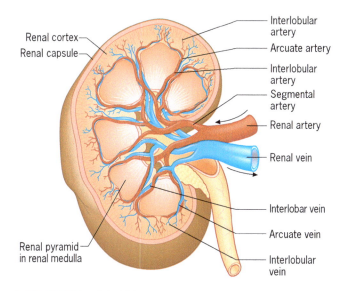

Renal cortex
Renal capsule
Interlobular artery
Arcuate artery
Interlobular artery
Segmental artery
Renal artery
Renal vein
Interlobar vein
Arcuate vein
Renal pyramid in renal medulla
Interlobular vein

△ **Fig 3.30** The right kidney.

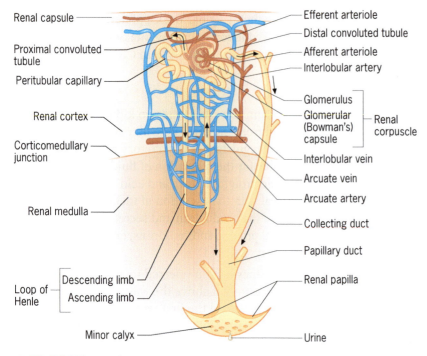

Renal capsule

Proximal convoluted tubule

Peritubular capillary

Renal cortex

Corticomedullary junction

Renal medulla

Loop of Henle — Descending limb / Ascending limb

Minor calyx

Efferent arteriole

Distal convoluted tubule

Afferent arteriole

Interlobular artery

Glomerulus

Glomerular (Bowman's) capsule — Renal corpuscle

Interlobular vein

Arcuate vein

Arcuate artery

Collecting duct

Papillary duct

Renal papilla

Urine

△ **Fig 3.31** The nephron.

The physiology of the kidneys is dependent on their internal anatomy, and **nephrons** (Figure 3.31) are the functional units. These carry out three basic processes that are the primary functional roles of the renal system:

- **glomerular filtration** – the process where water and solutes move from the blood into the glomerular capsule and then into the renal tubule;

- **tubular reabsorption** – the process whereby most of the filtered water and useful solutes are reabsorbed and returned to the blood as it flows through the peritubular capillaries and the vasa recta; and

- **tubular secretion** – in which the tubule and duct cells secrete materials such as drugs, excess micronutrients and ions into the fluid in the renal tubule.

The nephron starts in the renal cortex where the glomerular capillary bed is invaginated into it. Specialized cells form a Bowman's capsule, lining a Bowman's space. Fluid drained from the Bowman's space enters the proximal convoluted tubule, leading to the thin descending, thin ascending and thick ascending limbs of the loop of Henle. The tubule returns to the renal cortex and makes contact with the afferent arteriole of its glomerulus, forming the **juxtaglomerular apparatus** (the juxtaglomerular apparatus includes modified smooth muscle cells in the wall of the efferent arteriole). The tubule then becomes convoluted again (the distal convoluted tubule) and joins with other nephron segments to form a collecting duct that runs from the cortex (cortical collecting duct) to the

medulla (medullary collecting duct). These ducts merge further, become progressively larger, and terminate in the ducts of Bellini that drain into the ureter.

The major solutes that are reabsorbed are glucose, amino acids, organic acids and phosphate. There is normally no glucose in urine, but when blood glucose levels remain high for long periods due to lack of insulin production or sensitivity (i.e. untreated diabetes mellitus), glucose can be detected in urine. Inorganic ions such as sodium, chloride, magnesium, phosphate, sulphate and bicarbonate are reabsorbed. Potassium is an inorganic ion that is both reabsorbed and secreted.

Inorganic ions such as hydrogen ions are excreted in order to maintain extracellular pH. Some solutes that would be toxic if allowed to accumulate are also secreted, such as organic anions and cations, and ammonium.

A question that you might well ask is what controls the degree to which water, solutes and ions are reabsorbed or secreted. At the first stage, glomerular filtration, the rate of filtration is regulated in a variety of ways, such as

- by autoregulation – its own cells reacting and feeding back information;

- by neural pathways – activity of renal sympathetic nerves; and

- by hormonal regulation – **angiotensin II** and **atrial natriuretic peptide**.

The rates of tubular reabsorption and tubular secretion are regulated by hormonal control. The most important regulators of electrolyte (e.g. sodium, potassium, chloride) reabsorption and secretion are the hormones angiotensin II and **aldosterone**. Angiotensin II is synthesized by liver cells (hepatocytes) and the zona glomerulosa of the adrenal cortex secretes **aldosterone**, a mineralocorticoid, increasing sodium retention in the kidney tubules. The major hormone responsible for water reabsorption is **antidiuretic hormone (ADH)**. **Atrial natriuretic peptide (ANP)** has a lesser role in both electrolyte and water reabsorption.

The juxtaglomerular apparatus was mentioned above and these cells secrete **renin** in response to a reduction in blood volume and blood pressure. Renin secretion is also stimulated by sympathetic stimulation of the juxtaglomerular apparatus. Renin is an enzyme that synthesizes angiotensin I from angiotensiogen produced in hepatocytes, and angiotensin I is converted to angiotensin

II by **angiotensin-converting enzyme (ACE)**. Circulating angiotensin II causes vasoconstriction and increases total peripheral resistance, stimulating the release of aldosterone which stimulates the cells of the collecting ducts to reabsorb more sodium. This stimulates thirst and it increases the secretion of ADH (from the posterior pituitary gland) to reduce the loss of water to urine and to enhance systemic vasoconstriction.

An endocrine role of the kidney is the production of **erythropoietin (EPO)**, a naturally produced polypeptide hormone of the peritubular interstitial cells. EPO attaches to receptors on the surface of erythrocyte precursors and is stimulated by tissue hypoxia, which can be brought upon by exposure to high altitude, amongst other stimuli (see **Time Out** *Erythropoietin* sport on page 65).

The ability of the kidneys to regulate the rate of water loss in urine is the key factor in maintaining a stable volume of fluid in the body. In order to do this, we produce anything from a large volume of dilute urine to a small volume of concentrated urine, depending on fluid intake and fluid loss. When intake is low or losses are high, ADH is responsible for stimulating reabsorption of more water into the blood, generally resulting in a concentrated urine. This usually results in a yellow colour to urine which sometimes can be an indication that fluid intake has not matched fluid losses. Diuretics are substances that slow renal reabsorption of water and cause diuresis. Diuretic drugs are used, for example, as a therapeutic means of lowering blood volume and hence blood pressure in hypertension. Other, more common, substances such as alcohol and caffeine have a diuretic effect and have implications for sport and exercise.

From collecting ducts in the kidneys, urine drains into calyces that join and become larger and form the renal pelvis. From here, urine drains into the ureters and into the urinary bladder and the expulsion of urine, sometimes known as micturition, takes place when urine passes through the urethra. The bladder can accommodate about 300 ml of urine before becoming uncomfortable. When we become aware of the sensation of fullness, the micturition reflex is triggered and urination occurs.

So, in this chapter we have explored the physiological systems, with reference to some applications to applied sport and exercise science. Each of the systems discussed could be explored in much greater detail, and as your interest in sport and exercise develops and specialises in later years of study, you might need to explore a particular system in greater detail, but at this stage you should have a strong foundation from which to build on, and the key points are summarised below, before we move on to consider skeletal muscle in Chapter 4.

Summary

- Homeostasis is the maintenance of a relatively constant internal environment and it is made possible by the coordination and integration of physiological systems.

- The nervous system receives sensory information from receptor organs such as the skin, the ears, and the eyes. It processes this information and generates responses, it regulates and coordinates so as to maintain homeostasis, it is responsible for higher functions such as emotion, memory, thinking, and it also controls muscular movement.

- The endocrine system acts through chemical messengers (hormones) and has a key role in homeostasis. It is involved in the regulation of growth, development and reproduction, and, control of the storage and utilisation of energy substrates.

- The heart and blood vessels comprise the cardiovascular system, and its function is to circulate blood around the body – a transport system that serves other systems to maintain homeostasis.

- The respiratory system provides an infrastructure to ventilate the lungs and exchange gas, linking with the cardiovascular system to transport blood gases and maintain acid-base balance.

- The lymphatic system returns excess interstitial fluid to the vascular system, is a pathway for the absorption of fats from the intestine, and has a role in defence/immune responses.

- A complex array of offensive organisms and environmental agents require a complex defence or immune system, and there exists varying degrees and types of immune system specificity and components to protect us.

- Ingested materials pass by orderly and controlled means from the mouth to the anus in the digestive system, aided by smooth muscle activity that is responsible for gastric and intestinal motility, and absorptive processes that utilise a leaky epithelium – the small intestine, to take up ions, water, minerals and vitamins, monosaccharides, peptides and lipids.

- The liver is the largest gland in the body and hepatic cells are responsible for functions such as bile production, the storage and interconversion of nutrients, and detoxification of harmful metabolites and substances.

- The kidneys regulate the volume and composition of the body fluids, and excess intake of water and ions are excreted in the urine.

Review Questions

1. What is meant by the term homeostasis? Describe the role of the hormones insulin and glucagon in maintaining homeostasis.

2. Make a route map of the divisions and subdivisions of the nervous system.

3. Make a list of five hormones (in one column) and the metabolic effects of that hormone (in a second column).

4. Explain the route that blood takes through the body, making reference to the systemic and pulmonary circulations.

5. Compare and contrast the adaptive (specific) components and the less specific components of the immune system.

6. Explain the lung volumes and capacities that can be measured in respiratory physiology.

7. What are the functions of the liver and biliary system?

8. Think of a hormone produced by the kidney, and a hormone that acts on the kidney, and explain their roles.

References

Barcroft J (1934) *Features in the Architecture of Physiological Function*. Cambridge: Cambridge University Press.

Gleeson M, Blannin AK, Sewell DA, Cave R (1995) Short-term changes in the blood leucocyte and platelet count following different durations of high-intensity treadmill running. *Journal of Sports Science* 13: 115–123.

Gyntelberg F, Rennie MJ, Hickson RC, Holloszy JO (1977) Effect of training on the response of plasma glucagon to exercise. *Journal of Applied Physiology* 43: 302–305.

Johansson BW (2001) A history of the electrocardiogram. *Dan Medicinhist Arbog* 163–176.

Mahler DA, Moritz ED, Loke J (1982) Ventilatory responses at rest and during exercise in marathon runners. *Journal of Applied Physiology* 52: 388–392.

Nieman DC (1990) The effects of moderate exercise training on natural killer cells and acute, upper respiratory tract infections. *International Journal of Sports Medicine* 111: 467–473.

Sullivan FM, Swann IR, Donnan PT, Morrison JM, Smith BH, McKinstry B, Davenport RJ, Vale LD, Hammersley VS, Hayavi S, McAteer A, Stewart K, Daly F (2007) Early treatment with prednisolone or acyclovir in Bell's Palsy. *New England Journal of Medicine* 357, 1598–1607.

Further reading

Galbo H (1983) *Hormonal and Metabolic Adaptation to Exercise*. New York: George Thieme Verlag.

Hampton JR (2003) *The ECG Made Easy*, 6th edn. Edinburgh: Churchill Livingstone/Elsevier Science.

Warren MP, Constantini NW eds (2000) *Sports Endocrinology*. Champaign, IL: Humana Press.

Seeley RR, Stephens TD, Tate P (2008) *Anatomy and Physiology*, 9th edition. NY: McGraw-Hill.

Tortora GJ, Grabowski SR (2003) *Principles of Anatomy and Physiology*, 10th edn., IN: Wiley.

4 Skeletal Muscle Physiology and Metabolism

Chapter Objectives

In this chapter you will learn about:

- The mechanism of excitation and contraction in skeletal muscle in relation to its structure.
- The structure and function of the contractile proteins in skeletal muscle.
- How the integrity of the sarcomere is maintained by structural proteins.
- The energetics of, and pathways in, energy metabolism.
- Muscle cell/fibre types and their distribution in relation to exercise performance.

Introduction

Skeletal muscle is a distinctive and specialized organ that accounts for around half of the body mass of most humans, and indeed mammals, regardless of size. The gross anatomical features of the muscle groups and muscles of the body were introduced in Chapter 2, and as we saw, skeletal muscles differ in their visual appearance. In this chapter we will explore the different organizational levels – the basic physiology of skeletal muscle structure, its excitation and contraction, the characteristics of skeletal muscle – and look at molecular, cellular, metabolic and mechanical aspects of skeletal muscle function.

One aspect of the learning about muscle in Chapter 2 was the three types of muscle that we find in the body – smooth, cardiac and skeletal, in particular, the cells of these muscle types. We considered the basic characteristics of the muscle types, and made comparisons between them. In Chapter 3 we took a closer look at the structure and function of cardiac muscle, in relation to the cardiovascular system, in particular at the structure of the organ that it forms – the heart. We shall now take a close look at the structure of skeletal muscle. In Chapter 2 we also discussed the anatomical make-up of the skeletal musculature of the human body, particularly muscle groups, and muscles in relation to joints and their movement. As sport and exercise scientists we need to be familiar with a wide range of locomotory muscles because they are responsible for movement, acting around the major joints of the body.

We can now think about what constitutes a single muscle, the specialized cell type that is the skeletal muscle cell or fibre, and the mechanism of skeletal muscle excitation and contraction. Then later on in this chapter, we will take a closer look at, for example, the metabolism of skeletal muscle, the types and subtypes of skeletal muscle fibre that differentiate between the sprint/strength and the endurance type athlete, and the molecular physiology of skeletal muscle.

From the outside of the body, muscle is covered by skin, separated from it by a superficial fascia/subcutaneous layer of, for example areolar connective tissue and adipose tissue. A muscle is attached to the bone via a tendon and is surrounded by a muscle fascia (epimysium) (Figure 4.1). The **epimysium** or deep fascia allows for movement of muscles, carries nerves, blood vessels and lymph vessels, and fills spaces between muscles of muscle groups. Another connective tissue layer, the **perimysium** surrounds and binds together **fasciculi** – bundles of muscle fibres. A further connective tissue layer called the **endomysium** surrounds each muscle fibre (muscle cell).

Muscle fibres contain hundreds of contractile myofibrils, which contain structures such as mitochondria and the **sarcoplasmic reticulum**, and myofilaments, composed of the contractile proteins actin and myosin. The number of myofibrils in a muscle fibre is dependent on the size of the fibre, but is thought to be around 2000 in adult muscle (Figure 4.2).

The principal components of the muscle cell, also known as a fibre, are:

▷ Proteins (both contractile and structural)

▷ Sarcoplasmic reticulum (SR: enveloping the myofibrils)

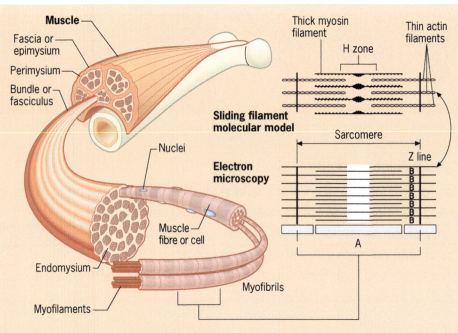

△ **Fig. 4.1** Skeletal muscle structure.

▷ T-tubular system (an invaginating network, running through the fibre and making contact with every myofibril)

▷ Mitochondria, nuclei, muscle spindles, glycogen granules, lipid droplets.

Proteins of the muscle cell will be discussed later, as part of our consideration of the molecular features of the muscle cell. The sarcoplasmic reticulum is a system of channels spreading out over the surface of the myofibrils. The calcium ions involved in initiating contraction are localized in the SR. Transverse tubules (T-tubules: you might recall that these are absent in smooth muscle) provide electrical links to the SR in skeletal muscle. They open to the extracellular space at the sarcolemma, forming a network or grid at the level of the junction between the A- and the I-bands in each sarcomere (discussed later in this chapter). The roles of mitochondria and nuclei have been introduced in Chapter 1, and the presence of glycogen (the storage form of carbohydrate) and lipid in skeletal muscle

and their roles in energy pathways and metabolism will be discussed later in this chapter. In Chapter 5 we discuss fuel sources for exercise from the perspective of whole body energy balance.

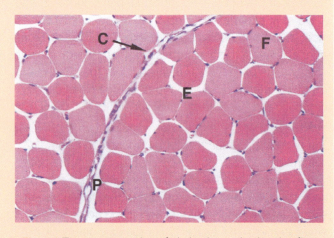

△ **Fig. 4.2** Transverse section of skeletal muscle showing fibres. P, perimysium separating two fascicles; E, endomysium; C, capillaries containing red blood cells; F, fibre.

Muscle fibres and types

Skeletal muscles differ in their visual appearance and they also have different contractile properties. Such differences exist due to, for example, the degree of capillarization to the muscle fibres that constitute each muscle fibre bundle, and the mitochondrial density and contractile protein content of each fibre. Muscle fibres can be characterized histochemically according to microscopic differences

between muscle fibres demonstrated using histochemical 'staining' reactions carried out on either thin sections of frozen, unfixed material or on fragments of single fibres dissected from freeze-dried tissue. Such techniques demonstrate acid lability and enzyme activity of muscle fibres and the organelles and substrates within them.

There are two principal types of muscle fibre – type I and type II. One of the most widely used histochemical procedures to characterize muscle fibre types in exercise, health and disease exploits the differential acid lability of the myosin ATPases contained within each fibre (Figure 4.3). 'Fast' myosins are inhibited at a pH of around 4.55 whilst 'slow' myosins are inhibited at a pH of around 9.5. Thus, it is possible to inhibit the 'slow' myosins by incubating muscle sections or fragments at pH 9.6 and then visualize microscopically the activity of the 'fast' myosins by incubating the section with ATP and an excess of calcium (Figure 4.3(a)). Fibres appearing very dark are classed as type II or fast twitch fibres whereas very light fibres are classified as type I or slow twitch fibres.

Type II fibres can be further subdivided and demonstrated by incubating sections in acid before following the normal alkaline histochemistry procedure. As a result of acid pre-incubation, the pattern of 'staining' intensity is reversed, type I fibres appear very dark, type IIa fibres appear very light, and type IIb fibres appear as an intermediate intensity (Figure 4.3(b)). More recently, characterization of muscle fibre types has been made clearer using myosin heavy chain isoform composition as molecular markers. Type IIb fibres are now referred to as type IIx fibres based on molecular identification. Despite the fact that the classification of fibre types still raises issues when different muscles in different physiological states and in different animal species are considered, or when histochemical criteria are combined with physiological and biochemical data (Hainaut *et al.,* 1981), classification systems can be combined. For practical purposes it is possible to consider fibres as:

- **Type I** – S or slow contracting, highly oxidative and highly fatigue resistant
- **Type IIa** – FR or fast contracting, oxidative/glycolytic, fatigue resistant
- **Type IIx** – FF or fast contracting, glycolytic, rapidly fatiguable.

(a)   **(b)**

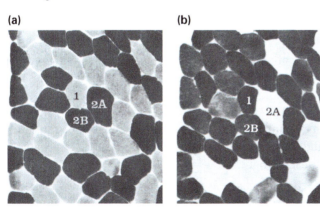

△ **Fig. 4.3** Muscle sections (TS) demonstrating different fibre types after myosin ATPase 'staining': (a) pH 9.4 (b) acid pre-incubation (pH 4.6). From Dubowitz (1985).

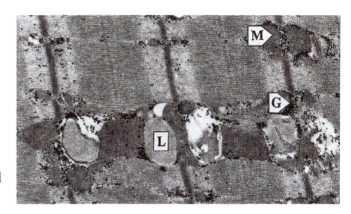

△ **Fig. 4.4** Mitochondria, glycogen and lipid (triglyceride) stored inside a muscle fibre. From Jones, Round and de Haan (1990).

We shall consider muscle fibre characteristics (Figure 4.4) a little later after we have looked at contraction, contractile proteins and muscle metabolism in more detail.

Nerve and muscle connection

Now that we have a good idea of the overall structure of skeletal muscle, the key components of the fibre and basic fibre types, let us think about the single muscle and the connection that the nervous system makes with it. You may remember from the description of the nervous system in Chapter 3 that an action potential, the basis of the signal-carrying capacity of nerve cells, is propagated along the length of a nerve axon. The type of nerve we are thinking about in this chapter is one which belongs to the peripheral nervous system (PNS), thereby providing an interface between the central nervous system (CNS) and the environment.

This somatic subdivision of the PNS consists of specialized cells (**motor neurons**) receiving information and transmitting signals to an effector cell (skeletal muscle cell). Each motor neuron has its origin (cell body) in the brainstem or the spinal cord grey matter, and projects its axon out of a cranial nerve or a ventral root respectively. The dendrites – extensions of the cell body – can be numerous (about 10) and quite long in motor neurons (i.e. more than 1 mm long) and receive many synaptic connections. A synapse is a site at which an impulse is transmitted from one cell to another. The neurons innervating skeletal muscle fibres (sometimes known as alpha motor neurons) distribute through peripheral nerves to the appropriate skeletal muscle, terminating synaptically on the surface of a muscle fibre at a **neuromuscular junction**, sometimes known as an endplate.

Near the neuromuscular junction the motor nerve loses its myelin sheath and divides into fine terminal branches

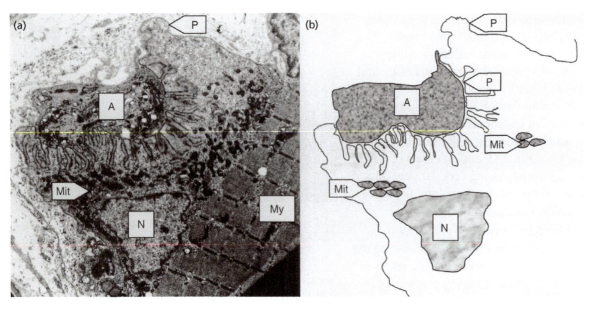

△ **Fig. 4.5** The neuromuscular junction in skeletal muscle. A, nerve axon; N, nucleus; My, myofibrils; P, muscle fibre plasma membrane; Mit, muscle mitochondria. From Jones, Round and de Haan (2004).

that sit in synaptic pits on the surfaces of the muscle cells (Figure 4.5). The neuromuscular junction is an example of a chemical synapse, the other type of synapse being an electrical synapse. Motor neurons and their axons are one of the few types of cells able to synthesize acetylcholine, the chemical neurotransmitter involved in synaptic transmission between nerve and muscle. The action potential is conducted down the motor axon to the presynaptic axon terminals, causing an increase in Ca^{2+} permeability and an influx of Ca^{2+} into the axon terminal. Synaptic vesicles fuse with the plasma membrane to empty their acetylcholine (ACh) into the synaptic cleft by exocytosis. Acetylcholine then diffuses to the post-junctional membrane where it combines with specific receptors to increase the permeability of the post-junctional membrane to Na^+ and K^+ to cause an endplate potential. The endplate potential is transient because the action of ACh is ended by its hydrolysis to form choline and acetate, catalysed by the enzyme acetylcholinesterase, which is present in high concentrations on the post-junctional membrane.

Before discussing what happens in the muscle fibre when an action potential reaches the neuromuscular junction, we should consider structures known as muscle **stretch receptors** that are important for both spinal reflexes and for proprioception (position sense and joint movement), and the concept of the motor unit. Ultimately, the cerebellum in the brain receives sensory input from stretch receptors.

Stretch receptors come in two types – the **muscle spindle** (or neuromuscular spindle) and **Golgi tendon**

organs. Muscle spindles are found in all skeletal muscles but are particularly concentrated in muscles that are controlled by small motor units (see below), such as those used for fine motor control (e.g. in the hand). Muscle spindles have their own sensory and motor nerve supply and specialized muscle fibres that are called **intrafusal** fibres. This distinguishes them from the regular fibres or extrafusal muscle fibres that we are mostly concerned about in this chapter. Muscle spindles lie between extrafusal fibres in a parallel direction, and are attached to connective tissue within the muscle (the endomysium). Their role is to respond to muscle stretch – when extrafusal fibres contract, the muscle spindle lying between them shortens, in contrast to when the whole muscle is lengthened and the muscle spindle is stretched because of its connection to the tissue framework of the muscle. In these ways, the muscle spindle is able to sense changes in length and the rate of changes in the length of that muscle (Figure 4.6).

Golgi tendon organs are the second type of stretch receptor found in skeletal muscle, wrapped around bundles of collagen fibres and arranged in series in the tendon of a muscle. The stimulus for a Golgi tendon organ is the force that develops in that tendon, compared with the muscle spindle that detects muscle length and the rate of change of muscle length.

A particular muscle is connected to the nervous system by a group of alpha motor neurons located in a **motor nucleus,** for example in the ventral horn. The set of alpha motor neurons that innervates a skeletal muscle is known as the **motor neuron pool** of the muscle. Each

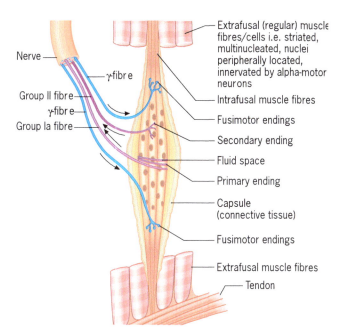

△ **Fig. 4.6** Longitudinal section of a muscle spindle.

skeletal muscle fibre (or cell) is supplied by only one alpha motor neuron, however, one alpha motor neuron may supply several muscle fibres. The number of fibres that a single motor neuron innervates is dependent upon the control required in that muscle. A small muscle that is required to execute fine motor control such as the first dorsal interosseus of the hand will be controlled with only about five or six muscle fibres per motor neuron, whilst a large muscle responsible for gross anatomical movements such as the *m. vastus lateralis* will be innervated by motor neurons that connect to more than 500 muscle fibres.

All of the varied reflex and voluntary contractions of a muscle are achieved by different combinations of active **motor units**. A motor unit is the basic 'quantal' unit of muscular contraction and represents the smallest number of fibres that can be activated by the CNS at any one time. The motor unit is the combination of motor neuron axon, terminal branches and the muscle fibres that are innervated by terminals from a single parent axon. The variation in size of motor units (with respect to the number of muscle fibres that they innervate) is similar across all mammalian species. The threshold of motor unit recruitment is that level of contraction strength at which the motor unit begins to fire as a muscle slowly increases its force of contraction. All the fibres in a particular motor unit will contract at the same time. This is produced by a muscle action potential and, as a consequence, muscle force is generated.

Innervation

The following concept of the action potential was introduced in Chapter 3 when we discussed the nervous system. Like nerve cells, plasma membranes of muscle cells are polarized, that is, they have a voltage/electrical charge difference across the membrane before the action potential can be generated. In the resting state, the charge difference, known as the **resting membrane potential** (RMP) of muscle (and nerve) cell membranes is around −75 mV (the inner surface is negative compared with outside the cell). An action potential is the reversal of the RMP, such that the inside of the plasma membrane becomes positive. This change in membrane potential is brought about by changing the permeability of the membrane to ions and hence allowing movement of ions across the membrane. Two types of channels allow for this – **ligand-gated ion channels** and **voltage-gated ion channels**. Unlike in cardiac muscle (see Chapter 3), the permeability changes during an action potential in skeletal muscle has only two phases – depolarization and repolarization (Figure 4.7):

● **Depolarization** – reversal of the resting membrane potential (RMP) is brought about by a change in the permeability of the membrane. There is a diffusion of Na^+ into the cell – the inside becomes positively charged relative to the inside as a result of opening of voltage-gated ion channels. The initial depolarization results in full depolarization if a **threshold** is reached, and many gated Na^+ channels open rapidly. Voltage-gated K^+ channels begin to open.

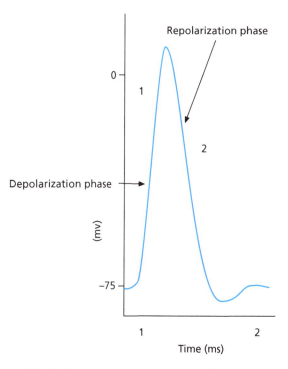

△ **Fig. 4.7** Phases of the action potential in skeletal muscle.

● **Repolarization** – this re-establishes the resting potential (the negative charge on the inside of the cell membrane surface is restored). The principal ionic events are the closing of the gated Na^+ channels (i.e. Na^+ ions stop diffusing in) and the gated K^+ channels continue to open (i.e. K^+ ions diffuse out). When the gated K^+ channels close, the action potential ends and the RMP is re-established.

Skeletal muscle repolarizes rapidly compared with cardiac muscle because there is no plateau phase during the action potential. You might like to compare the action potentials of skeletal muscle with cardiac muscle by referring back to Figure 3.14 in Chapter 3.

The endpoint of events that occur during neuromuscular transmission is the depolarization of areas of muscle membrane adjacent to the endplate, which initiates an action potential. This starts excitation–contraction coupling, a series of steps that describes the events from the generation of the muscle action potential to the binding of the myosin heads to actin (Figure 4.8).

The endplate potential is propagated across the surface of the muscle membrane and down the T-tubules of the muscle fibre. The change in membrane potential causes a release of Ca^{2+} from the sarcoplasmic reticulum.

Calcium ions are released from the lateral sacs and bind to troponin on the thin, actin filaments. Tropomyosin is physically moved aside to uncover cross-bridge binding sites on actin. The composition and structure of thin

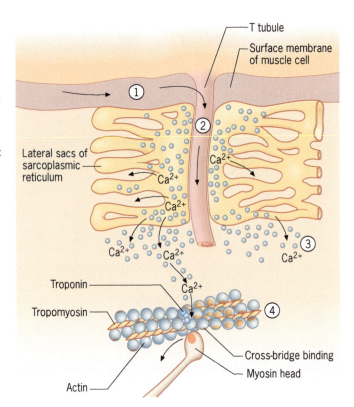

△ **Fig. 4.8** The four steps that constitute excitation–contraction (E–C) coupling.

(and thick myosin) filaments will be discussed later. The final step in the process of E–C coupling results in cross-bridge formation and movement – myosin heads attach

TIME OUT

TWITCH, TREPPE AND TETANUS

A single action potential may release sufficient Ca^{2+} to fully activate the contractile machinery in the fibres of the motor unit, however, the Ca^{2+} is rapidly pumped back into the sarcoplasmic reticulum before the muscle has had time to develop its maximal force. Such a submaximal response to a single action potential is called a twitch but repetitive action potentials can cause a partial or complete tetanus as Ca^{2+} pulses summate to maintain high concentrations of Ca^{2+} in the sarcoplasm. When stimulation is repeated with maximal stimuli, at a frequency that allows relaxation between stimuli, tension increases in steps (treppe) until maximal tension is developed by each twitch. Tension (force) produced is dependent on the number of motor units recruited and the size of the motor unit (i.e. fibre number).

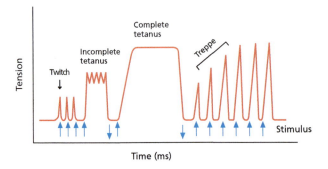

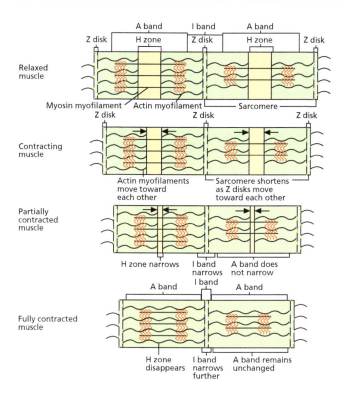

△ **Fig. 4.9** The sarcomere and sarcomere shortening. From Seeley *et al.* (2008)

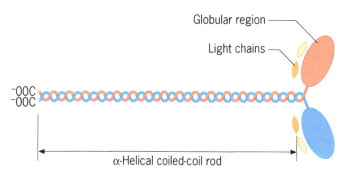

△ **Fig. 4.10** Schematic diagram of a myosin molecule.

to actin, pulling actin filaments towards the centre of the sarcomere. The energy for this process is provided by the breakdown of ATP to ADP.

At the end of contraction, calcium release channels are closed and calcium active transport pumps restore the level of Ca2+ in the sarcoplasmic reticulum, using ATP. Relaxation occurs when the troponin–tropomyosin complex slides back into position, blocking the myosin binding sites on actin.

When a muscle fibre acts, its basic functional unit, the **sarcomere** (a repeating unit in each myofibril) shortens. The thick and thin filament lengths, however, do not change because they slide past each other, so that the extent of overlap changes. The sliding filament model of contraction accounts for the way in which sarcomeres shorten without requiring actin and myosin filaments to change length (Figure 4.9). The A-band of the sarcomere remains a constant width, as it represents the length of myosin filaments, however, the I-band changes width when the sarcomere shortens and force is generated, as does the H-zone.

Molecular features of the sarcomere and muscle cell

The myofibrils are surrounded by sarcoplasm containing glycogen, ATP and other key substrates and enzymes. Many mitochondria are also found, mitochondrial

density being dependent upon muscle (and muscle fibre) type and training status. In the myofibril, the sarcomere repeats itself along the myofibril axis every 2.3 μm. The two types of protein filament ('actin' and 'myosin') overlap and interact in the space between the Z-lines.

Myosin, the main component of the thick filament, is a very large molecule (540 kDa) made up of six polypeptide chains (two heavy chain subunits [MyHCs]) each with a molecular weight of around 200 000 and four light chains (Figure 4.10). The muscle MyHC is a large protein of about 2000 amino acids and is split into two structural domains: the globular head at the N-terminus and the α-helical rod at the C-terminus. Two of the light chains are alkali/essential light chains (aMLC) and two are regulatory light chain subunits (regulatory MLC).

In addition to spontaneously assembling into thick filaments of skeletal muscle in optimal physiological conditions, myosin has other important biological activities: it is an enzyme (an ATPase, providing the energy to drive contraction) and it binds to actin, an interaction critical for the movement of filaments past each other in order to generate force.

Actin is a globular protein with a molecular weight of around 42 000 and is the major constituent of thin filaments (Figure 4.11). The other major component of the thin, twisted and two-stranded filaments is tropomyosin, which itself is composed of two separate polypeptide chains. Troponin (subunits: TnC, TnT and TnI) is the regulatory protein in skeletal muscle.

In addition to the contractile and regulatory proteins, there are a number of structural proteins that maintain

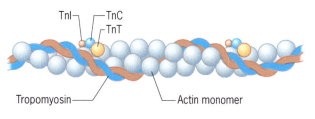

△ **Fig. 4.11** Actin and thin filament structure.

the integrity of the sarcomere (and of the muscle cell itself). Titin constitutes around 10 per cent of muscle mass, making it the third most abundant protein in muscle behind actin and myosin. It is the largest protein in the body, composed of 27 000 amino acids and has a molecular weight of 3-4 MDa. Despite this, its structure was not elucidated until the 1990s, but its importance in muscle tissue has rapidly emerged, particularly as abnormalities of titin are associated with disorders such as muscular dystrophy.

Titin is thought to act as a 'rule', controlling the relative positioning of actin and myosin, and it contributes to the extensibility and passive force development of muscle. It provides a continuous link between the Z-disc and the M-line of a sarcomere, with a single titin molecule spanning half the sarcomere, anchoring its NH_2 terminus in the Z-disk and its COOH terminus in the M-band. For a review of the structure and function of titin, nebulin and obscurin see Kontrogianni-Konstantopoulos *et al.* (2009).

Nebulin, another large, modular sarcomeric protein, is one of the most indistinct components of skeletal muscle. Like titin, it is thought to act as a 'rule' to specify the precise lengths of the actin filaments in skeletal muscle. It is ~1.0μm in length and its NH_2 terminus extends to the pointed ends of thin filaments, whereas its COOH terminus is partially inserted into the Z-discs (Figure 4.12). Interest in nebulin has suggested extensive isoform diversity, and its expression in cardiac tissue. It is proposed that it is a multifunctional filament system, possibly playing roles in signal transduction, contractile regulation and myofibril force generation (McElhinny *et al.*, 2003), and a growing number of mutations of nebulin are associated with human muscle pathology.

Obscurin is a more recently discovered muscle specific protein. Whilst titin and nebulin are integral components of sarcomeres, obscurin intimately surrounds sarcomeres, primarily at the level of the Z-disc and M-band, but is not present within the sarcomere. It is suggested that collectively the structural properties and subcellular locations of titin, nebulin and obscurin function as molecular scaffolds during myofibrillogenesis - facilitating the integra-

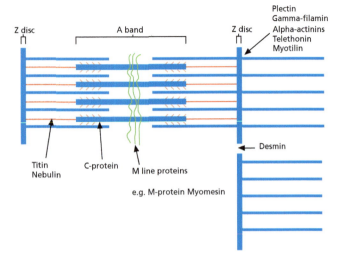

△ **Fig. 4.12** Some of the structural proteins in the sarcomere. Schematic diagram of the sarcomere and the location of some of the structural proteins.

tion of filaments of the contractile proteins actin and myosin, providing binding sites for a host of sarcomeric proteins and coordination of the alignment of structures like the SR (Kontrogianni-Konstantopoulos *et al.*, 2009).

C-protein is a myosin-binding protein seen in the A-band of striated muscle. You might recall from the previous sub-section that the A-band remains a constant width when the sarcomere shortens and muscle contracts. C-protein may have both structural and regulatory roles in the sarcomere, such as the aggregation of myosin monomers and regulation of thick filament length.

Alpha-actinins (ACTN) are major structural components of the Z-line, involved in securing actin-containing thin filaments. There are two isoforms in skeletal muscle, ACTN2 and ACTN3, which form a multigene family of four actin-binding proteins related to dystrophin. It appears that in human skeletal muscle ACTN2 is expressed in all muscle fibres, whilst ACTN3 is only found in type II. In some individuals ACTN3 can be absent, and in a recent study of six endurance athletes, in which only one subject showed typical type I fibre predominance, ACTN3 was absent in the *m. vastus lateralis* of a further

---

TIME OUT

THE DALTON

Biochemists and physiologists refer to the mass of a protein in a unit of mass called the dalton – named after John Dalton (1766–1844). Dalton developed the atomic theory of matter and the dalton was named after him as a unit of mass, equal to 1.000 on the atomic mass scale, nearly equal to the mass of a hydrogen atom. The kilodalton (kDa) is equal to 1000 daltons.

subject, suggesting a redundancy of this isoform of the protein, even in highly trained muscle (Zanoteli *et al.*, 2003). Those individuals lacking the stabilising influence of ACTN3 may experience more muscle damage in response to eccentric exercise (Vincent *et al.*, 2010).

Other Z-line proteins that have been discovered more recently include **telethonin**, **myotilin**, **plectin** and **gamma-filamin**, some of which have been linked to muscular dystrophies.

Dystrophin is thought to have a structural role in the surface membrane of the muscle fibre, acting as membrane 'scaffolding', and to also have a signalling role (by stabilizing nitric oxide synthase). It forms a link between the actin cytoskeleton and the extracellular matrix and is linked to the protein desmin via a novel protein called syncoilin. **Spectrin** is thought to have a similar role to dystrophin. **Desmin** is expressed in smooth and cardiac muscle tissue, as well as in skeletal muscle. In mature striated muscle fibres the lattice of desmin filaments surrounds the Z-discs, interconnects them to each other and links the whole contractile apparatus to the sarcolemmal cytoskeleton, cytoplasmic organelles and the nucleus.

In addition to these proteins, some of which can be seen in the schematic figure of the structural proteins in the sarcomere (Figure 4.12), there are thought to be proteins that keep myosin filaments in the correct spatial alignment in order for actin filaments to slide between them (M-line proteins) such as M-protein and myomesin. When muscle contracts, it is thought that the M-band buckles (Figure 4.13), and the Z-disc changes from a small square lattice to a basket weave lattice (Gautel, 2011).

The pursuit of a greater knowledge of structural proteins in skeletal muscle and their interactions is a target for research and is essential for a better understanding of muscle structure and function. This target will be helped enormously by proteomic techniques in the coming years. There are many other important proteins in muscle that have a variety of functions, including metabolic functions, many of which are still being identified through the work of the Human Genome Project and proteomics.

Practical proteomics involves the solubilization and isolation of proteins, fractionation by electrophoresis or chromatographic techniques, and identification by mass spectrometry and database searching. In the twenty-first century, proteomics is important for the understanding of cellular processes, to identify all of the proteins expressed in a cell and to find out about the relative degree of expression of each protein, its function(s), how it interacts with other proteins and how it might be modified post-translation. You might remember from Chapter 1 that translation is the mRNA-directed synthesis of a polypeptide. By understanding changes in protein expression and the up- and/or down-regulation of protein synthesis in disease and health states (such as muscle during exercise training or rehabilitation), a better understanding of physiology is being developed. Molecular physiology and molecular exercise physiology is set to have a significant impact on our lives in the future, fuelled by technological advances such as microarrays and antibodies for newly identified proteins.

Some further proteins with particular relevance to muscle and exercise metabolism that have emerged in recent years include uncoupling proteins, heat shock proteins and glucose transporter proteins.

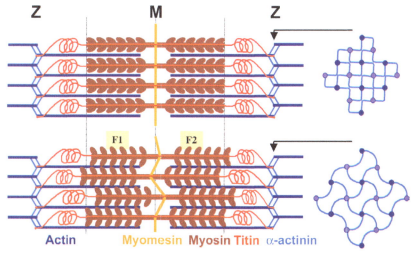

△ **Fig. 4.13** Schematic representation of a sarcomere at rest (top) and under isometric contraction (bottom). From Gautel (2011).

Uncoupling proteins

Uncoupling proteins (UCPs) are found in a variety of tissues, including brown adipose tissue (for example UCP1), and are responsible for heat generation (thermogenesis). UCPs uncouple oxygen consumption from ATP production, thereby dissipating energy as heat without ATP synthesis. More recently, further UCPs have been demonstrated, including UCP2 which has a wide tissue distribution, primarily in white adipose tissue and skeletal muscle. It may therefore have a role in regulating energy expenditure.

A UCP2 homologue called UCP3 is present in brown adipose tissue and skeletal muscle. Its mRNA exists as two isoforms: UCP3 long ($UCP3_L$) and UCP3 short ($UCP3_S$), though it is not known whether the two isoforms have different physiological functions. Expression of UCP3 varies between fibre types, with greater expression in glycolytic type II fibres than in oxidative type I fibres (Hesselink *et al.*, 2001). It has been demonstrated that whilst both endurance and sprint type training down-regulate UCP3 expression and protein content, strength training reduces UCP3 expression significantly less in type I fibres and significantly more in type IIx fibres (Russell *et al.*, 2003), suggesting that such an adaptation might enhance the efficiency of endurance-trained muscle by reducing mitochondrial uncoupling, whilst for sprint-trained muscle this is relatively less important.

The role of UCP3 appears to be linked to energy metabolism but it might not necessarily be a primary function. It may protect mitochondria from an overload of long-chain (non-metabolizable) fatty acids (Hoeks *et al.*, 2003). Knowledge of the expression and function of these novel uncoupling proteins in humans is still developing.

Heat shock proteins

Heat shock proteins (HSPs), also known as 'molecular chaperones', are expressed in skeletal muscle (and in other tissues and cells) and they play essential roles in protein homeostasis, including their folding, and membrane translocation. Oxidative stress, the imbalance between the production of chemical species called free radicals and their removal can induce the expression of HSPs, and in turn they can stimulate adaptive immune responses, with the ability to form a link between innate and adaptive immunity. You might recall that in Chapter 3 we discussed the immune system using the 4–4–2 soccer formation analogy, within which innate immunity is considered our second, semi-specific line of defence in midfield and adaptive immunity, provided by specific lymphocytes and antibodies, is the back line.

Immune function can be altered by exercise and training, and an increased level of free radical production, for example in intense exercise, may be a mechanism of exercise-induced immune function depression, mediated by HSPs. A role for HSPs in cardiovascular disease has also been suggested, and recent evidence suggests that higher circulating levels of HSP70 predict the future development of cardiovascular disease in people who already have a high blood pressure. Furthermore, as cardiovascular disease is a major cause of morbidity and mortality in the Western world, a role for HSPs is proposed in ageing, as a consequence of decreasing HSPs with advancing age. It has been shown that in a small sample of the offspring of centenarians (people aged 100 or more), circulating serum levels of HSP70 were significantly lower than in spousal controls, suggesting that this might be a marker for longevity (Terry *et al.*, 2004).

Glucose-transporter proteins

Glucose (GLU) transporter (T) or carrier proteins inside cells provide a means of access of glucose into cells, a process that is facilitated by the action of insulin. In response to insulin stimulation, glucose transporter (GLUT)-containing vesicles migrate to the surface, fusing with the plasma membrane. The increase in GLUT molecules at the cell surface causes a large increase in cellular glucose uptake. Different cells have different isoforms of glucose transporter, varying in their response to insulin and to the glucose concentration, and indeed to the extent of contractile activity. There are more than 12 glucose transporter (GLUT) protein isoforms, most of which have been discovered since the late 1990s. Several of them, such as GLUT4, GLUT8, GLUT11 and GLUT12 have been detected in human skeletal muscle (**Box 4.1**).

Box 4.1

Some of the human glucose transporter (GLUT) proteins and examples of their tissue distribution

GLUT-1	Red blood cells, brain, kidneys, placenta, small blood vessels
SGLUT-1	Sodium-glucose transporter found in the small intestine
GLUT-2	Liver, pancreatic beta cells, kidneys, small intestine
GLUT-3	Kidneys, brain, placenta
GLUT-4	Skeletal muscle, adipose tissue, cardiac muscle
GLUT-5	Small intestine
GLUT-8	Skeletal muscle
GLUT-11	Skeletal muscle – type I fibres only?
GLUT-12	Skeletal muscle

The expression of the isoforms can differ in a fibre type specific pattern, for example GLUT4 is generally expressed in all fibre types, whereas GLUT11 may be present in type I (slow twitch fibres) but not in type II fibres, suggesting a specialized function for GLUT11 with a regulation independent from that of GLUT4 (Gaster *et al.*, 2004).

Energetics and metabolism

Muscle cells require energy at all times, and therefore need to consume ATP for basic cellular activities, for example to maintain ion gradients. But over and above this resting metabolism, energy is required for exercise, to maintain cross-bridge cycling at rates much higher than resting metabolism. As described in Chapter 1, metabolism can be subdivided into **anabolism**, the biosynthesis of more complex molecules from simpler ones, and **catabolism**, involving the network of chemical pathways in which complex molecules are broken down into smaller molecules. Those metabolic pathways whose name ends with 'genesis', meaning to create, are anabolic, and those whose name ends in 'lysis', meaning to break down, are catabolic.

Catabolism generates precursor molecules for biosynthesis, reducing potential necessary for biosynthesis (e.g. in the form of the co-enzyme NADPH), heat and chemical potential energy in the form of ATP for the performance of muscular work and for biosynthesis.

$$\text{Fuel} + \text{ADP} + \text{P}_i \rightarrow \text{ATP}$$

The ATP present in muscle provides the energy to perform exercise. Its hydrolysis by myosin ATPase allows cross-bridge formation and myosin head attachment to actin, pulling actin filaments towards the centre of the sarcomere, and thereby enables the muscle's attempt to shorten. The rate of energy production and the need to transform chemical into mechanical energy is dependent on the load on the muscle and on the type of myosin. During contraction ATP is hydrolysed:

$$\text{ATP} + \text{H}_2\text{O} \rightarrow \text{ADP} + \text{P}_i + \text{H}^+$$

Producing relatively low forces at a high velocity requires high rates of ATP generation and consumption and muscle fibre shortening is dependent on the myosin isoform present. Fast fibres have an isoform of myosin present that is capable of a high rate of ATP synthesis compared with slow fibres.

A further energy requirement for muscle cells is in the phase of relaxation, as ATP is required by the Ca^{2+} pumps in the sarcoplasmic reticulum and the muscle cell membrane. The way in which skeletal muscle cells generate their energy (ATP) is the same as in all nucleated cells, but the relative importance of the different pathways varies in different muscle fibre types as a result of the myosin present and the density of contractile proteins and organelles such as mitochondria. There are three routes of ATP production: ADP rephosphorylation, glycolysis and oxidative phosphorylation.

ADP rephosphorylation by creatine kinase

As the ATP that is present at the onset of contraction is utilized by being dephosphorylated to ADP, some ADP can be immediately rephosphorylated as a result of the presence of phosphocreatine (PCr):

$$\text{PCr}^{2-} + \text{MgADP}^- + \text{H}^+ \xrightarrow{\text{creatine kinase}} \text{MgATP}^{2-} + \text{Cr}$$

The creatine kinase reaction rapidly controls the cytosolic ATP concentration and its availability to mitochondria. Utilization of phosphocreatine causes absorption of H^+ and may account for a quarter of the uptake of H^+ in human skeletal muscle during sprint-type exercise (see Table 4.1 on page 105.). The reversible ADP-dependent dephosphorylation of phosphocreatine to produce ATP and the guanido compound creatine was discovered by Lohmann in the early 1930s. Phosphagen (ATP and PCr) stores thereby act as an immediate buffer to the maintenance of ATP concentration whilst other routes for ATP resynthesis are building to full capacity. Phosphagen stores are, however, very limited and the total amount of energy available is small. In recent years, it has been hypothesized that a higher intracellular concentration of phosphocreatine and creatine can potentially aid muscle function.

The first published demonstration of an increase in the phosphocreatine and creatine content of human skeletal muscle as a result of oral creatine supplementation was described in 1992 by Harris and co-workers, and since then there has been a plethora of interest in the area of applied sport and exercise and some good basic research on creatine metabolism that is of benefit to muscle in health and disease.

The inability of muscle fibres to maintain the required rate of ADP rephosphorylation at times of very high energy demand (i.e. when the capacity of the PCr/Cr buffer is exceeded) results in an increase in ADP concentration. This leads to a displacement of the near equilibrium myokinase reaction, resulting in an increase in the formation of AMP. The second phosphate group of ADP can be transferred to another ADP molecule via the myokinase reaction, forming ATP and AMP:

$$2\text{ADP} \xrightarrow{\text{myokinase (adenylate kinase)}} \text{ATP} + \text{AMP}$$

Although this helps to maintain the ATP/ADP ratio, the cost is an elevated AMP concentration. This method of ATP regeneration is least satisfactory to the cell since AMP is deaminated to inosine monophosphate (IMP), with the concomitant production of ammonia (NH_3), resulting in a net loss of adenine nucleotide:

$$\text{AMP} + \text{H}_2\text{O} \xrightarrow{\text{AMP (adenylate) deaminase}} \text{IMP} + \text{NH}_3$$

TIME OUT

CREATINE SUPPLEMENTATION

The importance of the role of creatine (Cr) and phosphorylcreatine (PCr) in skeletal muscle has been known since 1911. PCr is the immediate phosphate buffer to support the resynthesis of ATP after its breakdown to ADP. The majority of the body Cr pool is found in skeletal muscle, however, the site of utilization is separated from the site of biosynthesis – liver and pancreas (Walker, 1979).

Vegetarians, who ingest no Cr in their diet, can adequately synthesize their needs endogenously from the precursor amino acids glycine, arginine and methionine. On the other hand, meat and fish eaters will downregulate their endogenous Cr synthesis and absorb Cr.

Creatine can also be ingested in its pure chemical form, and there has been an exponential interest in Cr as a supplement and commercial exploitation in products marketed for sport and exercise since 1992 – when the first of more recent scientific Cr ingestion work was published (Harris et al., 1992). It is possible to increase the total creatine content (TCr, i.e. PCr + Cr) of human skeletal muscle from about 120 mmol/kg dry muscle to about 150 mmol/kg dry muscle in those individuals who respond to oral Cr supplementation. Co-ingestion of large amounts of glucose during Cr supplementation augments muscle Cr accumulation, and removes some of the variability in uptake between individuals (Green et al., 1996). Furthermore, Cr supplementation has the potential to elevate muscle glycogen above a conventional glycogen 'loading' regimen (Robinson et al., 1999).

There is evidence to suggest that Cr supplementation can be of benefit to muscle metabolism in sport and exercise but also in health and disease, based on, for example, its effect on glucose disposal and tolerance. In exercise, the large number of research studies that have investigated performance enhancement are equivocal, but the lack of performance enhancement suggested by some may be due to variability in individual subject responses to Cr ingestion and other metabolic and nutritional individual differences.

The European Food Standards Agency (EFSA) have begun publishing opinions on supplement related health claims, creatine being one of them.

Sustained sprint exercise in humans and other species (e.g. that which results in fatigue in about 2 minutes) can result in a rapid fall in the muscle adenine nucleotide pool and is associated with fatigue. As a phosphorylated molecule, the IMP remains inside the muscle cell, but the extent of adenine nucleotide degradation can be gauged by the appearance of NH_3 in plasma, peaking at around 5 minutes after exercise – this is known as exercise **hyperammonaemia**. Plasma NH_3 concentration can rise from around 40 µmol at rest, to around 250 µmol in some individuals after short-duration, intense exercise (e.g. Sewell et al., 1994).

Glycolysis

As described more fully later, the overall principle of glycolysis is the conversion of a six-carbon glucosyl unit (i.e. from glucose or glycogen) to two three-carbon molecules of pyruvic acid. **Pyruvic acid** has two fates – conversion to **lactic acid** or continuation to oxidative means of energy production. Although inefficient in terms of net yield of ATP molecules per glucosyl molecule, glycolysis provides a rapid means of producing ATP and meeting most of the ATP demand in the short

term. One metabolic cost of a high rate of glycolytic ATP production is the production of lactic acid, which immediately dissociates to lactate and H^+, thereby decreasing the pH of the muscle cell.

Oxidative phosphorylation

This is the most efficient means of producing ATP, in the sense that the generation of reduced co-factors and hence ATP, per molecule of substrate such as glucose, is maximized. This is as a result of the pyruvic acid from glycolysis being converted to acetyl CoA and the oxidation of fatty acids to **acetyl CoA** within the mitochondrion (beta-oxidation) being utilized in the **tricarboxylic acid (TCA; Krebs or citric acid) cycle**, producing carbon dioxide, and the reduced co-enzymes produced (reduced forms of nicotinamide adenine dinucleotide (NADH) and flavin adenine dinucleotide (FADH)) being used in the **electron transport chain** to produce ATP and water. The products of the breakdown of dietary and tissue protein (and of alcohol) can also be oxidized to carbon dioxide and water, but these are not major sources of fuel in healthy, exercising individuals.

Pathways of energy metabolism in skeletal muscle

Where do we get all of the energy (ATP) from to continue the contractile process? Our resting muscle ATP content is around 21 mmol/kg dry muscle, but for intense muscular contraction (e.g. sprinting) we need about 10 mmol ATP/kg dry muscle per second. The total capacity of ATP production from glycolysis is greater than that provided by immediate phosphagen stores, however, the rate at which it can produce ATP is lower. Maximum speed can therefore only be sustained for about 10 seconds, then phosphagen levels fall and metabolic fatigue occurs. In Chapter 3 we looked at the assimilation of food, that is, from ingestion to its breakdown into basic chemical units (glucose, fatty acids and amino acids). We shall now consider the major metabolic pathways that are important in our understanding of the use of these substances. An overview can be seen in Figure 4.14.

Glycolysis

For many years, the desire of biochemists to find out how skeletal muscle cells obtain their energy for contraction ran alongside the question of how yeast cells ferment sugars to produce ethanol. By 1937, the metabolic pathway of glycolysis had been elucidated, explaining how glucose or glycogen is converted to pyruvic acid, and how, in the absence of oxygen (relatively), the pyruvic acid is reduced to lactic acid in muscle cells or is converted to alcohol in yeast cells. This was the work of Meyerhof and colleagues in Germany.

Glycolysis takes place in the cytoplasm and both glucose and its storage polymer form, glycogen, are substrates. The overall strategy is to convert a glucosyl unit (six-carbon molecule) to two molecules of pyruvate (three-carbon molecule). In the beginning, glucosyl units from glycogen (e.g. in muscle cells) enter the process as glucose 1-phosphate, due to the hydrolysis of glycogen (glycogenolysis). Molecules of glucose 1-phosphate are then converted to glucose 6-phosphate. Glucose from the bloodstream enters the cell, stimulated by insulin and aided by glucose transporter proteins (see earlier in this chapter), and is rapidly phosphorylated by hexokinase, forming glucose 6-phosphate. The eventual net yield of energy in the form of ATP from a glucose molecule is

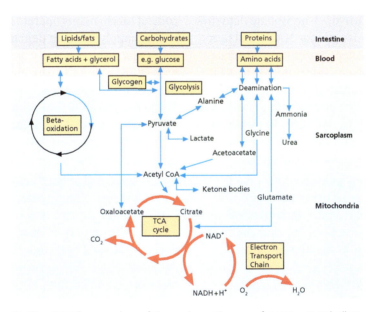

△ **Fig. 4.14** An overview of the major pathways of energy metabolism.

slightly lower due to the donation of phosphate from ATP for glucose 6-phosphate formation. Phosphorylation of glucose traps it within the cell as it can no longer cross the cell membrane, nor can it be transported back into the bloodstream. Glucose 6-phosphate is then further metabolized, that is, oxidized, beginning with the process of glycolysis (catabolism; or alternatively glycogen storage – anabolism). The series of reactions that occur in glycolysis are summarized in Figure 4.15.

The incorporation of energy into the process of glycolysis continues at a subsequent stage – when fructose 6-phosphate is converted to fructose 1,6-diphosphate. At this point, the net yield of ATP is minus two molecules. Things get better after the six-carbon molecule is split into two three-carbon molecules. The overall glycolytic yield of ATP molecules from glucose is four (steps 7 and 10), less the two required to phosphorylate glucose (step 1, Figure 4.15) and to phosphorylate fructose 6-phosphate (step 3). A further amount of ATP can eventually be produced from the mitochondrial oxidation of glycolysis (cytoplasmic)-generated NADH (conversion of glyceraldehyde 3-phosphate to 1,3-diphosphoglycerate; step 6).

You might well ask where the extra phosphate comes from in this step, if ATP is not required – this reaction, step 6, catalysed by glyceraldehyde-3-phosphate dehydrogenase, has an absolute requirement for NAD^+ and inorganic phosphate. In the next step, one of the phosphate groups of 1,3-diphosphoglycerate is transferred to ATP, one of the sources of the small amount of energy produced directly by glycolysis. The endpoint of glycolysis is pyruvic acid/pyruvate, which has two fates in metabolism in humans – conversion to acetyl CoA, the start point of the TCA cycle, or conversion to lactic acid/lactate.

In glycolysis, the breakdown of glucosyl units (glucose from glycogen breakdown) to pyruvic acid results in NAD^+ being reduced to NADH in the glyceraldehyde-3-phosphate dehydrogenase reaction. The availability of NAD^+ can become a problem, as the muscle cytosolic NAD^+ content is only a fraction of that needed to maintain high-intensity exercise. For this reason, rapid reoxidation of NADH is necessary in order for glycolysis to proceed and to contribute to the supply of energy. In addition to reoxidation of NADH through the mitochondrial respiratory chain (using oxygen), and through the conversion of dihydroxyacetone phosphate (DHAP) to glycerol 3-phosphate (an electron shuttle), NADH is also reoxidized by the conversion of pyruvic acid to lactic acid – whereby NADH is oxidized by pyruvic acid to form NAD^+ (NAD^+ being the oxidized form). This is of significance in metabolism because it is one of the essential functions of lactic acid production. In other words, lactic acid is not necessarily the bad metabolite of exercise metabolism.

By causing a change in oxygen affinity (the Bohr effect; shifting the oxygen dissociation curve to the right, see the section 'Oxygen transport in blood' in Chapter 3), circulating H^+ may also enhance the facilitation of oxygen diffusion into muscle by increasing P_{O_2}, thereby aiding oxygen movement from red blood cells to the tissue. Lactate can also be used as a fuel source that can be distributed by diffusion from muscle into other fibres, muscles and tissues (see the section on 'Gluconeogenesis' in this chapter). Obviously, the less beneficial effects of lactic acid production are the accompanying acidosis which it causes when it dissociates to lactate and H^1, and the rapid utilization of glycogen.

Glycolysis utilizes both circulating glucose and glucose molecules released from glycogen. Glycogenolysis (lysis, from the Greek meaning dissolution) is the hydrolysis of muscle glycogen to glucose 1-phosphate which is then degraded to pyruvic acid. The accumulation of products of ATP and PCr hydrolysis already mentioned, such as ADP, AMP, IMP and NH_3, all act as stimulators of glycogenolysis and glycolysis, making sure that ATP production is further maintained for a short while longer. Regulatory control mechanisms operate in glycolysis to ensure that energy supply matches demand. One of the key regulatory steps is the reaction catalysed by phosphofructokinase (PFK; step 3), which adds a phosphate molecule to fructose 6-phosphate to form fructose 1,6-diphosphate. The activity of PFK is inhibited by high concentrations of ATP (e.g. when skeletal muscle is at rest) and is stimulated by high concentrations of ADP and AMP (e.g. during moderate intensity exercise).

The pyruvic acid formed as a result of glycolysis can either undergo oxidative decarboxylation to form acetyl CoA or, when the rate of formation of pyruvic acid is too great for the amount of oxygen available, can be reduced by NADH to form lactic acid. Sprint type exercise demands a high ATP turnover rate and results in the production and accumulation of large amounts of lactic acid in skeletal muscle. Because it is a strong organic acid, at physiological pH values lactic acid dissociates almost completely to form a lactate anion ($C_3H_5O_3^-$) and a proton (hydrogen cation, H^+; Figure 4.16).

The increase in muscle lactate concentration that can be measured in muscle samples after exercise suggests an accumulation of so many H^+ ions that there could potentially be a decline in pH from 7.1 to around 1.0. Despite this, the fall in muscle pH from 7.1 at rest to around 6.5 following high-intensity exercise such as running 800 m corresponds to a relatively small increase in free $[H^+]$.

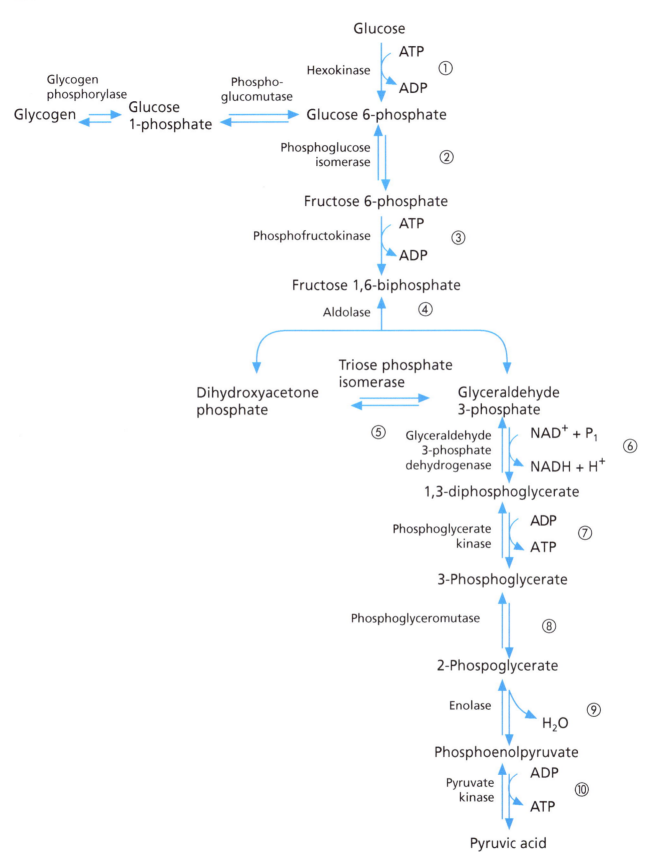

△ **Fig. 4.15** The glycolytic pathway.

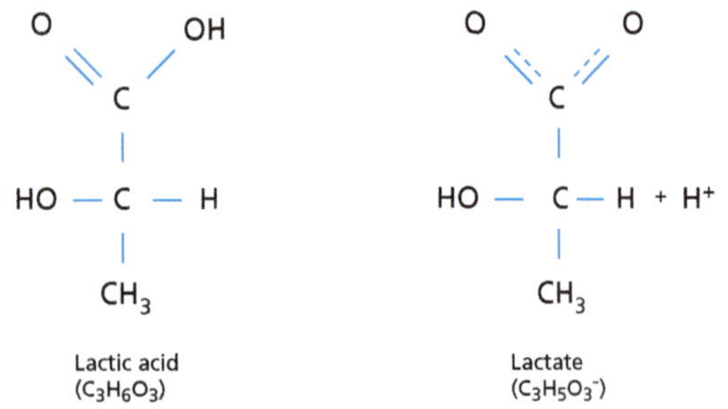

△ **Fig. 4.16** Dissociation of lactic acid to lactate and H^+.

Some of the H^+ produced will diffuse from the muscle to be buffered in the circulation whilst a proportion will be buffered within the muscle.

Although mechanisms exist to buffer H^+ produced during intense exercise, the capacity to neutralize H^+ may ultimately be saturated, leading to the development of metabolic acidosis. It is generally accepted that the development of metabolic acidosis during intense exercise is related to metabolic fatigue. A high intracellular $[H^+]$ may directly affect contraction in a number of ways, for example, by inhibiting the release of calcium from the sarcoplasmic reticulum or by impairing myosin ATPase activity, hindering the ability of myosin to detach from actin thereby slowing the contraction/relaxation process. Muscle acidosis may indirectly affect the fatigue process

by reducing the effectiveness of PCr to buffer the rise in $[ADP]$, due to the dependence of the equilibrium position of the creatine kinase-catalysed reaction on pH.

Acid–base control

The control of acid–base status in the whole body depends on the response and integration of a number of systems and in Chapter 3 we considered the contribution of extracellular buffering. We shall now consider the contribution of intracellular buffering mechanisms to acid–base homeostasis. It is vitally important that cells have mechanisms for appropriately regulating their cytoplasmic pH (pH_i). The acid–base insults that cells are subjected to might be acute or chronic. Acute loads are eventually corrected by the acid–base transport mechanisms that regulate pH_i. In the longer term, pH_i reaches a steady state when a balance is achieved between the rates of acid production and removal. Its regulation is therefore not dependent on buffering power and is unaffected by acute acid–base disturbances, but the ability to buffer the effects of high-intensity exercise in the short term may delay metabolic disturbance. For example, a greater propensity for the intracellular environment of muscle to buffer H^+ during exercise may result in a delay in fatigue. The enhancement of extracellular buffering capacity may also accelerate the removal of H^+ from the cytoplasm (thereby reducing $[H^+]$). Such

---

TIME OUT

BICARBONATE LOADING IN SPORT AND EXERCISE

Disturbance of acid–base status as a result of high-intensity exercise is a major metabolic consideration and is one of the factors related to fatigue in such exercise. Various physiological mechanisms have been put forward to explain how H^+ accumulation (muscle acidosis) might inhibit the contraction process, and as a result, inducing a metabolic alkalosis has been used both as a research tool to study muscle metabolism and fatigue, and as a potential ergogenic aid to sports performance. Bicarbonate loading has been used by athletes and has been administered to racehorses (and probably other racing animals!).

Sodium bicarbonate ($NaHCO_3$) might normally be ingested at a dose of 0.3 g per kg body mass – about 20 g for most people – 3 hours prior to exercise. Favourable and maximal acid–base changes, such as base excess (a clinical indicator of metabolic acid–base disturbance), do not take place until at least 3 hours after $NaHCO_3$ ingestion. The dose used is a large one and can result in gastrointestinal problems such as flatulence, diarrhoea and vomiting.

Sodium citrate is another alkalinizing agent that has been used experimentally and can have a similar effect on acid–base status without the same risk of side-effects.

Some sodium bicarbonate loading research studies have been able to demonstrate an improvement in performance in certain types of high-intensity exercise of short duration. ATP loss and adenine nucleotide degradation are associated with muscle fatigue, and it has been suggested that $NaHCO_3$ ingestion might favourably modify these metabolic consequences of high-intensity exercise. This will be of greater importance in those who have a high proportion of type II muscle fibres in key muscles, and who will, as a consequence, be potentially more capable of high-intensity (e.g. sprinting, repeated jumping) exercise performance.

an effect *in vivo* has been suggested by attempts to bicarbonate load prior to exercise (see **Time Out** *Bicarbonate loading in sport and exercise*).

Approximately 94 per cent of the H^+ released within muscle during exhaustive exercise is the result of the production of lactic acid (Hultman and Sahlin, 1980). Other acids account for a portion, such as pyruvic acid (0.3 per cent), malic acid (3 per cent), whilst the accumulation of glucose 6-phosphate and glycerol 1-phosphate contribute 2 per cent and 1 per cent respectively.

The pH of the cytoplasm can be buffered by three classes of processes: weak acids and bases (**physicochemical buffering**), biochemical reactions (**biochemical (metabolic) buffering**) and transport of acid–base equivalents across organellar membranes (**organellar buffering**). Of greatest value to us is the distinction between physicochemical and metabolic buffering (Table 4.1). Buffers can also be described as being intrinsic or extrinsic to the cell. Intrinsic buffers are those confined to the cell, including closed-system physicochemical buffers (e.g. histidine, carnosine), biochemical buffers and organellar buffers. Extrinsic buffers include open-system physicochemical buffers such as the CO_2–HCO_3 system referred to in Chapter 3.

Physicochemical buffering

In the cytoplasm, physicochemical buffers can be divided into those that cannot permeate the cell membrane (closed-system) such as the weak acid–base moieties of proteins, and those whose neutral form freely permeates the cell membrane such as the CO_2–HCO_3 system. Carbonic acid is the neutral form. In terms of the closed-system buffers, those that are considered most important within skeletal muscle are the protein-bound histidyl residues, histidine-containing dipeptides (e.g. carnosine in human skeletal muscle) and free histidine. The buffering potential of histidine is due to the acid–base behaviour of its imidazole ring structure. When the cytoplasmic $[H^+]$ rises, the imidazole group will bind protons, reducing the pH change that might occur. In the case of carnosine (β-alanyl-L-histidine), the pK_a (negative log of the dissociation constant) of the molecule is higher than that of histidine, increasing its effectiveness as a buffer in the physiological pH range.

Metabolic buffering

As described earlier in this chapter (see the section on 'Energetics and metabolism'), the creatine kinase reaction rapidly controls the cytosolic ATP concentration and its availability to mitochondria. Utilization of phosphocreatine (PCr) causes absorption of H^+ and may account for 27 per cent of the uptake of H^+ in skeletal muscle during exhaustive short-term exercise (Table 4.1). As a result of high-intensity exercise, it is possible that PCr will be almost totally depleted, which may be a cause of fatigue. Phosphocreatine depletion, specifically in type II fibres will result in a fall in the rate of ADP rephosphorylation and the development of fatigue. Phosphocreatine is resynthesized during recovery, and returns relatively quickly to pre-exercise values in comparison with metabolites such as ATP and IMP. The synthesis of PCr liberates H^+. Muscles with a greater proportion of type II fibres will have a higher content of PCr and individual differences in fibre composition, and hence PCr content, will affect buffering capacity.

The theory behind creatine supplementation is based on increasing muscle total creatine content and increasing resting [PCr]. Such alterations may delay the onset of fatigue caused by PCr depletion by increasing PCr resynthesis and improving the buffering capacity of skeletal muscle. As a result of adenine nucleotide degradation (Figure 4.17), ATP decreases, accompanied by an equal rise in IMP and the release of NH_3 (exercise hyperammonaemia).

$$ATP + nH^+ \rightarrow IMP + 2P_i + NH_4^+$$

The total amount of H^+ taken up by this reaction during exercise is small, but will increase as the amount of IMP formed rises with increasing high-intensity exercise duration.

Buffering process/substance	Uptake of H^+ (mmol/l muscle water)	% contribution
Physicochemical buffering		
Proteins (inc. free histidine), carnosine, glutamine	17.4	40.2%
$HCO_3 + H^+ \rightarrow H_2CO_3 \rightarrow H_2O + CO_2$	7.2	16.6%
Metabolic buffering		
$PCr \rightarrow Cr + P_i$	11.6	26.8%
$ATP \rightarrow IMP + 2P_i + NH_4$	0.8	1.8%
$H^+ + HPO_4 \rightarrow H_2PO_4$	3.2	7.4%
NH_4^- production during catabolism of amino acids	3.1	7.2%

1.30 mmol/l muscle water = 4.3 mmol/kg dry muscle = 1 mmol/kg wet muscle.
Modified from Hultman and Sahlin (1980) and Poortmans (1986)

△ **Table 4.1** Calculated uptake of H^+ in human skeletal muscle – contributions of various systems during exhaustive short-term exercise.

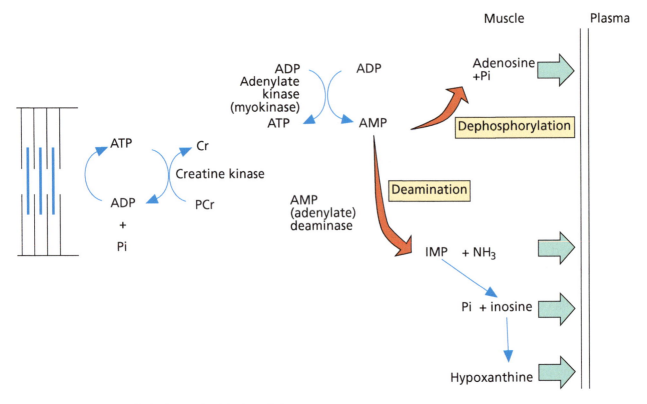

△ **Fig. 4.17** Pathway of adenine nucleotide degradation.

A certain proportion of protons are taken up by inorganic phosphate (P_i). The basic reaction scheme for this buffer system is as follows:

$$H^+ + HPO_4^{2-} \rightarrow H_2PO_4^-$$

At rest, and even at high rates of CO_2 production, normal circulation, ventilation and respiration will balance the CO_2 by elimination and keep [H^+] within normal range. When energy expenditure is too great to be met through oxidative means, the accumulation of lactic acid will ensue. The contribution of various systems to the uptake of H^+ during exhaustive short-term exercise can be seen in Table 4.1.

The contribution of the CO_2–HCO_3 system to intracellular buffering is less than that to extracellular buffering during exercise, maybe because the intracellular CO_2–HCO_3 system freely permeates the cell membrane and acts as an open-system buffer. The largest contribution to the uptake of H^+ inside the muscle cell comes from proteins, amino acids and dipeptides.

The intracellular pH of skeletal muscle becomes transiently alkaline within seconds after the onset of exercise. This can be entirely accounted for by H^+ consumption via net PCr hydrolysis, prior to the production and accumulation of large amounts of H^+. After the initial rise in intracellular pH, there is a decrease to around 6.5 as a result of intense muscular contraction. The recovery in muscle

pH after brief cycle exercise follows an exponential time course with a half-time of about 10 minutes (Sahlin *et al.*, 1976). Muscle pH is restored to around pH 7.1 (pre-exercise value) after a 20-minute rest. Occlusion of (blocking off) the blood supply (for example with a pressure cuff) results in no recovery in muscle pH, emphasizing the importance of the circulatory system for recovery of metabolites to pre-exercise levels.

The main factor determining intracellular pH during exercise is the production of lactic acid, though many other metabolic processes will affect acid–base balance. The hydrolysis of ATP, for example, releases H^+:

$$ATP + H_2O \rightarrow ADP + P_i \; 1 \; nH^+ + energy$$

in which the amount of H^+ released is small compared with other changes. The ATP–ADP reaction is coupled to the breakdown of PCr:

$$PCr + nH^+ \rightarrow Cr + P_i$$

which results in an uptake of H^+. The stoichiometry of H^+ uptake (n) is dependent on the pH of the system; at rest, when the intracellular pH is around 7.1, the uptake of H^+ per mole of PCr hydrolysed is only about half that when at fatigue, when intracellular pH is around 6.5. Metabolic buffering by the PCr system is quantitatively the second most important process involved in the uptake of H^+ during intense exercise.

The tricarboxylic acid cycle

Unlike in the relative absence of oxygen, when oxygen is available to muscle cells (and yeast and other cell types) glucose is oxidized completely to CO_2 by a combination of glycolysis followed by the tricarboxylic acid (TCA) cycle. This was elucidated by Hans Krebs, who called it the citric acid cycle, but it was also referred to as the Krebs cycle in his honour. Not only is it the pathway for the oxidative catabolism of glucose, as originally realized, but it is also necessary for the complete oxidation of lipids and amino acids. The cycle has a central position in metabolism because many of its intermediates also provide the starting points for anabolic or biosynthetic processes.

The TCA cycle occurs in the mitochondria in eukaryotic cells such as ours, providing most of the reduced co-enzymes for electron transport and is therefore responsible for the majority of ATP production.

Pyruvic acid molecules produced in glycolysis can be utilized in oxidative metabolism by moving into mitochondria. You will recall from Chapter 1 when we discussed the typical organelles of cells, that mitochondria have an inner and an outer compartment. In the inner compartment, a carbon atom and two oxygen atoms (CO_2) are removed from the three-carbon pyruvic acid molecule to form a two-carbon acetyl group, reducing NAD^+ to NADH. The acetyl group then combines with co-enzyme A (CoA) to form acetyl CoA. Enzymes that make up the pyruvate dehydrogenase complex (three enzymes) are responsible for this series of reactions. Control of the activity of the pyruvate dehydrogenase complex is important in the integration of the metabolic response to exercise.

The entry of lipids into the TCA cycle is a little different. Triglycerides are broken down into glycerol and free fatty acids. After fatty acids enter the muscle cell they are converted to a CoA derivative by fatty acyl CoA synthetase, in preparation for beta-oxidation into acetyl CoA. Fatty acyl CoA molecules in the sarcoplasm have to be transported into the mitochondria and this is done by forming an ester of the fatty acid with carnitine (carnitine substitutes the CoA group). The fatty acyl is carried into the mitochondria and once inside it reverts to acyl CoA, a reaction regulated by carnitine acyltransferase. This enzyme exists in two forms in skeletal muscle, one located on the outer surface of the membrane in order to generate acyl-carnitine, and the other is located on the inner surface of the inner mitochondrial membrane to regenerate the acyl CoA and free carnitine (Figure 4.18). Carnitine is a vitamin-like substance that we endogenously synthesise from the amino acids lysine and methionine, and is also obtained in the diet (e.g. meat, dairy products). There is little evidence for ergogenicity of carnitine supplementation for exercise performance.

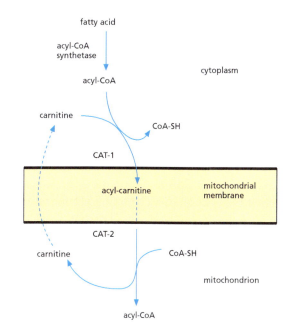

△ **Fig. 4.18** Entry of fatty acids into the mitochondria by utilizing carnitine. From Maughan *et al.*, 1997.

Beta-oxidation is a series of reactions in which two carbon atoms at a time are cleaved from the end of a fatty acid chain to form acetyl CoA. Once inside the mitochondria, fatty acyl CoA is oxidized to enoyl CoA, then to hydroxyacyl CoA (catalysed by enoyl CoA hydratase) to ketoacyl CoA and finally to a fatty acyl CoA molecule that is two carbon atoms shorter plus acetyl CoA. This process continues until all of the fatty acid chain is converted to acetyl CoA molecules, for subsequent entry into the TCA cycle for use in the generation of NADH.

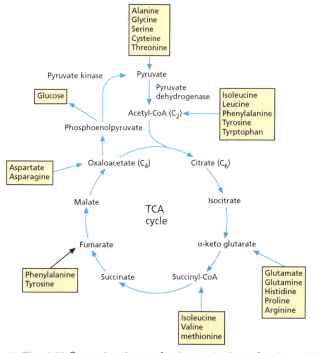

△ **Fig. 4.19** General pathways for the catabolism of amino acids.

Amino acids enter metabolism via the TCA cycle at various points (Figure 4.19). Fat and carbohydrate are, however, the major substrates that fuel ATP synthesis and amino acids are relatively unimportant as a fuel substrate for muscle for sport and exercise. Protein synthesis and breakdown that occurs as a result of muscle use and disuse is served by the free amino acid pool (Figure 4.20).

The TCA cycle is a series of eight enzyme-catalysed reactions that take place in the inner matrix of the mitochondria to oxidize acetyl CoA to carbon dioxide. Two carbon atoms enter the cycle as an acetyl unit and two leave the cycle in the form of two molecules of carbon dioxide. Citrate synthetase catalyses the transfer of the acetyl residue of acetyl CoA (a two-carbon molecule) to oxaloacetate (a four-carbon molecule) to form citrate (a six-carbon tricarboxylic acid; step 1: Figure 4.21). The reactions continue

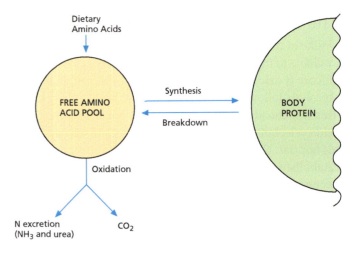

△ **Fig. 4.20** General scheme of amino acid/protein metabolism.

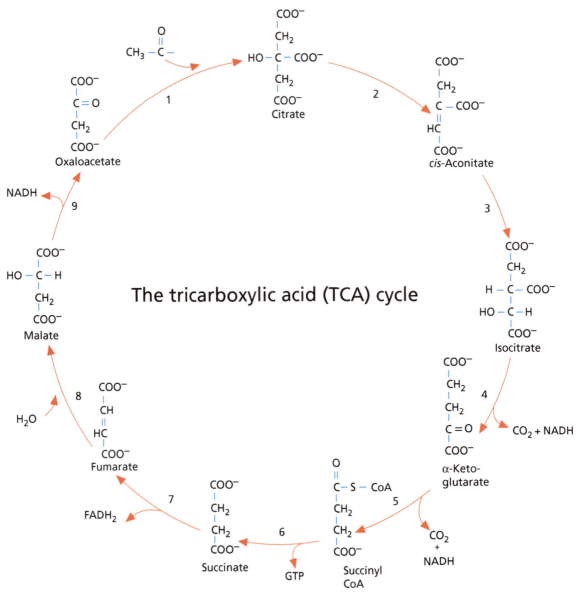

△ **Fig. 4.21** The TCA cycle.

through aconitase, isocitrate, alpha-ketoglutarate, succinyl CoA, succinate, fumarate, malate and oxaloacetate – the further intermediates of the TCA cycle. An acetyl group is more reduced than carbon dioxide, so oxidation–reduction reactions must take place in the TCA cycle. These occur at steps 4, 5 and 9 (to produce a molecule of NADH at each step), whilst step 7 produces a molecule of FADH. Three hydride ions are transferred to three NAD^+ molecules to form NADH, whereas one pair of hydrogen atoms (hence two electrons) are transferred to FAD (to form $FADH_2$).

In the electron transport chain (see below) when these electron carriers undergo oxidative phosphorylation, 11 molecules of ATP are produced, and one high-energy phosphate bond (in the form of guanosine triphosphate; GTP) is formed in each round of the TCA cycle itself. The phosphate group of GTP can be readily transferred to ADP to form ATP (in a reaction catalysed by nucleotide phosphate kinase) or can be used in aspects of metabolism such as protein synthesis and signal transduction.

Activity 4.1

Independent learning task: Enzyme examples from metabolic pathways

The following description of examples of enzyme types appeared in Chapter 1. See if you can complete the missing enzyme names now that they have been introduced to you again as part of the metabolic cycles described here in Chapter 4:

Dehydrogenases are oxido-reductase enzymes that add or remove hydrogen atoms (H). An example is _____ dehydrogenase which converts _____ to fumarate in the tricarboxylic acid cycle by the removal of H. Transferase enzymes transfer a chemical group from one substrate to another. In glycolysis, _____ transfers/adds a phosphate molecule to glucose to form glucose 6-phosphate. Lyases add a group across a carbon–carbon double bond, an example being _____ CoA _____ which reacts with water to convert acyl CoA to hydroxyacyl CoA. This is a step in beta-oxidation – the breakdown of fatty acids to produce acetyl CoA which can then enter the TCA cycle. Isomerases rearrange molecules and _____ isomerase is found in glycolysis where it converts dihydroxyacetone phosphate to glyceraldehyde 3-phosphate.

Electron transport chain/oxidative phosphorylation

Oxidative phosphorylation accompanies electron transport in mitochondria, involving the release of energy in the form of ATP in a process in which the reduced forms of the co-enzymes NAD^+ and FAD, namely NADH and $FADH_2$, donate their hydrogens and electrons to a series of carriers (the electron chain). The process involves the pumping of these hydrogens (also called protons) across the mitochondrial membrane. The carriers are cytochromes, flavoproteins and quinones, and the electron acceptor at the end of the chain is molecular oxygen – this is why the body must oxygenate tissues and consume oxygen.

The process is sometimes likened to that of a cascade, with different levels of the cascade 'catching' electrochemical energy. The potential energy in NADH and $FADH_2$ is transferred/coupled ('coupled respiration') from ADP to reform ATP at up to three 'energy' levels. NADH yields three molecules of ATP because it enters the respiratory chain at the highest possible energy level, whereas $FADH_2$ yields only two ATP for each hydrogen pair oxidized because it enters the chain at a lower energy level, beyond the synthesis site of the first ATP molecule.

The mitochondrial membrane is leaky, and the leakage of protons will increase oxygen consumption and instead of being used to produce ATP, heat is produced ('uncoupled respiration') which is a significant component of basal metabolic rate (see Chapter 5). Proton leak can occur through membrane proteins called uncoupling proteins (UCPs).

The energy yield (i.e. of ATP) of each glucosyl unit released from glycogen stored in muscle when oxidized to carbon dioxide and water can be summarized in an 'Energy balance sheet'. A total of 39 molecules of ATP are resynthesized (see Table 4.2).

Source	ATP molecules produced
Loss from phosphorylation of fructose 6-phosphate in glycolysis	−1
Cytoplasmic glycolytic ATP generation	4
Mitochondrial oxidation of NADH produced in glycolysis (2 NADH)	6
Mitochondrial oxidation of NADH produced in the TCA cycle (8 NADH)	24
Mitochondrial oxidation of $FADH_2$ produced in the TCA cycle (2 FADH)	4
Mitochondrial GTP produced in the TCA cycle	2
TOTAL	39

Or 38 if a glucose molecule (phosphorylation of glucose; see Figure 4.15).

△ **Table 4.2** Energy balance sheet showing the energy yield of each glucosyl unit released from glycogen stored in muscle when oxidized to carbon dioxide and water.

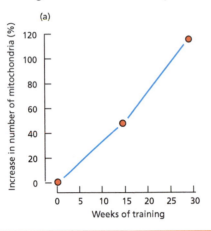
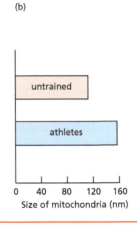
Gluconeogenesis

Gluconeogenesis provides an opportunity for the body to generate glucose from non-carbohydrate sources. It is the main source of glucose in fasting and in starvation for tissues such as the brain and red blood cells that cannot use any other fuel, and substrates such as lactate, pyruvate and glycerol are the precursors that are utilized. A small amount of lactate is continually being produced by skeletal muscle and by red blood cells, and gluconeogenesis provides a means by which this lactate and increased amounts produced during and after exercise can be recycled, particularly by the liver to resynthesize glucose (Figure 4.22).

Muscle cell fibre types and exercise performance

The heavy chain component of the myosin molecule determines the functional characteristics (e.g. velocity of contraction) of the muscle fibre. Three different isoforms of the MyHC can be identified in adult human muscle – these are known as I, IIa and IIx. Each muscle fibre has a predominance of one type of MyHC, and thus, fibres can also be designated as I, IIa and IIx. Fibres can also be categorized in other ways, for example, according to speed of contraction, metabolic properties or myosin ATPase activity (Table 4.3). Although there are similarities between the different classification systems, the schemes cannot be assumed to be interchangeable. As each classification reflects a different physiological characteristic of the fibre, the appropriate classification should be used when referring to a particular fibre property.

Classified according to:	Fibre types			
MHC isoform composition	I	IIa		IIx
Contractile properties	S	F-FR	F(Int)	FF
Metabolic properties	SO	FOG		FG
Myosin ATPase activity	I	IIa	IIc	IIb
ATPase activity related to twitch properties	ST	FTa		FTb

△ **Table 4.3** Muscle fibre type classifications.

As described earlier in this chapter, there are two types of muscle fibre, sometimes defined by their speed of contraction – slow and fast. Fast contracting fibres are subclassified further into fast IIa (fast oxidative, fast fatigue resistant) and fast type IIx (fast glycolytic, fast fatiguable). The different contraction speeds of muscle fibres is related to the way in which ATP is produced to obtain the energy required for contraction. 'Oxidative' fibres rely on the relatively efficient means of producing ATP, ultimately using available oxygen through mitochondrial respiration (hence the higher mitochondrial density in such fibres). Glycolytic fibres rely more heavily on ATP production by means of a high glycolytic flux, whereby pyruvic acid is converted to lactic acid rather than being converted to acetyl CoA and proceeding to the TCA cycle (Tables 4.4 and 4.5).

Slow contracting fibres are more important for endurance-type physical activities, such as long-distance cycling,

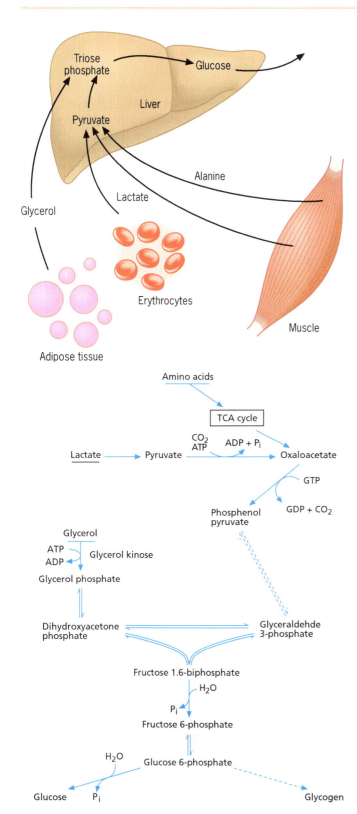

△ **Table 4.4** Contractile properties of the major fibre types.

Characteristic	Fibre type		
	I	IIa	IIx
Twitch contraction time	Slow	Fast	Fast
Maximum tetanic tension	Low	Intermediate	High
Resistance to fatigue	Very high	High	Low
Motor unit firing frequency	Low	Intermediate	High

running and swimming, and are recruited at low stimulation frequencies. At low exercise intensities, only slow/type I fibres are predominantly recruited. Fast contracting/type II fibres are more important for producing optimal power output, in activities such as sprinting, lifting a heavy weight and jumping, requiring high stimulation frequencies. At higher intensities, fast/type II fibres are recruited, adding to the type I fibre recruitment.

There are many important differences between the quality of the characteristics of muscle fibres, and the distribution of fibres between different muscles of the same person, and between different people. There are ranges of values for the fibre type distribution for human limb muscles (Table 4.6).

Unlike other mammalian species, where a particular muscle can be composed of mainly one fibre type (e.g. rat soleus), in humans virtually all muscles are a heterogeneous mixture of fibre types. There is, however, a large range of fibre distribution variability between human individuals for a given muscle. The variability in most muscles represents a normal distribution throughout the population. As a result of the re-introduction of the

△ **Fig. 4.22** Gluconeogenesis – the formation of glucose and glycogen from non-carbohydrate precursory such as lactate, glycerol and amino acids – involves inter-conversions between muscle and liver glycogen. Adapted from Maugham and Gleeson (2004).

Characteristic	Fibre type		
	I	IIa	IIx
Enzyme activity			
Myofibrillar ATPase	Low	High	High
Oxidative enzyme activity	High	Int–High	Low
Glycolytic enzyme activity	Low	High	High
Substrate concentration			
ATP	Int	High	High
PCr	Int	High	High
Glycogen	Low	Int	Int
Triglycerides	High	Int	Low

△ **Table 4.5** Histochemical and metabolic properties of the major fibre types.

Muscle	Human	Rat
Lateral gastrocnemius		
% type I	50–64	28
% type II	40–52	72
Soleus		
% type I	70–90	96
% type II	10–30	4
Tibialis anterior		
% type I	55–65	2
% type II	35–45	98
Vastus lateralis		
% type I	40–50	9
% type II	50–60	91

Modified from Rice *et al.* (1988)

△ **Table 4.6** Comparison of human and rat muscle-fibre type distributions.

Box 4.2

Fibre type proportions
The chart below shows the proportions of the different muscle fibre types (expressed as a percentage of total) for seven different human 'types', from average couch potato to extreme endurance athlete.

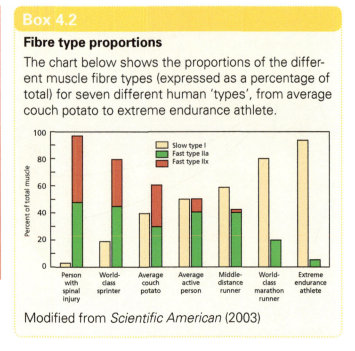

Modified from *Scientific American* (2003)

needle biopsy procedure by Jonas Bergstrom in the 1960s, exercise science and medicine has since used the technique in the study of muscle physiology and metabolism. The Bergstrom needle was in fact a modification of the needle first used by Duchenne in 1861. Small samples of muscle (for example from the *vastus lateralis*) can be obtained (see Figure 2.2) and the fibre type distribution determined from such samples after thin sections of the tissue are cut and histochemically treated.

In humans, samples can be obtained from the *vastus lateralis* that contain only 20 per cent type I, slow contracting fibres (and therefore 80 per cent type II, fast contracting fibres). Such a person might have the potential to become a good power athlete (many other factors are, however, required), whilst it would be unlikely that such a person could become a good marathon runner. The opposite would be true of someone with 80 per cent type I, slow contracting fibres. The 'average' healthy human adult has approximately 50 per cent of each type of slow and fast contracting fibres in such a muscle (Box 4.2).

In the 1970s particularly, some work was put into the muscle fibre type profiling of successful sports performers. This work consistently demonstrated a selective dominance, dependent upon the principal type of sporting activity, of either slow contracting or fast contracting fibre types. (For a good review of this topic, see Saltin and Gollnick, 1983.) The debate continues as to whether the preponderance of fibre types in such individuals is a result of genetic predisposition or of several years of training and participation in a type of activity. It is most probably a mixture of both, though studies of identical twins have demonstrated a strong genetic influence in fibre type distribution. There is no conclusive evidence that athletic success can be predicted on the basis of fibre type distribution. As well as the three distinct fibre types, there are hybrid fibre types that contain two different myosin isoforms, thereby creating a continuum of type, ranging from those containing only the slow isoform to those containing only the fast isoform.

Aspects of development, growth and ageing of muscle

In developing foetal muscle the number of myofibrils that form a muscle fibre can be relatively small (circa 50), however, in adult muscle, this increases to about 2000 myofibrils per muscle fibre.

During myogenesis three classes of myotube can be identified: primary, secondary and tertiary. Slow heavy-chain myosin synthesis is present in the earliest primary myotubes at around nine weeks' gestation but is not detected in secondary and tertiary myotubes until the third trimester (at around 29 weeks). The number of muscle fibres in each muscle is thought to be established by late gestation, so in order for muscle mass to increase there is hypertrophy (an increase in the size of muscle fibres) without hyperplasia (an increase in fibre number). At the other end of the lifespan, ageing results in a decline in skeletal muscle mass (sometimes referred to as sarcopenia), particularly if physical activity and exercise training is not continued into later years. The major reasons for the decline in muscle mass with ageing are the reduced muscle fibre size (atrophy), which is accompanied by a decrease in strength and a slowing of the contractile properties of muscle due to the changing dominance of myosin heavy-chain isoforms, i.e. a blurring of the distinction between slow and fast isoforms. This change in the slowing of the contractile properties of ageing muscle results in a loss of power.

Muscle fibres can be 'converted' from one type to another – this occurs as a result of a change in expression of the myosin isoform. When, for example, healthy muscle is resistance trained, some IIx fibres will convert to IIa fibres by expressing the IIa gene. It can be demonstrated (primarily in animal models) that re-innervation of a muscle fibre from a different motor unit type can transform the fibre, e.g. from a fast to slow, or, a slow to fast. Chronic electrical stimulation (again, primarily in animal models) can also result in structural (e.g. capillary density), metabolic (e.g. mitochondrial density) and contractile (e.g. myosin isoform expression) changes. Large shifts in characteristics (e.g. type IIx to type I) are, however, unlikely, and it appears easier experimentally to stimulate faster fibres to become slower, than to make slower fibres become faster.

Generating force – Length, tension and power characteristics

The length of a muscle is one of the determinants of the speed of shortening, and, as mentioned earlier, the heavy-chain component of the myosin molecule determines the functional characteristics (e.g. velocity of contraction) of the muscle fibre. Furthermore, the capacity of muscle to generate force is proportional to the cross-sectional area of the muscle. The concepts of length–tension and force–velocity are discussed further in the Biomechanics section of this book (Chapter 11 'Muscle contractions'), as are types of muscle contractions, namely isometric, isokinetic, isotonic, concentric and eccentric contractions. By way of introduction, and from a physiological perspective, we should first be clear about what we mean by the terms force, strength, work and power.

Most forces cannot be seen, but we often see and measure the effect of a force, or take for granted a force such as gravitational force. The gravitational force on a body is the product of the mass of the body and the acceleration due to gravity, so, a person with a 60-kg body mass will exert a **force** of 589 N (60 kg × 9.81 m/s = 589 N). **Strength** is a term that we use to describe the ability to generate a force. Lifting the force that the 60-kg person exerts on the ground requires work to be done, and **work** is the product of force and the distance it is moved through. The unit of measurement of work is the joule,

after James Prescott Joule (1818–1889), who showed the equivalence of heat, mechanical works and other forms of energy. **Power** is the term used to describe the rate of doing work and can be calculated either by dividing the work done (force multiplied by distance) by the time taken, or by multiplying force by velocity (because velocity = distance divided by time).

Length–force (tension) relationship in skeletal muscle

It can be demonstrated that there is an optimal muscle length at which tension (force) can be seen to be greater than when the muscle is shorter or longer than this optimum (see Figure 11.22). The two components of tension are the active tension that is produced by contracting myofibrils in the muscle fibre, and passive tension, produced by the connective tissue and structural protein elements. The force generated by muscle is low at short and stretched muscle lengths due to the extent of overlap and interaction of the actin and myosin filaments, as per the sliding filament mechanism of muscle contraction. Force (tension) decreases more rapidly when muscle shortens compared with when it lengthens because at very short lengths the thick filaments come into contact with the Z-lines.

Force–velocity relationship in skeletal muscle

The force generated by a muscle is dependent on filament overlap and muscle length, and it also varies with the velocity at which it is shortening, or lengthening (stretching). When muscle length is not changing, i.e. an isometric contraction, the force generated is high, however, as the velocity of shortening increases, the force rapidly declines until no force can be sustained – this is the point known as the maximum velocity of shortening (V_{max}; see Figure 11.21). Power, as mentioned above, is the product of force and velocity, and as force is proportional to the cross-sectional area of muscle, and velocity is proportional to length, power will be proportional to the product of area and length, that is, volume.

The volume of a muscle and the possibility of altering force and/or velocity has implications for exercise training. For example, a longer, thinner muscle will produce less force but shorten rapidly compared with a shorter, fatter muscle, which will generate more force but have a lower maximum velocity of shortening. The maximum power of a muscle is typically around one-third of V_{max}. If we are able to increase the maximum velocity of shortening and/or the maximum isometric force of a muscle, then we can increase the resultant power.

Summary

- Skeletal muscle structure, function and metabolism is a key area in our understanding of sport and exercise science, with this dispersed 'organ' accounting for around 50 per cent of the body mass of most individuals.

- Muscle and nerve meet at the neuromuscular junction, a chemical synapse that allows action potentials to cross the synaptic cleft, the endplate potential depolarises adjacent areas of muscle membrane, and the action potential initiates muscle excitation-contraction (EC) coupling.

- Sarcomeres are the basic functional units of the muscle fibre and contain structural and contractile proteins which maintain integrity of the functional unit and allow for the molecular mechanism that is the basis of the sliding filament model of muscle action.

- ATP production is maintained as a result of ADP rephosphorylation (using PCr), through glycolysis, and by oxidative phosphorylation, and the relative contribution of these routes is dependent on the physiological state of the body (e.g. rest, light exercise, high-intensity exercise).

- Key pathways of energy metabolism are glycolysis, the TCA cycle, and the electron transport chain/oxidative phosphorylation.

- The body has a number of mechanisms to control acid–base balance, and these are challenged severely as a result of high-intensity exercise, which results in a decline in muscle pH.

- There are two principal muscle cell or fibre types, type I and type II, with different contractile, histochemical and metabolic properties.

- Muscle fibre type proportions have been compared with exercise performance capability to indicate a dominance of type I muscle fibres in endurance-trained humans, and a dominance of type II fibres in sprint-trained individuals.

- Tension (Force) is generated by muscle and there are distinct relationships between the length of muscle and tension, and, between force and velocity.

Review Questions

1. What are the sequence of events in the contraction of skeletal muscle that are categorised as 'excitation-contraction' coupling?
2. What are the differences between intrafusal and extrafusal fibres in skeletal muscle?
3. Describe the structure and function of the contractile proteins in skeletal muscle.
4. Outline a scheme for the metabolism of carbohydrate, fat and protein in the cell.
5. How does skeletal muscle cope with the large amounts of lactic acid produced during high-intensity exercise?
6. What are the contractile properties that skeletal muscle fibre types might be classified under?
7. How can the product of force and velocity be altered in a muscle or muscle group?
8. How do fibre type proportions differ between individuals?

References

Gaster M, Handberg A, Schurmann A, Joost HG, Beck-Nielsen H, Schroder HD (2004) GLUT11, but not GLUT8 or GLUT12, is expressed in human skeletal muscle in a fibre type-specific pattern. *Pflugers Archive* 448: 105–113.

Gautel M (2011) The sarcomeric cytoskeleton: who picks up the strain? *Current Opinion in Cell Biology* 23: 39–46.

Green AL, Hultman E, MacDonald IA, Sewell DA, Greenhaff PL (1996) Carbohydrate ingestion augments skeletal muscle creatine accumulation during creatine supplementation in humans. *American Journal of Physiology* 271 (Endocrinol. Metab. 34).

Hainaut K, Duchateau J, Desmedt JE (1981) Differential effects on slow and fast motor units of different programs of brief daily muscle training in man. In: Desmedt JE, ed. Motor unit types, recruitment and plasticity in health and disease. *Progress in Clinical Neurophysiology* 9: 241–249.

Harris RC, Soderlund K, Hultman, E. (1992) Elevation of creatine in resting and exercised muscle of normal subjects by creatine supplementation. *Clinical Science* 83: 367–374.

Hesselink MK, Keizer HA, Borghouts LB *et al.* (2001) Protein expression of UCP3 differs between human type I, type IIa and type IIb fibres. *FASEB Journal* 15: 1071–1073.

Hoeks J, Hesselink MKC, van Bilsen M *et al.* (2003) Differential response of UCP3 to medium versus long chain triacylglycerols; manifestation of a functional adaptation. *FEBS Letters* 555: 631–637.

Hultman E, Sahlin K (1980) Acid–base balance during exercise. In: Hutton RS, Miller DI, eds. *Exercise and Sports Science Reviews*, Vol 8. New York: Franklin Institute Press, pp. 41–128.

Jones DA, Round J, de Haan A (2004) *Skeletal Muscle from Molecules to Movement*. Edinburgh: Churchill Livingstone.

Kontrogianni-Konstantopoulos A, Ackermann MA, Bowman AL, Yap Sv, Block RJ (2009) Muscle giants: molecular scaffolds in sarcomerogenesis. *Physiological Reviews* 80, 1217–1267.

Maughan R, Gleeson M (2004) *The Biochemical Basis of Sports Performance.* Oxford: Oxford University Press.

Maughan R, Gleeson M, Greenhaff PL (1997) *Biochemistry of Exercise and Training.* Oxford: Oxford University Press.

McElhinny AS, Kazmierski ST, Labeit S, Gregorio CC (2003) Nebulin: the nebulous, multifunctional giant of striated muscle. *Trends in Cardiovascular Medicine* 13: 195–201.

Poortmans J (1986) Use and usefulness of amino acids and related substances during physical exercise. In: Benzi G, Packer L, Siliprandi N eds *Biochemical aspects of physical exercise.* Elsevier Science Pub. pp 285–294.

Rice CL, Pettigrew FP, Noble EG, Taylor AW (1988) The fibre composition of skeletal muscle. In: Poortmans JR ed. *Principles of Exercise Biochemistry.* Medical Sport Science, Karger 27, 22–39.

Robinson TM, Sewell DA, Hultman E, Greenhaff PL (1999) Role of submaximal exercise in promoting creatine and glycogen accumulation in human skeletal muscle. *Journal of Applied Physiology* 87: 598–604.

Russell AP, Somm, E, Praz M *et al.* (2003) UCP3 protein regulation in human skeletal muscle fibre types I, IIa and IIx is dependent on exercise intensity. *Journal of Physiology* 550: 855–861.

Sahlin K, Harris RC, Nylind B, Hultman E (1976) Lactate content and pH in muscle samples obtained after dynamic exercise. *Pflugers Archiv* 367: 143–149.

Saltin B, Gollnick PD (1983) Skeletal muscle adaptability: significance for metabolism and performance. In: Peachey *et al.*, eds *Skeletal Muscle; Handbook of Physiology,* Section 10, pp. 555–663.

Seeley WW *et al.* (2008), Unravelling Bolero: progressive aphasia, transmodel creativity and the right posterior neocortex. *Brain* 131: 39–49.

Sewell DA, Gleeson M, Blannin AK (1994) Hyperammonaemia in relation to high-intensity exercise duration in man. *European Journal of Applied Physiology* 69: 350–354.

Terry DF, McCormick M, Anderson S *et al.* (2004) Cardiovascular disease delay in centenarian offspring: a role of heat shock proteins. *Annals of the New York Academy of Science* 1019: 502–505.

Vincent B, Windelinckx A, Nielens H, Ramaekers M, Leemputte MV, Hespel P, Thomis MA (2010) Protective role of α-actinin-3 in the response to an acute eccentric exercise bout. *Journal of Applied Physiology* 109, 564–573.

Walker JB (1979) Creatine: biosynthesis, regulation, and function. *Advances in Enzymology* 50: 177–242.

Young B, Heath JW (2000) *Wheater's Functional Histology – A Text and Colour Atlas*, 2nd edn. Sydney: Churchill Livingstone, p.83.

Zanoteli E, Lotuffo RM, Oliveira AS, Beggs AH, Canovas M, Zatz M, Vainzof M (2003) Deficiency of muscle alpha-actinin-3 is compatible with high muscle performance. *Journal of Molecular Neuroscience* 20: 39–42.

Further reading

Greenhaff PL, Harris RC, Snow DH, Sewell DA, Dunnett M (1991) The influence of metabolic alkalosis upon exercise metabolism in the thoroughbred horse. *European Journal of Applied Physiology* 63: 129–134.

Hunt CC (1990) Mammalian muscle spindle: peripheral mechanisms. *Physiological Reviews* 70: 643–663.

Sewell DA, Harris RC (1992) Adenine nucleotide degradation in the thoroughbred horse with increasing exercise duration. *European Journal of Applied Physiology* 65: 271–277.

Smith CA, Wood EJ (1991) *Energy in Biological Systems.* London: Chapman and Hall.

Tullson PC, Terjung RL (1990) Adenine nucleotide degradation in striated muscle. *International Journal of Sports Medicine* 11: S47–S55.

5 Energy Balance and Body Composition

Chapter Objectives

In this chapter you will learn about:

- Concepts relating to energy balance, namely energy intake and energy expenditure, and body composition.
- The methods and practice of measuring body composition.
- Energy and nutrient intake principles and methods of estimating energy intake.
- The components of daily energy expenditure, the determinants of energy expenditure and the principles of energy expenditure measurement through direct and indirect calorimetry.
- Concepts behind the use of fuels for exercise.

Introduction

In this chapter we turn our attention to nutritional energy balance. On the face of it, energy balance is a very simple concept:

Energy balance = Energy intake – Energy expenditure

Energy **im**balance manifests itself in a change in the body mass, and a change in body composition. We therefore have three major themes that this chapter will deal with: the theory and practice of body composition, energy intake and energy expenditure, and their respective measurements.

These themes are of importance to sport and exercise science because whatever aspect of sport, exercise, physical activity, health or disease that we wish to be involved in and think about (Figure 5.1), knowledge of and the study of energy balance enables us to ask a number of questions and seek answers. We need to establish concepts concerning energy requirements in health and disease (e.g. for optimal growth; during injury rehabilitation); we need to understand the mechanisms of depletion and repletion (e.g. how we refuel after exercise or illness); we need information about tissues and energy substrates of the body that might be lost or gained (e.g. as a result of injury; during recovery); and we need to know how we adapt to changes in energy intake and/or energy expenditure (e.g. in efforts of physical endurance; in an attempt to gain lean body mass).

A broad definition of energy requirements is defined by the leading world agencies (FAO/WHO/UNU, 1985) as:

The energy intake that will balance energy expenditure when the individual has a body size and composition and level of physical activity consistent with long-term good health; and that will allow for maintenance of economically necessary and socially desirable physical activity. In children and pregnant or lactating women, the energy requirement includes the energy needs associated with the deposition of tissues or the secretion of milk at rates consistent with good health.

This definition helps to set the content of this chapter in the context of sport, exercise and health, encompassing our three major themes and including the notion that economically necessary (and socially desirable) physical activity includes elite and professional sport.

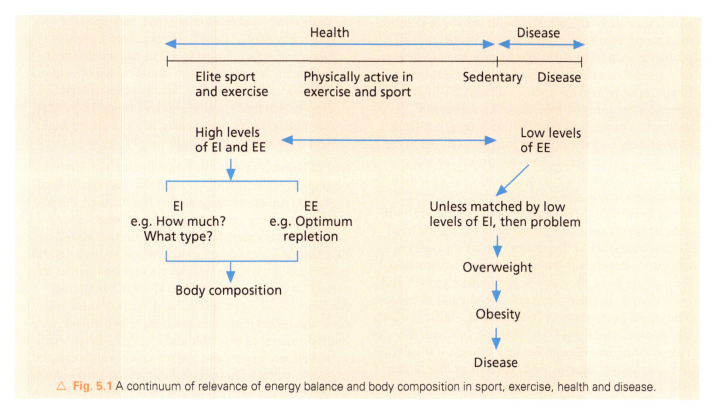

△ **Fig. 5.1** A continuum of relevance of energy balance and body composition in sport, exercise, health and disease.

Food and the diet

By way of introduction to energy balance, it is important to briefly consider the recent historical perspective of nutritional science, with particular reference to social, health and political change, and also to see how exercise and sport and nutrition has been perceived in the past.

Some of the landmarks in the fields of health, food and nutrition are described by James (2000). In the UK, since 1900 we have seen rapid growth of urbanization, and the standards of housing and sanitation, the wealth and the nutrition of the population have all improved overall. Despite this, at the beginning of the twenty-first century in the UK there are still great inequalities in wealth, health and diet, and these factors are inextricably linked, as has been recognized since the 1930s (for example Orr, 1936). The public health crisis that prompted the British government to take action in the early 1900s was the realization that Britain had nearly been defeated in the Boer War. It was thought that the cause of ill-health and poor physical and mental health of young men was malnutrition. As a result, a Schools Meals Service was introduced in 1906, to provide food for those children with the most need and to encourage them to take advantage of the education provided. By 1920 milk consumption was being promoted and the Milk Marketing Boards were set up in 1933 to improve the transport and efficiency of milk. Milk was

seen as both growth promoting and energy providing, even being promoted at the time as being a suitable alternative to a whole meal. Before the Second World War, fruit and vegetables were also recognized as being 'protective foods', and the 'food factors' contained in protective foods (vitamins) started to be identified.

The Second World War was to see a coherent strategy of UK-based food production, distribution, storage and sales (including rationing based on scientific principles) as a consequence of the German forces cutting off food supplies from overseas (which had provided 70 per cent of the UK needs). After the war, food rationing gave way to an expanding UK food industry to increase the availability of food at the lowest possible cost. Rationing, however, was not completely ended until 1954. Free school milk became universally available from 1944.

Since the Second World War food production has been a priority throughout Europe and North America. Cereal production over the last 50 years has increased, aided by, for example, improved seeds, fertilizers, pesticides, machinery and management. Compared with the rationing of the War years, dairy and meat products have now become plentiful as a result of intensive livestock management, and fish have become plentiful as a result of intensive trawling (at the expense of a reduction in sea fish stocks). These developments are not without their problems, which include overproduction and subsequent 'food mountains' (both cereal and animal based), health problems including outbreaks of salmonella, the risk of variant Creutzfeldt–Jakob disease (vCJD) from bovine spongiform encephalopathy (BSE)-infected meat, *Escherichia coli* infection, the overuse of antibiotics in animals, and foot-and-mouth disease in livestock. Understandably, this has resulted in calls from both governments and consumers for more sustainable agricultural practices (driven by policies) and more accountable food standards. Consumers in the twenty-first century are also concerned about genetic modification of food, emphasizing the need for the education of the general public on issues of nutrition, dietetics and food. In recent years, world food prices have risen due to population growth and poor climate in the major cereal producing nations. The world population in 2012 is circa 7 billion, which has doubled since 1961. In 2008, the U.K. was more at risk of rising food prices because it had a trade deficit in food, unlike some its peers such as France which had a surplus and the USA which was balanced.

A more detailed discussion of aspects of food safety, labelling and standards, the food industry, and the wider aspect of nutrition in developing worlds is beyond the scope of this text but might be dealt with as advanced aspects of your studies, particularly if you have the opportunity to take options in the areas of nutrition, dietetics and food science.

Examples of nutrition for sport and exercise will be discussed at various points in this chapter, however, it is interesting to take this opportunity to briefly discuss how nutrition for work, exercise and sport has been perceived since the 1900s. Anecdotal recollections of the diet of athletes in the late 1800s and early 1900s emphasizes the belief that a high protein/fat diet including a lot of meat was an important part of an athlete's diet. In his autobiography *Running Recollections and How to Train*, first published in 1902, champion sprinter Alfred Downer wrote:

Meals never trouble me at any time. So long as the cooking is good … I don't care what I eat. Chops, or bacon and eggs, followed by a little marmalade, used to form my breakfast. Dinner would consist of a cut off a joint, or some chicken, followed by a milk pudding, stewed fruit etc; while for tea I would have either fish, poached eggs, sweetbreads, or – in fact, anything light (Downer, 1902).

One of Downer's rivals was Edgar Bredin, the 'Half-mile Champion of the World' who said that 'As regards food, most men can eat as much as they feel inclined, but one had better make a rule to take no liquid whatsoever between meals unless feeling unduly thirsty'. He felt that 'the meat should be varied as much as possible, and poultry is a great aid in this direction; toast is better than bread'. For longer distances, Len Hurst 'Champion of the World at 20 miles' feasted on dinners of roast beef, roast and boiled mutton or chicken, helped out with a limited amount of vegetable and bread, the latter ought to be stale and crusty, and washed down with half a pint of good bitter ale. Downer's autobiography even contains a hint of the use of drugs in sport – according to Bredin 'Of opening medicines which are necessary to many men during hard work, three pills every Saturday night, and a seidlitz powder every Sunday morning, is the best possible aperient'!

Drinking and eating during exercise was shrouded in myth through the nineteenth and twentieth centuries. Noakes (1998) cites a number of examples, including how the Oxford University rowing crew of 1860 was restricted to a maximum of 1.2 litres of fluid each day during training, and the advice given to marathon runners of the early 1900s included the caveats 'don't take any nourishment before going seventeen or eighteen miles. If you do you will never go the distance. Don't get into the habit of drinking and eating in a Marathon race; some prominent runners do, but it is not beneficial'. Jackie Meckler, a world record holder at 30, 40 and 50 miles in 1954 is quoted as saying:

TIME OUT

ALFRED DOWNER 'CHAMPION SPRINTER OF THE WORLD'

During his time, Alfred Downer was called 'Champion sprinter of the world' and though there were no official world championships, he won the 100 yards, 220 yards and 440 yards at the SAAA championships for three consecutive years from 1893 to 1895. The facsimile edition of Downer's original autobiography is a fine leisure read! It 'contains a fascinating and uniquely rare insight, not only into Downer's own career but also into the whole structure of athletics in the latter part of the 19th century' (Downer, 1902/Balgownie Books, 1982).

In those days it was quite fashionable not to drink, until one absolutely had to. After a race, runners would recount with pride 'I only had a drink after 30 or 40 kilometres'. To run a complete marathon without any fluid replacement was regarded as the ultimate aim of most runners, and a test of their fitness (Noakes, 1998).

Apparently Meckler once competed in a 160-km race in which he first drank after only 120 km, and confirmed that this approach was still popular when he ran the 1969, 90 km Comrades Marathon, his last competitive ultramarathon. Bearing in mind the thoughts and practices that have prevailed in the minds of athletes over the last 100 years or so, we will discuss the modern, evidence-based approach to nutrition in sport and exercise at the end of this chapter.

Activity 5.1

Independent learning task: Nutritional changes in the twentieth century

Briefly describe the social, health and political changes that influenced nutritional change in the twentieth century, and the changes in the perception of nutrition for work, exercise and sport.

Why do we eat?

We eat about a tonne of food every year (Bender, 2002). Why? We need to generate energy to enable the body to carry out muscular work (about a quarter of energy expenditure, excluding that for sport and exercise) and to continue the normal metabolic processes, including those that maintain homeostasis (about two-thirds of energy expenditure). Not only are there physiological and psychological demands associated with meeting the need for metabolic fuels and nutrients, there are appetite needs that are also related to the pleasure of eating. The reason we eat, why we eat what we do, and the pleasure of eating also involves taste and flavour satisfaction, the availability and cost of food, religious, ethical and traditions associated with food, the luxury status of foods, and the social functions of food.

Some of these factors are associated with overnutrition, for example, cheap food and the social function of eating out/with a group of people. Conversely, some factors are associated with malnutrition and undernutrition, for example the unavailability of certain foods, apathy and/or the lack of incentive to prepare meals (this can be the case for people living alone). Accurate measurement of the nutritional status of an individual is very difficult, particularly if only a single type of assessment is attempted. For this reason, several parameters in tandem can be used to assess nutritional status, including anthropometry, body composition, energy intake and energy expenditure measurements. The long-term control of energy (food) intake and energy expenditure can result in changes in anthropometric and body composition parameters, and how these are measured, and their limitations, will be discussed next.

Body composition measurement

Medium to long-term imbalances in energy intake and energy expenditure result in changes in the composition of the body. **Anthropometry** is the term used for the measurement of physical characteristics and basic body composition estimations. Body height and mass (weight) are measured frequently during growth and maturation, and whilst height changes little during adult life, body mass often does – usually upwards in affluent countries. Longevity is significantly related to body mass, and a simple ratio of body mass to height can be a good indicator of relative risk of diseases related to overweight and obesity, such as cardiovascular disease and diabetes. Another reliable predictor of relative risk of these diseases, based on simple measurement, is the waist-to-hip ratio.

Many anthropometry measurements can be carried out using simple and inexpensive pieces of equipment such as a flexible measure and callipers, whilst more reliable estimates of body composition can involve expensive or cumbersome equipment only available in exercise and health laboratories or in hospitals. The methods used, essentially from inexpensive/low technology to expensive/high technology, and their implications, will be described after considering the chemical and biological levels at which body composition can be viewed.

The six most abundant elements of the body contribute about 99 per cent of our body mass (the 'chemical' level of composition). Oxygen, carbon and hydrogen account for around 94 per cent of that mass, with nitrogen, calcium and phosphorus the next most abundant elements. Knowing the total body content of an element can help to estimate macromolecular content of the body. For example, if we know the body nitrogen content, and assume a constant conversion factor of 6.25 from nitrogen to protein, then this is representative of body protein. Body cell mass can also be estimated using total body potassium (indirectly from ^{40}K measurement).

	Approximate percentage of total body mass
Water	
Intracellular	26
Extracellular	34
Lipid	
Essential	2
Non-essential	17
Protein	14
Mineral	5
Carbohydrate	<1

Based on Deurenberg and Roubenoff (2002).

△ **Table 5.1** Molecular and macromolecular ('biochemical') composition of the body.

At the level of molecules and macromolecules, we might consider the composition of the body being made up primarily of water (60 per cent; Table 5.1), and the remaining consisting of lipid, protein, mineral and carbohydrate.

We might consider the 'physiological' or cellular level of body composition as the mass of the body cells, including their water, protein and mineral content, plus extracellular fluids (i.e. plasma in the intravascular space and interstitial fluid in the extravascular space), plus a third component, that of the extracellular solids – proteins and minerals.

At an 'anatomical' or tissue/organ level some of the most sophisticated clinical methods of body composition measurement can be combined to estimate around 83 per cent of the total body mass of an individual (see Table 5.3). At the 'whole body' level of composition estimation, the most simple measurement techniques can be used as predictive tools, and this is where we will start our consideration of the basis of a whole range of techniques.

Simple physical characteristic measures

Height (*H*), the distance from the floor to the vertex of the head, measured to the nearest 0.01 m is perhaps the simplest measure, but care must be taken to use a properly mounted or free-standing, good-quality stadiometer. Specific details of anthropometry techniques and standards to use when measuring height (e.g. the need to stand without shoes and the posture to be adopted) should be available in your laboratory and can be obtained and learnt through courses organized by the International Society for the Advancement of Kinanthropometry (ISAK). Such courses might be organized for you by your tutors

as part of your course, or you may wish to pursue such a course if you see your career path involving the accurate and reliable collection of descriptive subject/client data, especially girth and skinfold measures.

Seated height (*SH*) can also be used in some subject/client groups where leg length and the seated height to stature ratio might be of interest.

Body mass (*M*; or weight), measured to the nearest 0.1 kg, again can be a very simple matter, but one in which significant inaccuracies can be introduced if care is not taken. Scales of poor quality can vary in their readings from one weighing to the next, and there can certainly be little guarantee of reliable repeated measures of body mass when standing on different scales (for example the scales in the home versus the scales in the sports centre).

If body mass and height are obtained, an index first introduced by Quetelet and known today as the body mass index (BMI) can be calculated. This is simply body mass (in kilograms) divided by the square of the height in metres. The BMI can be used as a guide to whether someone is below or above a desirable body mass for health (Box 5.1). Try it on yourself. If, for example, you are 1.74 m tall and have a mass of 65 kg your BMI is 21.5.

> **Box 5.1**
>
> **Basic adult body mass index categories and descriptions**
>
> | | |
> |---|---|
> | **BMI below 19** | Underweight |
> | **BMI 19–24.9** | Normal/Ideal weight range |
> | **BMI 25–29.9** | Overweight/Pre-obese |
> | **BMI 30–34.9** | Mildly obese/Obese class 1 |
> | **BMI 35–39.9** | Moderately obese/Obese class 2 |
> | **BMI above 40** | Severely obese/Obese class 3 |

Caution has to be exercised if calculating the BMI of a heavily muscled individual as this can result in BMI seemingly in the overweight category. Descriptors (based on World Health Organization guidelines) from underweight to severely obese can be assigned, as indicated in Box 5.1. Children and pregnant women are two further populations for whom BMI is not a valid or appropriate index.

If properly measured, subtracting seated height from height can give an estimate of leg length, and a seated height/stature ratio can also be ascertained (SH divided by H, multiplied by 100). Body surface area can also be estimated using the following equation (Haycock *et al.*, 1978):

Body surface area (m^2) = (M$^{0.5378}$ × H$^{0.3964}$) × 0.024265

Girth (circumference) measurements

Girths should be measured using a flexible steel measure, taking care not to alter the contour of the skin by compressing soft tissue. The pattern of body fat distribution is an important predictor of the health risks of overweight and obesity. **Waist-to-hip ratio** is a simple but significant predictor of the risk of hypertension, diabetes and coronary heart disease – it can define two patterns of adipose tissue distribution:

- upper body segment/trunk, the classical male pattern of overweight/obesity ('apple-shape'); or
- lower body segment, the classical female pattern ('pear-shape').

The measurements of waist (smallest circumference below the rib cage) and hip girth (largest circumference at the posterior extension of the buttocks) are therefore key measures in a health-related setting. Ratios above 0.86 for women and 0.95 for men suggest a significantly increased health risk, the greater the ratio, the greater the disease risk.

Limb girths are perhaps of greater interest in a sport and exercise setting. The commonly measured girths are the forearm, the upper arm (relaxed and flexed), the thigh (can you remember the quadriceps and hamstring muscles of the thigh learnt in Chapter 2?) and the calf. These measures can be useful in training and rehabilitation settings (e.g. muscle development) to monitor improvements, or in a detraining setting (e.g. restoration of muscle mass after long-term injury). It is possible to use height (H) and selected limb girths to estimate total limb skeletal muscle mass (TLSMM).

Using 12, older adult, male cadavers, Martin *et al.* (1990) generated an equation based on the assumption that limb muscle girth (MG) is equal to the total limb girths less the skinfold thickness (bone was accounted for in the model):

$$\text{Muscle girth (MG)} = \text{girth (G)} - 2\pi \times \text{skinfold}$$

But the skinfold measure is twice the thickness of adipose tissue, so:

$$\text{MG} = \text{G} - (\pi \times \text{skinfold})$$

Muscle limb girths should be corrected (C) for the thigh (CTG), and the calf (CCG). One other girth, the forearm (FG) is also required.

Total limb skeletal muscle mass (TLSMM):

$$\text{TLSMM} = H \times ([0.0553 \times \text{CTG}^2] + [0.0987 \times \text{FG}^2] + [0.0331 \times \text{CCG}^2]) - 2445$$

Using sliding callipers, measurements can also be made across specific bone marks. You will hopefully remember from Chapter 2 the skeletal landmarks used in the measurement of the following commonly used sites; the **biacromial distance** (between the tips of the acromial processes); the **bicristal distance** (between the most lateral points of the iliac crest); the **elbow distance** (between the medial and lateral epicondyles of the humerus); and the **knee distance** (between the medial and lateral epicondyles of the femur). These measures are perhaps most useful in monitoring change in a developmental setting (e.g. children and youths in sport and exercise training).

Skinfold thickness measurements

Skinfold thickness measures enable an estimate of body fat to be calculated based primarily on work carried out by Durnin and Rahaman (1967), and Durnin and Womersley (1974). The latter involved making measurements on nearly 500 adults with a wide age range. The Durnin and Rahaman (1967) paper was reprinted in the *British Journal of Nutrition* as a 'citation classic', thought to be the most highly cited article that the *BJN* has published.

On the face of it, measuring skinfolds appears to be a simple procedure, but in order to obtain accurate and reliable skinfold measures, a great deal of care and systematic standardization needs to be practised. The appropriate methodology will no doubt be demonstrated to you in your course, and you will be given the opportunity to practise skinfold measurements. There is at least one international society with a keen interest in kinanthropometry – the measurement of human body composition. The International Society for the Advancement of Kinanthropometry (ISAK) has developed standards for anthropometric measurement and a training and accreditation scheme (www.isakonline.com).

Four sites are commonly used to obtain a sum of skinfolds: biceps, triceps, subscapular and suprailiac. An estimate of percentage body fat, based on the sum of these four skinfolds can be obtained from the table of Durnin and Womersley (1974). Further sites that might be measured are an abdominal site, the medial calf and the anterior thigh. These seven sites are illustrated in Figure 5.2.

The same skinfold thickness in women relates to a higher body fat content than in men. Also, at equal skinfold thickness, older people have a higher body fat content due to internal fat. A table can be used to relate the sum of four skinfold thicknesses to the percentage body fat content of men and women of various ages (Table 5.2).

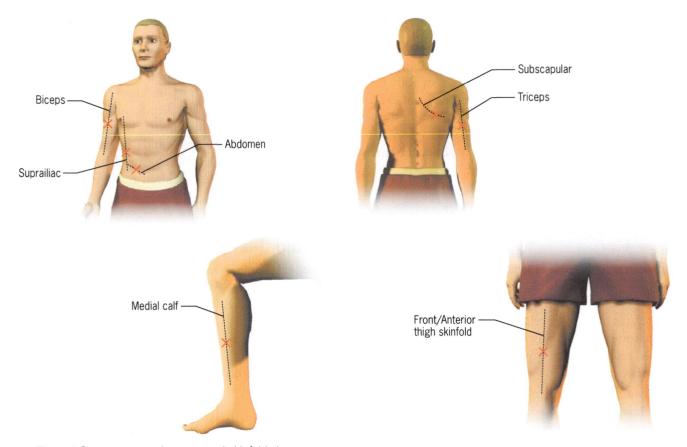

△ **Fig. 5.2** Seven commonly measured skinfold sites.

Bioelectrical impedance

This is a technique that measures electrical conductivity using a pair of electrodes attached to the left hand and the left foot. Current is impeded less through fat-free tissue and extracellular water (due to the greater electrolyte content) than through fat tissue. The impedance of the electric current measured is therefore postulated to be relative to the total fat content of the body. Interestingly, however, the equations on which the model is based is also dependent upon the physical characteristics of body weight, height and age. For example, based on Heitman (1990):

for women $F = (0.819W) - (2790H^2/R) - (23.1H) + (0.077A) + 14.941$

for men $F = (0.755W) - (2790H^2/R) - (23.1H) + (0.077A) + 14.941$

where F is fat (kg), W is weight (kg), H is height (m), R is impedance (ohms) and A is age (years).

For example, for a male, $A = 20$ years, $W = 67$ kg, $H = 1.73$ m, $R = 450$ ohms, then $F = 8.5$ (equivalent to 12.7 per cent body fat). Some typical percentage body fat estimates for various adult population groups are shown in **Box 5.2**.

Box 5.2

Typical percentage body fat estimates of different adult population groups

Sedentary male	20 per cent
Sedentary female	30 per cent
Elite male endurance athlete	10 per cent
Elite female endurance athlete	15 per cent
Professional rugby football forward	18 per cent
Professional soccer player	14 per cent

An alternative formula for the calculation of fat-free mass using bioelectrical impedance can be seen in the work of Deurenberg *et al.* (1990), particularly with reference to working with older people.

All of the methods described so far have in common the fact that they are doubly indirect techniques of assessing body composition, that is, the techniques are not measuring the body tissue that we wish to consider, but instead make a variety of assumptions and utilize statistical relationships between anthropometric characteristics that are easily measured.

Sum of four skin-folds (mm)	Percentage body fat (Males)				Percentage body fat (Females)			
	17- 29 yrs	30-39 yrs	40-49 yrs	>50 yrs	16-29 yrs	30-39 yrs	40-49 yrs	>50 yrs
15	5				11			
20	8	12	12	13	14	17	20	21
25	11	14	15	16	17	19	22	24
30	13	16	18	19	20	22	25	27
35	15	18	20	21	22	24	26	29
40	16	19	21	23	23	26	28	30
45	18	20	23	25	25	27	30	32
50	19	22	25	27	27	28	31	33
55	20	23	26	28	28	29	32	35
60	21	24	27	29	29	31	33	36
65	22	24	28	30	30	32	34	37
70	23	25	29	32	31	33	35	38
75	24	26	30	33	32	33	36	39
80	25	27	31	34	33	34	37	40
85	26	27	32	35	34	35	38	40
90	26	28	33	36	35	36	38	41
95	27	28	34	37	36	37	39	42
100	28	29	34	37	36	37	40	43
105	28	30	35	38	37	38	40	43
110	29	30	36	39	38	39	41	44
115	29	31	36	40	38	39	42	45
120	30	31	37	40	39	40	42	45
125	31	32	38	41	40	40	43	46
130	31	32	38	42	40	41	43	46
135	32	32	39	42	41	41	44	47
140	32	33	39	43	41	42	44	47
145	33	33	40	44	42	42	45	48

Based on Durnin and Womersley (1974).

△ **Table 5.2** Body fat content as a percentage of body weight, for the sum of four skinfold measures (biceps, triceps, subscapular, suprailiac) for male and female adults of different age groups.

Techniques in body composition estimation are improved by making fewer assumptions and relying less on statistical relationships, and more on sound physical science concepts – what might be termed 'indirect techniques'. We shall start by considering densitometry.

Estimating body fat and lean body mass using densitometric principles

A relatively cheap but not particularly practical or accessible method of more accurately estimating body composition that has been used for some time is hydrostatic or underwater weighing. The method is dependent, for example, on the Siri equation which assumes that the density of body fat is 0.9 kg/l, whilst that of lean tissue is 1.1 kg/l. Using the Siri equation:

Body fat (per cent) = (495/body density) – 450

There are other predictive equations because one set of assumptions does not fit all genders and ethnicities, so for example in the literature, separate equations have been developed and recommended for caucasian men and women, and for black men or women.

Body density is a function of body mass and body volume (namely mass divided by volume).

$$\text{Body density} = \frac{\text{body mass (kg)}}{\text{body volume (litres)}}$$

It is easy to ascertain body mass, but body volume is rather more difficult. It can be measured using underwater weighing. The measurement is based on Archimedes' principle that an object immersed in a liquid receives an upthrust equal to the weight of the liquid displaced by the object. The principle explains the nature of buoyancy, and the buoyant force of water is dependent on its density, that is, its weight per unit volume. Measures of body mass in air, and body mass submerged in water can be used to calculate body density. Hydrostatic weighing can be used to calculate body density. Hydrostatic weighing can be carried out in either a sitting or supine position and a detailed methodological description of hydrostatic weighing is described by Pollock *et al.* (1995). In order to account for the air in the lungs, the measurement requires the subject to completely breathe out just before immersion. The remaining air – the residual lung volume – can be estimated and therefore a correction can be made for the buoyancy of this air in the lungs (Figure 5.3).

△ **Fig. 5.3** Underwater weighing.

As you can imagine, there are a number of disadvantages of the use of this technique.

A newer, but relatively expensive method of measurement of body composition based on air displacement has become available (Dempster and Aitkins, 1995). One commercially available system using air displacement plethysmography (measurement of the size, usually volume, of a body) is called the BOD POD (Life Measurement, Inc, Concord, CA, USA). The BOD POD essentially uses the same physical principles and assumptions as underwater weighing but uses air instead of water. The volume of the body is measured indirectly by measuring the volume of air it displaces inside an enclosed chamber. The air inside the chamber is measured by applying relevant physical gas laws (those of Boyle and Poisson). (**Box 5.3** *Air displacement plethysmography apparatus*)

A good review of body composition assessment using air displacement plethysmography by Fields *et al.* (2002) might be consulted for more background and assessment of BOD POD measurements compared with other methods. Some of the advantages of the BOD POD system in comparison with underwater weighing are that measurements are less time consuming, it is more subject friendly, and it can accommodate a wider range of subjects.

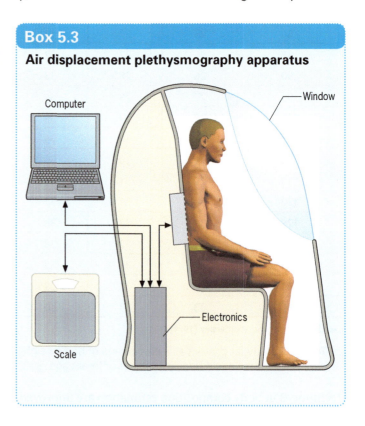

Box 5.3

Air displacement plethysmography apparatus

Computer

Window

Electronics

Scale

Other indirect methods of body composition assessment include dual-energy X-ray absorptiometry (DEXA) (sometimes abbreviated as DXA), a medical imaging technique, and, total body water using stable isotope dilution, using the same principles as those described in Chapter 1 (**Time Out** *Stable isotopes and their use in health and exercise research*). A DEXA image of the whole body can also be seen in **Time Out** *Bone mineral density measurement and bone health* in Chapter 2.

For the purposes of the evaluation of body composition, the body has most commonly been viewed as being partitioned into two basic compartments: fat mass (FM) and fat-free mass (FFM). This is the model relevant to body composition analysis using skinfold measures or densitometry. It assumes that the densities of these compartments are constant within the tissues, and among individuals and populations. Whilst fat is a relatively homogeneous compartment, fat-free mass is not, as it includes water, bone, protein and carbohydrate. More recently, methods have been developed that try to measure each of these compartments, and then 'piece' the body back together again without assuming that one constituent is related to another.

Multicompartment models, for example three- and four-compartment (3-C and 4-C) models, have become the 'gold standard' against which other body composition methods might be validated. In the 3-C model, the compartments are FM, total body water (TBW: measured using a stable isotope dilution technique) and the 'dry' component of protein (P) and mineral together. In a 4-C model, TBW is estimated, plus mineral (i.e. bone mineral density; measured by DEXA), plus FM and P compartments.

Another multicompartmental model has been suggested by Deurenberg and Roubenoff (2002) – see Table 5.3. At the tissue level of body composition, the model might be:

$$\text{Body mass} = \text{adipose tissue} + \text{skeletal muscle} + \text{bone} + \text{organs} + \text{the rest}$$

Using sophisticated clinical techniques, adipose tissue mass can be estimated by computerized tomography or magnetic resonance imaging (MRI), skeletal muscle mass can be estimated using 24-h urinary creatinine or N-methyl-histidine excretion, bone mineral density can be estimated using DEXA, blood volume using a dilution technique and specific organs might be imaged and mass estimated using ultrasound or MRI. Estimation of skeletal muscle mass using creatinine excretion is based on the principle that urinary creatinine (measured in a 24-h urine sample collection) relates to muscle mass, and that 1 g excreted creatinine is equivalent to about 20 kg muscle in the body.

There are considered to be only two direct measures of body composition – carcass analysis and *in vivo* neutron activation analysis (IVNAA). **Carcass analysis** can equip us with knowledge post-mortem but is obviously of no benefit in the intact, free-living person. In contrast, information obtained at the chemical composition level using **IVNAA** is a relatively new technique for body composition. A number of elements in the body can be quantified, such as nitrogen, calcium, phosphorus and potassium and if, for example, it is assumed that body protein consists of 16 per cent nitrogen, then total body protein can be calculated as 6.25 multiplied by the total body nitrogen content. The cost of **IVNAA** is prohibitive, and the radiation dose varies from low to high depending on the element being measured.

Most methods of body composition analysis have been developed and validated on 'normal adult' populations. The validity of some measures is reduced depending on the nature of the population, such as pregnant women (who have more body water), and elderly people (who have less bone mineral), and for some populations (e.g. children, pregnant women), the use of some techniques or validating them is less easily justified because of risks (often very small) inherent in the measurement procedure itself (e.g. radiation doses).

Energy balance over time

The joule (J) is the unit of energy of the International System of Units (Système International: SI), the name being after James Prescott Joule (1818–1889) who demonstrated the equivalence of heat, mechanical work and other forms of energy. Many people still use the old unit of energy, the calorie (1 calorie is equal to 4.184 J).

As stated in the introduction to this chapter, energy balance is a simple concept, described by the equation:

$$\text{Energy balance} = \text{Energy intake} - \text{Energy expenditure}$$

and energy imbalance results in a change in the body mass, but it is important to understand that body mass gains are made generally (e.g. in the developed world) over a long period of time (e.g. months and years), and energy imbalance on a day-to-day basis is normally only very slight. Using a modest example of body mass gain, if an individual between the age of 30 and 60 years of age gains 10 kg, and this can all be attributed to an increase in adipose tissue, then the energy density of this tissue is around 300 MJ (30 MJ/kg – adipose tissue is not all lipid). This represents a positive energy imbalance of 300 MJ per 30 years, or approximately 27 kJ/day (the energy content of

	Approximate percentage of total body mass
Skeletal muscle	40
Adipose tissue	21
Blood	8
Bone	7
Skin	4
Liver	2.5
Other tissues/organs	17.5

△ **Table 5.3** Typical adult body composition using a multicompartment/tissue model.

one bite/one-third of a chocolate biscuit), or approximately 9 kJ per meal. It is therefore important to realize, in the context of overweight or obese individuals, that the imbalance may have only been slight, but for a long period of time (months, years), and that body mass is stable in the short term (daily, weekly) and such individuals are very close to energy balance in the short term.

The way that nature intended fat stores to be used, that is, fluctuating depending upon the seasonal availability of food, is still evident in some areas of the developing world (Figure 5.4).

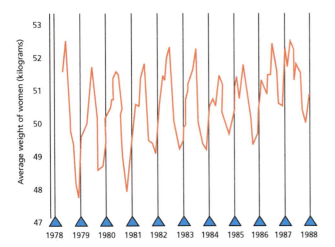

△ **Fig. 5.4** Stores of body fat in women from Keneba village in The Gambia nearly double after the harvest is gathered each year to provide them with enough energy to see them through the lean times. From Prentice and Jebb (1995).

Over a much shorter period of time (weeks), in sport, we can see how energy is balanced when levels of intake and expenditure, and of performance, are close to the limits of human tolerance. Energy intake and energy expenditure are nearly matched in elite, competitive cyclists taking part in the most demanding cycling competition in the world (Figure 5.5). The components of energy balance, intake and expenditure, and the concepts and practice behind their estimation will now be discussed.

Energy and nutrient intake and its estimation

Energy or nutrient intake might be considered from one of at least three perspectives:

- At a population/epidemiological level (e.g. people living in one country eat more of a particular food product than those in another country)
- In relation to a disease or clinical condition (e.g. people with coeliac disease must avoid food products containing gluten)

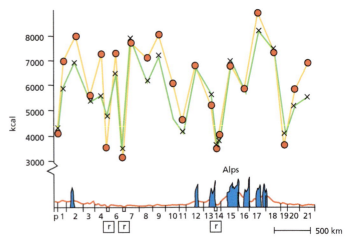

△ **Fig. 5.5** Energy intake (crosses) and expenditure (red circles) during the Tour de France. From Saris *et al.* (1989).

- At the individual level (e.g. eating 20 MJ per day in order to meet normal energy needs plus that to match the expenditure resulting from 2 hours of exercise training).

The majority of discussion in this part of the chapter will relate to the individual and the way in which their intake of energy and nutrients might be considered, and assessed.

Broad, healthy eating guidelines for an individual are the territory of public advertising campaigns on hoardings and in the media. A summary of the current major goals of healthy eating are listed in **Box 5.4**. A balanced diet, of the ideal macronutrient proportions, where the macronutrients obtained are from a variety of sources (particularly different protein sources), and where simple sugars and saturated fats are kept to a low proportion (compared with complex carbohydrate (starch) and unsaturated fats respectively), should obviate the need for any food supplements in healthy, active individuals. Food pyramids in public health nutrition education are being replaced by plates, such as 'The eatwell plate' used by the Foods Standards Agency in the UK and 'MyPlate' (ChooseMyPlate.gov) in the USA Further aspects of diet and health are considered in Chapter 6.

In order to know how well we adhere to healthy nutritional principles and guidelines, we have to have methods of assessing our energy intake. We each might have an idea as to how close we match the general healthy eating guidelines, but how might we 'measure' this in a more robust way? What techniques can be employed to better inform ourselves, and others, of whether these goals are being achieved? The ideal single tool to measure the nutritional status of a person does not exist, but

Box 5.4

General healthy eating guidelines and goals

- Look at and learn from the ingredient and nutritional information on food labels
- Eat as wide a variety of food as you can, from as many food groups as possible
- Eat at least five portions of fruit or vegetables each day
- Eat less fat, for example, by trimming off visible fat from meat, eating vegetarian dishes more often, use semi-skimmed or skimmed milk
- Eat a variety of carbohydrate sources (e.g. bread, pasta, rice, potatoes) and avoid simple sugars
- Drink alcohol in moderation, try to regularly have alcohol-free days.

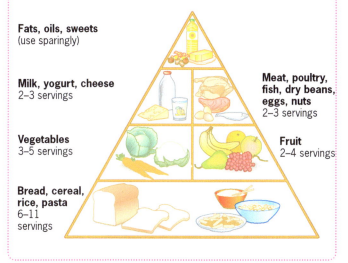

Fats, oils, sweets (use sparingly)

Milk, yogurt, cheese 2–3 servings

Meat, poultry, fish, dry beans, eggs, nuts 2–3 servings

Vegetables 3–5 servings

Fruit 2–4 servings

Bread, cereal, rice, pasta 6–11 servings

dietary evaluation (and anthropometric, biochemical and clinical/physical assessment) can help us to create a profile. Dietary evaluation might range from simple basic qualitative information, to semi-quantitative, to quantitative measurement.

Simple food questionnaires help us to evaluate whether an individual might at least be adhering to some of the most basic nutritional guidelines. Questions such as 'Do you eat portions of fruit and vegetables every day?', 'Do you have alcohol-free days?' can create some idea of the healthiness of food intake. Quantifying phrases such as 'How many portions of fruit and vegetables do you eat, each day, on average?' can help us start to build a quantitative picture, and if individuals are given a list and asked to tick food items or types they have consumed each day, we can find out more. Paper or computer-based food frequency questionnaires, on

their own or combined with some sort of food record can be utilized. Assessment techniques such as these are mainly retrospective, but there are limitations to such measures, such as the ability to accurately recall information, underestimation and overestimation of certain foodstuffs, and seasonal variations (i.e. excess at holiday periods).

Examples of food frequency questionnaires include the computer-based Food Feedback (National Dairy Council, www.nationaldairycouncil.org) and a paper-based one (but capable of being optically scanned) produced by the Anti Cancer Council of Victoria, Melbourne, Australia.

Prospective dietary assessment such as a food diary with weighed intake of foodstuffs has the potential to provide a very accurate measure of intake. This advantage, however, can be outweighed when, for example, people modify their diet for the convenience of recording, or, they understate some of their intake for fear of embarrassment. Nevertheless, the concept of a food diary, constructed by recording and, where necessary, weighing the food consumed, is an important one for research and for personal use, in sport, exercise, health and disease, so let us look at the fundamental principles of this in more detail.

The principles of energy and nutrient intake are to maintain a desirable body composition and mass by ingesting suitable quantities (as a percentage of total energy intake) of the macronutrients (carbohydrate (CHO), fat, protein); suitable quality of macro- and micronutrients for optimal daily living and performance; and an adequate fluid intake, normally monitored by thirst. If we exercise or are in a non-temperate environment, energy and/or fluid requirements may well be increased. When considering macronutrient intake, we have to take into consideration a fourth, and often significant contributor to energy intake, that of alcohol. Fat is the most energy-dense nutrient (37.8 kJ/g), followed by alcohol (29 kJ/g), then carbohydrate (16.9 kJ/g) and protein (16.8 kJ/g). These figures are known as the biological fuel values of nutrients.

We can calculate the energy intake from foodstuffs by multiplying the amount of macronutrient in grams by the biological fuel value. For example, if we eat an apple weighing 130 g, then 9 g per 100 g of apple will be carbohydrate (from information readily obtained from **food composition tables**) and therefore the energy provided from carbohydrate will be 11.7 g × 16.9 = 198 kJ, representing 100 per cent of the total energy provided by the apple (there is no fat or protein in an apple).

Activity 5.2

Independent learning task: Food intake

Compare the following meals:

- Beef steak 250 g; broccoli 100 g; boiled potatoes 125 g
- Beef steak 125 g; broccoli 100 g; boiled potatoes 250 g

From food composition tables you will find that the content of carbohydrate, fat and protein (per 100 g) are typically as follows:

- Beef steak (grilled): carbohydrate 0 g, fat 14 g, protein 19 g
- Broccoli: carbohydrate 1.6 g, fat 0 g, protein 3 g
- Boiled potatoes: carbohydrate 20 g, fat 0 g, protein 1.4 g.

Using biological fuel values, calculate the energy content (as percentage of energy from carbohydrate, fat and protein) of the two meals. Furthermore, can you find out how many grams of fibre are contained in the two meals? The meals are essentially the same (food type and total quantity), but what effect does cutting down on the meat content, and increasing the potato content have in nutritional terms? What difference is there in the saturated fat content of the two meals?

In addition to food composition tables (e.g. Roe *et al.*, 2002), much detailed information about the food and drink we consume can be obtained from the packaging and labelling of the food we buy. If the precise information about the amount of carbohydrate, fat, protein, fibre, etc. contained in the product is not provided, we can weigh the amount we eat and look up the nutritional information in food composition tables, or utilize electronic databases that contain similar information. Not only is macronutrient information available in these ways, but the amounts of vitamins and minerals are available for thousands of food ingredients and foodstuffs.

When we have obtained detailed information about our food intake from a food diary and the weighed intake of that food, we can carry out an analysis of energy intake (**Box 5.5**), and even micronutrient intake, using data contained in detailed food composition tables or computer packages.

The total amount of energy provided by a food can be measured in a **bomb calorimeter**, an instrument which measures heat released during the combustion of a known amount of food inside a sealed chamber. The amount of carbohydrate, fat, protein (and the building blocks of these, such as particular fatty acids or amino acid content), and each vitamin and mineral, or the fibre content in food is measured using a whole variety of chemical and biochemical analysis procedures. Armed with food composition data, we can set about making

changes to our meals and overall diet using sound scientific data and principles. The energy intake and proportions of energy from different macronutrients of a standard or regular meal can, for example, be considered by changing portion sizes.

The recording of a food diary, incorporating the weighing of food when necessary, is time consuming, but despite this, it is an essential tool for the modification, optimization or manipulation of diet in an elite or professional athlete, if a scientific approach to training and competition is to be taken. Food intake will vary quite substantially from one day to the next, but making a seven-day continuous record can 'smooth' out the variable daily intake to give a reasonable estimation of the general trends, particularly the total daily average energy intake. Despite this, the average intake over seven days can differ by as much as 20 per cent from an overall mean measured over several weeks (Rutishausen and Black, 2002).

The daily energy intake of around 9 MJ for a typical adult with a sedentary lifestyle can then be looked at in terms of how much energy is obtained from the different macronutrients. The oft-quoted 'Official Government' recommended daily energy intake for 'average' adults is 8372 kJ (2000 kcal) for women and 10 465 kJ (2500 kcal) for men. These values are of limited value at the level of the individual as they are dependent on a number of factors, including physical activity and body mass. Gender is of less importance in comparison. As a population, in the UK, the energy obtained from carbohydrate, fat, protein and alcohol is currently out of balance. We derive too much energy from fat, too little from carbohydrate, about the correct amount from protein, but too much from alcohol. Three examples of macronutrient energy proportion intake can be seen in Table 5.4, where 'Poor' is typical of the current situation in the U.K., 'Better' is the goal for individuals in the general population, and 'Exercise' is typical of what might be required to sustain increased levels of physical activity either in work and/ or leisure.

What is most surprising is how many people become teetotal when they are asked to complete a food diary! Alcohol metabolism and the impact of alcohol on energy intake and on health will be discussed in Chapter 3.

Protein intake has been discussed earlier in Chapter 3 (see section on 'Assimiliation of fuels', page 82), however, at this point we need to consider, within the general goals of macronutrient intake, the type of fat and carbohydrate that is desirable for good health. There are two principal types of dietary fat: saturated (found mainly in animal and dairy products) and unsaturated (monounsaturated and polyunsaturated, found mainly in vegetable oils and oily fish). The problem with fat is that it is energy dense and therefore an easier weight gain substrate, and dietary cholesterol carries with it a heart disease risk.

The incidence of premature death from a variety of conditions is increased with a higher fat intake, particularly saturated fat intake. The total intake of fat as well as the ratio of saturated fat to unsaturated fat affects cholesterol concentrations in the blood, and elevated serum cholesterol levels (above 5.0 mmol/l) start to increase coronary heart disease mortality. For these reasons, it is recommended that fat should only constitute around 30 per cent of the energy content of the diet, and that no more than 10 per cent of this should be saturated fat.

If we are to reduce saturated fat intake and the proportion of total energy from fat (and alcohol), this has to be made up from elsewhere, and the considered nutritional opinion is that this should be compensated for by an increase in carbohydrate (Table 5.4). This should be obtained from 'complex' carbohydrates of starch, rather than from simple sugars. The average intake of simple sugars is already considered to be too high, contributing to dental decay, obesity, diabetes mellitus, atherosclerosis

	Poor	Better	Exercise
Carbohydrate	42	55	65
Fat	40	30	22
Protein	15	14	12
Alcohol	3	1	1

△ **Table 5.4** Macronutrient distributions as a percentage of energy intake.

and coronary heart disease. An increase in complex carbohydrate intake (at the expense of reduced fat intake) should aid the maintenance of an increase in physical activity, as carbohydrate is the major substrate for exercise. If we are to address the positive energy balance that is causing an increase in overweight and obesity, we need to increase our energy expenditure. There is little evidence to suggest we are eating more (though food intake estimates may be subject to problems of under-reporting), but some evidence that we are less physically active due to the increasing use of cars, other labour-saving devices and an increase in television viewing (Prentice and Jebb, 1995; see also Figure 6.1).

Box 5.5

Page of a typical food diary with calculation of energy content
Assessing percentage energy intake from different nutritional fuels using a food diary – an example
Name: A.N. Other   Dates: From Monday 17/05/2011 to Sunday 23/05/2011

Day/Time	Food	Amount (g or ml)	CHO (g)	Fat (g)	Protein (g)	Alcohol (g)
Monday						
07.30	Cornflakes	61	49.8	0.5	4.3	0
	Milk (semi-skimmed)	400	18.8	0.4	12.8	0
	etc.	xx	xx	xx	xx	xx
Tuesday						
	TOTALS		2605	480	657	74

Total energy intake = total energy from:

Carbohydrate   2605 g × 16.9   = 44,024 kJ
+
Fat   480 g × 37.8   = 18,144 kJ
+
Protein   657 g × 16.8   = 11,037 kJ
+
Alcohol   74 g × 29.0   = 2,146 kJ
TOTAL   = 75,351 kJ (weekly total)
              Average 10,764 kJ per day (10.8 MJ)

where 58% energy from CHO, 24% from Fat, 15% from Protein, and 3% from Alcohol.

Fuel	Energy stored (kJ)	Approximate time at 75 per cent of maximum work
ATP, PCr and glycolysis	80	1 min
Blood glucose (oxidation)	350	5 min
Liver glycogen (oxidation)	1 500	20 min
Muscle glycogen (oxidation)	6 000	1.5 h
Plasma FFA and TG (oxidation)	170	2 min[a]
IMTG (oxidation)	9 000	2 h[a]
Fat stores (oxidation)	360 000	3–4 days[a]
Protein	200 000	1–2 days[b]

PCr, phosphocreatine; FFA, free fatty acids; TG, triglycerides; IMTG, intramuscular triglycerides.

For lipid, it is assumed that provision is 30 kJ/g (not all fat mass is available for conversion to energy).

[a]Rate of liberation of energy is only about 50 per cent of that of CHO oxidation and therefore the times allocated are theoretical as 75 per cent of maximum work cannot be achieved with fat metabolism alone.

[b]Only used in prolonged fasting/starvation.

△ **Table 5.5** Approximate available fuel reserves.

Fuels for exercise

So what energy sources do we have for exercise? We know that the universal energy currency is ATP, but do we have enough to sustain exercise for long periods? How do we (re)generate ATP? Having read Chapter 4 you will have gained an insight into the answer to these questions. The endogenous energy substrates of the body and the amount of time that these sources are able to sustain exercise can be seen in Table 5.5.

Energy substrates of the body tissues and their metabolism

There are a number of situations that require rapid and/or quantitatively high rates of fuel mobilization. Exercise is one of them, but, as examples in human biology, starvation and stress states might also be considered. We will only consider exercise, as this is an introductory text for the sport and exercise sciences.

The major fuel for exercise is carbohydrate, and this is why the amount of carbohydrate in the diet is important for both normal daily living – that should incorporate more physical activity than it currently does – and for exercise participation and training. Ultimately, the molecules that the muscle cell requires for contraction are the high-energy phosphates (HEP; e.g. ATP, PCr) and the rate of formation and reformation of HEPs is dependent upon the source and type of molecular substrate (Figure 5.6).

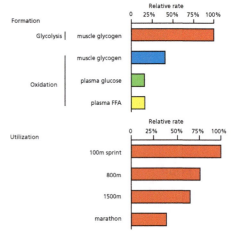

△ **Fig. 5.6** Relative rate of formation and utilization of high-energy phosphates (HEPs) from metabolic routes/activities in adults. The maximal rate of HEP formation is around 2.4 mol HEP per minute, and maximum rate of utilization is around 2.6 mol HEP per minute. Adapted from van Loon (2001), Van der Vusse and Reneman (1996) and Hultman and Harris (1988).

At this point, it would be a useful exercise for you to be able to recall, and be able to sketch, an overview of muscle energy metabolism, in the way that was described in Chapter 4 (see Figure 4.14). If we take the major athletic running distances from sprint to marathon, we can see how the contributions to total energy production of 'oxidative' and 'glycolytic/HEP' sources are provided (Table 5.6).

One of the major factors that affects substrate utilization during exercise is intensity. At low levels of exercise intensity, we use relatively little carbohydrate but can be very much

Running distance	World record holder as at February 2012	Year	Duration (min:sec)	~ Per cent oxidative	~ Per cent glycolytic
100m	Usain Bolt	2009	0:9.58	10	90
	Florence Griffith-Joyner	1988	0:10.49		
200m	Usain Bolt	2009	0:19.19	20	80
	Florence Griffith-Joyner	1988	0:21.34		
400m	Michael Johnson	1999	0:43.18	30	70
	Marita Koch	1985	0:47.6		
800m	David Lekuta Rudisha	2010	1:41.01	60	40
	Jarmila Kratochvilova	1983	1:53.28		
1500m	Hicham El Guerrouj	1998	3:26.0	80	20
	Yunxia Qu	1993	3:50.46		
1 mile	Hicham El Guerrouj	1999	3:43.13	80	20
	Svetlana Masterkova	1996	4:12.56		
5000m	Kenenisa Bekele	2004	12:37.35	95	5
	Tirunesh Dibaba	2008	14:11.15		
10,000m	Kenenisa Bekele	2005	26:17.53	97	3
	Junxia Wang	1993	29:31.78		
42,200m	Patrick Makau Musyoki	2011	2 hrs 03:38	99	1
	Paula Radcliffe	2003	2 hrs 15:25		

Based on outdoor world records (www.iaaf.org).

△ **Table 5.6** Approximate contribution of glycolytic and oxidative energy sources to total energy production of maximal work over different running distances.

dependent on our fat sources (Figure 5.7). The price we pay for this is the low rate of formation of HEP (see Figure 5.6). As we increase the intensity of exercise, we become increasingly reliant on our carbohydrate sources to match the rate of HEP formation required, in particular our muscle glycogen stores (Figure 5.7 and Figure 5.8). Training status is another factor that affects substrate utilization during exercise and Christensen and Hansen (1939) were the first to demonstrate that endurance training leads to an increased capacity to utilize fat as a substrate and reduce reliance on our carbohydrate stores during exercise. Of course, this can only work to an extent – we still have the problem that the maximum rate of formation of HEP from fat alone is insufficient to maintain moderate to high exercise intensities.

The availability of carbohydrate to the body, from stores in the liver and in muscle, is a key to our exercise capacity, particularly for longer periods of exercise and for manual work. This has also been recognized since the work of Christensen and Hansen in the 1930s. The ability for the liver to store glycogen and to supercompensate after starvation followed by a carbohydrate-rich diet was

described in Chapter 3. Skeletal muscle also has 'plasticity' in its capacity to store carbohydrate in the form of glycogen. The pioneering work in this area was made possible by the re-introduction of the needle biopsy technique by Jonas

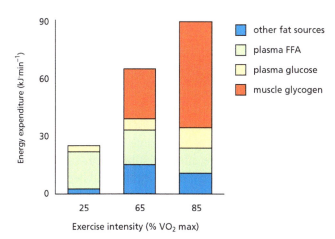

△ **Fig. 5.7** Energy expenditure and substrate sources compared with exercise at different intensities. Adapted from van Loon (2001) and Romijn et al. (1993).

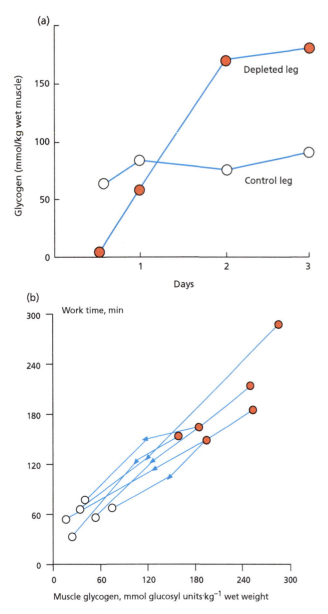

(a)

(b)

△ **Fig. 5.8** Pioneering work demonstrating the importance of (a) a high carbohydrate diet to restore and supercompensate muscle glycogen storage, and (b) initial muscle glycogen content and capacity endurance in exercise performance. From Bergstrom and Hultman (1966) and Bergstrom *et al.* (1967).

Bergstrom (1962) (see Chapter 1), and by the work of Eric Hultman (e.g. 1967). Together, Bergstrom and Hultman (1966) demonstrated that muscle glycogen could be depleted in one exercising leg whilst the glycogen content of the non-exercising leg of the same individual remained the same. Furthermore, after ingesting a high-carbohydrate diet, the glycogen content in the leg that had been exercised increased to a level far above that of the non-exercising leg (Figure 5.8(a)). This supercompensatory effect was then used to demonstrate the relationship between initial muscle glycogen content and cycle ergometer work time at 75 per cent of maximal oxygen consumption (Figure 5.8(b)) when exercise followed three days of isocaloric diets: mixed

normal diet (triangles), followed by a carbohydrate-free diet (open circles), followed by a carbohydrate-rich diet (closed circles).

Energy expenditure and its measurement

We have now dealt with energy intake in some detail, so it is time to turn our attention to the other side of the equation – energy expenditure. What do we use all of this energy for? If we are in energy balance, how do we expend an amount of energy equivalent to our intake? A short-term, daily perspective of this was illustrated earlier in the chapter – energy intake and energy expenditure are quite well matched in elite, competitive cyclists taking part in the most demanding cycling competition in the world (Figure 5.5). This happens to be one of the highest recorded sustained levels of energy expenditure in humans. At this extreme, the proportion of energy expenditure through physical activity becomes the major single component of total daily energy expenditure, with other components seemingly insignificant in comparison.

Such high levels of energy expenditure are, however, most uncommon. The components of total energy expenditure (TEE) are basal metabolic rate (BMR), thermogenesis (heat production, primarily the thermic effect of food (TEF)) and physical activity (ACT) (Figure 5.9):

$$TEE = BMR + THERM + ACT$$

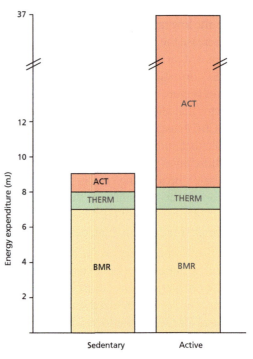

△ **Fig. 5.9** Components of total energy expenditure.

The largest single component of TEE is normally BMR, accounting for about 7 MJ in a healthy 70-kg person. Thermogenesis (THERM) accounts for about 10 per cent or about 1 MJ of TEE, composed mostly of post-prandial thermogenic processes (heat produced by eating food), sometimes referred to as either dietary-induced thermogenesis (DIT) or TEF. There can be other, generally smaller, thermogenic contributions such as shivering. Habitual activity such as personal care, walking around the house, walking around at work can contribute about 1 MJ. So if we are sedentary, our TEE will be around 9 MJ. If we are active in work or in leisure, then we will add to this TEE depending on the level of activity (e.g. up to the extremes of a Tour de France cyclist expending around 37 MJ on the most arduous and gruelling mountain day).

Basal metabolic rate (BMR, sometimes known as the average daily metabolic rate: ADMR) is the daily amount of energy expenditure required to maintain basic physiological functions, such as the work of breathing and the circulation of blood and fluids. It does not include the energy for the digestion and absorption of food, or for temperature regulation. Measurement of BMR can be carried out using calorimetry (to be discussed shortly) but to do this the subject must be lying at physical and mental rest, in a comfortably warm (thermoneutral) environment, and have not eaten for at least 12 hours. In practice, this needs to be early in the morning, before breakfast, after an overnight fast, having not exercised vigorously or taken in any stimulants such as caffeine, nicotine and capsaicin, for 24 hours.

Alternatively, an estimate of BMR can be calculated using the following formula examples (Department of Health, 1991):

For men aged 18–29 BMR (MJ per day) = 0.063 (body mass (kg)) + 2.896
For women aged 18–29 BMR (MJ per day) = 0.062 (body mass (kg)) + 2.036

If any of the conditions for BMR measurement are not met, the measured energy expenditure should be referred to as resting metabolic rate (RMR), and this will be a little higher than BMR, for example as a result of an increase in energy expenditure due to a recent meal. During sleep, energy expenditure is about 5–10 per cent lower than BMR. As a ball-park figure, average adult BMR might be about 4.5 kJ/min over the course of a day but there are numerous determinants of energy expenditure that render this value very approximate.

It is interesting to consider the relative contribution to BMR of some organs and tissues, particularly in relation to their mass. In an average 70-kg adult, small organs such as the liver (about 1.8 kg) and the brain (about 1.4 kg) each account for about 20 per cent of BMR, reflecting their importance in metabolism. The relatively large mass of skeletal muscle (about 28 kg) accounts for about 22 per cent of BMR, whilst about 15 kg of adipose tissue is relatively metabolically inactive and only accounts for about 4 per cent of BMR. The remaining organs and tissues account for the rest of the 34 per cent of BMR (about 2400 kJ).

Principles of the measurement of energy expenditure at the level of the whole body can be traced back to the 'father of modern chemistry' Antoine Lavoisier (1743–1794). He recognized that candles only burnt in the presence of oxygen and was to demonstrate that living organisms produced heat in a similar way – they required oxygen for life and combusted food as they released heat. Thus:

Food + oxygen → water + carbon dioxide + heat

This is the basis of **calorimetry**, and if we are able to directly measure the heat liberated by the body, by inviting a person to spend time inside an insulated, energy-tight chamber whilst measuring the change in temperature (of air or water) passing through the chamber over a period of time, we will have a measure of energy expenditure. Facilities for such direct calorimetric measurement of energy expenditure do exist (first established by American scientist Wilbur Atwater at the end of the nineteenth century), but they are not widely available (Figure 5.10).

Of greater convenience and availability is the principle of **indirect calorimetry** – energy expenditure resulting from the oxidation of fuels used by the body – assessed by the consumption of oxygen (and the concomitant production of carbon dioxide). Simply stated, oxygen consumption is the difference between the amount of oxygen going into the body and that coming back out. The calculation of oxygen consumption is possible if we can measure the volume of air inspired and expired (V_I and V_E) and the fractions of oxygen in those volumes (F_IO_2 and F_EO_2) in a known amount of time:

$$V_{O_2} = (V_I \times F_IO_2) - (V_E \times F_EO_2)$$

One litre of oxygen consumed corresponds to approximately 20 kJ of energy expenditure. The heat produced per litre of oxygen consumed is almost the same, whether carbohydrate, fat or protein is being oxidized. At rest an adult might consume 250–300 ml/min of oxygen (the O_2 consumption; VO_2) and we produce in exchange about 200–250 ml/min of carbon dioxide (the CO_2 production; VCO_2). The concept of the ratio of carbon dioxide produced to oxygen consumed is discussed a little later. At rest this ratio is dependent on the chemical nature of food or endogenous energy source that is being oxidized. During exercise, it is dependent on both catabolism of energy substrate and on ventilation.

There are a wide variety of techniques that can be used to measure oxygen consumption, and these techniques and methods of respiratory gas collection and analysis

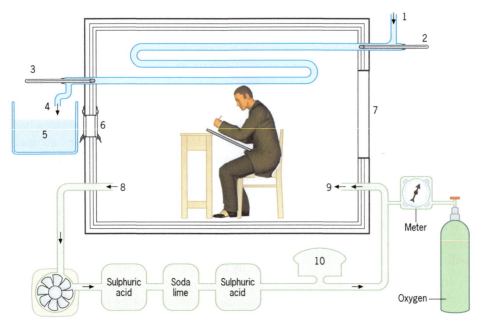

△ **Fig. 5.10** Schematic picture of a whole body calorimeter room (Atwater chamber). Such facilities are not widely available – circa 2004 there were only six in the UK, three of these at the MRC Dunn Nutrition Centre (Cambridge).

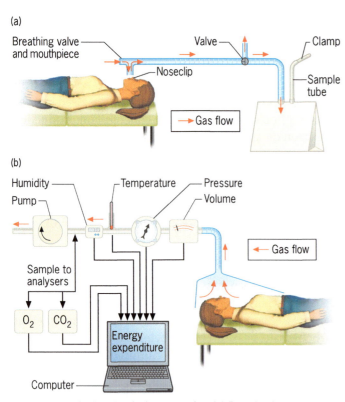

△ **Fig. 5.11** Indirect calorimetry using (a) Douglas bags or (b) a ventilated hood/canopy.

a timed collection has been made, the contents of the bag can be analysed for oxygen and carbon dioxide content using stand-alone analysers, and the volume of air in the bag measured. Corrections are made for air pressure and moisture content (correction to standard temperature and pressure dry (STPD)) and oxygen consumption can then be calculated using the formula presented earlier. Douglas bags can be used for resting and exercising measures of oxygen consumption, but require the use of a noseclip and mouthpiece or a mask covering the mouth and nose. A more subject-friendly and comfortable approach to respiratory gas analysis is the use of a 'ventilated hood' or canopy. This is used in resting metabolic studies, primarily for the measurement of oxygen consumption and carbon dioxide production to determine, for example, BMR, RMR or TEF, and requires the subject to sit or lay quietly with a hood over the head and shoulders, whilst a constant flow of air is pulled past the mouth and nose to dilute expired air. The air passes into the mixing chamber of an on-line analyser and is sampled by integrated oxygen and carbon dioxide analysers (Figure 5.11(b)).

For longer periods of measurement, it is possible to construct a chamber, much like an Atwater-type chamber, where the subject can stay inside for a day or more, eating, sleeping, working at a desk and exercising (normally using a cycle ergometer). Rather than measuring heat production, such a chamber (a whole body respiration chamber) is essentially a large ventilated hood with air passing through and being constantly sampled and analysed, using indirect calorimetry principles.

are appropriate for both the measurement of whole body metabolism, and for exercise responses. You will hopefully have the opportunity to become familiar with at least one of the measurement procedures and the hardware involved.

The most established technique for the collection of expired air is that of Douglas bags (Figure 5.11(a)). Once

△ **Fig. 5.12** Ambulatory indirect calorimetery.

Over the years, it has been a goal to measure oxygen consumption outside of the constraints of a laboratory or a specially designed room. The Kofrani–Michealis respirometer enabled physiologists from the 1950s to study work and exercise capacity during everyday tasks such as working at a desk or walking up a flight of stairs. The modern-day equivalent for ambulatory indirect calorimetry measurement is an instrument like the CosMed respirometer (Figure 5.12).

Energy metabolism in the cell results in the consumption of oxygen and the production of carbon dioxide. The ratio of the amount of carbon dioxide produced to the amount of oxygen consumed by tissue substrate utilization is known as the **respiratory quotient (RQ)**:

$$RQ = \frac{\text{carbon dioxide production}}{\text{oxygen consumption}}$$

If the cell is utilizing carbohydrate, i.e. glucose ($C_6H_{12}O_6$), the following chemical equation will apply:

$$C_6H_{12}O_6 + 6O_2 \rightarrow 6CO_2 + 6H_2O$$

In this situation, an equivalent amount of carbon dioxide is produced compared with the oxygen consumed, so

$$RQ = 1.$$

The oxidation of 1 mole of glucose (molecular weight 180) releases 2.8 MJ of heat, therefore 1 g of glucose liberates 2800 kJ/180 = 15.6 kJ. Compared with carbohydrate, proportionally more oxygen is required for the oxidation of fat and RQ is about 0.707 (depending on the fatty acid being oxidized).

In sport and exercise physiology, we are primarily concerned with whole body measures of physiological function, and instead of measuring oxygen consumption and carbon dioxide production at the level of the cell, we do so at the interface with the atmosphere – the air that we breathe in and out. At this, the ventilatory level, we can apply the same principle of the ratio of the amount of carbon dioxide produced to the amount of oxygen consumed, but instead this is known as the **respiratory exchange ratio (RER)**. When we are in a steady state, for example in a resting, post-prandial state, or if exercising at an intensity at which heart rate remains constant, RER is equivalent to RQ and we can use RER data to calculate the macronutrient substrates that are being utilized during such periods. If, however, we are in a non-steady state, we cannot make such assumptions – for example, at high exercising workloads that cannot be sustained, we hyperventilate, 'blowing off' carbon dioxide which artificially increases the apparent carbon dioxide production in relation to the oxygen consumed.

Substrate oxidation at rest or during exercise can be calculated from RER data if the measurement was carried out during steady state. Using a simple example, if C is the energy due to carbohydrate oxidation and F is the energy due to fat:

$$C + F = 1$$

(assuming protein is a negligible energy source and no alcohol is being metabolized).

If RER = 0.83 = $(1 \times C) + (0.707 \times F)$

then $(1 \times C) + (0.707 \times [1 - C]) = 0.83$

Per cent carbohydrate oxidation =

$$\frac{(0.83 - 0.707)}{(1 - 0.707)} \times 100 = 42$$

If $C = 42$, then $F = 58$, i.e. 42 per cent of energy was being provided by carbohydrate oxidation.

Furthermore, if we know the amount of oxygen consumed, after a series of derivations the rate of substrate oxidation (e.g. g/min of carbohydrate) can be calculated, and then the total amount of substrate over any given period can be calculated.

When we exercise, the amount of oxygen consumed by the body (VO_2) increases. At lower workloads such as 50W or 100W on a cycle ergometer, VO_2 increases before reaching a new steady-state. At higher workloads, VO_2 continues to increase beyond the 2 min or so to reach steady state at low intensities. It is at such higher intensities (e.g. 150 W) that the slow component of oxygen uptake kinetics is therefore seen. When we exercise at ever increasing exercise workloads in steps of around 3 min, we eventually reach a point where VO_2 does not increase and exercise cannot be continued. This is termed

VO_{2Peak}. In a small number of cases, the same VO_2 might be recorded despite being able to exercise at one further increase in workload before exhaustion – this is a true plateau in VO_2, termed VO_{2Max}. In Figure 5.13 we can see these two alternative scenarios in a plot of VO_2 compared with running speed. A trained athlete is more likely to be able to continue an incremental test to exhaustion and achieve a plateau in VO_2, whereas an untrained person might not. Using this approach, we can then extrapolate from the relationship workloads (e.g. Watts on a cycle ergometer, speed on a treadmill) that correspond to a percentage of that person's maximum or peak VO_2.

but for monitoring and research in exercise training and in health, there is a requirement for longer term energy expenditure measurements that allow subjects to be 'free-living'. Non-calorimetric methods exist that fulfil such objectives, but there is still a need to refine such methods or find new methods. Doubly labelled water can be used to obtain an integrated measure of all components of daily energy expenditure over one to two weeks and was introduced in Chapter 1 (see **Time Out** *Stable isotopes and their use in health and exercise research*). Accelerometery instruments have been refined in recent years, such that earlier uniaxial models (that measure movement in one plane only) have now been superseded by triaxial devices (Figure 5.14).

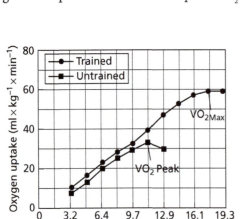

△ **Fig 5.13** Oxygen consumption increases with increasing workload until a peak or maximum (plateau) is attained.

When exercising for a prolonged period at a moderate intensity i.e. an intensity of around 70% of VO_{2max}, which can normally be sustained for 1–2 hours, a phenomenon known as VO_2 drift can be seen – a slow increase in VO_2 during prolonged, submaximal, constant intensity exercise.

Oxygen consumption and carbon dioxide production data collected over longer periods can be used to estimate **total energy expenditure (TEE)**. A formula often used in human indirect calorimetry is that of Weir (1949):

$$TEE \text{ (kJ)} = 16.489\, VO_2 \text{ (litres)} + 4.628\, VCO_2 \text{ (litres)} - 9.079\, N \text{ (g)}$$

where N is nitrogen excretion in g. If N excretion is not measured but it is assumed that protein turnover accounts for 15 per cent of total energy expenditure, then the formula becomes:

$$TEE \text{ (kJ)} = 16.318\, VO_2 \text{ (litres)} + 4.602\, VCO_2 \text{ (litres)}$$

The methods of measurement of energy expenditure so far described have been dependent on specially constructed rooms and respiratory gas analysis in laboratories, or ambulatory equipment. These methods are appropriate for many of the procedures required for sport and exercise,

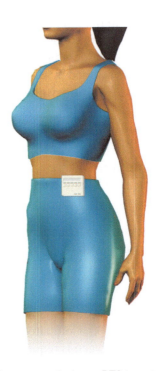

△ **Fig. 5.14** Accelerometer device – RT3 type triaxial accelerometer with docking/downloading unit

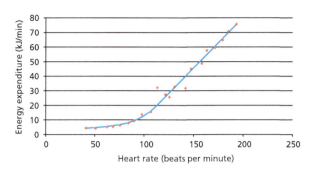

△ **Fig. 5.15** Calibration curve example of energy expenditure plotted against heart rate. Energy expenditure can be calculated from oxygen consumption data based on the assumption that 1 litre of oxygen consumed corresponds to 20 kJ.

Heart rate monitors (HRM) can also be used to estimate energy expenditure, particularly when combined with a calibration curve obtained from indirect calorimetry for the subject (Figure 5.15). Although heart rate rises linearly with oxygen consumption at exercising workloads, the relationship is not linear over the range of physiological heart rates. Some currently available models of HRM enable the recording of heart rate for over 24 hours and the subsequent download of the information, including each recorded data point, heart rate profile and a frequency distribution of the data. By plotting oxygen consumption (converted to energy expenditure) against heart rate, a rate of energy expenditure can then be assigned to particular frequency sets of heart rates and the total energy expenditure for the period summated.

Physical activity level, and determinants of energy expenditure

As mentioned earlier, there are numerous factors or determinants of energy expenditure that render a 'typical' average value of energy expenditure (e.g. 4.2 kJ/min) very approximate at the level of the individual. We have also seen how the major component of energy expenditure, normally BMR, can become much less significant if a person engages in high levels of physical activity. Levels of physical activity can be related to (multiples of) BMR known as **metabolic equivalents (METS)**. This provides us with a basic way of expressing the metabolic cost of exercise: the energy cost of running at 10 km/h is about 10 METS, and at 15 km/h is about 17 METS. MET intensities for a whole range of exercise and everyday physical activities have been estimated (Ainsworth *et al.*, 2000).

We can also relate our RMR to the total amount of energy expended on a daily basis. This comparison of TEE to RMR is referred to as the **physical activity level (PAL)** factor. For example, if TEE is around 12 MJ/day and RMR is around 6 MJ/day, then the PAL factor is 2.0, indicating that TEE is twice the RMR. In Tour de France cyclists, this can reach 5.0, and it is thought that migrating birds reach a PAL of around 20. A low PAL indicates a sedentary lifestyle, a high PAL an active one. People with a sedentary occupation with little leisure that involves physical activity, will have a PAL in the region of 1.4, and those with light activity in work or leisure, around 1.6. Active occupations and/or leisure lifestyles tend to raise PAL to above 1.75. It is suggested that the optimal PAL to protect against overweight and obesity is around 1.8 or higher and in order to increase PAL from around 1.6 to 1.8 requires an hour of light activity per day. This might be achieved by sport or exercise or by making adjustments to a normal daily activity such as the means of travelling to work – is it possible to walk or cycle to work and back (or part of the way) instead of using the car or bus?

There are a number of other determinants of energy expenditure in addition to physical activity. These affect BMR and the energy cost of physical activity. **Body size** is one of the major determinants of energy expenditure in humans, accounting for about half the variability in BMR between individuals. A difference in body mass of 10 kg accounts for a difference of 0.5 MJ/day in adult BMR or about 0.7 MJ in TEE of someone cycling to work. You might recall that the contribution to BMR of various organs and tissues is not related to their mass. Two of the largest contributors to body mass have contrasting energy demands, so **body composition** is an important determinant of energy expenditure. Adipose tissue has a lower metabolic demand than other tissues such as 'fat-free' or lean tissues. For this reason, BMR is sometimes expressed relative to fat-free mass (e.g. as 'MJ per kg fat-free mass'). The lifespan has contrasting demands on energy expenditure, so **age** is another determinant. Young infants require more energy to maintain body temperature, the cost of growth in children is about 20 kJ/g of normal gain in body mass, and in adulthood there is often a decline in fat-free mass and an increase in fat mass/adiposity which can account for a decline in BMR with older age. This is not inevitable, however, and the maintenance of lean mass and functional capacity has the potential to be maintained well into older age.

Gender is a further determinant of energy expenditure as a result of an average difference in both body size and composition between males and females. The BMR of an adult male of around 70 kg will be about 1 MJ/day higher than that of a female of the same age and body mass. Diet is a further determinant as a result of post-prandial thermogenesis. More chronically, a change in BMR of around 5–10 per cent will occur if, for example, we severely restrict our calorific intake, reducing the TEF component of energy expenditure, and if we overfeed we will increase our BMR.

The **environment** around us influences energy expenditure by altering the energy cost of maintaining body temperature. Factors such as our clothing, the radiant temperature of the room we are in, or if outside, the wind speed and temperature all come into play. At low room temperatures (e.g. below 18°C) energy expenditure might be increased due to cold-induced thermogenesis, or at high temperatures (e.g. above 28°C) energy expenditure might be increased due to the energy cost of sweating.

It has been demonstrated in studies of subjects matched for age, gender, body mass and fat-free mass that BMR can vary by about 10 per cent between individuals, implying a **genotype** determinant to energy expenditure. **Hormonal status** can also be influential either in disease (e.g. hyper- or hypothyroidism) or in health (e.g. there are cyclical changes in BMR over the menstrual cycle, with a rise shortly after ovulation).

The importance of measuring BMR when the subject is 'lying at physical and mental rest' brings attention to the fact that **psychological status**, particularly acute status, is a determinant of energy expenditure, as anxiety is a potent stimulator of adrenaline secretion and thereby slightly elevates energy expenditure. It is not known whether chronic psychological stress influences energy expenditure. Another feature of BMR measurement raised earlier in the chapter was the absence of any dietary, pharmacological or social **stimulant agents** such as caffeine, amphetamines or nicotine, as these increase energy expenditure by small amounts.

It can be seen that a whole range of factors determine energy expenditure in apparently healthy individuals. It is also important to remember that in **disease**, processes such as fever and tumours can also increase energy expenditure. Clearly reliable and valid information relating to energy expenditure (and therefore energy balance) can only really be obtained by considering the individual and by comparing and contrasting a wide range of factors.

If a person is in energy balance, then energy intake will be equal to energy expenditure. When considering energy requirements for that person, this gives us an option to base requirement on either intake or expenditure measures. It can be argued that intake estimates are less reliable because of a tendency for people to at best forget about some of their intake, or at worst, be economical with the truth! Energy expenditure methodologies can be used to validate energy intake; if, in doubly labelled water experiments, expenditure is greater than intake over a period of a week or so, but body mass and composition remains the same, an underreporting of food intake is likely.

Energy balance – techniques in perspective

How concepts of energy homeostasis can be obtained from different types of energy balance studies has recently been discussed by Elia *et al.* (2003). Energy balance does not necessarily mean that we have optimum 'body size and composition and level of physical activity consistent with long-term good health'. When considering how we look at the energy balance of a person, we can take a number of routes, based on the principles and concepts discussed in this chapter. If we were to simplify the energy balance equation and restrict energy intake (starvation), then energy balance would become equal to energy expenditure. This is less relevant to sport and exercise because we need to eat often to perform well, but as a methodology in health, this can be appropriate.

Whole body direct and indirect calorimeters are the most accurate and precise means of assessing and monitoring energy expenditure, but being monitored for a day or more in a small room has the major disadvantage of not representing the 'free-living' situation. We might instead glean energy balance information from measurement of body composition. The tissues of greatest interest in sport, exercise and health are arguably fat and fat-free tissues. We can assign energy values to the loss or gain of either of these tissues, but the major disadvantage of most of the methodologies widely available is the lack of precision, particularly in detecting changes over a period of a few days.

A further approach to energy balance assessment and monitoring is the combination of both energy intake and energy expenditure measurement. Measures of both intake and expenditure are more accurate and reliable in controlled conditions (e.g. we can determine how much food is being given to a person in a chamber) but again, measures of energy balance in controlled conditions do not reflect those in free-living conditions. As described in the previous section of this chapter, if a person is in energy balance, then any major discrepancies between intake (food diary/weighed intake) and expenditure (measured by doubly labelled water tracer) can be attributed to underreporting of intake.

You should now have an understanding of the main issues relating to energy balance as summarized in **Box 5.6**.

Box 5.6

Energy balance measurement and study in perspective – Summary of major points

- The most accurate and precise measures of energy balance can be obtained in artificial environments.

- Measures of energy expenditure are improving but they lack accuracy and precision.

- Energy intake measures have their problems depending on the method and the population group.

- There have been recent advances in body composition techniques but they lack short-term precision (and most are clinical and expensive techniques).

- Body composition changes provide little knowledge of dynamics and interrelationships between energy intake and expenditure, their components and their determinants.

- It is recognized that there is a need to develop better methodology for the study (e.g. assessment and monitoring) of energy balance in free-living subjects. This would be of benefit in sport, exercise, health and disease.

Diet and physiological status/Exercise nutrition

Nutrition and energy balance is of importance to a wide range of types of physiological status. The sport and exercise scientist may well find themselves dealing with exercise support and thereby working with a wide variety of client groups, including pregnant and lactating women, children and adolescents, or the elderly. Each of these groups have different energy and nutrient balance needs that will interrelate with their exercise needs and prescription. These are the types of subject groups that may well be the focus of specialist interest for later parts of a sport and exercise science course. In the meantime, we will summarize here the key features of energy balance that relate to the physiological status of the exercise performer, as highlighted in this chapter.

In the example of cycling in the Tour de France race (Figure 5.5), we have seen how exercise can demand high levels of energy intake and energy expenditure. In such an example, exercising energy expenditure becomes the major component, making BMR and TEF seem relatively unimportant. Different sporting performance events demand different proportions of energy from glycolytic and oxidative sources (see Table 5.6). Such high levels of energy expenditure in training and competition are

rare, however, and for most people BMR is the major component of energy expenditure (see Figure 5.9), and a wide variety of general determinants of BMR and energy expenditure were briefly discussed.

For most of the athletic running events from 1500 m to 42 km, the major source of energy is carbohydrate. Energy substrates for exercise were discussed and the importance of stored carbohydrate for exercise in terms of the relatively high rate of liberation of energy from this source (see Figure 5.7), the ability to manipulate muscle glycogen stores, and the dependence of prolonged, moderate intensity exercise on glycogen stores has been illustrated (see Figure 5.8).

Despite most people having a lot of fat stored in the body, the rate of liberation of energy from fat is relatively low. Although fat is metabolized to provide energy alongside carbohydrate, the ingestion of a high-fat diet does not increase the capacity to oxidize fat and improve exercise performance. Protein requirements of exercise are easily met by an increase in energy intake (assuming it is for the support of an exercise-induced increase in energy expenditure), providing the diet is balanced (e.g. around 14 per cent of energy is derived from a range of dietary protein foods). Reducing body mass or fat mass for health or performance should be done at a modest rate, addressing both sides of the energy balance equation where possible. A sensible rate of loss of body fat is around 0.5 kg/week and should never be more than 1 kg/week. One kg of fat mass loss is equivalent to about 27 MJ, so, if attempted by dietary restriction alone, to forego around three days' worth of energy intake in a seven-day period is not advised.

One further aspect of dietary provision that you were introduced to briefly in Chapter 3 is **hydration**. Maintaining hydration status for optimal exercise capacity and sports performance can be combined with replacement of energy, and a further introduction to this specialized area of sport and exercise nutrition is therefore required.

Hydration and rehydration in sport and exercise

Some of the earliest work to demonstrate the importance of fluid replacement during exercise in the heat was carried out in industry and in the military by Adolph (1947). That dehydration results in an increase in body core temperature was also demonstrated by Wyndham and Strydom in 1969. Athletes who became dehydrated by more than 3 per cent in 32-km races had elevated post-race rectal temperatures, suggesting that preventing an excessive rise in core temperature (and subsequent heat injury) during prolonged exercise was dependent upon the avoidance of dehydration. Despite these early findings being difficult to replicate, research studies over the last 40 years have established the benefits of optimal hydration prior to exercise and fluid ingestion during exercise (**Box 5.7**).

The current debate is to what extent the rate of fluid replacement influences exercise performance. Guidelines suggest replacing the same amount of fluid lost during exercise, though this extent of replacement might not be necessary to maintain performance during exercise. Two important considerations for fluid intake in sport and exercise are the maintenance of fluid balance and the need to provide energy. Most rehydration strategies include drinks with electrolyte and carbohydrate, and the presence and quantity of these in the drink has an effect on gastric emptying (see Chapter 3). The amount of fluid required and the desirable amount of carbohydrate will depend on a number of factors, including the rate of fluid loss (determined by training status, exercise intensity and environmental conditions), and the duration of exercise.

- Energy intake can be estimated, for example, at population, disease and individual level.
- It is often important in sport and exercise to gauge energy intake that will sustain 'economically and socially desirable physical activity', and of importance for health is the estimation of energy intake that will prevent over-consumption.
- Energy expenditure has three primary components, including BMR and physical activity, and several determinants such as body size, body composition, gender, psychological status.
- Much of the measurement of energy expenditure in sport, exercise and health is dependent on indirect calorimetry, assessed by measuring oxygen consumption in the laboratory or in the field.
- Most exercise efforts lasting more than five minutes are highly dependent on carbohydrate as a substrate source, which is stored as glycogen in muscle and the liver and its storage can be manipulated by diet and exercise.
- There is a need to develop better methodology for the study of energy balance in free-living subjects, that would be of benefit in sport, exercise, health and disease.
- Hydration and rehydration in sport and exercise has received greater attention in recent years and is important in reducing the deleterious physiological effects that affect performance and health of the athlete.

Summary

- Energy balance is represented by the difference between energy intake and energy expenditure, and imbalance manifests itself as a change in body mass and/or composition.
- Political, social, economic, perceptual and scientific influences have all had a role to play in general dietary/nutritional patterns and those in sport and exercise over the last 150 years.
- Body composition can be considered at a number of levels, from chemical to tissue to organ and these all have a role to play in the compartment modelling of human body composition, of which three or four compartment models are becoming 'gold standards'.
- There are few available direct methods of measuring body composition, and most techniques for estimating body composition are therefore based on indirect or doubly-indirect anthropometric/physiological/imaging measurements with varying assumptions.

Review Questions

1. Describe what our general energy requirements are and what they require us to be able to do.
2. What are the biological levels at which the composition of the body might be considered?
3. How can body fat content be estimated using skinfold callipers?
4. How can densitometric principles be utilised to estimate body composition?
5. In what ways can energy intake be estimated?
6. What factors determine energy expenditure?
7. How can the storage of carbohydrate by body tissues be manipulated as a result of diet and exercise?
8. What is the major component of energy expenditure for a) a sedentary individual and b) a Tour de France cyclist?
9. Describe the main macronutrient and energy requirements for endurance exercise.

References

Adolph EA (1947) *Physiology of Man in the Desert*. New York: Interscience Publishers.

Ainsworth BE and 11 others (2000) Compendium of physical activities: an update of activity codes and MET intensities. *Medicine & Science in Sports and Exercise* 32: S498–S516.

Bender DA (2002) *Introduction to Nutrition and Metabolism*, 3rd edn. London: Taylor and Francis.

Bergstrom J (1962) Muscle electrolytes in man. Determined by neutron activation analysis on needle biopsy specimens. A study on normal subjects, kidney patients and patients with chronic diarrhoea. *Scandinavian Journal of Clinical and Laboratory Investigation* 14 (Suppl 68): 1–110.

Bergstrom J, Hultman E (1966) Muscle glycogen synthesis after exercise: an enhancing factor localised to the muscle cells in man. *Nature* 210: 309–310.

Bergstrom J, Hermansen L, Hultman E, Saltin B (1967) Diet, muscle glycogen and physical performance. *Acta Physiologica Scandinavica* 71: 140–150.

Christensen EH, Hansen O (1939) Respiratorischer quotient und O$_2$-aufnahme. *Scandinavian Archives of Physiology* 81: 180–189.

Dempster P, Aitkins S (1995) A new air displacement method for the determination of human body composition. *Medicine and Science in Sports and Exercise* 27: 1692–1697.

Department of Health (1991) *Dietary Reference Values for Food Energy and Nutrients for the United Kingdom*. London: HMSO.

Deurenberg P, van der Kooij K, Evers P, Hulshof T (1990). Assessment of body composition by electrical impedance in a population aged >60 years. *American Journal of Clinical Nutrition* 51: 3–6.

Deurenberg P, Roubenoff R (2002) Body composition. In: Gibney MJ, Vorster HH, Kok FJ eds *Introduction to Human Nutrition*. The Nutrition Society Textbook Series. Oxford: Blackwell Publishing, pp. 12–29.

Downer AR (1902) *Running Recollections and How to Train*. Gale and Polden. Facsimile edition published in 1982 by Balgownie Books, Aberdeenshire.

Durnin JV, Rahaman MM (1967) The assessment of the amount of fat in the human body from measurements of skinfold thickness. *British Journal of Nutrition* 21: 681–689 (reprinted *British Journal of Nutrition* 2003; 89: 147–155).

Durnin JV, Womersley J (1974) Body fat assessed from total body density and its estimation from skinfold thickness: measurements on 481 men and women aged from 16 to 72 years. *British Journal of Nutrition* 32: 77–97.

Elia M, Stratton R, Stubbs J (2003) Techniques for the study of energy balance in man. *Proceedings of the Nutrition Society* 62: 529–537.

FAO/WHO/UNU (1985) Energy and protein requirements. Report of a Joint FAO/WHO/UNU Expert Consultation. *World Health Organization Technical Report* Series 724: 1–206.

Fields DA, Goran MI, McCrory MA (2002) Body composition assessment via air-displacement plethysmography in adults and children: a review. *American Journal of Clinical Nutrition* 75: 453–467.

Haycock GB, Schwartz GJ, Wisotsky DH (1978) Geometric method for measuring body surface area: a height–weight formula validated in infants, children, and adults. *Journal of Pediatrics* 93: 62–66.

Heitman BL (1990) Evaluation of body fat estimated from body mass index, skinfolds and impedance: a comparative study. *European Journal of Clinical Nutrition* 44: 831–837.

Hultman E (1967) Muscle glycogen in man determined in needle biopsy specimens. *Scandinavian Journal of Clinical Laboratory Investigation* 19: 209–217.

Hultman E, Harris RC (1988) Carbohydrate metabolism. In: Poortmans JR ed. *Principles of Exercise Biochemistry*. Basel: Karger, pp. 78–119.

James WPT (2000) Historical perspective. In: Garrow JS, James WPT, Ralph A eds *Human Nutrition and Dietetics*, 10th edn. Edinburgh: Churchill Livingstone, pp.3–12.

Martin AD, Spenst LF, Drinkwater DT, Clarys JP (1990) Anthropometric estimation of muscle mass in men. *Medicine and Science in Sports and Exercise* 22: 729–733.

Noakes TD (1998) Hydration and rehydration in exercise and sports. In: Arnaud MJ ed. *Hydration throughout Life*. London: John Libbey and Company, pp. 95–103.

Orr JB (1936) Food, Health and Income. London: Macmillan. Cited in: Garrow JS, James WPT, Ralph A eds *Human Nutrition and Dietetics*, 10th edn. Edinburgh: Churchill Livingstone, p. 6.

Prentice AM, Jebb SA (1995) Obesity in Britain, gluttony or sloth? *British Medical Journal* 311: 437–439.

Roe MA, Finglas PM, Church SM eds (2002) McCance and Widdowson's *The Composition of Foods*, sixth summary edition. London: Royal Society of Chemistry.

Romijn JA, Coyle EF, Sidossis LS, Gastaldelli A *et al.* (1993) Regulation of endogenous fat and carbohydrate metabolism in relation to exercise intensity and duration. *American Journal of Physiology* 265: E380–391.

Saris WHM, van Erp Baart MA, Brouns F *et al.* (1989). Study on food intake and energy expenditure during extreme sustained exercise: The Tour de France. *International Journal of Sports Medicine* 10: S26–S31 (Supplement).

Van Loon LJC (2001) The effects of exercise and nutrition on muscle fuel selection. PhD thesis, Maastricht University, The Netherlands.

Van der Vusse GJ, Reneman RS (1996) Lipid metabolism in muscle. In: Rowell LB, Sheperd JT eds *Handbook of Physiology*, section 12: Exercise: Regulation and integration of multiple systems. Bethesda, MD: American Physiological Society.

Weir JB (1949) New methods for calculating metabolic rate with special reference to protein metabolism. *Journal of Physiology* (London) 109: 1–9.

Wyndham CH, Strydom NB (1969) The danger of an inadequate water intake during marathon running. *South African Medical Journal* 43: 893–896.

Further reading

Maughan R, Gleeson M (2004) *Biochemical Basis of Sports Performance*. Oxford: Oxford University Press.

6 Lifestyle Factors affecting Health

The health of the people is really the foundation upon which all their happiness and all their powers as a state depend.

Benjamin Disraeli

Chapter Objectives

In this chapter you will learn about:

- Health and how it is measured.
- The factors that affect health.
- The physiological effects of major behavioural, modifiable factors, such as physical inactivity, nutrition, alcohol, smoking and drug use.
- The ways in which these lifestyle factors can be controlled by the individual.
- The impact of physical inactivity and the resulting rise in chronic diseases in populations.
- How different dietary patterns have an influence on disease development and progression.
- The patterns of consumption of alcohol and its short- and long-term effects on the body and on health.
- How the use of tobacco and illicit drugs affects health and social well-being.

Introduction

One of the potential roles that exercise biologists have in the early part of this century is to use the increased understanding of the scientific bases of health to improve the health of the world's population. Exercise is only one of the factors that affect health, and the sciences of physiology and psychology have a particularly important role to play, with different slants applied depending on where in the world we live. In the industrialized world, we can continue to be dependent on expensive medical and pharmacological interventions to treat the consequences of what might be considered modern day excesses. Alternatively, we might make a serious effort to reduce our dependency on expensive intervention by looking to promote and achieve the goal of a healthier lifestyle – taking more responsibility for our own health (primary prevention), at the same time as being aided by primary care, and being less reliant on acute/secondary care. Key contributors to ill-health include physical inactivity, poor diet, tobacco smoking, alcohol consumption and drug abuse; however, the effects of some of these patterns of behaviour can be reversed.

Lifestyle, however, is a difficult thing to change; it turns slowly, more like an oil tanker than a racing car, and lifestyle changes seem to be alien to our quick-fix mentality and consumption philosophy. Altering the physiology and metabolism of the body is dependent on changes in the psychology of the person: they have to want to effect a lifestyle change for the benefit of their health. Educators, public health and medical authorities, national and international regulators, and industry also have major roles to play in applying the wealth of knowledge we have, in facilitating change and providing the resources needed. At the same time we would hope that the world will address the more fundamental needs of the developing nations (where major problems remain political and/or nutritional and preventable-disease based) and enable such nations to learn from our mistakes.

Nutrition has an important role in a healthy lifestyle, and the basic tenets of energy balance and nutrition have been introduced in the previous chapter. The public health implications of good and poor nutrition will be discussed in this chapter, as will the case for physical inactivity as a public health concern, and the case for exercise as part of a healthy lifestyle from an epidemiological perspective. How we find out about and what we know of the physical activity and nutritional status of the populations that policymakers target will also be considered.

Finally, social factors affecting health that border on nutrition/consumption, such as alcohol, smoking and drug use, will also be discussed. These are issues that pervade society generally, including the world of sport, where amongst other things, celebrity status can occasionally bring excess and temptation.

Health

As individuals we are affected by ill health randomly, though it can also be systematic as some disease states predispose to other illnesses (for example, diabetes predisposing to peripheral circulatory diseases). At the population level, social and economic circumstances and lifestyles are closely associated with health, both positively and negatively. Assessing the health needs of a population is not, however, straightforward. Health is not easily measured, except in broad terms, and it cannot be measured with precision. A wide range of sources of health information are available and by piecing these together we can develop our understanding of the health of a nation.

At the population level, we can obtain information from a large group of people by way of a survey, and depending on the sample size and its ability to represent a bigger population (e.g. that of a nation), we can attempt to construct policies and practice to influence the outcome of 'health'. Particular surveys that have been carried out in the U.K. in the past and are worthy of mention in the context of the current chapter are the Health and Lifestyle Survey (HALS), the National Fitness Survey and the Health Survey for England. The HALS was a large (9003 participants) national survey of the population of England, Wales and Scotland carried out in 1984/5 and 1991 and in it, people were asked in detail about their health and lifestyles, certain aspects of their 'fitness' were measured and participants were invited to express their opinions and attitude towards health and health-related behaviour (Blaxter, 1990). The National Fitness Survey (1992), funded by Allied Dunbar, the Health Education Authority and the Sports Council, was an attempt to describe the different patterns of physical activity and levels of activity prevalent among the adult population of England. The Health Survey for England (carried out annually from 1991 to 1997), was used in 1995 to monitor some of the targets of a Health of the Nation strategy. Two documents – *The New National Health Service:*

Modern, Dependable (Department of Health, 1997) and *Saving Lives: Our Healthier Nation* (Department of Health, 1999) set out a package of measures intended to improve the quality of services to patients and to reduce unacceptable variations in health. These measures included the introduction of National Service Frameworks (NSFs), key new tools for tackling major health issues and important diseases (Department of Health, 2001).

National Service Frameworks in the areas of Coronary heart disease, Diabetes, Cancer, and Older people, for example, were originally developed in an effort to improve health services by setting national standards to improve service quality and tackle variations in care. In addition to the aforementioned, there have also been developed NSFs or strategies for Children, Chronic obstructive pulmonary disease, Long-term conditions, Mental health, and Stroke. Physical activity has a potential role to play in the areas of health for which NSFs have been developed, and therefore this is where graduates of the exercise sciences have a role in health promotion.

What exactly does 'health' mean to people? Take a look at the ten tips from Liam Donaldson, then the Chief Medical Officer (CMO) of the UK, in **Box 6.1**. They might help to introduce you to the breadth of the topic. In his most recent annual report (Department of Health, 2010), the outgoing CMO addressed five new public 'health' topics – Physical activity, Cold weather, Rare diseases, Grandparents, and Climate change, just to add to the diversity, broad interpretation and perceptions of health.

Box 6.1

What does health mean? (Department of Health, 1999)

Ten tips – your guide to better health

1. Don't smoke. If you can, stop. If you can't, cut down
2. Follow a balanced diet with plenty of fruit and vegetables
3. Keep physically active
4. Manage stress by, for example, talking things through and making time to relax
5. If you drink alcohol, do so in moderation
6. Cover up in the sun, and protect children from sunburn
7. Practise safer sex
8. Take up cancer screening opportunities
9. Be safe on the roads
10. Learn the First Aid ABC – airways, breathing and circulation

There is no simple or obvious way in which health can be defined; some people prefer a biomedical or scientific model, others a more holistic model. Concepts for describing the health of oneself might be different from those used for describing the health of others. Without getting in too deep with schools of thought on defining health, discussions of which can be found in texts more specific to health than this, interesting anecdotes of how an individual might view health are described by Blaxter (1990): a Scottish woman said of her husband that 'he had a lung taken out, but he was aye healthy enough'. In another study, on the death of an elderly parent, the woman's daughter said that her mother 'had been an active woman before this with no previous restriction apart from general old age, deafness, loss of sight in one eye and loss of memory'. This demonstrates how health and 'fitness' have different dimensions in the minds of the populace.

Measurement of health

Blaxter (1990) identified four dimensions of health:

1. Unfitness/fitness – whereby physiological measurements are used
2. Disease presence/absence – based on reported medical conditions
3. Experience of illness or freedom from illness – based on reports of symptoms suffered
4. Psycho-social malaise or well-being – based on reports of psychological symptoms.

These are aspects of health that can be experienced independently of the others and are often used in health surveys.

Unfitness/fitness can be measured by collecting relatively easily measured and non-invasive physiological data such as body mass to height ratio (i.e. body mass index; BMI), or waist to hip ratio, blood pressure and lung function variables. Such measures represent health objectively and have a general significance for health and longevity. The terms 'unfitness' and 'fitness' are less appropriate, particularly in relation to sport and exercise, and a better term might be **physiological health**.

The **presence or absence of disease** might be considered the medical dimension and can be ascertained by asking questions related to whether a respondent has any longstanding illness or disability. Answers may be declared as specific diseases and conditions, such as type 1 diabetes, and be recognized and under treatment, whilst others such as back pain, migraine or arthritis might be untreated and possibly self-diagnosed.

The **experience of illness or freedom from illness** is based on short-term reporting of common symptoms such as headaches, colds, menstrual pain, etc. These can be added together and a score allocated.

Psycho-social health is based on the frequency of symptoms such as disturbed sleep, feelings of stress, for example, to assess depression. Again, reported symptoms can be added together and a score allocated. An example of this is the Patient Health Questionnaire (PHQ-9), which scores each of the nine questions (e.g. 'Little interest or pleasure in doing things') as 0 – not at all, through to 3 – nearly every day. A maximum score of 9x3 = 27 can be obtained. A total score of 5–9 is given a depression severity score of mild, whereas 20–27 is considered severe.

Patient-reported outcome measures (PROMs) are increasingly being used to measure health status or health related quality of life at a single point in time, and have been introduced particularly before and after NHS surgical treatment interventions, such as hip and knee replacements, varicose vein and hernia treatments.

One way of combining the dimensions of health is to generate an overall health index. This can be constructed empirically by an examination of possible combinations of the health scores described above. In the Health and Lifestyle Survey eight summary health categories were created, as listed in **Box 6.2**.

> ## Box 6.2
> **An example of basic descriptors of health ranked from good to poor**
> - Excellent health
> - Good health
> - Good health but 'unfit'
> - Good physical health but poor psycho-social health
> - High rate of illness without disease
> - 'Silent' disease
> - Non-limiting disease and ill
> - Limiting disease and ill

Factors affecting health and health promotion strategies

Whilst health is closely associated with social circumstances, the factors affecting health that are most relevant to the student of sport and exercise science are those that can be controlled by the apparently healthy person or patient through behaviour. Behaviour or voluntary lifestyles (as represented by physical activity, diet, alcohol, smoking and drugs) affect health, and health also affects behaviour. We can do nothing about some factors that affect health ('non-modifiable' factors) such as genotype, ageing and ethnic origin. Current population health trends for the UK were published and discussed in a report called 'Securing Good Health for the Whole Population' (Wanless, 2003).

Broadly speaking, there are two complementary approaches to improving health (Department of Health, 2000, Chapter 1):

- in the **population approach** the aim is to lower the average level of risk factors in the population and
- in the **high risk approach** the aim is to identify those people at high risk and offer appropriate advice and treatment.

These two approaches are not mutually exclusive and both are important. The 'high risk approach' will increasingly lend itself to our new-found knowledge of the human genome and to methods for screening of individuals susceptible to disease. This should not, however, lead us into a sense that as individuals, the development of disease X or condition Y is inevitable. The genetic component of disease is a non-modifiable factor, however, the modifiable factors are, to use a sporting analogy, balls that are in our own court. The extent to which we look after ourselves can be affected by health promotion and education strategies.

A number of models can be identified. A **behavioural change model** is aimed at the individual in an attempt to change their behaviour (e.g. to eat more healthily) but does not take into consideration socio-economic factors such as availability and affordability of particular foods. An **educational change model** provides the individual with knowledge about health with the aim of making an informed choice, providing sound scientific argument and assuming that as individuals we take personal responsibility. This does not take into account, however, intellectual, peer pressure and socio-economic factors. A **political model** is aimed at taking some degree of control away from the individual, by applying legislation or taxation or policy to change behaviour, for example raising taxation on cigarettes and tobacco, banning smoking in public places (negative) or providing milk for under 5s at nursery, school meals (positive). This can have an influential effect but is authoritarian, causing resentment and greater resolve in some individuals not to adopt a healthier behaviour if the promotion is seen to take away individual choice. A **medical model** targets the positive benefit to health based on the absence of disease, for example taking folic acid during pregnancy to reduce the risk of foetal spinal cord defects, or also on the dangers of a lack of change in behaviour, for example

not practising safe sex, increasing the risk of HIV transmission, emphasizing the negative message. The model uses scientific information to support the message but some individuals might not have the choice or intellect to adopt healthier practice.

Behavioural change models of the mid-1980s in the UK included the 'Look After Your Heart' programme, one of the major objectives of which was to increase public participation in physical activity. Campaigns promoted by the then quasi non-governmental organization, the Sports Council, carrying straplines such as 'What's Your Sport?' and 'Ever Thought of Sport?' were an attempt to encourage people to exercise, and to do so more often, a goal that is still evident as we enter the twenty-first century. With hindsight, we can question how effective such strategies have been, as in the past 20 years the British population has become 'fatter and unfitter'.

Physical inactivity, health and disease

It is not possible to keep politics out of sport. A report in the *Guardian* newspaper (Chavdhary, 2003) after the England Rugby World Cup victory in 2003 announced that 'Politicians are to examine how best the country can cash in on the success of the England rugby union team and get more of us to take part in sport after figures were released showing that, despite sport receiving more publicity and funding than ever before, Britain is still essentially a nation of couch potatoes'. A government source was also quoted as saying 'The figures for participation in sport are pathetic and quite embarrassing, given how much money has been put into it . . . we need to do something about this because the figures also show that the population is getting fatter and unfitter' (Chaudhary, 2003). The 2012 Olympic Games is a further example of the notion of political gain from a sporting event, with a 'legacy' expected – 'inspiring a new generation of young people to take part in volunteering, cultural and physical activity'. This was one of the promises presented to the International Olympic Committee by the Olympics Minister in 2007. The U.K. Government had a target of 2 million more adults active by 2012.

Much of what we can do to be physically active and to increase our energy expenditure can be through habitual daily activities and physical activity such as walking and cycling. As pointed out in the previous chapter, there is little evidence to suggest we are eating more, but there is some evidence that we are less physically active due to the increasing use of cars, other labour-saving devices and an increase in television viewing (Figure 6.1). This may

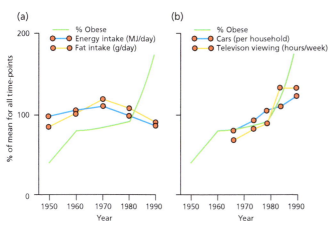

△ **Fig. 6.1** Trends in (a) diet and (b) alternatives to physical and habitual activity in relation to obesity in Britain. Reproduced with permission from Prentice and Jebb (1995).

not, however, be an entirely true picture because of the problem of underreporting of food intake.

Although it is common knowledge that habitual physical activity such as cycling and walking is beneficial to health, this knowledge is not always acted upon. It is particularly ironic that in Shanghai, where cycling has been a popular mode of transport for decades, the city authorities proposed to ban bicycles from major roads to ease congestion caused by a surge in private car ownership (Figure 6.2), putting itself in danger of repeating the mistakes made in the West. In 2005 there was one car for every 70 people in China, whereas there was one car for every 2 people in the U.S.A. China is the second largest consumer of oil, with the U.S.A. the largest. With a huge population of about 13.4 million, Shanghai and other rapidly growing Chinese cities could be giving themselves a potential time bomb for the diseases of affluence, which could exacerbate health problems for the twenty-first century and beyond.

△ **Fig. 6.2** Habitual physical activity: Come rain or shine Shanghai residents cycle around – banned from major roads to allow for a surge in private cars.

One of the most common diseases contributed to by a lack of physical exercise is **cardiovascular disease**. Responsible for substantial morbidity and mortality – amounting to some 17 million deaths a year worldwide – the term cardiovascular diseases covers a range of conditions that affect the functioning of the heart and blood system and thus restrict the supply of oxygen to vital organs such as the brain and the heart itself. The most common cardiovascular diseases are myocardial infarction (MI or heart attack), heart failure, congenital heart disease, cardiomyopathies, cerebrovascular disease (e.g. stroke), transient ischaemic attacks and peripheral vascular disease.

Many of the deaths due to cardiovascular causes can be attributed to tobacco smoking, which increases the risk of dying from cardiovascular disease or coronary heart disease by 2 to 3 times. Other major risk factors are physical inactivity and inappropriate diets. The risk of cardiovascular disease increases with age and, despite the widespread misconception that it is mainly a disease of middle-aged men, affects as many women as men (WHO/CDC, 2004).

Coronary heart disease (CHD), also known as ischaemic heart disease, is caused by the narrowing or obstruction of coronary arteries by atheromatous (fatty) deposits and thrombosis, resulting in the impairment of the blood supply to the myocardium. Such pathologies alone are a leading cause of death, killing over 110 000 people in England in 1998, including more than 41 000 under the age of 75. Although death rates for CHD have been falling for the past 20 years, UK death rates remain high in international terms and have fallen less than in other comparable countries. Physical inactivity is one of the major risk factors for CHD (others being smoking and overweight/obesity, alongside the co-morbidities of diabetes, hypertension).

The epidemiological evidence for an association between higher levels of habitual physical activity and a lower incidence of CHD includes the early work of Morris and co-workers, who compared London bus drivers with bus conductors (Morris *et al.*, 1966). Jerry Morris died in 2009, aged 99, and was described as a titan of twentieth century public health by the then Chief Medical Officer. His work on preventing heart disease, on physical activity and health and on health inequalities made a significant impact on population health.

One should also note the work of Paffenbarger *et al.* (1978) on dockworkers (longshoremen) and Harvard alumni, and a host of subsequent studies conducted by other workers up to the present day. Attempts to encourage populations into being more physically active have included such public campaigns as the Canada Fitness

> **TIME OUT**
>
> SOME DEFINITIONS
>
> - Epidemiology is the study of the distribution and determinants of disease, injury, death or disability. It sets out to define and determine contributory factors and explain geographical patterns. It makes observations on existing disease patterns and their statistical association with, for example, lifestyle factors. The information obtained is used to plan and evaluate strategies to prevent ill-health.
> - In a case–control study, subjects or patients ('cases') with a particular disease are compared with respect to their activity or dietary pattern or environmental exposure with 'control' subjects or patients. Controls do not have the disease and 'matching' of as many characteristics as possible such as gender and age help to minimize confounding factors between the case and control groups.
> - In a prospective study, characteristics of lifestyle or baseline measures are recorded at a chosen time, in a group of people (a cohort) that do not have any disease. Some time later (months, years) clinical episodes such as heart attacks or strokes that have occurred are recorded, and the strength of the statistical association between the baseline measures and the clinical episode being investigated are examined.
> - Intervention or experimental studies introduce a treatment on a group of subjects for a period of time (days, weeks). A control group who do not receive the treatment are monitored and, after time, the outcome measures of interest are examined in both the treatment and control groups. In well-designed studies the treatment and control groups and the investigators do not know (are 'blind' to) which subjects are receiving the treatment, and the treatment and control groups might also be matched.

Survey, subsequently rolled out as a home fitness test, and Heartbeat Wales, building on the earlier epidemiological and survey data. Despite the evidence that has built up over the last 50 years of the positive effect of physical activity on health, levels of physical activity have not grown in tandem. A lot of Britons are in denial about their health, with claims of good 'fitness' levels and low drinking (alcohol) levels, despite evidence of low activity and high levels of drinking. The Health Survey for England (2008) found over 60 per cent of the adult population failed to meet the minimum recommendation of 30 minutes of physical activity five times a week.

Much of the emphasis on the protective effect of physical activity on health has focused on prevention of coronary heart disease. In recent years, a health problem for women has been highlighted, particularly in the popular media, that of breast cancer. In comparison, women are around four times more likely to die from heart disease than breast cancer. The risk of CHD among sedentary people is nearly two-fold. Sixty per cent of men and 70 per cent of women can be classed as sedentary, making the population-attributable risk from physical inactivity very high (Department of Health, 2000, Appendix C). A commonly reported measure of health outcomes is potential years of life lost (PYLL), which calculates the total number of years by which people dying are failing to reach a specified age – often 70 or 75 years (Wanless, 2003). The main contributors to PYLL to age 75 are CHD and cancer, based on the main cause of death for each individual (Table 6.1).

Contributing factor	Percentage of total life years lost
CHD	18
Cancer	17
Injury/poisoning	9
Suicide and undetermined death	6
Stroke	6
Respiratory disease	6
Liver disease (alcohol related)	2
Road traffic accidents	2
Diabetes	1
Others	33
Total	100

From Wanless (2003).

△ **Table 6.1** Causes of years of life lost up to age 75 in England (1999).

The health benefits of physical activity are many and varied, all with sound physiological and psychological bases. These include the benefits for improved cardiovascular, skeletal muscle, connective tissue, skeletal, joint, metabolic and psychological functions (Table 6.2).

Regular physical activity of an appropriate form (i.e. a combination of type, intensity, duration and frequency) is associated with beneficial blood lipid profiles, for example, a higher ratio of high-density lipoprotein (HDL)

Function enhanced by physical activity	Example of preventative benefit
Cardiovascular	Blood pressure reduction, delayed ageing processes
Skeletal muscle	Maintenance of lean muscle mass
Tendon/Connective tissue	Reduced injury risk, increased muscle function
Skeletal	Maintenance of bone mineral density
Joint	Improved stability
Metabolic	Body mass regulation, muscle substrate turnover
Psychological	Improved mood, memory, self-esteem

△ **Table 6.2** Health benefits of regular physical activity.

to low-density lipoprotein (LDL), and a lower total cholesterol level. These are concomitant with a reduced deposition of fat in coronary arteries, reduced atherosclerosis, and therefore reduced blood pressure and risk of CHD. Benefits also include the management of disease risk factors such as hypertension, diabetes and the related factor of overweight/obesity. Physical activity is a strategy for the prevention of loss of bone mineral density (alongside other modifiable factors, such as dietary factors), thereby reducing osteoporosis and subsequent fracture risk, and for the health and maintenance of tendon, connective tissue and joint function, thereby improving mobility, flexibility and limb range of motion, and delaying osteoarthritis. Physical activity increases the insulin sensitivity of tissues and therefore glucose disposal, potentially improving blood glucose control. Lastly, but not least, physical activity is of benefit to psychological and mental health, including behavioural states and cognitive performance.

All of these benefits of physical activity far outweigh any risk that may be associated with exercise. Sudden death in sport and exercise, whilst tragic and often of high profile, is very rare. There is the risk of acute (e.g. traumatic) and

chronic injury associated with some sport and exercise activities, and this serves to emphasize the importance of improving exercise science and sports medicine support and service provision in the big picture of physical activity promotion. An important factor in keeping people active is not to let momentum and enthusiasm for exercise be diminished by injury. One would hope that, in the future, this is considered by policymakers as an integral part of strategies to promote physical activity, and such strategies should include the professional recognition of sport and exercise medicine as a distinct discipline in medicine.

Whilst a lot of attention is given to the beneficial effects of physical activity on cardiovascular health, it should not be forgotten that physical activity also protects against **cancer**. The current general consensus is that physical activity protects against colon and breast cancer, and possibly other cancers such as those of the prostate, lung and endometrium. In the USA 133 500 new cases of colon and rectal cancer were diagnosed in 1996, with colon cancer being the third leading cause of cancer incidence and mortality (Colditz et al., 1997). No association has been found between physical activity and cancer of the rectum (Batty and Thune, 2000). The physiological basis for the protection against gastrointestinal tract cancers is thought to be the favourable effect of physical activity on insulin, bile acid and prostaglandin levels. Physical activity may also modulate the production and metabolism of sex hormones and thus reduce the risk of development of cancers in the breast, endometrium and prostate. The improvement in pulmonary ventilation and perfusion provided by physical activity may favourably modify the exposure and interaction of carcinogenic agents to the lung. It is not thought that physical activity increases the risk of any cancer.

Figure 6.3 shows the mortality from cancer in various countries. This gives us a perspective on cancer mortality in the light of the possible protective effects of physical activity.

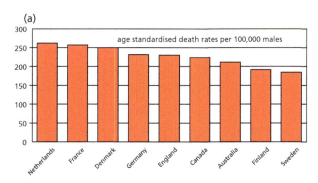

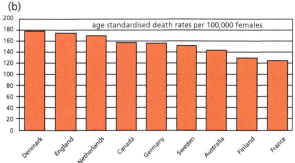

△ **Fig. 6.3** Mortality from cancer (a) males (b) females, by country. Reproduced with permission from Wanless (2003).

Measuring population physical activity

Attempts have been made to assess and monitor change in health-related physical activity status in populations. For example, the National Fitness Survey of 1990 in England (Sports Council and Health Education Authority, 1992) involved two sets of measurements, one using an interviewer-administered questionnaire ('home interview') and one using a mobile laboratory to carry out body composition, physiological health and exercise capacity measures. A probability sample of 6000 addresses across 30 parliamentary constituencies eventually resulted in interviews (average duration 1 hour 15 minutes) being obtained from a resident in 75 per cent (4316) of effective addresses. From those interviewed, 70 per cent took part in the mobile laboratory physical appraisal, this being around 2700 participants. The participant numbers are being emphasized here to illustrate the typical attrition (drop-out) rates associated with such work, and to allow the reader to judge the extent to which this data might be representative of a 'nation', despite the considerable effort and dedication of the research team over several years to set up, pilot and carry out the work, and then analyse the data.

There were nearly equal numbers of men and women involved in the questionnaire and laboratory appraisals, ranging in age from 16 to 74 years. One important component of health-related 'fitness' and the risk of CHD is the aerobic capacity of an individual. The test used in the National Fitness Survey involved treadmill walking, as walking is the most common daily activity for the majority of people. This was thought to have advantages over cycle ergometry (used in an Australian Fitness Survey) and stepping (used in the Canadian Fitness Survey). Despite the common nature of walking, unless we are used to walking on a treadmill, there is a degree of familiarization required, and ironically, this took up a large amount of the total time needed to conduct the physical appraisal of the subjects.

Some of the conclusions from the National Fitness Survey were that activity levels and, consequently, the aerobic capacity of most of those surveyed were too low (i.e. to protect against the diseases of affluence) and that body mass was too high and the prevalence of overweight and obesity was rising. Twenty years since this survey, things have only become worse. One barrier to overcome is that of self-perception – the sample interviewed felt that the population in general did not do enough exercise to keep 'fit', but when asked about themselves, the majority thought that they did engage in sufficient physical activity.

Such fitness surveys aim to establish baseline measures to devise health policy, and in the future to assess change against such baseline data to refine health policy, however, the financial and/or political will required to carry out subsequent identical surveys appears to have been lacking or deemed not cost-effective. Inactivity remains a major public health threat, for example, affecting more people in England than the combined total of those who smoke, misuse alcohol or are obese (Table 6.3).

Nutrition, health and disease

The fact that nutrition and physical activity are complementary in promoting health and well-being is increasingly being recognized by policymakers and incorporated under the label 'public health nutrition'. Good nutrition and physical activity working together can remodel physiology and metabolism and improve health. Life scientists, including physiologists and social and behavioural psychologists, have a major role to play in investigating complementary and interactive relationships between nutrition and physical activity.

The National Diet and Nutrition Survey (NDNS; Great Britain) looks at a relatively small sample of 2000 individuals and measures weighed food intake, anthropometric and physiological data (e.g. height, weight, blood pressure) and biochemical data. Each survey covered a different age group of the population and the cycle was repeated every 12 years. These surveys provide data on trends in diet and nutrition, as well as allowing comparison with recommended intakes, and with biochemical

	Alcohol misuse	Smoking	Obesity	Inactivity
Percentage of adult population affected in England	6–9%	20%	24%	61–71%
Estimated cost to the English economy every year	£20 billion	£5.2 billion	£15.8 billion	£8.3 billion
Estimated cost to the NHS per year	£2.7 billion	£2.7 billion	£4.2 billion	£1–1.8 billion

Source: 2009 Annual Report of the Chief Medical Officer (Department of Health (2010)) – adapted from a number of reports of official statistics and other published work.

△ **Table 6.3** The public health threat of physical inactivity in England.

or other health-related reference ranges (Donaldson and Donaldson, 2003). The NDNS has evolved into a rolling programme, the latest results of which (fieldwork for which was carried out in 2008/2009) are publicly available via the Food Standards Agency website.

Through the diet we can reduce the risk of ill-health from cardiovascular disease (CVD), diabetes and cancer, particularly by preventing overweight and obesity. An unhealthy diet along with physical inactivity, for example, is thought to be responsible for about a third of premature deaths due to cardiovascular disease in Europe. The case for physical activity in cardiovascular disease prevention has been outlined earlier; the other major risk factors for cardiovascular disease include excessive alcohol intake, smoking (see following sections) and poor diet. The negative dietary influences on cardiovascular disease are primarily the fat content as a proportion of the energy intake of the diet, particularly that obtained from saturated fat (potentially leading to high serum cholesterol levels; Figure 6.4), the low intake of fruit and vegetables, and a high salt intake (particularly in individuals with a genetic predisposition to hypertension). The physiological mechanisms for these factors include the accumulation of atherosclerotic plaques in coronary arteries and the low level of antioxidant protection against oxidative stress.

The objective of healthy eating is to promote the consumption of a balanced diet by everyone in the community, and one of the major goals of the National Health Service in the UK is to encourage healthier eating in order to reduce heart disease in the population. A balanced diet is one which emphasizes more fruit and vegetables, fish (especially oily fish), starchy foods such as bread, rice, pasta and potatoes, and includes a smaller proportion of foods containing fat and saturated fat, and less salt (COMA, 1994). These principles are associated with a reduced risk of both cardiovascular disease and cancer.

Saturated fat intake has a more positive association with cardiovascular disease than total fat intake. This is clearly illustrated by a comparison between Crete and Finland (Gibney and Gibney, 2004). The incidence of coronary events is much lower (age-standardized rates) in Crete than in Finland despite similar total fat intakes (circa 40 per cent) however, the intake of saturated fat in Crete is around 8 per cent of total energy intake, whereas it is 22 per cent in Finland.

The food energy from total fat and saturated fatty acids in Great Britain, 1974–2000, is shown in Figure 6.5.

Diabetes mellitus

A higher risk of cardiovascular disease is one of the consequences of type 2 diabetes. Significant inequalities exist in the risk of developing type 2 diabetes mellitus: those who are overweight or obese, physically inactive or have a family history of diabetes are at increased risk of developing diabetes. People of South Asian, African, African-Caribbean or Middle Eastern descent have a higher than average risk of type 2 diabetes, as do less affluent people (Department of Health, 2001). Diabetes mellitus is a condition in which the ability of the pancreatic beta cells to produce insulin is lost (type 1 diabetes) or the amount of insulin that is produced or the sensitivity of the tissues to insulin is insufficient (type 2 diabetes). This results in a disturbance of glucose homeostasis – a high concentration of circulating blood glucose if untreated, resulting in urinary excretion of glucose.

The goal of diabetes management is to maintain normal blood glucose concentrations through lifestyle or therapeutic interventions. There is evidence that tight control of blood glucose (and blood pressure) increases life expectancy and improves the quality of life for people with both type 1 and type 2 diabetes (Department of Health, 2001). Regular physical activity can lower the risk of

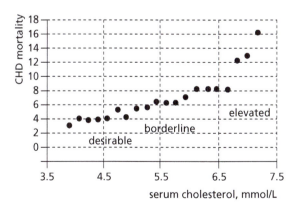

△ **Fig. 6.4** Serum cholesterol and CHD mortality. Reproduced from *The Journal of the American Medical Association*.

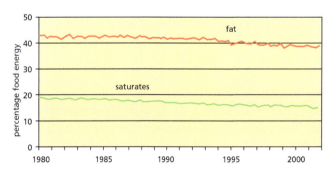

△ **Fig. 6.5** Food energy from total fat and saturated fatty acids, Great Britain, 1974–2000. Reproduced with permission from Wanless (2003).

developing diabetes and has a key role in its management through the maintenance of an ideal body mass and also by increasing insulin sensitivity. Physical inactivity, which is higher amongst 'at risk' groups such as the elderly and black and ethnic minorities, compounds the problem.

At present there is a worldwide diabetes epidemic. For example, estimates are of 346 million diabetics world-wide, in the USA around 16 million type 2 diabetics and about a million type 1 diabetics, and in China there may be 90 million diabetics. Many diabetics remain undiagnosed for some time, for example in Scotland it was estimated in 2004 that 210 000 people had diabetes, but 87 000 of them did not know it. In England in 2010 there were around 2.3 million diabetics. Most diabetics are non-insulin-dependent (NIDDM/type 2 diabetics), whilst about 15 per cent are insulin-dependent (IDDM/type 1), requiring doses of insulin each day. Whilst type 1 diabetes commonly occurs in childhood, type 2 diabetes was once known as 'late-onset' diabetes but it is becoming increasingly common in children and adolescents as a result of the rise in overweight and obesity.

People with diabetes are at considerable risk of excessive morbidity and mortality from coronary heart, cerebrovascular and peripheral vascular disease, leading to myocardial infarction, strokes and amputations. There can, however, be considerable gain from reducing risk factors. The effect of reducing blood pressure, as indicated by the UK Prospective Diabetes Study (a 20-year trial of over 5000 type 2 diabetes patients in clinical centres in England, Northern Ireland and Scotland) demonstrated that 'tight' blood pressure control (mean 144/82 mmHg) compared with 'less tight' control (mean 154/87 mmHg) reduced heart failure by 56 per cent, strokes by 44 per cent and combined myocardial infarction, sudden death, stroke and peripheral vascular disease by 34 per cent (Watkins, 2003).

Excessive adipose tissue mass reduces insulin sensitivity and is therefore a risk factor for type 2 diabetes. The 1995 Health Survey for England indicated an increase in levels of obesity in both men and women from the 1991 survey. *The Health of the Nation* (a Green Paper published in 1991) target was to reduce the levels of obesity in women from 12 per cent to 8 per cent by 2005, however, obesity in women had risen to 16.5 per cent in 1995 (compared with 15 per cent in 1991). At the start of the twenty-first century in England, it is estimated that one in five adults are obese and 40 per cent are overweight. Obesity is rarely a result of a metabolic disorder, though there are, for example, a small number of people with a higher susceptibility to the development of obesity, including children with the Prader–Willi syndrome.

Obesity increases the risk for other chronic diseases such as cardiovascular disease, including coronary heart disease, hypertension and stroke. Risk of a variety of other diseases is increased by obesity, including gallbladder disease, some types of cancer (i.e. breast, ovarian, cervical cancer in postmenopausal women, colorectal), respiratory illnesses (e.g. sleep apnoea, Pickwickian syndrome) and reproductive problems (Bandini and Flynn, 2003). Whilst addressing the problem of diet and health is a huge struggle, the evidence is there for all to see. Exercise scientists have a major role to play in the tackling of physical inactivity and nutrition problems – key determinants for the prevalence of obesity, which continues to rise in Europe among children and adults.

Cancer

The focus of this section on nutrition health and disease has been the causality of cardiovascular and related diseases, however, diet, together with smoking and physical inactivity, is also one of the main determinants of cancer. It is estimated that physical inactivity and an unhealthy diet could be responsible for up to one third of cancers. Some of the dietary factors thought to have effects on cancer incidence are discussed below.

Dietary fibre is considered to be a protective component of the diet. The effects of dietary fibre on the colon and rectum that have the potential to reduce cancer risk include the dilution of carcinogens, the provision of a surface for adsorption of carcinogens, a faster transit time, altered bile salt metabolism and favourable consequences of fermentation, such as lower pH and an altered microbial metabolism. Fibre increases faecal output and decreases the amount of time taken for materials to pass, resulting in stools being softer and more voluminous with a fibre-rich diet. Constipation is reduced, as is the risk of colon cancer. In northern-European countries more of the total fibre in the diet comes from bread compared with southern European countries where fruit and vegetables are the predominant sources.

Oxygen is an essential component of human life, but ironically, chemically reactive forms of oxygen that arise out of normal metabolism (collectively known as 'reactive oxygen species') have the potential to be detrimental to the body. Most reactive oxygen species are 'free radicals' – these are chemical species capable of independent existence that contain one or more unpaired electrons. Most free radicals are unstable and highly reactive. The goal of normal healthy metabolism is to achieve a balance between pro-oxidant factors and antioxidant defence, such that the body does not experience undue oxidative stress. Some examples of pro- and antioxidants are listed in Table 6.4.

Potential pro-oxidants/ Pro-oxidant contributors	Potential antioxidants
Exercise	Endogenous tissue antioxidants
Tobacco smoke and air pollution	Vitamins (e.g. C and E)
Carcinogens	Carotenoids
Inflammation	Glutathione

△ **Table 6.4** Some potential pro- and antioxidants.

Food sources of **antioxidants** that are present in a normal, healthy and balanced diet exert a protective effect against several diseases. Many of these antioxidants are of plant origin, possibly because of their own role in protecting themselves against radiation.

Epidemiological studies show that fruit and vegetable intake is associated with a lower risk of cardiovascular disease and cancers, and that in the case of carcinogenesis, the potential mechanisms for their protective effect occur in the earlier rather than later stages. This emphasizes the need for good dietary habits (as well as good physical activity habits) to start at an early age. An oft-quoted target for fruit and vegetable intake is at least five portions a day, more than the current UK average of about three portions. The 'five a day' recommendation is, however, only based on a simplistic survey study carried out by the World Health Organization in 1990. The average fruit and vegetable intake in countries where consumption is considered to be a factor in the low rate of heart disease and cancer (e.g. Greece, Spain) was thus taken as an 'arbitrary ideal'. The 'at least' might now be borne in mind, considering that the average Greek is thought to eat around 10 portions a day.

Table 6.5 shows what is meant by a 'portion' and Figure 6.6 shows the average weekly consumption of fruit and vegetables in Great Britain from 1974 to 2000.

Fruit	Vegetable
One apple	Seven cherry tomatoes
A handful of grapes	Three celery sticks
Half a grapefruit	Fourteen button mushrooms
Two clementines	Eight spring onions
Two kiwi fruits	Small tin of baked beans
A large slice of pineapple	Chick peas (in a pulse-based recipe)
Three apricots	Lentils in a meal (in a pulse-based recipe)
One glass of fruit juice	

△ **Table 6.5** What is a portion?

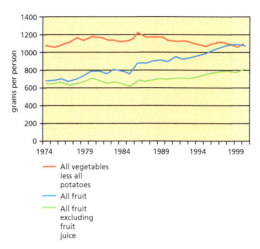

△ **Fig. 6.6** Average weekly consumption of fruit and vegetables, Great Britain, 1974–2000. Reproduced with permission from Wanless (2003).

Alcohol, health and disease

Alcohol has a longstanding place in human culture, and perhaps a shorter one in sporting culture. Increasingly, however, those who take their sport, exercise and health seriously are waking up to the detrimental effects that excessive alcohol consumption, particularly binge drinking, can have on exercise performance. For example, too much alcohol can lead to poor match performance, reduced effectiveness of post-exercise rehydration and replenishment of glycogen stores. It also has an effect on energy balance (the energy content of alcohol in the diet was discussed in the previous chapter) and on longer term health.

Alcohol consumption has increased in the UK, particularly in the form of binge drinking by young adults (Figure 6.7)

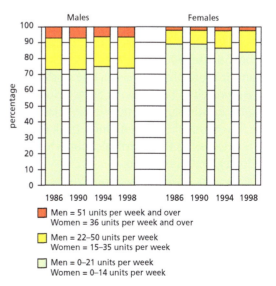

△ **Fig. 6.7** Alcohol consumption, people aged 18 and over, Great Britain, 1986–1998. Reproduced from Wanless (2003).

in recent years, and it has been reported that around 25 per cent of all accident and emergency cases in the UK are related to alcohol consumption. This is one measure of the acute effects, but we are now seeing the chronic effects of behaviours in alcohol consumption, and the health costs (personal and economic) of these, for example, liver damage. In 2008 there were nearly 5,000 deaths due to alcoholic liver disease in England and Wales (75 per cent of them men), and death rates linked to alcoholic liver disease have risen by 69 per cent from around 1980–2010.

In a social model of health a socially motivated moderate drinking message/definition is provided by Haggard and Jellinek (1942): the moderate drinker is someone who 'does not seek intoxication and does not expose himself to it. He uses alcoholic beverages as a condiment and for their milder sedative effects. Alcohol constitutes neither a necessity nor a considerable item in his budget'. Alcohol, formed during the fermentation of carbohydrate-containing substrates by yeasts, has been consumed for centuries, and in its purest form, ethanol, has a variety of other commercial and industrial uses. As a constituent of the diet, alcohol and alcohol-containing beverages are a source of energy and might be considered a nutrient, but with little added health benefit. The health benefit gained is mainly in the form of a lower risk of CHD as a result of very modest consumption but this is soon lost as the quantity of alcohol consumed increases.

Alcohol has powerful effects on the central nervous system that are not consistent in nature between individuals, nor within the same individual in different circumstances (for example, emotional, prandial). Consumption patterns range from complete abstinence, sometimes for life, sometimes for months during a training period, to moderate and/or occasional consumption, to occasional but heavy (binge) consumption – arguably harmful misuse – to alcohol abuse and dependency. In a UK government report on alcohol misuse (2003), in which binge drinking is defined as four or more pints of beer in one session, British men are reported to be the worst binge drinkers in Europe, despite consuming less total alcohol than most European men. It is thought that men binge drink on 40 per cent of the occasions they drink alcohol, whilst women do so 20 per cent of the time.

Alcohol abuse by British women in the 18- to 24 year-old age range was said to have increased by 70 per cent from 1992–2004.

In excess, alcohol has the potential to breakdown physiological and psychological function and lead to 'the extreme dependence on excessive amounts of alcohol associated with a cumulative pattern of deviant behaviours', this being one definition of alcoholism (Gurr, 1996).

Although 'moderate' drinking of alcohol is difficult to define, there is a perception that it is acceptable and causes limited damage to performance and health. Moderate drinking is considered to be perhaps more than 'one drink a day', in other words, 10–40 g alcohol per day. Those who totally abstain have a higher all-cause mortality than those who drink modest amounts of alcohol (e.g. one unit occasionally). Health campaigns in recent years have attempted to educate us about how we might account for, or measure, the amount of alcohol we consume in order to know whether it is 'moderate' or excessive. When measuring alcohol consumption for scientific purposes (e.g. by questionnaire or interview) data tend to produce an underestimation of consumption, whereas anonymous questionnaires can be more reliable.

Measurement is not helped by the amount of alcohol that constitutes a 'unit' varying from country to country; for example one unit is equivalent to 12 g alcohol in the USA, but 8 g in the UK. Modest or light consumption might be regarded as up to one unit per day, moderate consumption might be regarded as up to 3 (UK) units per day. Some drinks and their alcohol contents are shown in **Box 6.3**.

> ## Box 6.3
> ### How many units am I drinking?
> - A pint of beer (alcohol content around 5 per cent) contains about 20 g alcohol (over 2 units)
> - 100 ml of wine (alcohol content around 12 per cent) contains about 10 g alcohol (over 1 unit)
> - A 275-ml bottle of 'alcopop' (alcohol content around 5 per cent) contains about 14 g of alcohol (2 units)

Alcohol is absorbed into the body rapidly, but this is delayed when a meal, particularly a high-fat meal, has been consumed before drinking. Meals will slow absorption and also stimulate gastric metabolism of alcohol by stomach alcohol dehydrogenase. The metabolism of alcohol by the stomach and the liver is briefly discussed in Chapter 3. The effects of alcohol on the central nervous system result from specific interactions between ethanol and major neurotransmitter systems (dopamine, serotonin, gamma-aminobutyrate, glutamate and endogenous opioids) as well as second-messenger systems, the pharmacological effects of ethanol being the sum of its interaction with these systems. Interactions between these systems have been reviewed by Eckhardt *et al.* (1998), but here we will concentrate on health and disease issues related to alcohol consumption.

Alcohol can be metabolised as a source of energy. The energy constituents of some alcoholic drinks in relation to the measures they are usually consumed in is indicated in Table 6.6.

Drink, measure, volume	Alcohol (g)	Carbohydrate (g)	Energy (kJ)
Beer (bitter, 4.5 per cent) 568 ml (1 pint)	20.4	13.1	813
Beer (lager, 4 per cent) 568 ml (1 pint)	17.9	8.5	662
Cider (sweet, 5 per cent) 568 ml (1 pint)	22.4	24.4	1062
Sherry (medium) 70 ml	10.4	2.5	343
Red wine 188 ml (glass/quarter bottle)	18.1	0.6	535
White wine (dry) 188 ml (glass/quarter bottle)	18.1	1.1	544
White wine (sweet) 188 ml (glass/quarter bottle)	18.1	11.1	713
Bacardi/vodka-based (5 per cent) FAB drink 275 ml	13.8	0	400

1 ml of pure alcohol has a mass of 0.79 g.

FAB, flavoured alcoholic beverage or 'Alcopop'.

△ **Table 6.6** Approximate energy constituents of some alcoholic drinks.

Estimates in 1990 suggested that on an average daily basis, alcohol may account for nearly 7 per cent of energy intake in British men, and nearly 3 per cent in women (MAFF, 1995).

The amounts of alcohol drunk by young people under the age of 20 in Britain has risen dramatically in recent years, possibly doubling in the decade to 2002. One likely cause of this was the increased popularity of flavoured alcoholic beverages ('alcopops'). Whilst the number of young people who drink has not risen dramatically, according to the UK Department of Health in a survey of 285 schools in 2002, the average consumption by children aged 11–15 increased from about 5 units in 1990 to about 10 units in 2001. Consumption tends to be concentrated into one or two nights of drinking, that is, binge drinking.

The adverse effects of alcohol on tissues of the digestive tract include the development of cancers. An association was recognized many years ago between the heavy consumption of alcohol and the development of cancer of the oesophagus. Epidemiological methods have improved our knowledge, particularly to exclude confounding factors, and it has been shown that cancers of the mouth, oesophagus and larynx are associated with alcohol consumption in a dose-related but exponential manner (i.e. the incidence of cancer resulting from consumption is small except in heavy drinkers), except in the presence of smoking, which multiplies the risk substantially. There is no evidence that alcohol consumption is associated with cancers of the stomach or pancreas. Alcohol itself is not a carcinogen but may be co-carcinogenic – it may enhance the process without having a capacity to initiate cancer itself.

Excessive consumption of alcohol increases the risk of developing several diseases of the liver, primarily 'fatty liver', hepatitis and cirrhosis. Whether or not cancer of the liver is caused by alcohol is less certain. An early sign of liver damage caused by alcohol is the accumulation of fat in hepatocytes. Triacylglycerols accumulate as a result of the inhibition of oxidation of fatty acids by alcohol. Acetaldehyde, a major product of alcohol metabolism, may also reduce the availability of proteins needed to carry fat. Alcohol may also cause liver damage by the promotion of lipid peroxidation, causing oxidative stress. This might normally be kept under control by antioxidants, but excessive alcohol consumption in combination with a poor diet (e.g. lacking in antioxidants) may reduce the ability of hepatocytes to cope with oxidative stress.

Substantial evidence exists of the impact of dietary factors on blood pressure, and a proportion of hypertension (high blood pressure) cases are believed to be due to alcohol intakes of more than 30 g, either daily, or frequently on individual occasions (i.e. binge drinking). Above the 30 g per day threshold increments of 10 g per day can raise blood pressure by about 1–1.5 per cent on average. A significant proportion of cases of hypertension in developed countries is due to excessive alcohol intake and a direct relationship has been shown to exist between alcohol intake (and sodium intake and body mass) and hypertension (i.e. the higher the intake, the higher the blood pressure). In some individuals these relationships are exacerbated by genetic traits.

Individuals who have a higher than normal (i.e. 120/80 mmHg) blood pressure are recommended to adopt a healthier lifestyle, including a reduction of alcohol intake, particularly if, for example, it constitutes more than 1 per cent of the average daily energy intake. Other complementary changes, if appropriate, would be an increase in fruit and vegetable intake, a decrease in fat and sodium intake, a reduction in body mass and an increase in physical activity. Such measures to adopt healthier lifestyle habits are part of an American set of guidelines known as DASH – Dietary Approaches to Stop Hypertension. It is estimated that in China more then 160 million people have a higher than normal blood pressure.

Despite the adverse effects on blood pressure in some people, an overall reduction in risk of CHD has been associated with sensible drinking, particularly in men over the age of 50 in whom CHD risk is greatest. One possible reason could be an effect on lipoprotein profile (e.g. increasing HDL), therefore slowing the atherogenic process. During atherosclerosis deposits (atherosclerotic plaques) build up on the inside walls of arteries, narrowing the vessel. Amongst the constituents of plaques are lipids originating from LDLs in the blood. Whilst alcohol can raise HDL levels (helping to remove cholesterol for disposal by the liver), high blood pressure caused by alcohol consumption higher than the 30 g threshold encourages plaque formation. Thus, there appears a delicate balance between the small positive and larger negative effects of alcohol.

Secondhand effects of alcohol are increasingly of concern. Such effects include fights, sexual assaults and domestic violence and an attempt has been made recently to quantify such effects (Langley *et al.*, 2003). Amongst university students in this study, a tenth of women and a fifth of men were assaulted at least once in the four weeks preceding the survey. Damage to property is also a problem (Wechsler *et al.*, 2002).

In the UK, the medical and social consequences of alcohol consumption are enormous (**Box 6.4**). In recognizing that there is clear evidence of an increasing burden of harm from alcohol misuse, the UK government developed a national alcohol-harm reduction strategy that has been implemented since 2004.

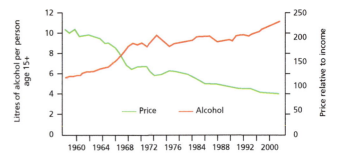

△ **Fig. 6.8** Consumption of alcohol in the UK relative to its price. From Academy of Medical Sciences (2004).

The relationship between cost of alcohol and consumption has changed dramatically since 1960 (Figure 6.8) and the Academy of Medical Sciences (2004) has called on the UK government to take measures to stop the rise in alcohol consumption and, furthermore, reduce levels. It has been highlighted that the 'personal' allowance of alcohol imported from the European Union in 2004 equates to a 2-year supply for a male drinking the maximum 'sensible' level of alcohol, compared with only a 40-day supply of cigarettes to sustain a 20-per-day habit.

Smoking, health and disease

Of the four behavioural habits discussed in this chapter, tobacco smoking is considered to be the one showing the clearest association with ill-health of all kinds (Blaxter, 1990). Smoking includes cigarette, cigar and pipe smoking, and in researching the effects of smoking on health, scientists are concerned with the effects of **active smoking or mainstream smoke** (MSS), that is, smoke inhaled directly into the smoker's upper airway and lower respiratory tract, and/or **environmental tobacco smoke** (ETS), which is passive.

Smoking may well be declining overall in the UK population, but smoking in young people (aged 16–24) and men (aged 25–34) was found to be increasing in the mid-1990s in England (Prescott-Clarke and Primatesta, 1995). In 1996 the proportion of young people starting to smoke was around 13 per cent but since 1998 the figure has remained around 10 per cent. The prevalence of smoking amongst young people under the age of 20 has serious implications for public health, particularly as young smokers are more likely to also drink alcohol or take illicit drugs.

Smoking is increasingly seen as antisocial and unhealthy, as indicated by the increasing number of smoking bans in the workplace and in public places. A ban on smoking in public (indoor) places took effect in Scotland in March 2006, and in England and Wales in 2007. Despite the ban, the UK's largest cigarette company, Imperial Tobacco

Box 6.4

Medical and social consequences of alcohol consumption in the UK (Academy of Medical Sciences, 2004)

- The annual alcohol-related costs of crime and public disorder are estimated at £7.3 billion, workplace costs £6.4 billion and health costs £1.7 billion
- There are 150 000 hospital admissions per year, accounting for up to one-third of all accident and emergency attendances
- About 2.9 million, or 7 per cent of the population, are dependent on alcohol
- In a poll, 7 out of 10 respondents saw drinking in public places as a problem in their locality
- 47 per cent of victims of violence believed that their assailant was under the influence of alcohol
- Between 1993 and 2001 the total number of casualties from road accidents involving alcohol rose by one-fifth
- Between 30 per cent and 60 per cent of child protection cases involve alcohol. Up to 1.3 million children may be adversely affected by family drinking

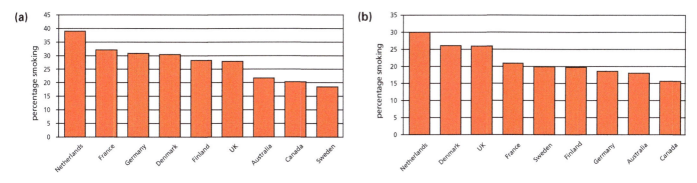

△ **Fig. 6.9** Smoking prevalence (a) males (b) females, by country. Reproduced with permission from Wanless (2003).

appeared to crassly celebrate that their sales in Scotland fully recovered in the period from March to October 2006. Smoking prevalence figures for 2003 in some Western countries can be seen in Figure 6.9. The overall decline in smoking in the UK and some other developed countries may now be coming to an end, as since 2007 smoking prevalence in the UK has remained at about 20 per cent.

King James I (1566–1625) was not complimentary about smoking, saying that it was: 'lothsome to the eye, hatefull to the Nose, harmful to the braine, dangerous to the Lungs, and in the black stinking fume thereof, nearest resembling the horrible Stigian smoke of the pit that is bottomlesse', and there are many today that would agree with him. The effect of smoking is to reduce physical activity capacity and tolerance due to effects on the blood, the cardiovascular system and the lungs. One such effect is carboxyhaemoglobinaemia, which results in a reduced oxygen-carrying capacity without the reduced blood viscosity found in anaemia (Wasserman *et al.*, 1994). When carboxyhaemoglobin levels increase, the arterial oxygen content of blood is reduced. This causes a shift to the left of the oxyhaemoglobin dissociation curve (see section 'Oxygen transport in blood' in Chapter 3), and reduces oxygen consumption. The effect on exercise immediately after smoking is to increase heart rate and blood pressure.

In a well-controlled case study of an overweight subject, comparing exercise under two conditions – smoking in the 5 hours prior compared with no smoking in the 5 hours prior – minute ventilation (see section 'Oxygen transport in blood' in Chapter 3), maximum oxygen consumption (the concept of oxygen consumption is discussed in section 'Energy expenditure and its measurement in Chapter 5) and anaerobic threshold (an exercise physiology concept that you will become familiar with in the future) were all reduced by smoking (Wasserman *et al.*, 1994).

Smoking *per se* tends to affect health markers such as illness, psycho-social health and 'fitness' with little difference between those who are light/occasional smokers compared

with regular smokers. The pathophysiological effects of smoking that become apparent after time are primarily linked with the cardiovascular system and subsequent disease. Specifically, smoking has pathophysiological effects on the heart, blood vessels, coagulation system and on lipoprotein metabolism.

There is strong evidence for the link between smoking and blindness, the most common cause of smoking-related blindness being age-related macular degeneration, which results in severe irreversible loss of central vision. This has been reported in a pooled analysis of data from three cross-sectional studies (Kelly *et al.*, 2004).

Smoking has been identified as the single greatest cause of preventable illness and premature death. In 1995 it was estimated that over 120 000 people in the UK and 400 000 in the USA died because of smoking-related causes. Cardiovascular disease (including hypertension, heart disease and strokes) accounts for the largest number of smoking-related deaths, and smoking increases the risk of dying from cardiovascular disease by 2 to 3 times. Adverse cardiac events fall by around 50 per cent after stopping smoking, and the risk of cardiovascular disease, including acute myocardial infarction (MI), stroke and peripheral vascular disease, also decreases significantly over the first 2 years. Aggressive promotion of smoking by the tobacco industry in newer markets such as Africa and Asia is contributing to a new worldwide health problem, that of tuberculosis. Smokers are about twice as likely to get TB and die from it, and it has recently been estimated that 40 million smokers could die from TB by 2050 (Basu *et al.*, 2011).

In the UK about 25 per cent of women who smoke continue to do so during pregnancy. Those that do so tend to be young, single, of lower educational achievement, and in manual occupations (Coleman, 2004). Smoking has substantial adverse effects on the unborn child, including increased risk of miscarriage, preterm birth and growth retardation. Ideally, women should stop smoking before getting pregnant, however the foetus will still benefit from cessation of smoking at the onset of pregnancy, and even during pregnancy. There are benefits to both the mother and foetus soon after stopping smoking.

Unfortunately, two-thirds of women who stop smoking while pregnant restart afterwards and this is of particular concern to paediatric respiratory health. Many young children are exposed to adults who smoke and have detectable levels of cotinine, a metabolite of nicotine that is commonly used as a marker for smoke exposure. Those people most socially deprived are at an increased risk of developing CHD as other risk factors in these groups are greater. Smoking rates among the higher socio-economic classes in the UK have declined in the last 30 years, but the proportion of smokers from lower socio-economic classes (IV and V) has remained unchanged at around 50 per cent (Sharpe, 1998).

Smoking cessation has substantial immediate and long-term health benefits for smokers of all ages. In the short term, cessation results in a small increase in lung function and reverses the effect of an accelerated rate of decline in lung function with age. The rate and extent of reduction of risk depends on the disease, the risk of CHD decreasing quickly after stopping smoking. Within a year, the excess mortality due to smoking is almost halved, and within 15 years the risk is almost the same as for those who have never smoked. The risk for lung cancer falls more slowly, over about 10 years to 30–50 per cent of that of continuing smokers, but some risk remains even after 20 years of abstinence.

Recreational drugs, health and disease

The consumption of recreational drugs has become a phenomenon of recent social history, a problem and a challenge that society has to face up to. The consumption of recreational drugs has reached epidemic proportions – 45 million European Union citizens have used cannabis at some time, with proportionately higher use among younger people (Ghuran et al., 2001). Harder drugs such as cocaine and heroin are consumed by around 1.5 million problem users in the European Union, and cardiovascular complications of recreational drugs are an important cause of morbidity and mortality.

Cocaine, MDMA (3,4-methylenedioxymetham phetamine, 'ecstasy') and amphetamine have similar adverse effects on the cardiovascular system, activating the sympathetic nervous system and causing, for example, a dramatic rise in circulating catecholamine concentrations. Further effects of sympathetic activation include tachycardia and vasoconstriction.

Whilst recreational drugs are not particularly used by those that take sport and exercise seriously, a number of sports celebrities have succumbed to the use of recreational drugs for social or psychological reasons, and they are contained in doping regulation lists of banned substances. The avoidance of drug testing by sports people raises suspicion of the use of drugs for recreational and/or performance enhancement and the commitment for governing bodies to promote sport as healthy and purposeful means that the penalties for avoidance are necessarily severe.

Recreational drugs could be considered to fall into three main groups – **stimulants** (MDMA, amphetamine, cocaine, amyl and butyl nitrites), **depressants** (heroin, tranquillizers, methadone, volatiles) and **hallucinogens** (LSD, cannabis, magic mushrooms). It is not intended to discuss the detailed pharmacology of each of these drugs but to provide the reader with a brief description and to raise awareness of the detrimental effects of a number of those that are in use.

Stimulants

Amphetamine (also known as speed, and also uppers, whiz, sulph or sulphate) is most often sold as a white powder folded in paper. The subjective feeling is of high 'energy' and confidence combined with a loss of appetite followed by depression as the effects diminish. **MDMA** (also known as ecstasy, E, love hearts, doves, rhubarb and custard and disco burger) is a derivative of amphetamine. It is popular among young people and sold in the form of white or brown tablets or capsules. Like amphetamine, MDMA activates the sympathetic nervous system indirectly, by releasing noradrenaline, dopamine and serotonin. It can cause hyperthermia, leading to heat illness, dehydration and sometimes death and is more dangerous if combined with alcohol, which increases the risk of dehydration. Frequent use of amphetamine and MDMA can produce irritability and longer term effects include a reduced dose/effect, anxiety, confusion and sleep disturbance, cardiac arrest with high doses, mental illness and liver damage.

Cocaine or **crack** (also known as coke, snow and base) is sold as a white powder and is usually sniffed but can be smoked. Crack comes as small crystals and is smoked. It is a habituating drug that exerts its sensations – feelings of confidence and high 'energy' – by acting on chemical synapses. It acts by blocking the re-uptake into vesicles of the neurotransmitter dopamine. As a consequence, the dopamine receptors remain saturated with the neurotransmitter for a long time, maintaining a signal instead of responding to repeated releases. Higher doses run the risk of heart failure and the longer term effects include dependency, physical and mental ill-health, and nasal/breathing problems due to snorting the drug.

Amyl and **butyl nitrites** (also known as poppers) belong to the group of drugs known as alkyl nitrites and are vapours that are inhaled through the mouth or nose. Amyl nitrite is a medicine used to treat angina and is controlled in the UK under the Medicines Act (i.e. legal only on prescription), but there are no controls on sales of butyl nitrite at present. Effects include a 'rush', caused by vasodilation. Side effects include headaches, vomiting and dermatitis, and, in severe cases of amyl nitrite overdosing, clinical shock, unconsciousness and death. Longer term use leads to a reduced dose/effect and therefore the dangers of the side effects mentioned.

Depressants

Heroin (also known as smack, brown, junk, scag and H) is a semi-synthetic analogue of morphine and sold as a light brown powder. It can be injected, sniffed or smoked, injection being the most dangerous route because of the danger of overdosing and the transmission of human immunodeficiency virus (HIV) and other infectious diseases through the use of non-sterile/shared syringes. Regardless of how it is taken, it is a very addictive and dangerous drug, along with morphine accounting for around half of all drug-related deaths. Heroin slows physiological and psychological functioning, for example heart rate and breathing rate, causing drowsiness and analgesia. Ignorance about the purity and strength of the sample creates a high risk of overdose resulting in unconsciousness and sometimes death, exacerbated when used with other drugs. Longer term effects include dependency, insensitivity to its effects (requiring more drug to obtain the same effect), and the physiological and psychological problems associated with withdrawal.

Tranquillizers (also known as tranx, benzos, eggs, jellies and norries) are primarily prescription drugs, often obtained from legitimate users (prescribed for mild depression, sleep disturbance or anxiety). Benzodiazapines such as diazepam (tradenames such as Valium), chlordiazepoxide (e.g. Librium), lorazepam (e.g. Ativan) and temazepam when used as prescribed and under medical supervision by those who have clinical symptoms have effects (and side effects) that are manageable but when used as recreational drugs, sometimes mixed with alcohol or injected, they are dangerous drugs. A calm and less anxious state can be produced when taken normally (orally), but problems have occurred with the injection of dissolved gel filled capsules (such as temazepam) when they have re-solidified, causing severe circulatory problems. Gangrene has followed abuse

by injection. Dependency is a long-term problem associated with tranquillizers, as is depression and aggression, unpredictability of behaviour and the withdrawal effects of restlessness and anxiety. This emphasizes the nature of the use of these types of drugs even in those who might genuinely benefit from them and the need for more clinical research into these drugs.

Methadone is used as a treatment for dependency on opiate drugs such as heroin. It is a green/yellow liquid, swallowed (but sometimes injected), and its effects resemble those of heroin but only less intense and longer lasting. It can disturb sleep, despite causing drowsiness, and cause nausea. Very small doses are required to achieve an effect and these can be lethal in children and even adults unaccustomed to it. The longer term problems associated with methadone are the dependency and withdrawal problems.

Volatile substances such as solvents, fuels and glues are readily available, butane gas lighter refills being the most commonly misused substance. The effects are similar to alcohol, sometimes producing hallucinations, and can result in poor judgement and being involved in or causing an accident. There is a risk of suffocation if a plastic bag is used to inhale, and vomiting, unconsciousness, choking and heart failure are further risks. Longer term effects include memory loss and the inability to concentrate. As with a number of drugs, intense exercise (for example chasing after someone) whilst under the influence can result in heart failure.

Hallucinogens

Lysergic acid diethylamide (LSD) (also known as acid and trips) is normally sold in small amounts soaked onto blotting paper. It is a hallucinogen, causing changes in the way in which the environment is perceived (colours, sound) and feelings such as having the identity of another person. These can also occur as flashbacks after the initial effects have subsided. The experiences that people have can seem good or bad, and, although there is little evidence of long-term change or damage to personality or behaviour, flashbacks can be disturbing.

Cannabis (also known as shit, dope, draw, smoke, grass or hash) is the product of the cannabis plant, when its leaves are dried and a resin or oil produced. It the most widely consumed recreational drug, normally smoked with tobacco in cigarettes or without tobacco in a pipe. Skunk is a variety of cannabis of higher strength. Cannabis is a mild hallucinogen, causing relaxation and sometimes analgesia, but there is concern about cannabis use predisposing to higher rates of depression and anxiety, particularly in young adulthood. For some time, in surveys and studies of long-term users, frequent recreational use has been linked to high rates of depression and anxiety but there is uncertainty as to why this is so. A recent study demonstrated a strong association between frequent cannabis use and later depression and anxiety in teenage girls (Patton *et al.*, 2003). Other studies have linked cannabis use to a substantially increased risk of having psychotic symptoms and usage is reportedly becoming the leading cause of psychosis in the UK.

Magic mushrooms (psilocybin) (also known as mushies) grow wild in Britain and are collected most often in the autumn, then dried and eaten or soaked in water to make a 'tea'. The effects are similar to those experienced with LSD, but not as strong. The danger of picking a poisonous mushroom by mistake is one problem, but there are no known long-term effects of imbibing. At least some exercise is usually involved in the collection of mushrooms!

Possession, supply or production of drugs other than by registered practitioners or manufacturers is illegal and the basis of the legal position in the UK is The Misuse of Drugs Act of 1971 (amended 1986). Penalties depend on the type or class of drug and range, for example, from up to 5 years imprisonment or an unlimited fine or both (for class C drugs, e.g. temazepam) to up to life imprisonment (class A drug). It is really not worth getting involved.

In this chapter we have discussed factors that affect health and disease and are modifiable. There is strong evidence pointing to the potential health benefits of being active, with adults who are physically active having up to a 30 per cent reduced mortality risk compared with those who are inactive. Good physical activity habits, like good nutritional habits, are best started at a young age. Nutrition is one of the major determinants for cardiovascular disease and cancer. It is estimated that an unhealthy diet and a sedentary lifestyle might be responsible for up to a third of the cases of cancer and a third of premature deaths due to cardiovascular disease in Europe.

Factors affecting health that often start as habits in late childhood are alcohol consumption, smoking and drug use. Alcohol is one of the key health determinants in Europe that needs to be tackled, there being a significant burden of disease and injury attributable to alcohol in the European Union. Tobacco is the single largest cause of avoidable death in the European Union, accounting for over half a million deaths each year and over a million deaths in Europe as a whole. In 2003 there were up to 8000 drug-related deaths reported in 15 member countries of the European Union.

It is increasingly common that people with one disease develop further chronic conditions. This is termed co-morbidity or multi-morbidity. It has been estimated that of the 90 million Americans with a chronic condition in the early 1990s, 45 per cent had more than one (Starfield *et al.*, 2003). This can make treatment of disease much more complex and efforts should be made to reduce risk factors as much as possible to prevent co-morbidity from developing. Often, co-morbidities develop which can be restricting in terms of these modifiable factors, for example physical activity has a beneficial effect on reducing risk from CHD but osteoarthritis may limit physical activity. Medical evidence points to a cumulative disease state developing as a result of excessive alcohol intake and/or smoking and/or poor nutrition and/or physical inactivity. Increasingly the onus is on the individual to take responsibility for their own health, aided by the guidance of knowledgeable practitioners (for example, sport and exercise scientists) and local, national and international government directives and initiatives.

Summary

- Health lies in the medical domain, however, health is affected by physical activity and nutrition, both of which have a key role to play in primary prevention and in secondary care management.

- Health is a wide ranging concept, as evidenced by the UK Department of Health 'tips' that attempt to guide the population to better health.

- Dimensions of health might include physiological health, the absence of disease, the freedom from illness and psycho-social health.

- The population approach to improving health complements the high risk approach.

- The high risk approach has the potential to be influenced by nutrigenomics – the influence of nutrition and individual genetic variation on risk factors for chronic diseases.

- A major risk from physical inactivity is the incidence of coronary heart disease, the risk amongst sedentary people being nearly two-fold.

- There are numerous benefits to health of regular physical activity, including those that benefit the cardiovascular and musculoskeletal systems, and metabolism, and psychological health.

- Measurements of population physical activity levels have shown that in affluent countries such as the UK, the majority of the population are insufficiently active to optimally benefit health.

- Nutrition, working in tandem with physical activity can remodel physiology and metabolism and improve health.

- Disease risk, in states such as diabetes mellitus and cancer, can be reduced through optimal nutrition, and can also be used to manage aspects of disease.

- Alcohol has found a place in society and in culture but can be detrimental to both short-term and long-term health, contributing significantly but unhelpfully to energy intake, and also to hypertension.

Review Questions

1. What might 'health' mean to a cross-section of the British population?

2. How might health be measured?

3. Is there evidence that people in developed European countries such as the UK are active enough?

4. What health issues does physical inactivity have an impact on?

5. What impact does poor nutrition have on the risk of developing non-insulin dependent diabetes mellitus, and cancer?

6. On what social, physiological and health grounds should alcohol intake be low and never in large quantities?

7. What are the health problems associated with tobacco smoking and drug use?

References

Academy of Medical Sciences (2004) *Calling Time: The Nation's drinking as a major health issue.* Available at www.acmedsci. ac.uk.

Bandini L, Flynn A (2003) Overnutrition. In: Gibney MJ, Macdonald IA, Roche HM eds *Nutrition and Metabolism.* The Nutrition Society Textbook Series. Oxford: Blackwell Publishing, pp. 324–340.

Basu S, Stuckler D, Bitton A, Glantx SA (2011) Projected effects of tobacco smoking on worldwide tubercolosis control: mathematical modelling analysis. *British Medical Journal* 343, d5506.

Batty D, Thune I (2000) Does physical activity prevent cancer? *British Medical Journal* 321: 1424–1425.

Blaxter M (1990) *Health and Lifestyles.* London: Routledge.

Chaudhary V (2003) World Cup win prompts top talks. The *Guardian*, 29 November.

Colditz GA, Cannuscio CC, Frazier AL (1997) Physical activity and reduced risk of colon cancer: implications for prevention. *Cancer Causes and Control* 8: 649–667.

Coleman T (2004) Special groups of smokers. In: Britton J ed. *The ABC of Smoking Cessation.* London: BMJ Publishing.

COMA (1994) *The Nutritional Aspects of Cardiovascular Disease.* London: The Stationery Office.

Department of Health (1997) *The New National Health Service: Modern, Dependable.* London: The Stationery Office.

Department of Health (1999) *Saving Lives: Our Healthier Nation.* London: The Stationery Office.

Department of Health (2000) *National Service Framework for Coronary Artery Disease: Modern standards and service models.* London: The Stationery Office, chapter 1.

Department of Health (2001) *National Service Framework for diabetes: Standards.* (2001). London: Department of Health.

Department of Health (2010) *2009 Annual Report of the Chief Medical Officer.* http://www.dh.gov.uk/en/Publicationsandstatistics/Publications/AnnualReports/DH_113912.

Donaldson LJ, Donaldson RJ (2003) *Essential Public Health*, 2nd edn. Newbury: Petroc Press.

Eckhardt MJ, File SE, Gessa GL *et al.* (1998) Effects of moderate alcohol consumption on the central nervous system. *Alcoholism, Clinical and Experimental Research* 22: 998–1040.

Ghuran A, van der Wieken LR, Nolan J (2001) Cardiovascular complications of recreational drugs. *British Medical Journal* 323: 464–466.

Gibney MJ, Gibney ER (2004). Diet, genes and disease: implications for nutrition policy. *Proceedings of the Nutrition Society* 63: 491–500.

Gurr M (1996) *Alcohol – Health issues related to alcohol consumption*, 2nd edn. Washington DC: International Life Sciences Institute.

Haggard HW, Jellinek EM (1942) *Alcohol Explored.* Garden City: Doubleday, p. 12.

HM Government Cabinet Office Strategy Unit (2003) *Alcohol Misuse: How Much Does it Cost?* London.

Kelly SP, Thornton J, Lyratzopoulos G, Edwards R, Mitchell P (2004) Smoking and blindness (Editorial). *British Medical Journal* 328: 537–538.

Langley JD, Kypri K, Stephenson SCR (2003) Secondhand effects of alcohol use among university students: computerised survey. *British Medical Journal* 327: 1023–1024.

MAFF (1995) *Manual of Nutrition*, 10th edn. London: HMSO.

Morris JN, Kagan A, Pattison DC, Gardner MJ (1966) Incidence and prediction of ischaemic heart disease in London busmen. *Lancet* 7463: 553–559.

Paffenbarger RS Jr, Wing AL, Hyde RT (1978) Physical activity as an index of heart attack risk in college alumni. *American Journal of Epidemiology* 108: 161–175.

Patton GC, Coffey C, Carlin JB, Degenhardt L, Lynskey M, Hall W (2003) Cannabis use and mental health in young people: cohort study. *British Medical Journal* 325: 115–1198.

Prentice AM, Jebb SA (1995) Obesity in Britain, gluttony or sloth? *British Medical Journal* 311: 437–439.

Prescott-Clarke P, Primatesta P eds *Health Survey for England 1995*. London: The Stationery Office.

Sharpe I (ed.) (1998) *Social Inequalities in Coronary Heart Disease: Opportunities for Action*. London: The Stationery Office.

Sports Council and Health Education Authority (1992) *The Allied Dunbar National Fitness Survey: Main Findings*. London.

Starfield B, Lemke K, Bernhardt T, Foldes S, Forrest C, Weiner J (2003) Comorbidity: implications for the importance of primary care in 'case' management. *Annals of Family Medicine* 1: 8–14.

Wanless D (2003) *Securing Good Health for the Whole Population: Population Health Trends*. London: HM Treasury/HMSO.

Wasserman K, Hansen JE, Sue DY, Whipp BJ, Casaburi R (1994) *Principles of Exercise Testing and Interpretation*, 2nd edn. Philadelphia: Lea and Febiger.

Watkins PJ (2003) ABC of diabetes: cardiovascular disease, hypertension, and lipids. *British Medical Journal* 326: 874–876.

Wechsler H, Lee JE, Kuo M, Seibring M, Nelson TF, Lee H (2002) Trends in college binge drinking during a period of increased prevention efforts: findings from four Harvard School of Public Health college alcohol study surveys: 1993–2001. *Journal of American College of Health* 50: 203–217.

WHO/CDC (World Health Organization/Centers for Disease Control) (2004) *The Atlas of Heart Disease and Stroke*. Geneva: WHO.

Further reading

Hirsch GL, Sue DY, Wasserman K, Robinson TE, Hansen JE (1985) Immediate effects of cigarette smoking on cardiorespiratory responses to exercise. *Journal of Applied Physiology* 58: 1975–1981.

Wanless D (2004) *Securing Good Health for the Whole Population: Final report*. London: HM Treasury/HMSO.

II Biomechanics

7 Introduction to Sports Biomechanics

Chapter Objectives

In this chapter you will learn about:

- The basic principles of sports biomechanics.
- The field's history and developments.
- What sports do and their training.
- Career opportunities in the field.

Introduction

Human movement is subject, without exception, to the laws of mechanics. Biomechanics is derived from the Greek word *bios* meaning 'life' and *mekhaniki* meaning 'mechanics', and can be defined as the study of the mechanics of life forms (Yeadon and Challis, 2008). It is one of the disciplines of human movement and sport and exercise science and is the field that deals with the effects of forces acting on or produced by living bodies (Norman, 1975). It can be split into three general categories: **clinical biomechanics,** concerned with areas such as gait, neuromuscular control, tissue mechanics and movement evaluation during rehabilitation; **occupational biomechanics**, which looks at ergonomics and human growth or morphology and how they influence movement; and **sports biomechanics**, which investigates the movements of athletes using the laws of mechanics.

What is sports biomechanics?

Sports biomechanics has two main branches, (i) the study of sport-related injuries and (ii) the understanding of performance in sport. From developing playing surfaces and equipment to designing shoes, sports biomechanics plays an important role in recognizing which practices and techniques are less dangerous and effective and how performance is optimized by athletes. Research in sports biomechanics is diverse and multifaceted and may take the form of movement analysis or developing innovative equipment designs. For example, injury-related research may attempt to identify the causes of back injuries in cricket, whereas performance-related research may focus on the development of new sport-shoe sole designs, or the aerodynamics of ski-jumping and the influence of V-style on the flight parabola.

The subject of biomechanical investigation is thus mechanical movement, taking into consideration the mechanical characteristics and the prerequisites of the movement mechanisms (bones, joints and the neuromuscular system). These prerequisites are functionally dependent on the biological conditions of the organism.

Biomechanics therefore examines the internal and external forces acting on the human body and the effects of these forces. It helps us to understand why humans walk the way they do, the effects of gravity on the musculoskeletal system and how mobility impairment in elderly people can be improved. Sports biomechanics concerns itself with the study and evaluation of sports techniques and how the skills are performed. Sports biomechanists have contributed invaluably to improving performances in selected sports. Understanding the mechanical principles that underpin human movement can help to answer questions related to human movement and sport.

Sub-branches of mechanics

The two main sub-branches of mechanics are called statics and dynamics:

- **Statics** deals with systems in a constant state of motion. This includes systems that are at rest or moving with constant velocity.
- **Dynamics** deals with systems in which acceleration is present, or where velocity is not constant.

Other areas of research include **kinesiology**, which is the study of human movement from the point of view of the physical sciences. Its main areas are mechanical and anatomical kinesiology. In a broader sense it includes biomechanics, exercise physiology, sport psychology, motor behaviour and pedagogy. **Anthropometry** is the study of mechanical properties of body segments such as weight and dimension.

History of biomechanics

Biomechanics is an applied form of mechanics, and so the methods used to investigate it have also been derived from mechanics. If we are to truly understand the subject area, we need to have knowledge of its history.

Born approximately 2500 years ago, the Greek philosopher Socrates taught that in order to understand the world around us we need to understand ourselves and our own functions. He was executed, however, at the age of 70 because it was believed he was corrupting the minds of the young.

Another Athenian, Plato, 51 years younger, began the philosophical enquiries that were the blueprint for Western concepts of philosophy, psychology and politics. He considered observations and experiments worthless, but believed that the best tool for pursuing knowledge was mathematics. This created the framework for the growth of mechanics.

Aristotle went to Athens to study with Plato at his academy. He studied anatomy and wrote what might be considered the first book on the subject of biomechanics, called *De Motu Animalium (On the Movement of Animals)*. Aristotle considered questions such as the physiological difference between imagining performing a task and actually doing it. He also considered animals' bodies as being mechanical systems. He eventually moved away from Plato's mathematical philosophy and supported a qualitative method of enquiry. In the space of approximately 100 years the fundamental scientific tools of mathematical and deductive reasoning were established. Biomechanics had emerged, although there were no further developments in Western science until the Renaissance.

The history of modern day biolocomotion, which was later to become motion analysis/biomechanics, and of electrology, which today is known as kinesiological electromyography, can be traced back to the latter half of the seventeenth century. In 1679, Borelli, an Italian mathematician, was the first to establish the centre of gravity in humans by experimentation. He had already studied running, jumping and skating and is generally considered to be the pioneer of modern biomechanics (Clarys and Alewaeters, 2003). In 1836, the brothers Wilhelm and Eduard Weber, German physiologists, published *Mechanics of the Human Locomotor Organs*. Their theory

claimed, however, that locomotion of the legs alternated as a result of gravity.

This theory was opposed later by other scientists, first by the French researcher Etienne Marey, who developed chronocyclo-photography as early as 1880, and later, at the turn of the century, by the German scientists Christian Wilhelm Braune and Otto Fischer, who developed a new method for determining the centre of gravity through their work with cadavers. They also determined path-time progress of the human through experimental procedures. These, along with the work of Gheorghe Marinescu in 1898 and Friedrich Kohrausch in 1891 added the cinematographic or photographic dimension to biomechanical research (Clarys and Alewaeters, 2003). The Soviet scientist Bernstein continued the work of Braune and Fischer, perfecting the chronocyclo-photography methodology and simplifying the evaluation of measured path-time values.

Progress in biomechanics has been largely dependent on methodology and the state of science and technology. Du Bois Reymond is recognized as being one of the first sports scientists. Publication of Dally's book in 1857 entitled *Cinesiologie ou Science du Mouvement* made reference to Du Bois Reymond's work on concentric and eccentric muscle contractions and the relationship to muscle force and strength. A continuation in the development of equipment opened doors to new discoveries in biomechanical research. It has also been largely determined by the research objectives. Towards the end of the nineteenth century, biomechanical objectives were established by orthopaedics and the physiology of work and industry. Electromyography, or electrology as it was then known, became a scientific discipline and it became accepted as a diagnostic (clinical EMG) and functional/coordination tool (kinesiological EMG).

With the advent of the Olympic Games in Athens in 1896, however, interest grew into using the principles of mechanics to improve sports techniques and performance. Athletes tried to learn which movements led to the greatest performances. Mistakes were made as factors such as strength, speed and endurance were not considered. In most countries sports biomechanics developed through its use in education as a means to technique improvement. It also developed as a scientific discipline after the Second World War.

In 1967, the International Council of Sport and Physical Education (ICSPE) held its first international conference in biomechanics in Zurich. Also in 1967, the Working Group of Biomechanics (WGB) was set up. This was a specialist research committee group of the ICSPE. Following this, the International Society of

Biomechanics (ISB) was founded on 30 August 1973, to promote the study of biomechanics at international level (Hochmuth, 1984). In 1982 the International Society for Biomechanics in Sports was founded (ISBS) and it is now the only international society totally dedicated to biomechanics in sports.

During the late 1980s, great advances were made in technology, especially in computer hardware and real-time data acquisition interfaces. This resulted in a large volume of three-dimensional kinematic systems, although understanding and interpretation of human movement was still limited by single plane analysis. In their attempt to analyse human movement, contemporary biomechanists have attempted to improve their understanding of the laws that govern movement by developing incredibly sophisticated mathematical body models.

Sports biomechanics will continue to contribute to further knowledge but the presentation of research papers with application to the coaching environment and their ability to communicate effectively with coaches and athletes needs to be addressed. Future research directions will also include further advances in sports equipment and the development of appropriate testing methodology for this equipment.

Biomechanical models

Biomechanics can be defined as the science that examines internal and external forces acting on the human body and the effects produced by these forces. For example, a sprinter running down the track is generating internal forces in the leg muscles during each stride. The net result of these internal forces is a sequence of pushes against the ground with the foot. With each push against the track (note that the track or ground is an external force), the ground pushes back against the athlete and so the athlete moves forward. Other external forces also act against the athlete such as gravity and air resistance. Each of these forces has an effect upon the motion of the human body.

While we can visualize the human body running down the track, the actual body with all its bones, muscles, connective tissue and internal organs is too complex to consider for a biomechanical analysis. Human movement is also three-dimensional and the human body is multi-segmented. Therefore, to make the study of human movement possible, biomechanics has adopted three simplified models (Figure 7.1):

- The particle model
- Stick figure or free body diagram
- Rigid segment body model.

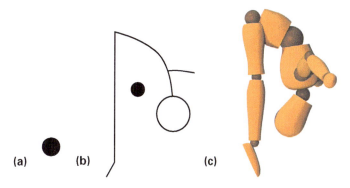

△ **Fig. 7.1** Three models are used to represent human movement. In this example of a diver, the movement is represented by: (a) the particle model (b) the stick figure model or (c) the rigid segment body model. The choice of model would depend on the biomechanical analysis to be made.

The **particle model** is a dot representing the centre of mass of the body or object. Because it involves the centre of mass only, its use is limited to bodies in flight (projectile). In sport these include any object that is thrown, hit or kicked, as well as the human body in flight such as diving or long jumping. When the body becomes a projectile, often the only force acting on it is gravity through its centre of mass. High-velocity objects such as javelins and cricket balls are also affected by air resistance (aerodynamics).

When objects or bodies are in contact with the ground a **stick figure model** is used to represent the body. Body segments are represented by rigid bars linked together at the joints. The stick figures (free body diagram) represent the position and size of the body segment and are used to represent the body configuration for two-dimensional gross motor skills such as the sprint start.

They do not, however, represent longitudinal rotational movements, segment rotations or three-dimensional movements.

A sequence of stick figures is known as a **composite diagram**. Composite diagrams give a picture of the body's actions throughout the full movement range, but they cannot be individualized to the athlete. Multiple force vectors can be drawn on the free body diagram wherever contact is made with the environment.

When undertaking more complex three-dimensional analyses, biomechanists use a **rigid segment body model**. Each body segment is represented as an irregularly shaped three-dimensional volume.

Three steps must be considered before human movement can be described using the biomechanical models detailed above.

1. Identify the system to be analysed and isolate the object from its surroundings.

2. Identify the reference frame (coordinate system) in which the movement takes place (Chapter 8). For example, a runner changes her body position relative to the ground and a starting position. Many sports are described in two dimensions but actually occur in three dimensions.

3. Identify the type of movement that is occurring, the planes of motion and the axes of rotation (Chapter 8).

Composite diagrams, such as the one shown in Figure 7.2, provide a pictorial overview of a performance. If the movement does not occur in a two-dimensional plane then accurate measurements of angular velocities, for example, are difficult.

△ **Fig. 7.2** Composite diagram of a golfer to give a pictorial overview of performance.

What do sports biomechanists do?

Sports biomechanists undertake varied roles. Many students decide that on graduating they would like to follow careers in physical education teaching, coaching, strength and conditioning, personal fitness training or physiotherapy. As undergraduates they often ask the question, 'Why do we have to study biomechanics?' The answer is that an understanding of movement analysis and the underlying principles is mandatory if the potential coach or PE teacher is to be successful in identifying faults in technique and correcting them.

The study is also closely linked to injury prevention both in the sports and the exercise environment, as well as performance enhancement. According to Norman (1975) a major problem faced by coaches, physical educationalists and fitness instructors when teaching basic skills is that of error detection. Inaccurate, poor or non-specific feedback can adversely affect the quality of learning of the athlete. Identifying the causes of poor technique can save time and frustration in the learning process for both the athlete and the coach. An understanding of the mechanics of human movement is the key to effective skills coaching and this is achieved through the study of sports biomechanics.

This knowledge can be utilized more effectively when an interdisciplinary approach to problem solving is adopted by the practitioner. An interdisciplinary approach involves more than one discipline and uses more than one area of sport science to solve a problem or challenge. For example, a PE teacher preparing an up-and-coming hurdler for an athletics championship may consider the implications of increased sprinting speed between hurdles on body position at take-off and the trail leg position during the flight phase over the hurdle and the athlete's subsequent confidence levels (British Association of Sport and Exercise Sciences, 1997). There are biomechanical, physiological and psychological considerations here.

To return to the question of what sports biomechanists do, it is clear that practitioners have three primary roles, involving teaching, research and consultancy (Figure 7.3).

The teaching role

As educators, sport and exercise biomechanists are primarily involved in informing the widest possible audience of the ways and methods in which biomechanics can be used to enhance performance, prevent injury, help in sport and exercise equipment development and explain the physical effects of the environment. Many people benefit from this education, including coaches, performers at all levels, teachers, exercise and health professionals, national governing body administrators and the media.

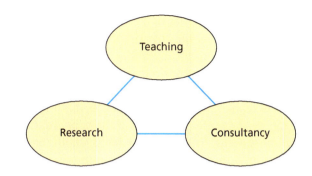

△ **Fig. 7.3** Tripartite roles of a sports biomechanist.

Sports biomechanists often have a teaching or lecturing role at universities or institutes of higher education. They predominantly teach biomechanics or kinesiology modules to undergraduate and postgraduate students, particularly if they teach on sport and exercise science courses. Some institutions also offer higher degrees in sports biomechanics and therefore would play a significant role in the delivery and development of these courses as well as providing dissertation support and supervision up to doctoral level. Many biomechanists have knowledge in specific areas, which may be of great interest to a limited audience, but of restricted interest to a more general audience.

The research role

Another major role of a sports biomechanist is to conduct research. Fundamental and applied research are both important for the investigation of problems in sport and exercise biomechanics. **Applied research** provides the necessary theoretical grounds to underpin education and sports-science support services. The understanding gained from applied research is, for example, a prime resource for coach education. **Fundamental research** such as the mechanics of skeletal muscle allows specific applied research to be developed. The nature of sport and exercise biomechanics requires a research approach based on a mixture of experimentation and theoretical modelling The majority of biomechanists based in universities undertake research aimed at advancing the knowledge within the field.

Biomechanists then share their findings with colleagues in the field. They might study the kinematics of tennis players to produce high velocity serves or develop new techniques for assessing ball speed and impact force in volleyball or baseball for example. The International Society of Biomechanics in Sports (ISBS) provides the platform on a world stage for discussion and debate to take place. The ISBS held its first full-scale conference in San Diego in 1982, however the first ISBS biomechanists were involved in field research at the 1976 Montreal Olympics and 1978 Commonwealth Games in Edmonton, Canada. Many research projects have been conducted since, along with symposia, seminars, clinics, exhibits, lectures, newsletters, research journals and annual conferences.

The consulting role

The third role is that of consultancy. Many biomechanists work with individual athletes or teams, providing biomechanical scientific support to ultimately enhance athletic performance. In this role, biomechanists use their scientific knowledge for the benefit of the client. This usually involves undertaking a needs' analysis to ascertain the

---

TIME OUT

EXAMPLE OF RESEARCH

Arabatzi *et al.* (2010) investigated vertical jump (VJ) biomechanics following plyomteric (PL), Olympic weightlifting (OL) and combined (weightlifting + plyometric) (WP) training. Thirty-six male physical education students were randomly assigned to 4 groups: PL group (n = 9), OL group (n=9), WP group (n = 10) and control group (C) (n = 8). All participants were physically active and had at least 1 year of experience in resistance training. The experimental groups trained 3 days/week for 8 weeks and sagital kinematics, VJ displacement, power and electromyographic (EMG) activity from rectus femoris (RF)and medial gastrocnemius (GAS) were collected during squat jumping and counter-movement jumping (CMJ) before and after training. The results showed that all experimental groups improved VJ height (p < 0.05) when compared with the control group, however the mechanisms for these improvements differed between the 3 training protocols. They concluded that Olympic weightlifting training may be more appropriate to achieve changes in VJ performance and power in the pre-competition period of the training season. Emphasis on the PL exercises should be given when the competition period approaches, whereas the combination of OL and PL exercises may be used in the transition phases from pre-competition to the competition period.

client's requirements, followed by the development and implementation of a support or intervention. The first stage involves the biomechanist seeking to understand the nature of the problem. In the second stage, the appropriate qualitative or quantitative analytical techniques are used to deliver the relevant scientific support. Careful interpretation of data from the analyses is carried out by the biomechanist, who then translates the science into a 'user friendly' language appropriate to each individual problem and client.

There are a limited number of full-time self-employed biomechanists working in the UK, with the majority being based at academic institutions and providing their services to athletes, coaches and National Governing Bodies on a part-time consultancy basis.

Along with the widespread acceptance of the provision of sport and exercise science support by coaches, athletes and clients, comes the accompanying problem of standards. Consequently, the British Association of Sport and Exercise Sciences (BASES), the professional body for those with an interest in sport and exercise science within the UK, offers an accreditation programme for sport and exercise scientists (research and scientific support). Those individuals who want to become accredited sport and exercise scientists (biomechanics/physiology/psychology) have the opportunity to undertake a rigorous peer-reviewed process that involves undergraduate and postgraduate training as well as a three-year accreditation process–namely BASES-supervised experience. There is also a specialist route to BASES accreditation for individuals providing sport science services to high performance athletes and sports programmes (i.e. High Performance Sport Accreditation). BASES accreditation is seen as the mark of distinction.

Future directions must continue to address the relatively poor supply of graduates with appropriate knowledge and experience within sports biomechanics. This has been addressed in part by BASES through its undergraduate endorsement scheme (BUES) by providing undergraduates with opportunities to develop knowledge and skills considered essential to enter into the profession. The scheme is concerned with the appropriateness of the curriculum, resources and opportunities that undergraduate programmes offer for training sport and exercise scientists. Programmes must demonstrate that at least 10 per cent of student time is spent in each of the subdisciplines of biomechanics, physiology and psychology, and that at least half their total time is spent in these three areas.

In addition, continued developments in technology will enable more accurate and varied methods of assessment. In particular, automatic motion analysis systems allow for increases in the number of trials analysed in both a laboratory and/or training environment. Similarly, force transducers can record forces produced by athletes and equipment in sports as diverse as weightlifting, skiing and throwing activities. Sports biomechanics may also benefit from developments in data processing systems (e.g., portable and accurate 3-dimensional camera calibration systems) (Yeadon and Challis, 2008). Encouraging new and innovative approaches to biomechanical methods and procedures will ensure continued progress within the discipline.

Summary

- Biomechanics can be used in many areas of sport and exercise: to improve sports performance through technique and equipment, to avoid sports injury by identifying safer techniques, and in the development of protective equipment.

- It is also used in occupational injury prevention (ergonomics), for example, in the study of low back pain, injury rehabilitation, the reduction of physical or functional declines, improving flexibility and product design.

- Sports biomechanics is the scientific study and evaluation of sports techniques and skills. Contemporary sports biomechanists play a number of different roles, including teaching, conducting research and consultancy.

- Sports biomechanists have undergraduate and postgraduate degrees in either sport and exercise science or related disciplines, or move into the area having studied mainstream physics or mathematics. All practitioners should be accredited.

- In the UK, the British Association of Sport and Exercise Sciences (BASES) offer an accreditation programme involving three years of supervised experience.

- Although sport science support is more accepted today than previously by coaches, athletes and their governing bodies, employment opportunities are still limited.

- The greatest gains in the field of sports biomechanics are to be made by educating coaches, athletes, physical educators, strength and conditioning coaches, fitness instructors and physiotherapists in biomechanical principles. An application of these principles through interdisciplinarity is the way forward. This will allow these professionals to achieve their specific goals whether in a sport or exercise setting.

Review Questions

1. What is sports biomechanics?

2. Describe the roles of the sports biomechanist.

3. Why is there a need for accreditation in contemporary sports biomechanics?

4. What career opportunities are available to the sports biomechanist?

5. Why should students of sport science study biomechanics?

6. Where have the greatest advancements in sports biomechanics taken place?

7. Where are future developments likely to occur?

References

Arabatzi, F, Kellis, E and De Villarreal, S-S (2010) Vertical jump biomechanics after plyometric, weightlifting and combined (weightlifting + plyometric) training. *Journal of Strength and Conditioning Research* 24: 9, 2440–2448.

British Association of Sport and Exercise Sciences (BASES) (1997) *Interdisciplinary Section Future Directions.* Leeds: BASES, December.

Clarys JP, Alewaeters K (2003) Science and sports: a brief history of muscle, motion and ad hoc organisations. *Journal of Sports Sciences* 21: 669–677.

Hochmuth G (1984) *Biomechanics of Athletic Movement.* Berlin: Berlin Sport Publishing House.

Norman RWK (1975) How to use biomechanical knowledge. In: Taylor JW ed. *How to be an Effective Coach.* Canada: Manufacturer's Life Insurance and the Coaching Association of Canada.

Yeadon MR and Challis JH (2008) The future of performance-related sports biomechanics research. *Journal of Sports Sciences* 12, 3–32.

Further reading

Bing Y, Broker J, Silvester LJ (2003) A Kinetic Analysis of Discus Throwing Techniques. *Journal of Sport Biomechanics* I: 25–46.

Blackwell J, Knudson D (2003) Effect of type 3 (oversize) tennis ball on serve performance and upper extremity muscle activity. *Journal of Sports Biomechanics* I: 187–191.

8 Human and Linear Kinematic Concepts

Chapter Objectives

In this chapter you will learn about:

- Anatomical reference positions and directional terminology.
- Reference planes and axes of rotation.
- Linear kinematic concepts such as displacement, velocity and acceleration, their mathematical basis and their importance to the analysis of human movement.
- The importance of obtaining velocities and accelerations from displacement data through graphical differentiation.
- Projectiles in terms of optimum release angles, velocities and ranges, and be able to carry out simple calculations of these concepts.

Introduction

In biomechanics, movement can be categorized in a number of ways. Movement can be linear (rectilinear or curvilinear), angular or general. **Linear motion** occurs in a straight line. All parts of an object or system move the same distance in the same direction at the same time. Rectilinear motion is linear motion along a straight line, for example a 100 m sprinter and curvilinear motion is linear motion along a curved line, for example a javelin throw. **Angular motion** or rotation involves movement of an object or body about an axis of rotation, e.g. a gymnast somersaulting (see Fig 8.1). All points move through the same angle in the same time. The axis of rotation is an imaginary line that the object spins about and is positioned perpendicular to the plane of rotation. Body segments also experience angular motion about their joints when they extend, flex and longitudinally rotate. **General motion** is a combination of linear and angular movement. Most human movement consists of general movement. Kent (1994) defines motion as the constant change of relative position of an object in space.

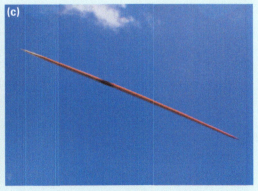

△ **Fig. 8.1** (a) A 100 m sprinter (rectilinear motion) (b) A gymnast somersaulting (angular motion) (c) An airborne javelin (curvilinear motion).

In order to describe body segment movements and to measure joint angles in biomechanics, it is essential to identify the reference position, planes and axes associated with the human body. **Body segments** in biomechanical terms are rigid bodies and include the foot, lower leg or shank, the thigh, pelvis, torso, hand, forearm, upper arm and head. There are many methods used in biomechanics to record kinematic information. These include goniometers, potentiometers, several forms of transducers and cameras. However, cameras are used more often than not because of their simplicity, accuracy and versatility. Normally, reflective markers are placed onto specific landmarks (usually the joints between adjacent body segments). The reflective markers reflect the light from the cameras and the data are recorded and analysed.

Kinematics versus kinetics

The study of **kinematics** describes spatial and timing characteristics of motion of the human body and its segments. These variables are used to describe both linear and angular motion. They answer four questions: How long? How far? How fast? and How consistent? **Kinetics** (Chapter 9), however, focuses on the various forces that cause a movement, in other words the forces that produce the movement and the resulting motion.

The forces acting on a human body can be internal or external. **Internal forces** refer to forces generated by muscles pulling on bones via their tendons, and to bone-on-bone forces exerted across joint surfaces. **External forces** refer to those forces acting from outside of the body such as gravity or the force from any body contact with the ground, environment, equipment or opponent. Generally speaking, internal forces cause individual body segment movements while external forces affect total body movements.

Anatomical reference position

The anatomical reference position Figure 8.2 (front (a), rear (b)) is used as the position of reference for anatomical nomenclature. For a person standing in the anatomical reference position all joint angles are zero (0). It involves:

- Erect standing
- Feet separated slightly and pointed forward
- Arms hanging at the sides
- Palms facing forward.

Directional terminology

In studying kinematics it is important to be familiar with the following terms:

- Superior – Closer to the head
- Inferior – Further from the head

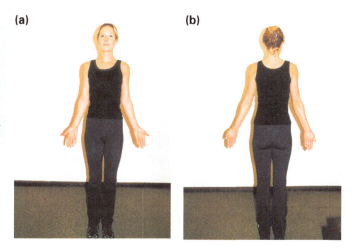

△ **Fig. 8.2** (a) Anatomical reference position – front. (b) Anatomical reference position – rear.

- Anterior – Toward the front of the body
- Posterior – Toward the back of the body
- Medial – Toward the midline of the body
- Lateral – Away from the midline of the body
- Proximal – Closer to the trunk
- Distal – Away from the trunk
- Superficial – Toward the surface of the body
- Deep – Away from the surface of the body.

Anatomical reference planes

These are three imaginary perpendicular planes that divide the body in half by mass (Figure 8.3).

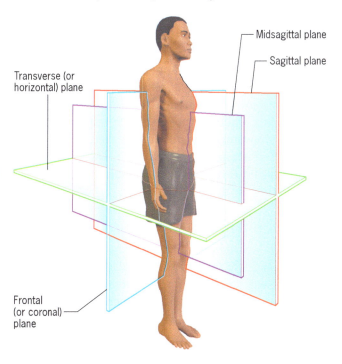

△ **Fig. 8.3** The three perpendicular planes of motion (sagittal, transverse and frontal).

Planar movements

Movement is said to occur within a plane if it is parallel to the plane.

- Sagittal movement can occur forward/back or up/down
- Transverse movement can occur forward/back or right/left
- Frontal movement can occur right/left or up/down.

Anatomical reference axes

The **axes of rotation** are imaginary lines about which rotation occurs. The axes pass through a joint's centre of rotation and in the reference position they are perpendicular to the anatomical planes.

- Transverse axis – The axis for sagittal plane rotations
- Longitudinal axis – The axis for transverse plane rotations
- Sagittal axis – The axis for frontal plane rotations.

Multiplanar movements

The structure of most joints allows movements in more than one plane simultaneously. For example, movement is in three planes in the hip (flexion/extension, abduction/adduction and internal/external rotation), two planes in the wrist (flexion/extension, abduction/adduction) and three planes in the subtalar joint: (pronation/dorsiflexion, abduction, eversion, supination/plantarflexion, adduction/inversion).

It is important to understand which sporting movements are primarily planar (e.g. running and cycling) and which are primarily multiplanar (e.g. tennis serve and karate kick). Planar movements are much easier to analyse. In practical terms, whenever possible, biomechanists try to represent movements as planar. The plane of motion provides information about the action of muscles. Muscles within plane act as prime movers and muscles out of plane act primarily as joint stabilizers.

Reference frames in biomechanics

Reference frames in biomechanics are needed to quantify kinematics of the body (e.g. position, displacement, velocity and acceleration). In order for descriptions of body segments to be meaningful, movement specialists must be able to communicate with one another using a standard set of nomenclature. Planes and axes of motion are used to describe the orientation of the body segments, while joint action terminology is used to describe the actual movement at a particular joint.

Spatial reference frames

A spatial reference frame is a set of coordinate axes (1, 2 or 3) positioned perpendicular to each other. It provides a means of describing and quantifying positions and directions in space.

- A one-dimensional (1D) reference frame quantifies positions and directions along a line. It requires only one number to describe position (Figure 8.4).

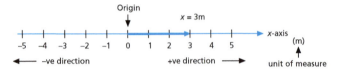

△ **Fig. 8.4** One-dimensional reference frame (x-axis).

- A two-dimensional (2D) reference frame requires two numbers (x, y) or coordinates to specify position and one angle (θ) to specify position. Linear motion (x), linear motion (y) and rotation are analysed separately (Figure 8.5).

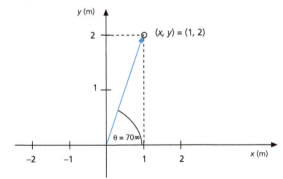

△ **Fig. 8.5** Two-dimensional reference frame.

- A three-dimensional reference frame quantifies positions and directions in space and requires three numbers to describe x, y and z positions and three numbers to describe direction (Figure 8.6).

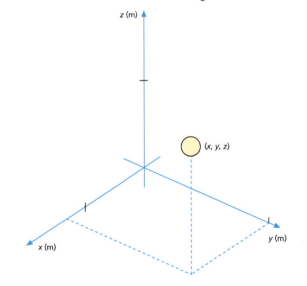

△ **Fig. 8.6** Three-dimensional reference frame.

Selecting a reference frame

The position or motion of a system does not depend on the choice of reference frame but the numbers used to describe them do. Always specify the reference frame used. Use only as many dimensions as is necessary. Link the reference frame with clearly defined directions (e.g. compass directions and anatomical planes).

For example:

- Sagittal plane: x = anterior, y = upward
- Frontal plane: x = left, y = upward
- Transverse plane: x = anterior, y = left.

The origin should have meaning and the reference frame should not be moved. For example in Figure 8.7:

$$x = 0 \text{ initial position}$$
$$y = 0 \text{ ground height}$$

Stoichiometry

Conversion between units can be made very simple if you use the principle of stoichiometry. For example, in an examination you are asked to calculate the speed of a sprinter over 100m. You are given the time recorded by the sprinter and nothing else. Unfortunately you cannot remember the units of measurement for speed, what do you do? Without the units the quantities calculated lose their meaning. As long as you can remember the units associated with speed in this example (distance/time), you can work out the formula:

Distance = 100m
Time = 9.95s
Speed = Distance/time
Speed = 100 m/9.95 s
Speed = 10.05 m/s

You can see how carrying the units can help. However there are other fundamental uses for stoichiometry, namely conversion between units. For example, referring back to the example above, the average speed is calculated in metres/second, but the sprinter's coach cannot grasp how fast 10.05 m/s really is. You agree to recalculate the speed in traditional English units (miles per hour). How is this done?

1. Always start with what you are given; in this case 10.05 m/s.
2. Work out what you need (i.e. what units do you want to end up with?) – miles/hour.
3. Write down what you know and establish relationships between units. Make sure all units, except those you want to end up with, cancel each other out.

- So 10.05 m/s = 10.05 m/1s × 60 s/1min × 60 min/1 hour × 1 km/1000 m × 0.621 mile/1 km
- Cancelling out units we are left with the following: 10.05 m/s = 10.05 × 60 × 60 × 1 × 0.621mile/1 × 1000 × 1 hour
- Which means that 10.05 m/s = 22467.78miles/1000 hour
- Therefore 10.05 m/s = 22.47miles/hour.

△ **Fig. 8.7** System identification and frame reference. The system is the runner and the reference frame is a two-dimensional x, y plane, with the start line being the origin (O).

Linear kinematics

Kinematics is the branch of biomechanics that studies the position of bodies through time without regard to the forces causing motion (without regard to Newton's second law). It is used to describe how a body changes position, orientation, shape and/or size over time. Knowledge of kinematics is an absolute prerequisite for understanding kinetics. Kinematic data provide the biomechanist with valuable information to analyse athletic ability (Briggs *et al.*, 2003).

Why study linear kinematics?

Linear kinematics should be understood by both coaches and athletes alike. Being able to measure an athlete's position, velocity, acceleration or displacement can provide valuable information as to the strengths and weaknesses of the athlete and about the effectiveness of a training programme. For example, it is essential to be able to calculate the release velocity of a javelin thrower or how long it takes to run a race. How long refers to the temporal characteristics of a performance, either of the total skill or of its phases. It is a time interval calculated as the difference between the beginning and end of two instants of time. The study of linear kinematics aims to put into place concepts involving movements in a single direction such as velocity and acceleration.

Vector algebra

To truly understand sports biomechanics, we must understand mathematical principles. Trigonometry can be used to resolve horizontal and vertical components of vectors and the angles these components make. For example, identifying the optimal take-off angle in the long jump will allow the horizontal and vertical velocities to be calculated to achieve the greatest possible distance. A knowledge of trigonometry and vector algebra will allow you to understand the concept and principles of vectors more effectively.

Many important quantities in biomechanics are meaningless without knowing both a magnitude and a direction:

- Speed and direction of travel (velocity)
- Applied force
- Position
- Change in position.

Trigonometric functions

Figure 8.8 shows the trigonometric functions used. They are sine, cosine and tangent.

$$\sin \theta = \text{opposite/ hypotenuse}$$
$$\cos \theta = \text{adjacent/hypotenuse}$$
$$\tan \theta = \text{opposite/adjacent}$$

Example 1

From Figure 8.8, let opposite = 1.5, adjacent = 2.1 and hypotenuse = 2.5. Find $\sin \theta$, $\cos \theta$ and $\tan \theta$. From the above definitions:

$$\sin \theta = \text{opposite/hypotenuse} = 1.5/2.5 = 0.6$$
$$\cos \theta = \text{adjacent/hypotenuse} = 2.1/2.5 = 0.84$$
$$\tan \theta = \text{opposite/adjacent} = 1.5/2.1 = 0.71$$

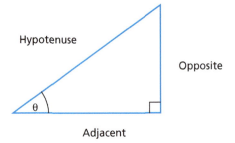

△ **Fig. 8.8** The appropriate trigonometric functions of sine, cosine and tangent can be obtained from a right-angled triangle.

TIME OUT

SINGLE MAXIMAL VERSUS COMBINATION PUNCH KINEMATICS

Piorkowski *et al.* (2011) determined the influence of punch type (Jab, Cross, Lead Hook and Reverse Hook) and punch modality (single maximal, 'In-synch' and 'Out of synch' combination) on punch speed and delivery time. Ten competitive-standard volunteers performed punches with markers placed on their anatomical landmarks for 3D motion capture with an eight-camera optoelectronic system. Speed and duration between key moments were computed. There were significant differences (p<0.05) in contact speed between punches with Lead and Reverse Hooks developing greater speed than Jab and Cross. There were significant differences in contact speed between punch modalities with the Single maximal higher than 'Out of synch' or right lead. Delivery times were significantly lower for Jab and Cross than Hook. The authors concluded that a defender may have more evasion-time than previously reported. The research could be of use to performers and coaches when considering training preparations.

Example 2

Use your calculator to find the following:

$\sin 25$ degrees $= 0.423$
$\cos 45$ degrees $= 0.707$
$\tan 60$ degrees $= 1.732$

To calculate the angles from the trigonometrical values we use the inverse function on the calculator. For example:

$\sin \theta = 0.2754 \Rightarrow \theta = \sin^{-1}(0.2754) = 15.99$ degrees
$\cos \theta = 0.5634 \Rightarrow \theta = \cos^{-1}(0.5634) = 55.71$ degrees
$\tan \theta = 0.8777 \Rightarrow \theta = \tan^{-1}(0.8777) = 41.27$ degrees

Vectors

Vectors represent both **magnitude** and **direction**. It is the convention to print vectors in boldface type to distinguish them from scalars (magnitude only). Graphically, an arrow can conveniently represent a vector (Figure 8.9(a)). The length of the arrow is proportional to its magnitude while the tip of the arrow shows its direction.

A vector can also be specified by its components:

$$\underline{\mathbf{v}} = [\underline{\mathbf{v}}_x, \underline{\mathbf{v}}_y, \underline{\mathbf{v}}_z]$$

The number of the components determines the dimension of a vector: two-dimensional (two components) versus three-dimensional (three components). The components are related to the coordinate system in use. Figure 8.9(b) shows vectors described in Cartesian coordinates.

The magnitude of the vector is the same as the length of the vector arrow shown in Figure 8.9(b).

$$\underline{\mathbf{v}} = /\mathbf{v}/$$
$$\underline{\mathbf{v}} = /\mathbf{v}/ = \sqrt{\mathbf{v}^2_x + \mathbf{v}^2_y + \mathbf{v}^2_z}$$
$$= (\underline{\mathbf{v}}^2_x + \underline{\mathbf{v}}^2_y + \underline{\mathbf{v}}^2_z)^{1/2}$$

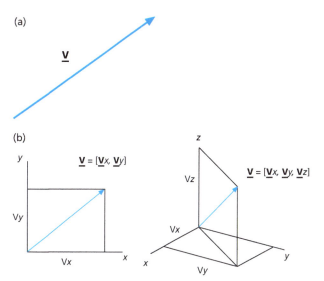

△ **Fig. 8.9** (a) The arrow represents a vector which possesses both magnitude and direction. (b) A vector can be specified by its components which determine dimension. The figure shows vector $\underline{v}$ and the components v_x and v_y.

A vector whose magnitude is unity (= 1) is called a unit vector. The unit vector of vector **v** shown in Figure 8.9(b) can be obtained by dividing the vector with its magnitude. The unit vector of the axes (unit coordinate vectors) can be expressed as $\underline{\mathbf{i}} = [1, 0, 0]$, $\underline{\mathbf{j}} = [0, 1, 0]$ and $\underline{\mathbf{k}} = [0, 0, 1]$, where $\underline{\mathbf{i}}$, $\underline{\mathbf{j}}$ and $\underline{\mathbf{k}}$ are the unit vectors of the x, y and z axes respectively.

Vector addition – the graphical method

Vectors may be added or subtracted to give the resultant or net vector. If vectors of the same nature are present in the same plane, they can be added using the tip-to-tail method. The tail of one vector is placed at the tip of the other vector giving a resultant.

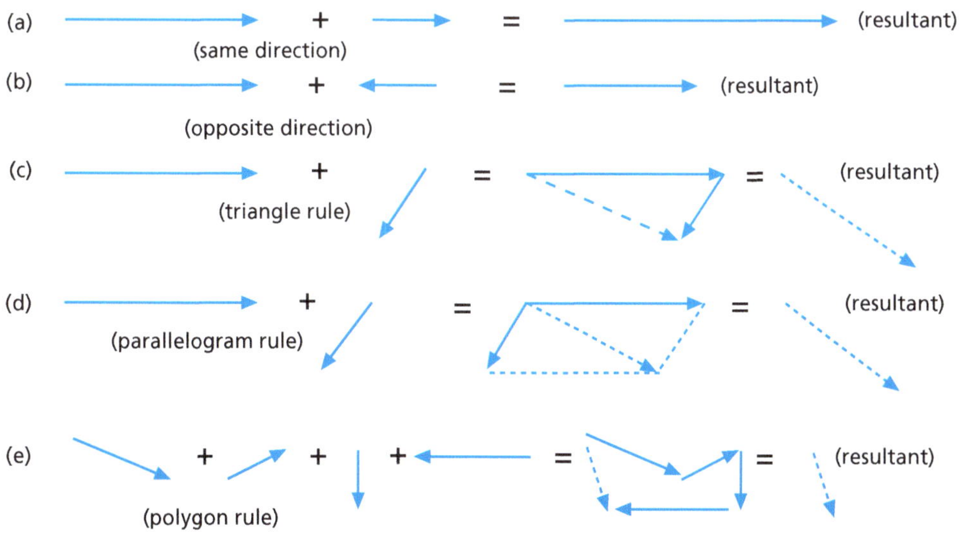

(a) ——————→ + ——→ = ————————→ (resultant)
(same direction)

(b) ———→ + ←——— = ——→ (resultant)
(opposite direction)

(c) ———→ + / = = (resultant)
(triangle rule)

(d) ———→ + / = = (resultant)
(parallelogram rule)

(e) ↘ + ↗ + ↓ + ← = = (resultant)
(polygon rule)

△ **Fig. 8.10** Vector addition – the graphical method.

During vector addition, vectors can be moved freely, yet the magnitude and direction remains unchanged, as shown in Figure 8.10.

Displacement versus distance

Displacement is defined as the distance moved in a specified direction, or the location of a particle with respect to the origin. For example, sprinters run 100 m in a westerly direction, volleyball players jump 0.7 m upward during a block. Distance, however, is concerned only with the length of the path moved. If an athlete runs a distance of 100 m the expression '100 m' is a scalar quantity. But if the athlete runs along a straight line and the direction of travel is mentioned, e.g. 100 m due south, then this expression is now a vector quantity, and this is called the displacement of the athlete. Displacement is therefore concerned with the change in position. Its standard unit is the metre (m).

For example, a runner completes three laps of a track. What is the distance run? What is the runner's displacement?

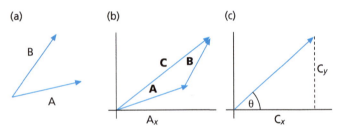

(a)   (b)   (c)

△ **Fig. 8.11** Vector C in (b) is the resultant of vectors A and B in (a). They have been added using the tip-to-tail method previously described. The component vectors of C are represented in (c).

Displacement vectors

Regardless of the number of component vectors used to describe displacement, they can always be represented by a single resultant vector. For example, a swimmer's initial position is (4.0, 5.0). She swims in a straight line to a final position (6.0, 2.0). Calculate the displacement $\underline{d}$:

$$\underline{d} = \underline{r}_2 - \underline{r}_1 = (x_2, y_2) - (x_1, y_1) = (x_2 - x_1, y_2 - y_1)$$

Using Pythagoras' theorem, the magnitude or size of $\underline{d}$ can be found:

$$\underline{d} = \sqrt{[(x_2 - x_1)^2 + (y_2 - y_1)^2]}$$
$$\tan\theta = (y_2 - y_1)/(x_2 - x_1)$$

Therefore the direction of $\underline{d} = \tan^{-1}[(y_2 - y_1)/(x_2 - x_1)]$. Substituting into the above equations:

$\underline{d} = (2, -3)$ (the negative sign being an indication of direction)

Hence, the magnitude $= \sqrt{(9 + 4)} = \sqrt{13}$

Therefore the direction of $\underline{d}$ is $\theta = \tan^{-1}(3/2)$ and the swimmer's displacement is 3.61 m in a direction of 56.3 degrees to the horizontal axis. This is shown in Figure 8.12.

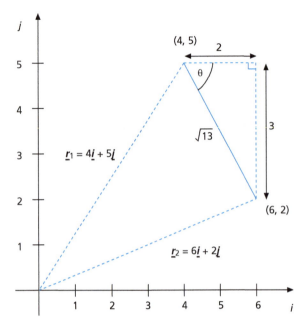

△ **Fig. 8.12** The diagram shows the displacement of a swimmer from her point of origin at A. Her position is represented in vector format.

Vector addition – the component method

The component method can be used to calculate the resultant vector or vectors of the same nature. Add the vectors graphically using the tip-to-tail method (see Fig. 8.10); visualize the resultant. Calculate the components of the individual vectors; then calculate the components of the resultant vectors. Finally calculate the magnitude and direction of the resultant vector using Pythagoras and the inverse trigonometrical function respectively.

Example

An athlete runs 400 m in an easterly direction and then 500 m at an angle of 45 degrees to the horizontal (northeast). Calculate the overall displacement (Figure 8.13).

Components of the individual vectors:

$$\underline{a}_x = 400 \quad \underline{a}_y = 0$$
$$\cos45° = \underline{b}_x/500 \Rightarrow \underline{b}_x = (500)(\cos45°) = 353.50$$
$$\sin45° = \underline{b}_y/500 \Rightarrow \underline{b}_y = (500)(\sin45°) = 353.50$$

Calculate the components of the resultant vector:

$$\underline{r}_x = \underline{a}_x + \underline{b}_x = 400 + 353.50 = 753.50$$
$$\underline{r}_y = \underline{a}_y + \underline{b}_y = 0 + 353.50 = 353.50$$

Calculate the magnitude and direction of the resultant vector:

$$\underline{r} = \sqrt{\underline{r}_x^2 + \underline{r}_y^2} = \sqrt{(753.50)^2 + (353.50)^2} = 832.30$$

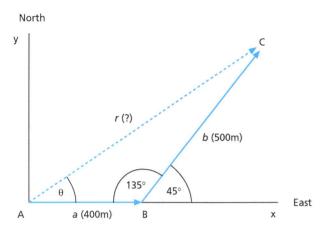

North

△ **Fig. 8.13** The component vectors run parallel to the *x*- and *y*-axes and are perpendicular to each other.

$$\tan \theta = \underline{r}_y/\underline{r}_x = 353.50/753.50 = 0.469$$
$$\theta = \tan^{-1}(0.469) = 25.13 \text{ degrees}$$

The resultant displacement is 832.30 m in a north-easterly direction of 25.13 degrees.

Velocity versus speed

In everyday conversation the word 'velocity' is often used in place of 'speed'. In biomechanics it is important to distinguish between these two terms. **Velocity** is defined as the rate of change of distance moved with change in time in a specified direction (rate of change of displacement). Velocity is therefore a vector.

$$\text{Velocity} = \text{displacement/time}$$
$$\underline{v} = \underline{d}/\Delta t$$

Velocities can be calculated as instantaneous, that is occurring over a very small interval, or as an average, where the time interval is longer (e.g. the time it takes to run a complete race). For example, the 1996 Olympic 100 m champion, Donovan Bailey, reached a peak velocity of 12.1 m/s, but his average velocity was 10.14 m/s.

Speed is defined as the rate of change of distance moved with change in time. Speed is therefore a scalar.

$$\text{Speed} = \text{distance/time}$$
$$v = d/\Delta t$$

A body is said to move with uniform velocity if its rate of change of distance moved with time in a specified direction is constant. The standard unit of measure for velocity is metres per second (m/s).

Example

A sprinter in the 100 m Olympic final records a time of 10.00 sec. What is his average velocity? Using $\underline{v} = \underline{d}/\Delta t$ we can say

$$\underline{v} = 100/10 = 10 \text{ m/s}$$

Why do we need to know how fast an object moves in a specific direction? As sports scientists, athletes and coaches we need to quantitatively measure strengths and weaknesses in athletic performance if further improvements are to be made. This information is then used in the short-, medium- and long-term planning for the athlete in terms of manipulation of training programmes, development of technique, psychological or nutritional preparation.

Vector resolution and composition

Vectors can be resolved into components. Vector $\underline{a}$ shown in Figure 8.14(a) can be resolved into the components $\underline{a}_x$ and $\underline{a}_y$ as illustrated in Figure 8.14(b). The tip-to-tail method states that the sum of vectors $\underline{a}_x$ and $\underline{a}_y$ is the same as vector $\underline{a}$. The two component vectors are parallel to the *x*- and *y*-axis and perpendicular to each other. They form a right-angled triangle.

The following equation can be used to calculate the magnitudes of the component vectors. Rightward and upward are positive and leftward and downwards are negative directions.

$$\cos \theta = \underline{a}_x/\underline{a} \Rightarrow \underline{a}_x = \underline{a} \times \cos \theta$$
$$\sin \theta = \underline{a}_y/\underline{a} \Rightarrow \underline{a}_y = \underline{a} \times \sin \theta$$

where θ is the vector angle to the horizontal axis.

(a)   (b)

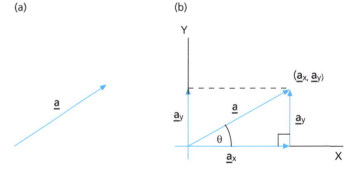

△ **Fig. 8.14** The vector in (a) can be resolved into the components in (b).

Example

A ball is kicked at an angle of 45 degrees to the horizontal. If the velocity of the ball is 15 m/s, find the horizontal and vertical velocities of the ball. Since velocity is a vector quantity it can be resolved into horizontal and vertical components as shown in Figure 8.15. Therefore from the equation above:

$$\underline{v}_x = \underline{v} \times \cos45° = (15)(0.7071) = 10.61 \text{ m/s}$$
$$\underline{v}_y = \underline{v} \times \sin45° = (15)(0.7071) = 10.61 \text{ m/s}$$

When component vectors are known, the actual vector can be derived from the components. This is called **vector composition**. In Figure 8.14(b), the components $\underline{a}_x$, $\underline{a}_y$

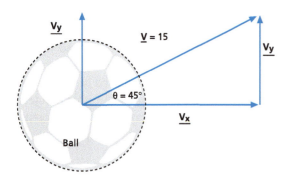

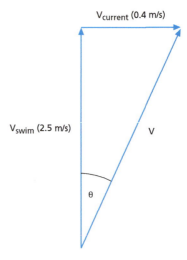

△ Fig. 8.15 As the velocity of the ball has a vector quality it can be resolved into horizontal and vertical components.

and the resultant vector **a** form a right-angled triangle. Using Pythagoras theorem:

$$\underline{a}^2 = \underline{a}_x^2 + \underline{a}_y^2 \Rightarrow \underline{a} = \sqrt{(\underline{a}_x^2 + \underline{a}_y^2)}$$

Example

What is the resultant velocity of a swimmer who swims in a perpendicular direction parallel to the two banks. His velocity is 2.5 m/s and the velocity of the current is 0.4 m/s (Figure 8.16).

The swimming and current velocities are perpendicular to each other and these can be regarded as the component vectors. Draw the component and resultant vectors, use Pythagoras' theorem to calculate the magnitude of the resultant, and then use the inverse tangent function to find the direction of the resultant vector:

$$\underline{v} = \sqrt{(\underline{v}_{swim}^2 + \underline{v}_{current}^2)} = \sqrt{(2.5^2 1 0.4^2)} = \sqrt{(6.41 = 2.53)} \text{ m/s}$$
$$\tan \theta = \text{opposite/adjacent} = 0.4/2.5 = \tan^{-1}(0.16) = 9.10$$
degrees

△ Fig. 8.16 The vectors in the diagram represent the swimming and current velocities which are perpendicular to each other. These are regarded as component vectors.

Activity 8.1

Independent learning task: Sample vector problems

1. Find the x and y components of the following vectors:
   (a) 240 N at 330 degrees
   (b) 34 m/s at 210 degrees
   (c) 15 m at 12 degrees.

2. From the x and y components given below, find the direction and magnitude of the resultant.
   (a) $\underline{F}_y = 120$ N, $\underline{F}_x = 345$ N
   (b) $\underline{v}_y = 31$ m/s, $\underline{v}_x = 8$ m/s.

3. A soccer ball is kicked with a horizontal velocity of 11.3 m/s and a vertical velocity of 3.5 m/s. What is the magnitude and direction of the ball's velocity?

4. A shot putter applies a force of 415 N to a shot at an angle of 37 degrees. What are the horizontal and vertical components of this force?

TIME OUT

KINEMATIC CHARACTERISTICS OF ELITE MEN'S AND WOMEN'S 20KM RACE WALKING AND THEIR VARIATION DURING THE RACE

Hanley *et al.* (2011) analysed important kinematic variables in elite men's and women's 20km race walking. Thirty men and thirty women were analysed from video data recorded during the World Race Walking Cup. Video data were also recorded at four points during the European Cup Race Walking and 12 men and 12 women analysed from these data. Two camcorders operating at 50Hz recorded at each race for 3D analysis. The two main performance determinants of speed were step length and cadence. Men were faster than women because of their greater step lengths but there was no difference in cadence. A reduction in cadence was the initial cause of slowing down with later decreases in speed caused by reductions in cadence. Shorter contact times were important in optimising both step length and cadence, and faster athletes tended to have longer flight times than slower athletes. It was less clear which other kinematic variables were critical for successful walking, particularly with regard to joint angles. Different associations were found for some key variables in men and women, suggesting that their techniques may differ due to differences in height and mass.

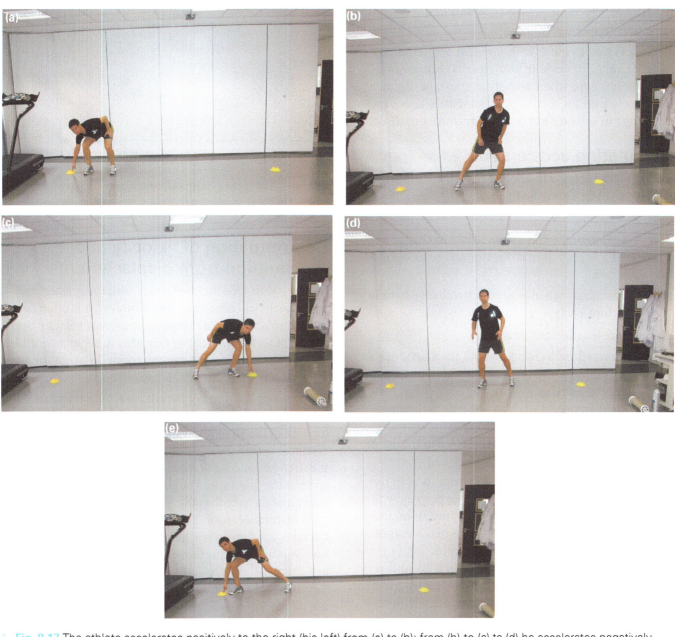

△ **Fig. 8.17** The athlete accelerates positively to the right (his left) from (a) to (b); from (b) to (c) to (d) he accelerates negatively, and acceleration again becomes positive from (d) to (e).

Acceleration

Acceleration is defined as the rate of change of velocity with respect to time. Success in many sporting activities depends upon the athlete's ability to increase or decrease speed or change direction rapidly such as in rugby, netball, putting the shot or decreasing ball velocity when catching.

Generally, the rate of change is regarded as **positive** if the velocity is increasing (acceleration) and **negative** if the velocity is decreasing (deceleration or retardation). However, we need to be more specific when referring to positive or negative direction. For example, when an athlete side steps at a *constant rate* to the right (positive

direction) as part of an agility drill, the acceleration is zero. If the athlete gets faster in the positive direction, he accelerates positively, and if he slows down then he accelerates negatively (see Figure 8.17, a, b, c, d). If the athlete then changes direction by accelerating back towards the starting point (or the origin in the negative direction) he accelerates negatively. Acceleration in the negative direction (negative acceleration or deceleration) is what happens if he continues to apply a force that opposes the original direction of movement. If the athlete again slowed down in the negative direction, he would be accelerating negatively in the negative direction – this is positive acceleration (two negatives make

181

a positive). The same principle applies in the vertical direction (Figur 8.17(d) to (e)).

A body is said to move with uniform acceleration if its rate of change of velocity with time is constant. Acceleration is zero when velocity is constant. Because velocity is a vector, a change in velocity is also a vector. A change in velocity has both magnitude and direction.

Calculating acceleration

To calculate a change in velocity (Δv), the following equation is used (see Figure 8.18):

$$\Delta \underline{v} = \underline{v}_{final} - \underline{v}_{initial}$$

Linear acceleration is therefore the rate of change of linear velocity. Acceleration is a vector and has magnitude and direction.

Acceleration = change in velocity/change in time
Therefore $\underline{a} = (\underline{v}_1 - \underline{v}_2)/(t_1 = t_0)$

Its units are length/time^2 or m/s^2.

Acceleration reflects a change in both the magnitude and direction of velocity and is defined as a change in velocity with respect to a change in time. The component vectors are shown in Figure 8.19, which presents the component velocity vectors (see Figure 8.19(a)) and the corresponding component acceleration vectors (reflected in the definition) – see Figure 8.19 (b).

Generally, acceleration reflects a change in both the magnitude and direction of velocity (e.g. a rugby player changing velocity and then side-stepping an opponent). Acceleration can also reflect a change in the magnitude of velocity without a change in direction (e.g. the change in speed of a 100-m sprinter in a linear direction during the acceleration phase of the run), or it can reflect a change in the direction of velocity without a change in magnitude (e.g. a distance runner moving at constant velocity around a circular track or route).

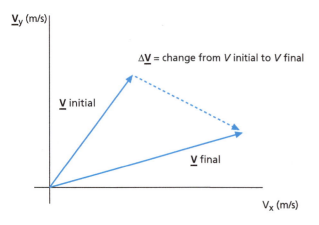

△ **Fig. 8.18** Indicating a change in velocity.

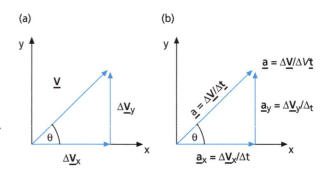

△ **Fig. 8.19** The component velocity vectors are represented in (a) and the corresponding component acceleration vectors are represented in (b).

Determination of velocity and acceleration from displacements

Over a 100 m race the competitor with the greatest average velocity will obviously win. However, it is often of interest to analyse the competitor's average velocity over shorter intervals within the race, that is, segments of 10m, 5m or even less. So what can the coach do to obtain more information about his athlete?

By recording the times of the athlete at 10m intervals, for example, the mean or average velocity can be found for each of the 10 m intervals. Velocity over these shorter distances within the race allows one to determine some of the strengths and weaknesses of a sprinter. For example, if a sprinter has good speed out of the blocks but slows towards the end of the race the athlete may require additional muscular endurance work. If the athlete has a good top speed but takes too long in reaching it, more strength/power training should be performed. The data can be recorded in a chart as shown in Table 8.1 and then represented as a displacement-time graph (Figure 8.20). To calculate average velocity and acceleration, use the following equations:

Velocity ($\underline{v}$) = Change in displacement/Change in time,

or:

$$\underline{v} = (\Delta \underline{d})/(\Delta t)$$

Similarly acceleration is defined as the rate at which velocity changes with respect to time. That is:

Acceleration = Change in velocity/Change in time

or:

$$\underline{a} = (\Delta \underline{v})/(\Delta t)$$

Example calculation

An athlete has performed two 10 m sprints, one using a crouched (c) start and the other with a standing (s) start. The following times were recorded electronically:

Crouch = 1.890 seconds   Standing = 1.927 seconds

Distance covered (m)	Time (t)	Change in distance (m)	Time interval Δt (s)	Mean velocity per 10 m interval
0		10		
10		10		
20		10		
30		10		
40		10		
50		10		
60		10		
70		10		
80		10		
90		10		
100		10		

△ **Table 8.1** Example of chart for recording an athlete's times at intervals of 10 metres.

Calculate the average velocity.

Crouch velocity = _____ Standing velocity =

Instantaneous Velocity and Acceleration_____

If the coach requires more information about his athlete a mathematical technique called *finite difference* technique allows the coach to obtain data for instantaneous velocity. The method is straightforward. Using the equation $\underline{v}_{xi} = (x_{i+1} - x_{i-1})/2\Delta t$, the finite difference technique is used to calculate the slope of the curve at the *i*th sample point (see Figure 8.21). This is used for the calculation of velocity-time data from displacement-time data.

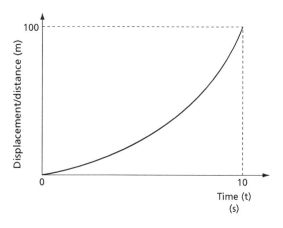

△ **Fig. 8.20** The relationship between displacement and time, which represents the gradient at time (*t*).

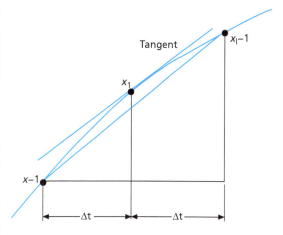

△ **Fig. 8.21** Instantaneous velocities can be calculated from displacement-time data.

Calculating instantaneous acceleration-time data

Acceleration at the *i*th point can be calculated from either displacement or velocity data:

From displacement data use the following equation:

$$\mathbf{a}_{xi} = (x_{i+1} - 2x_i + x_{i-1})/\Delta t^2$$

From velocity data use the following equation:

$$\mathbf{a}_{xi} = (\mathbf{v}_{xi+1} - \mathbf{v}_{xi-1})/2\Delta t$$

Interpreting graphs of kinematic data

Sports biomechanists can represent and interpret information graphically for displacement, velocity and acceleration as functions of time. This information can be used to better understand the factors influencing performance in many sports.

The kinematic *relationships* can be best illustrated using curves obtained from a force platform during a countermovement jump. A force platform measures the force exerted on it by the athlete, and according to Newton's third law of motion (covered in Chapter 9) equates to the force exerted by the platform on the athlete (i.e. a ground reaction force). During a countermovement jump, the jumper starts from an upright position and makes a preliminary downward movement by flexing at the knees and hips, followed by an immediate and vigorous extension at the knees, hips and ankles to jump vertically off the ground (Figure 8.22). A countermovement jump is an example of a movement that uses the stretch-shortening cycle (SSC),

Activity 8.2

Independent learning task: Displacement, velocity and acceleration.

The curves with (a) to (i) provided represent instantaneous profiles of displacement (d), velocity (v) or acceleration (a), with respect to time (t). Enter the most appropriate time (t_i) that represents the following for each curve:

- Peak positive displacement
- Peak negative displacement
- Peak positive velocity
- Negative acceleration
- Peak positive acceleration
- Peak negative acceleration

- Zero velocity
- Positive velocity
- Negative velocity
- Peak negative velocity
- Zero acceleration
- Positive acceleration

Note that *more* than one time may be appropriate for any of the parameters. Also, *no* time may be appropriate.

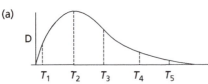
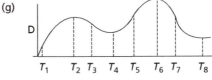
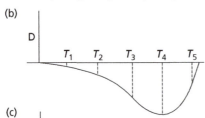
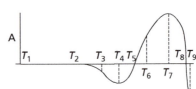
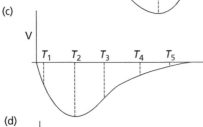
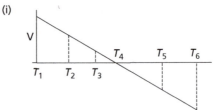
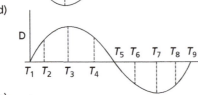
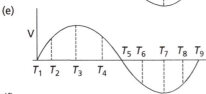
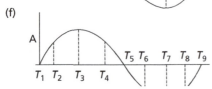

Activity 8.3

Independent learning task: Velocity–time graph

A weightlifter performs the clean and jerk. She lifts the bar from the floor and racks it across the front of her shoulders, pauses momentarily and then drives the bar overhead. She then lowers the bar back to the ground. The graph below shows the vertical height of the bar with respect to time.

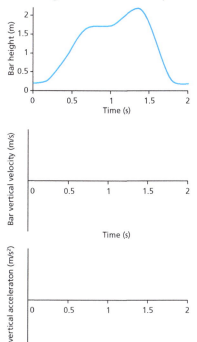

1. Sketch the vertical velocity of the bar with respect to time.
2. Sketch the vertical acceleration of the bar with respect to time.
3. On the velocity–time graph, clearly mark the point where the downward velocity of the bar is at its greatest.

whereby the muscles are pre-stretched before shortening in the desired direction. There are many actions from sport and human movement (e.g. walking, running, jumping, throwing) that utilise this mechanism. Figure 8.23 shows the key times and phases that can be identified during a vertical jump (adapted from Linthorne (2001b):

a: Point **a** is the start of the jump; the jumper stands upright and stationary.

a-b: The jumper relaxes the hip and leg muscles, allowing the knees and hips to flex under the effects of gravity. The resultant force on the jumper (F_{GRF}-mg) (where GRF = Ground Reaction Force, m = body mass and g = acceleration due to gravity) becomes negative, and the jumper's centre of mass moves and accelerates downwards – acceleration and velocity of the centre of mass are both negative.

b: Point **b** marks the maximum downward acceleration of the jumper's centre of mass.

b-c: The downward/negative motion of the jumper cannot continue forever. The jumper starts to increase activation of the extensor muscles, but the centre of mass is still moving downward. The resultant force on the jumper and the acceleration of the jumper's centre of mass are still negative.

c: At this point the GRF is equal to body weight, and so the resultant force on the jumper and the acceleration of the jumper's centre of mass are zero. Point *c* marks the maximum downward velocity. The region **a-c** is sometimes called the 'unweighting' phase because the ground reaction force is less than body weight.

d: Point **d** is the lowest point of the countermovement jump, where the centre of mass is momentarily at rest (velocity = 0). The leg muscles are now strongly activated and the ground reaction force is close to maximum.

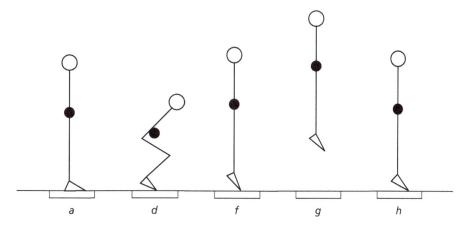

△ **Fig. 8.22** Sequence of actions during a counter-movement jump. The jumper's centre of mass moves in the vertical direction and is shown at key times during the jump.

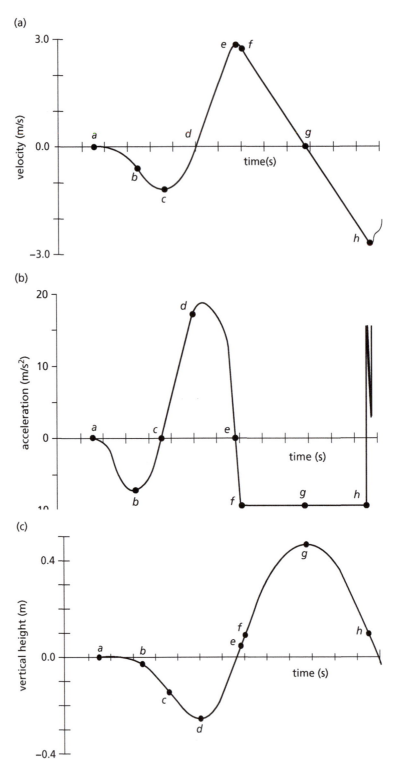

△ **Fig. 8.23** Kinematic relationships: (a) displacement-time, (b) velocity-time, and (c) acceleration-time graphs of the centre of mass during a counter-movement jump. Plots are aligned by the point of take-off beginning the flight phase where time equals zero.

A common error when examining force-time curves is to identify point **b** as the lowest point of the countermovement. Although the velocity of the centre of mass is zero, the acceleration is extremely high, as is the GRF.

d-e: This is the 'propulsion' phase; the jumper moves upwards (concentric muscle actions) by extending the knees and hips. The velocity is now positive (upwards). For many jumpers, the maximum ground reaction force occurs early in this phase, shortly after the lowest point of the countermovement.

e: The ground reaction force has dropped to become equal to bodyweight. The resultant force on the jumper and the acceleration of the jumper's centre of mass are therefore zero. Point **e** marks the maximum upward vertical velocity. The high jump and long jump also display this phenomenon, where the maximum upward vertical velocity is not at the instant of take-off, but at a short time before take-off.

e-f: The ground reaction force drops below bodyweight and so the resultant force on the jumper and the acceleration of the centre of mass are negative. The centre of mass is still moving upwards but it has started to slow due to the effects of gravity.

f: Point **f** is the instant of take-off, where the ground reaction force first becomes zero. The centre of mass is higher at take-off because the jumper has extended the ankle joints.

f-g: The only force acting on the jumper is the jumper's weight and so becomes a projectile in free flight. The reason **f-g** marks the ascent of the flight phase where the jumper's centre of mass is moving upward but slowing down due to gravity.

g: Point **g** marks the peak of the jump; the jumper's centre of mass is momentarily at rest.

g-h: This is the descent of the flight phase; the centre of mass is moving downwards (velocity is negative) and speed is increasing.

h: Point **h** is the initial point of landing, where the feet first contact the ground. The GRF shows a sharp impact peak and eventually becomes equal to bodyweight when the jumper is again standing motionless on the platform.

An introduction to gait analysis

Unnatural or abnormal forces placed upon the body due to alterations in running or walking technique can cause injury. During human movement, the limbs and trunk should be considered as movable segments or chains, linked together at the joints. **Open chain movements** involve the distal segment remaining free whilst the proximal bone segment remains fixed. During **closed chain movements**, both the proximal and distal segments remain fixed. Muscle function differs during closed and open chain movements. For example, open chain dorsiflexion when sitting down occurs as a result of a concentric contraction of the anterior tibialis. Closed chain dorsiflexion when walking involves an eccentric contraction of the calf muscles (Norris, 1999). The joint movements in walking or running involve the hip, knee and ankle joints. The hip joint is the articulation between the femoral head and the acetabulum; the knee joint articulation is between the condyles of the femur and the tibia; and the ankle or talocrural joint involves the trochlea surface of the talus and the distal ends of the tibia and fibula.

In actions such as walking or running, the body attempts to move horizontally across the ground. Any other movement, especially vertical, does not help this aim and uses up energy. If the body had wheels then we could avoid vertical movement altogether. However, we have legs and so there must be some vertical movement. This occurs because at the heel-strike and toe-off the two legs make up the sides of a triangle, while during mid-stance the leg is vertical. This has the effect of lowering the upper body at heel-strike and toe-off and raising it during mid-stance (Figure 8.24).

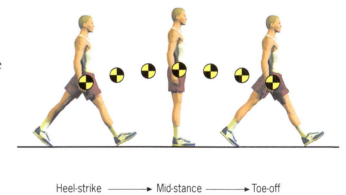

Heel-strike ⟶ Mid-stance ⟶ Toe-off

△ **Fig. 8.24** The path of the centre of mass during gait.

Research has shown that there are mechanisms used by the body to reduce this up-down path of the centre of mass during walking. These have been called the determinants of gait. The main goals of these determinants are to reduce the maximum height of the centre of mass during mid-stance and also to increase the minimum height of the centre of mass at heel-strike and toe-off.

The gait cycle

The gait cycle can be divided into two phases, the **stance phase** and the **swing phase**. The stance phase occurs when the foot is in contact with the floor supporting the bodyweight. Closed chain movements occur in this phase, such as during deceleration. The swing phase occurs when the foot leaves the floor and acceleration and open chain movement occur.

The foot goes through four phases during the stance phase. Initially the heel makes contact with the floor. As the centre of mass continues to move forward, the foot flattens and again as the centre of mass continues to move forward, the heel lifts off the floor and finally the toes push off the floor (toe-off) (Figure 8.25).

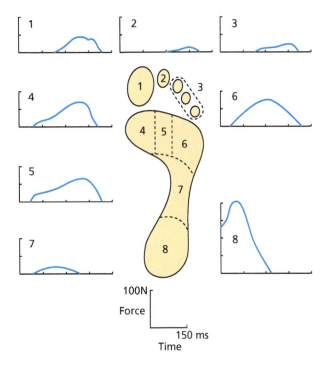

△ **Fig. 8.25** Forces produced at the foot during the stance phase vary from a peak at the heel (8) followed by a reduction in mid-support (7) an increase at the ball of the foot (4), (5) and (6) a significant reduction at the toes (3) and (2) and an increase at take-off or toe-off (1). From Reid (1992).

At the initiation of the swing phase the leg accelerates, during mid-swing the speed is constant and finally the leg decelerates as it is lowered to the ground. The cycle starts all over again when the heel strikes the floor (Figure 8.26).

Determinants of gait

The following actions all contribute to very small reductions in the amplitude of the centre of mass. Added together, however, they reduce it significantly – approximately 5 cm (up–down and side-to-side) in normal gait (Kirtley *et al.*, 1985).

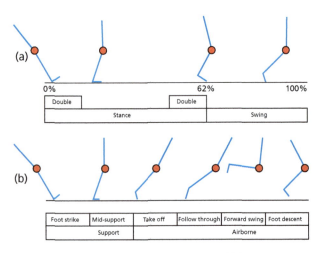

△ **Fig. 8.26** The walk cycle and run cycle. This highlights the differences between the walk and run cycles. The stance phase in walking accounts for approximately 60% of the total gait cycle. The living phase accounts for about 40%. In running, movements are more rapid. As a result, the stance phase takes up less of the total cycle time. From Subotnick (1989).

- **Pelvic rotation**: At heel-strike the pelvis rotates anteriorly and at toe-off it rotates posteriorly. This increases the effective leg length at these times.
- **Pelvic list**: The pelvis moves downwards to increase the effective leg length at heel-strike and toe-off.
- **Stance phase knee flexion**: Slight knee flexion lowers the centre of mass during stance.
- The ankle is dorsiflexed at heel-strike and plantarflexed at toe-off. Both of these actions increase the leg length.
- **Transverse rotation**: The lower leg is lengthened by external rotation and shortened by internal rotation.
- **Genu valgum**: The anatomical valgus at the knee allows a narrower walking base, and so a smaller lateral shift occurs.

The biomechanics of running

The biomechanical aspects of running are of extreme importance for a number of reasons. First, a knowledge of movements of the body segments may provide useful information with regard to the mechanisms of the neuromuscular system. Secondly, identifying optimal running mechanics may improve technique and consequently athletic performance. Finally, injury mechanisms specific to running may aid injury prevention, although further investigation into the relationship between biomechanical parameters and performance and injury are still needed to clarify this (Williams, 1985).

Sprint results occur due to many biomechanical factors. According to Mero and Komi (1994), the most important factors are start reaction time, technique, EMG or electromyographic activity, neural factors, force production,

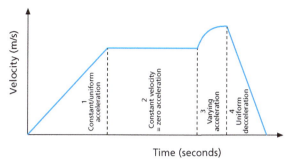

△ Fig. 8.27 An acceleration curve for an athlete sprinting 100m.

muscle structure and external factors like weather conditions, footwear and running surface. The sprint can be split into four phases (Figure 8.27) (Mero *et al.*, 1992):

1. The start/acceleration phase
2. Constant velocity phase
3. Varying acceleration phase
4. Uniform deceleration.

In terms of the gait cycle there are two major phases when running (Figure 8.28):

1. **The support phase:** This starts as the foot makes contact with the ground and continues until the foot leaves the ground (toe-off).

2. The **non-support phase** or **airborne phase:** This starts when the toe leaves the ground and continues until the same foot again makes contact with the ground.

Sprinting is a cyclical movement, and it is important to identify differences between walking and running. In walking, both feet are simultaneously in contact with the ground during the support (or stance) phase. One foot is touching the ground while another foot is leaving the ground in the non-support (or swing) phase. In running, no foot is in contact with the floor during the early part of the non-support (or swing) phase.

Maximal sprinting velocity is the product of stride length and stride rate. These are interrelated and dependent upon technique, morphology, contact time and force production in both the braking and propulsive phases. Studies have reported that the greater the velocity of the athlete the shorter the contact time (Mero *et al.*, 1992). Optimal contact time is essential if the greatest horizontal force is to be achieved in the propulsive phase, since this force moves the sprinter forward.

According to biomechanical principles, the longest stride and contact times should occur during the acceleration phase so the athlete can generate maximal force. There should be a large difference between the braking and propulsive phases during foot contact. Braking forces

△ Fig. 8.28 Model for running. (a) The athlete stands in the support phase, before moving to the non-support or airborne phase (b). Once the athlete's right leg makes contact with the ground, he will be back in the support phase (c).

should be kept to a minimum, with propulsive forces at their highest. Therefore to ensure the most economic phase of the forward support so that the loss in horizontal velocity of the centre of mass is the smallest possible, a high backward grabbing velocity of the foot under the centre of mass of the body must occur.

Activity 8.4

Independent learning task: Kinematics and speed assessment

Assessing speed is easy, kinematic data can be obtained from using either timing gates, a radar gun, video camera or a stopwatch; however, improving speed is the real challenge for both the coach and athlete. Testing considerations should include the distance covered, running surface, footwear and weather conditions if testing outside. In terms of the sub-qualities of speed there are a variety of ways of looking at the concept of speed assessment. These sub-qualities of speed are namely:

- Reaction: Considered the ability to react to an opponent or stimuli which may be auditory, visual or kinaesthetic – also known as 'quickness'.

- Agility/coordination: Defined as a rapid whole-body movement with change of velocity or direction in response to a stimulus. There are 2 main factors that are included in the development of agility: change of direction and cognitive. The ability to rapidly change direction whilst maintaining good body mechanics/coordination is trainable. The greater the number of changes in direction within a movement episode, the greater the agility demands and the lesser the maximum velocity demands.

- Acceleration: Already defined as the rate of change in velocity with respect to time, typically in sports it implies speed over the first 5–10m from a stationary start. However, it also includes the rate of change in velocity from different starting/moving positions; all need to be trained.

- Maximal velocity: This is considered the highest velocity attained during a speed episode. Typically occurs between 30–40m in a field sport athlete but maybe 40–50m in a track and field athlete.

- Speed-endurance: This is considered the ability to repeat speed efforts with limited degradation in performance. Very important for field and court sports athletes, however, speed endurance is the least studied of all speed sub-qualities.

N.B: The above qualities *are related*, but separate enough to warrant different training methods.

Equipment

Timing gate (or light) systems are becoming increasingly popular as a means of assessing acceleration and speed by

Activity 8.4 *(continued)*

measuring the time to cover certain distances. A timing gate consists of a light source (transmitter) and an optical pick-up (receiver). Visible, infrared, standard or monochromatic laser lights are the usual light source used. The light source is aimed across the track at the reflector and bounced back to the receiver. This forms a 'gate' that the athlete will pass through. Timing gate systems can vary from single, dual and three-beam reflector units. The disadvantage of single beam systems is that they can be triggered early by a swinging arm, and may produce measurement error of up to 80ms. In addition, three-beam systems are not necessary, as accuracy improvement is less than the system resolution of 0.01s.

Task

Although the use of timing lights is widespread, the descriptions of their use, in particular the methods used, are generally very limited. It is important that when measuring athletic abilities the procedures are standardised. This helps us to collect *reliable* data that we can use to monitor the progress of the athlete.

1. Define reliability.

2. When using timing gates, how would you test the reliability of your assessment procedures?

3. Name three factors that you would need to *standardise* when using timing gates to ensure that your assessment procedures were reliable.

Centre of mass and sprinting

The centre of mass is the point about which every particle of a body's mass is equally distributed. During the standing position it is located anterior to S2 at approximately 55 per cent of body height. In biomechanical analyses, the centre of mass represents the centre of the whole body in each posture. Its location changes when we change our posture. The changes in the centre of mass are used to calculate the velocity and acceleration of the whole body. This is discussed in greater detail in Chapter 9. However, as we are already aware, velocity is equal to the displacement of the centre of mass divided by the change in time or the time interval. The acceleration is calculated by dividing the change in the velocity of the centre of mass by the time interval.

Projectile motion

In this section we will look at projectile motion and how gravity influences it, discuss the effects of projection angle, speed and relative height and learn how to calculate maximum height, flight time and flight distance of projectiles.

Activity 8.5

Independent learning task: Stride length and stride rate

To understand how to develop speed technique we need to understand the mechanical components that affect running. In its simplest form, a runner's speed is determined by the stride length (distance travelled by consecutive foot contacts) and stride rate (number of strides taken per unit of time).

SPEED = STRIDE LENGTH × STRIDE RATE

Therefore to increase speed either or both the length or rate of stride can be increased (see Figure 8.29).

Task

A subject performs three runs across the gymnasium floor. The first of these is performed at a slow jog, the second at a moderate pace, the third at a fast pace. Once the subject reached a constant pace the time taken to perform 5 strides was recorded, as well as the distance of covered in the 5 strides. The time taken from starting heel strike to 5th heel strike was recorded using a video system. To do this the number of frames for the athlete to complete 5 strides was counted. Hint: Each video frame = 1/30 second = 0.033 seconds.

STRIDE RATE = NUMBER OF STRIDES/TIME TAKEN

Given the data for number of video frames counted for 5 strides and the distance covered, complete the table below.

Stride length and rate – changes with running speed

Speed	Frame count for 5 strides	Stride Time (s)	Distance for 5 strides	Stride length (m)	Stride rate (strides/s)	Speed (m/s)
Slow	121		6.25			
Medium	115		14.80			
Fast	79		15.45			

Note: m/s can also be written m.s^{-1}

On graphs (a) and (b) mark the values of stride length and stride rate above each point to indicate the speed.

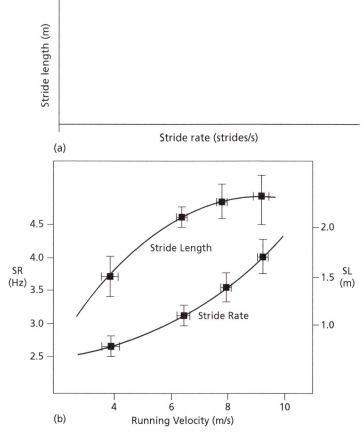

(a)

(b)

Stride length and rate as a function of running velocity. (Cronin, 2006).

Questions

1. Discuss the relationship between stride length and stride rate at different speeds.

2. If you wanted to improve the maximum velocity ability of your athlete, what component would you focus on developing and give an example how you might do this?

Performance in many sporting activities is dependent on the ability either to control or predict the motion of a projectile. In attempting to produce a particular trajectory of a projectile, we have the ability to manipulate three basic characteristics of the projectile at the instant it is released into the air: its **velocity**, the **angle** at which it is projected, and the **height** at which it is released above the landing area. Of particular interest in many examples of projectile motion is the horizontal distance travelled by the projectile (its range). In fact, this represents the performance measure in many sports (e.g. long jump, shot put).

Box 8.1

Considerations for filming human movement (see also Chapter 13)

- **Frame rate** – This is the number of samples per second (Hz) or the time between each still image of the performance (e.g. 25Hz, 1 frame = 0.4s). During sport and exercise, many activities or movements are carried out at a relatively high velocity. Therefore, 4–5 times the rate of the movement being recorded is appropriate, i.e. the frequency content of the signal.

- **Shutter speed** – This is the period that the film is exposed or the time period to capture each image. The shutter speed must be fast enough to prevent blurring, but slow enough to allow enough light to enter. Similarly, the light intensity must be bright enough to give sufficient contrast between markers and background. Therefore the aperture width must be large enough to allow light in through the lens.

- **Camera position** – This is important when analysing sport and exercise movements. The camera position must be perpendicular to the plane of motion to reduce perspective errors, and a large distance from the plane of motion to reduce parallax errors. For two-dimensional analyses only one camera is required, along with one plane of motion (usually sagittal plane – greater movements).

As can be seen from Figure 8.29, the term perspective error is used to refer to the error in the recorded length of a limb or body segment which is at an angle to the photographic plane of motion and appears to be shorter than it actually is. Typically, objects behind the plane of motion appear *smaller*; objects in front of the plane of motion appear *larger*. Perspective errors can be reduced by increasing the camera distance from the plane of motion (Figure 8.29(b)). In addition, movement parallel to the plane of motion does not affect measured angles, but rotation out of the plane of motion affects lengths and angles (Figure 8.30). Likewise, parallax errors are caused by viewing away from the optical axis across the plane of motion. The view is therefore not always side on. Bartlett (1997) suggests that the combined results of these errors is

that limbs nearer the camera appear to be larger and to travel further than those furthest from the camera. Errors then occur when digitizing coordinates.

△ **Fig. 8.29** Rotation out of the plane of motion affects both lengths and angles.

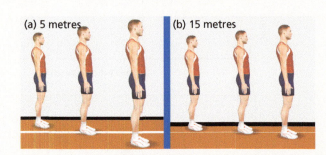

△ **Fig. 8.30** Placing the camera further away from the plane of motion can reduce measurement error.

- **Spatial reference** – A length scale such as a metre rule and a vertical reference such as a plumb line must be included in the field of view. Filming of known lengths allows the scaling of digitized coordinates to real life dimensions. The inclusion of horizontal and vertical scalings makes sure that the aspect ratio (video cameras scale horizontal and vertical coordinates differently) is maintained.

(Continues)

Box 8.1 (*continued*)

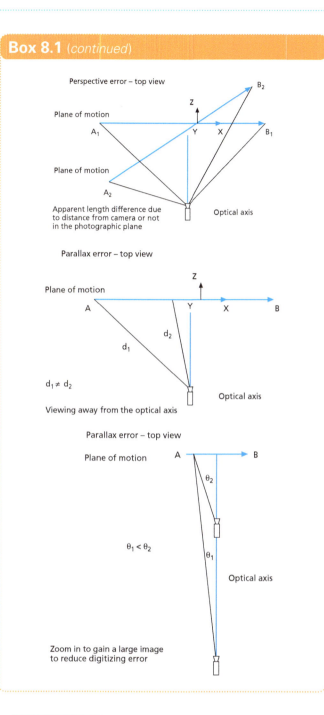

The javelin is an example of an event in which biomechanics research has assisted in fine-tuning the performances of elite athletes. Coaching used to concentrate on the release speed of the javelin, but through the use and application of quantitative biomechanical analysis, researchers have analysed speed of run-up and delivery, elbow angle during various stages of the run-up and the relationships between hips and shoulders.

What is a projectile?

A projectile is a body or object that is in the air, is subject only to the forces of gravity and air resistance and follows a parabolic path. Examples include balls, implements and the human body. The movement of the centre of mass of any object in free fall is governed by the laws of projectile motion.

Why is projectile motion important?

Success in many sports involves projectile motion. Objects acts as projectiles in basketball, football, rugby, tennis and badminton for example, whereas human bodies act as projectiles in the high jump, long jump, ski jumping and gymnastics, etc. Variables of interest include the flight distance, flight time and maximum height. There are four types of projectile motions:

1. **Vertical projectiles:** Where the object travels straight up and down.
2. **Horizontal projectiles:** Begin with only a horizontal velocity and are projected from a height.
3. **Classic projectiles:** Take-off and landing height are equal.
4. **General projectiles:** Take-off and landing height are not equal.

TIME OUT

KINEMATICS OF SPRINTING AND ANAEROBIC TESTING

Berthoin *et al.* (2001) studied the relationship between kinematic parameters measured during the first 10 seconds of a 100 m sprint and anaerobic tests in 22 male physical-education students. During the first 10 seconds of the sprint, the position of the runners was 'continuously' measured with a laser telemeter. Maximal acceleration, maximal velocity and time to reach the maximal velocity were derived from position data. In addition the subjects performed anaerobic tests: squat jump, countermovement jump, and a force–velocity test to measure maximal power, maximal theoretical cranking velocity, maximal theoretical isometric force and the 30-s Wingate anaerobic test. Since the countermovement jump was the anaerobic performance best correlated to the different kinematic parameters of the run ($r < 0.56$ and $p < 0.05$), the results failed to identify one anaerobic test that specifically explains one sprint kinematic parameter.

Why study projectile kinematics?

As a coach or athlete, knowledge of projectile release velocity and angle of projection is essential if maximum potential is to be achieved. Most individuals are aware that if you want to maximize the range of a projectile it must be released at an angle of 45 degrees. But this is only true if the release height and the landing height are the same.

Figure 8.31 shows the ranges a shot putter achieved at various release speeds in relation to projection angle. A shot putter whose maximal release velocity is 12 m/s reaches a distance of approximately 16 m at a release angle of 40–42 degrees. Any release angle greater than this produces a reduction in shot put performance at the same release velocity. This further reduction in performance is due to the relationship between a decrease in projection speed with increasing release angle. Having measured your athlete's release angle through video analysis, how can you improve performance? What are the implications for your athlete's strength and conditioning programme? In other words, how can you, as a coach, improve your athlete's release velocity?

The kinematic quantities of interest in projectile motion include:

- Horizontal displacement: shot put, javelin, long jump, discus
- Maximum vertical displacement: high jump, pole vaulting
- Trajectory (speed and accuracy): basketball, rugby, football.

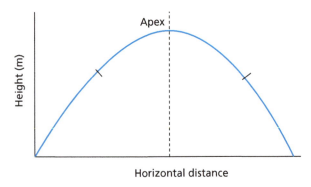

△ **Fig. 8.32** Gravity causes projectiles to move in a parabolic path that is symmetric about the apex.

Factors that affect projectile motion

- **Gravity:** This can be defined as the **pull** of the mass of the earth on a body. When objects move between the angles of 0° (i.e. in the horizontal direction) and 90° (i.e. in the vertical direction), gravity will only act on bodies that move with some vertical motion. For example, gravity will pull a projectile downwards, causing a parabolic trajectory that is symmetric about its apex (i.e. the highest point) (Figure 8.32). It has the effect of accelerating an object vertically downwards towards the centre of the earth. Gravitational acceleration or acceleration due to gravity (g) is always straight downward at a constant of 9.81 m/s² (Figure 8.33). However, there are occasions where the acceleration due to gravity (g) of an object will vary, and will be dependent upon its location on earth. Due to the radius of the Earth being slightly larger at the equator than at the poles, and its distorted shape caused by the earth's spin, gravity is slightly lower at the equator (9.78 m/s²) than it is at the north and south poles (9.83 m/s²). Gravity is also lower at altitude; it's approximately 0.2 per cent lower at the top of Everest. The combined effects

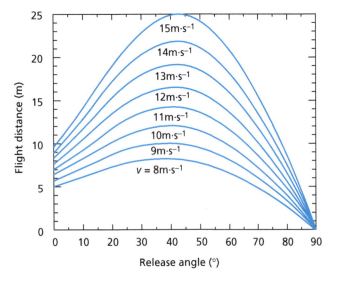

△ **Fig. 8.31** The ranges achieved by a shot-putter at various release speeds in relation to projection angle. From Linthorne (2001a).

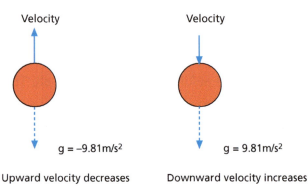

△ **Fig. 8.33** Gravity accelerates objects in a vertical direction and is represented by the letter g. When an object is projected upwards g is negative and when projected downwards g is positive.

of altitude and a near equatorial position (i.e. lower acceleration due to gravity) were evident at the 1968 Olympic Games in Mexico City where many world record performances were set.

- **Air resistance**: This affects the trajectory of the projectile. The extent of its influence is dependent upon the velocity and shape of the projectile. In many cases it is ignored for simplicity.

Let us put some equations in place here for the effects of gravity on projectile motion.

Gravity and vertical velocity

From the law of constant acceleration:

$$v = u + at$$

where v is the final vertical velocity, u is the initial vertical velocity, a is acceleration due to gravity (= g) and $t = \Delta t$.

Vertical velocity changes linearly with time (Figure 8.34).

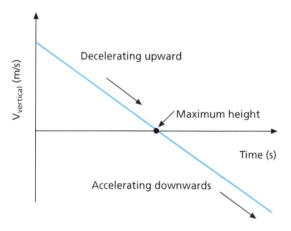

△ **Fig. 8.34** Vertical velocity changes linearly with time. When objects are projected upwards they eventually start to decelerate until they reach maximal height where $v = 0$ at time t. The object returns to earth by accelerating downwards.

Gravity and horizontal velocity

Gravity does not change the horizontal velocity of an object. From the law of constant acceleration:

$$v = u + at$$

where v is the final horizontal velocity, u in the initial horizontal velocity, a is the horizontal acceleration and $t = \Delta t$. Since for gravity, $a_{horizontal} = 0$, then $v_{horizontal} = u_{horizontal}$, which means that $v_{vertical}$ changes while $v_{horizontal}$ remains constant (Figure 8.35).

Since weight is proportional to mass, the acceleration produced by weight alone is always the same. This means that

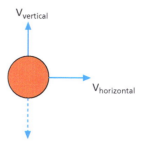

△ **Fig. 8.35** Gravity has no influence on horizontal velocity.

athletes or objects in flight have weight as the predominant force. Figures 8.36 (a) and (b) show that after take-off the flight of the long jumper and diver is governed by weight only.

Influences on projectile trajectory

There are three factors that influence projectile trajectory (Figure 8.37):

- Projectile angle
- Projection speed (the most influential)
- Relative height of projection (i.e. the projection height – landing height)

Although there are specific influences involved in projectile trajectory, research undertaken by Cushing and Robinson (2003), has derived a mathematical equation based on the 85 per cent kicking success rate of England rugby's fly half Jonny Wilkinson. The first part of the equation identifies the biomechanical factors in kicking a rugby ball between a set of posts. The second part looks at the external factors. Analysis of Wilkinson's performance using part 1 of the equation concluded that much of his success was due to his consistency – his ability to execute as near to the same length of run-up to the ball, with the same run-up length, angle and speed of approach at each attempt. This was combined with striking the ball with a similar foot position and force at every attempt. The second part of the equation included environmental conditions such as the wind and rain, the player's psychological state depending on the crowd or the score at that time, and physical factors such as fitness levels and fatigue.

The researchers concluded that the principles are applicable to any closed skill like javelin throwing and they highlight the interdisciplinarity that is involved in performing such activities successfully.

Influences of projection angle

Figure 8.38 shows the effects of various projection angles on trajectory, where projection speed and height remain

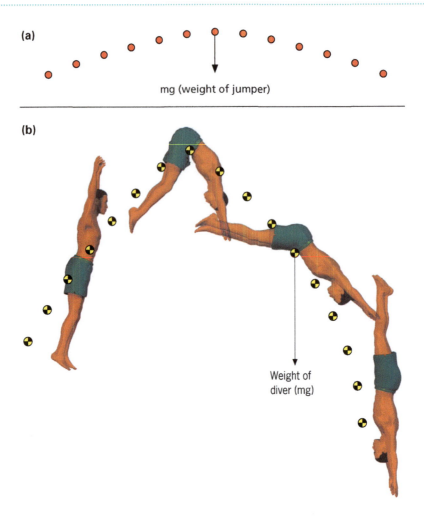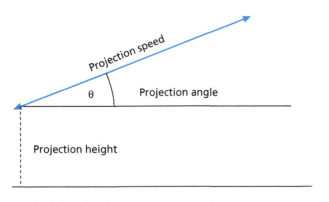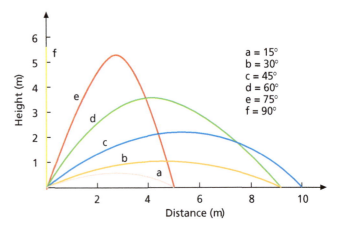

mg (weight of jumper)

Weight of diver (mg)

△ **Fig. 8.36** Flight phase of (a) a long jumper (b) a diver.

Projection speed

Projection angle

θ

Projection height

△ **Fig. 8.37** Major influences on projectile trajectory.

a = 15°
b = 30°
c = 45°
d = 60°
e = 75°
f = 90°

Height (m)

Distance (m)

△ **Fig. 8.38** The effects of various projection angles on trajectory, where projection speed and height remain constant.

constant (v = 10 m/s and projection height = 0). Trajectory shape depends only on projection angle.

Influences of projection speed

Figure 8.39 shows the effect of projection speed on object trajectory, where the projection angle = 45 degrees and the projection height = 0.

Influences of projection height

Figure 8.40 shows the effect of relative projection height on object trajectory, where v = 10 m/s and projection angle = 45 degrees. Relative projection height is the

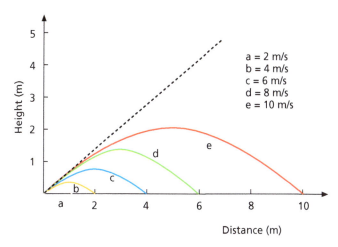

△ **Fig. 8.39** Effect of projection speed on object trajectory where the projection angle = 45 degrees and projection height = 0.

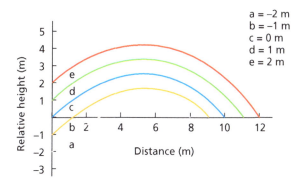

△ **Fig. 8.40** The effect of relative projection height on object trajectory where v = 10 m/s and projection angle = 45 degrees.

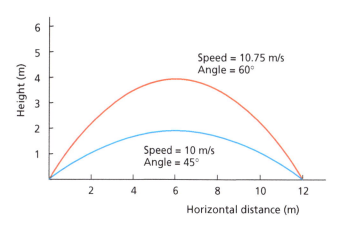

△ **Fig. 8.41** Trade-offs between factors make it possible to achieve the same height or distance with different combinations of projection speed, angle and height.

difference between the take-off or release height and the landing height. So, in shot putting the release height of the shot is higher than landing height, which is always at ground level.

There are, however, trade-offs between the factors and it is possible to achieve the same height or distance with different combinations of projection speed, angle and height (Figure 8.41).

Analysis of projectile motion

The strategies used to analyse projectile motion include:

● The projectile motion equation

● The resolution of projection velocity into horizontal and vertical components

● The treatment of the horizontal and vertical movements separately.

The projectile motion equation

The projected distance of an object is determined largely by release height, release angle and release velocity. The equation that governs projectile motion gives us an idea as to the relative importance of each parameter:

$$d_{projected} = v^2 \sin^2\theta/2g \left[1 + (2gh/v^2\sin^2\theta) \, 1/2\right]$$

where $d_{projected}$ is the projected distance, v is the release velocity, θ is the release angle, g is acceleration due to gravity (9.81 m/s²) and h is the release height.

In athletic events such as the shot put and javelin, release height is the least important factor in the equation, since little can be done to change it; it is predominantly determined by the athlete's anthropometric parameters. Release angle is the second most important factor in the equation and is determined by the angle of the throwing arm as well as the orientation of the trunk relative to the ground. As can be seen from the above equation, projected distance is a function of velocity squared. Release velocity is therefore by far the most important factor in determining projected distance.

Nevertheless, coaches must consider all parameters since release velocity is dependent upon all other factors that lead up to that point. It is important to examine all parts of the throw to determine what factors have a positive effect on the distance thrown.

Resolution of projectile velocities

The projectile velocities can be resolved using the following equations (Figure 8.42):

$$vx = v\cos\theta$$
$$vy = v\sin\theta$$

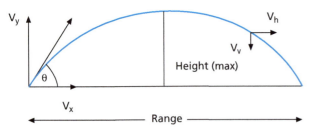

Fig. 8.42 Resolution of projectile velocities.

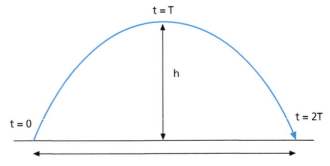

Fig. 8.43 Graphical representation of the soccer ball trajectory indicating maximal height and range.

For example, a football is kicked with a velocity of 15 m/s at an angle of 35 degrees to the horizontal. What are the horizontal and vertical component velocities?

$$v_x \, 5 \, v\cos\theta = 15 \times 0.819 = 12.29 \text{ m/s}$$
$$v_y \, 5 \, v\sin\theta = 15 \, \theta \, 0.574 = 8.60 \text{ m/s}$$

These data can then be fed into the equations of projectile motion:

- Vertical:

$$v_v = v_y \times gt$$
$$d_v = v_y \times t - 1/2gt^2$$
$$v_v^2 = v_y^2 - 2gd \text{ (where } g = 9.81 \text{ m/s}^2)$$

- Horizontal:

$$v_h = v_x$$
$$d_h = v_x \times t$$

- Ascending time:

$$t = v_y/g$$

- Height of the apex:

$$h = v_y^2/2g$$

- Horizontal distance:

$$d = v_x \times (2t)$$

Example

A soccer ball is projected with a horizontal velocity (vx) of 12.29 m/s and a vertical velocity (vy) of 8.60 m/s. Calculate the positions and velocities after 0.1s.

$$d_h = v_x \times t = (12.29)(0.1) = 1.30\text{m}$$
$$d_v = v_y \times t - 1/2gt^2 = (8.60)(0.1) - \times (0.5)(9.81)$$
$$(0.1)^2 = 0.81\text{m}$$
$$v_h = 12.29 \text{ m/s}$$
$$v_v = v - gt = 8.60 \, (9.81)(0.1) = 7.62 \text{ m/s}$$

Find also the maximal height (apex), ascending time and the maximal horizontal distance of the soccer ball where $v_x = 12.29$ m/s and $v_y = 8.60$ m/s (Figure 8.44).

$$h = v_y^2/2g = (8.60)^2/(2)(9.81) = 73.96/19.62 = 3.77\text{m}$$
$$t = v_y/g = 8.60/9.81 = 0.88\text{s}$$
$$d = v_x \, 3 \, (2t) = (12.29)(2)(0.88) = 21.63\text{m}$$

Optimum projection conditions

Projection angle for maximal distances depends on **relative projection height**. If the relative projection height is greater than 0 then the optimal angle is less than 45 degrees. If the relative projection height equals 0 then the optimal angle equals 45 degrees and if the relative projection height is less than 0, then the optimal angle of projection is greater than 45 degrees. The projection angle for maximal height is 90 degrees to the horizontal (Figure 8.44).

In the shot put the distance achieved is determined by the release parameters. Using the projectile motion equation you can find the optimal release parameters giving a specific distance, which in turn could then be included in the training goals. However, release parameters are not independent of one another as the equation suggests. Unlike non-human projectiles, the human body is a highly complex system of pulleys and levers that do not function with equal capacity in all positions and angles. Release velocity and release angle possess a strong inverse relationship, therefore it is clear that the simplest way to increase release velocity is to alter the release angle so that release velocity is optimized (above

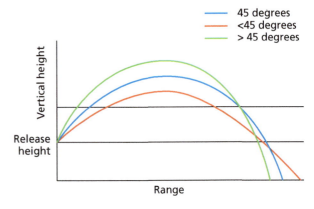

Fig. 8.44 Considerations of projection angle for maximal distance or range

13 m/s) while still maintaining an angle that will allow elite level performances. Release angles that allow this are considerably lower than the mathematically optimal range (42 degrees).

This is supported by Linthorne *et al.* (2001a) who suggest that athletes have their own individual release angles because of individual differences in the rate of decrease in release velocity with increasing release angle. All attempts should be made to maximize release velocity, but the coach should be aware that increasing athlete or implement velocity at any stage of the throw may not result in an increase at release. This can be due to technical changes, instability or inefficient sequencing of muscles.

Coaching application

Why do shot putters and javelin throwers train differently in terms of resistance training programmes? The aim of both activities is to generate maximum release velocities to the implement. However the differences occur due to implement weights. The male shot weighs 7.257 kg and the female shot weighs 4 kg. The male javelin weighs 0.8 kg and the female javelin weighs 0.6 kg. These differences in implement weight have considerable consequences in the release velocities. For elite athletes the release velocity of the shot is approximately 14 m/s whereas for the javelin it is approximately 30 m/s. The values correspond to different parts of the force–velocity curve (this is explained in more depth in Chapter 11).

Shot putters must generate higher maximal forces at release which is why 50 per cent of their conditioning involves heavy resistance training. Javelin throwers spend approximately 15–25 per cent of their time weight training since greater emphasis should be placed on high velocity training (Zatsiorsky, 1995).

TIME OUT

BIOMECHANICS OF JAVELIN THROWING

Range in javelin throwing is determined by the release parameters and aerodynamic factors. Research undertaken by Viitasalo *et al.* (2003) investigated the effects of release speed, release angle and uncorrected angle of attack measured at the foul line on the official javelin throwing result. Data were collated in international competitions for 26 elite male and 15 elite female javelin throwers. Data collection was carried out using a computerised photcell gate consisting of two invisible infrared walls 2 m apart, perpendicular to the throwing direction. Multiple regression models were used to predict the range of the throw for individual throwers, a group of throwers using the mean value for each thrower and all individual throws registered for each gender separately. An increase of 1 m/s in the release speed from 29 m/s to 30 m/s was calculated to increase the official result between 2.12 m and 6.14 m for males and the effects of an increase from 24 m/s to 25 m/s in release speed calculated an increase of 2.25 m to 3.68 m in females. The study emphasized the importance of investigating javelin throwing biomechanics on an individual thrower basis.

Activity 8.6

Practical investigation: Projectile motion

This investigation looks at the factors involved with projectile motion. These include height of release, velocity of release, angle of release, horizontal distance travelled or range and flight time.

Basic concepts

If the velocity of release is constant, then the angle of release will determine the vertical and horizontal velocity components. The vertical component is influenced by gravity and determines the height reached during flight and the time of flight. Assuming there is no air resistance, the horizontal velocity component is constant throughout the flight and for a given time of flight will determine the horizontal distance covered.

Method

This laboratory investigation involves the use of a digital video camera and either a laptop or computer with access to Siliconcoach software packages, or a pen and tracing paper.

1. Video a classmate putting the shot. Stand perpendicular to the thrower since the analysis is planar and two-dimensional. Remember to follow the considerations for filming motion set out earlier on in this chapter (see **Box 8.1** *Considerations for filming human movement*).

2. Record a number of trials with a variety of release angles. Instruct the athlete to use maximal effort for each throw.

3. Either download the video recording onto your software package and work out the release heights, velocities, angles of release, and the ranges of the shot or obtain the angles of release, release heights and ranges from tracing off the video recording.

4. Observe the changing effects of release angle of the shot on the trajectory of the flight path that it follows. Discuss these changes.

Questions

1. Increasing the release angle from the optimal value tends to increase the time of flight and the maximum height reached during flight, however, the distance thrown is reduced. Outline a sporting situation where such a scenario would be desirable.

2. Decreasing the release angle from its optimal value tends to decrease the time of flight, the maximum height reached during the flight and the distance thrown. Outline a sporting situation where such an occurrence would be desirable.

Summary

- This chapter has introduced the kinematic concepts that are essential if the reader is to achieve a full understanding of human movement.

- The differences between vectors and scalars are identified and the component, graphical and resolution methods of vector construction with examples are provided.

- Each kinematic concept is defined alongside the mathematical underpinning. Mathematical underpinning of human movement will provide a greater understanding of the area. The foundations of biomechanics stem from mathematics. Improved technology confirms this. However, in reality this helps to cut down the need for longwinded calculations.

- The student must be able to interpret graphical representations of one kinematic variable with another (displacement, velocity and acceleration).

- Finally, projectile motion is introduced and the equations to calculate vertical and horizontal components, the range and angles of projection are introduced. Working examples are provided for the student, along with practical investigations to further the understanding of the concepts particularly in an applied setting.

Review Questions

1. What factors are involved in the anatomical reference position?

2. Identify the anatomical reference planes and axes.

3. Explain the differences between speed–velocity and distance–displacement in terms of vectors and scalars. Give examples from sport.

4. A runner accelerates at 12.4 m/s^2 for 0.5s. If her final velocity is 9.7 m/s, what was her initial velocity?

5. A shot is released at a height of 1.90m and with a release velocity of 12.5 m/s at an angle of 35 degrees. Calculate the maximum height reached and the time when this occurs. What is the range and the time of flight?

6. A sprinter begins his race and accelerates out of the blocks to a velocity of 7 m/s over 1.5 s. What is his acceleration over this time period?

7. When studying projectile motion, what factors are involved in the optimum projection conditions? Which of these is the most influential?

8. A rugby player attempts a penalty 35 m out from the posts. The goal post crossbar is 3m above the field. If the ball is kicked from the ground what is its height when it reaches the crossbar given the following velocity components:

(a) $v_x = 15$ m/s and $v_y = 15$ m/s
(b) $v_x = 17$ m/s and $v_y = 12$ m/s?

9. Explain the difference between vertical, horizontal, classic and general types of projectile motions.

10. Why do you think it is important that coaches can interpret graphical representations of displacement, velocity and acceleration?

References

Bartlett RM (1997) *An Introduction to Sports Biomechanics.* London: E & FN Spon.

Berthoin S, Dupont G, Mary P, Gerbeaux M (2001) Predicting sprint kinematic parameters from anaerobic field tests in physical education students. *Journal of Strength and Conditioning Research* 15: 75–80.

Briggs SM, Tyler J, Mullineaux DR (2003) Accuracy of kinematic data calculated using SIMI motion. *Journal of Sports Sciences* 21: 235–365.

Cronin, J (2006) Resisted sprint training for the acceleration phase of sprinting. *Strength and Conditioning Journal,* 28(4): 42–51.

Cushing A, Robinson P (2003) Formula reveals kicking prowess. *BASES World.* December.

Dahlkvist NJ, Maya P, Seedhom BB (1982) Form during squatting and rising from a deep squat. *Medicine* 11: 69–76.

Hanley B, Athanassios B and Drake A (2011) Kinematic characterisics of elite men's and women's 20km race walking and their variation during the race. *Journal of Sports Biomechanics,* 10(2): 110–124.

Kent M (1994) *The Oxford Dictionary of Sports Science and Medicine.* Oxford: Oxford University Press.

Kirtley C, Whittle MW, Jefferson RJ (1985) Influence of walking speed on gait parameters. *Journal of Biomedical Engineering* 7: 282–288.

Linthorne NP (2001a) Optimum release angle in the shot put. *Journal of Sports Sciences* 19: 359–372.

Linthorne NP (2001b) Analysis of standing vertical jumps using a force platform. *American Journal of Physics,* 69(11): 1198–1204.

Mero A, Komi PV (1994) EMG, force and power analysis of sprint-specific strength exercise. *Journal of Applied Biomechanics* 10: 1–13.

Mero A, Komi PV, Gregor RJ (1992) Biomechanics of sprint running. *Sport Medicine* 13: 376–392.

Norris CM (1999) *Sports Injuries: Diagnosis and Management,* 2nd edn. Oxford: Butterworth Heinemann.

Piorkowski BA, Lees A and Barton, GJ (2011) Single maximal versus combination punch kinematics. *Journal of Sports Biomechanics,* 10(1): 1–11.

Reid DC (1992) *Sports Injuries: Assessment and Rehabilitation.* London: Churchill Livingstone.

Subotnick SI (1989) *Sports Medicine of the Lower Extremity.* London: Churchill Livingstone.

Viitasalo J, Mononen H, Norvapalo K (2003) Release parameters at the foul line and the official result in javelin throwing. *Journal of Sports Biomechanics* 2: 15–34.

Williams KR (1985) Biomechanics of running. *Exercise Sport Science Review* 13: 389–441.

Zatsiorski VM (1995) *Science and Practice of Strength Training.* Champaign, IL: Human Kinetics.

Further reading

Hay JG (1993) *The Biomechanics of Sports Techniques.* Englewood Cliffs, NJ: Prentice Hall.

9 Linear Kinetics: The Study of Force and Movement

Chapter Objectives

In this chapter you will learn about:

- Force and Newton's laws and their relevance to sport and exercise.
- The importance of the free body diagram in the solution of mechanical problems.
- The significance of such concepts as friction, centre of gravity and impact in human movement.
- The link between work, power, energy and performance.
- The application of the principles of linear kinetics to performance and training.

Introduction

Most of our movements ultimately rely upon our interaction with the ground. We are constantly pushing against the ground both vertically and horizontally as we initiate and modify movements of the total body and its segments. Consider just a few examples of movements, both simple and complex, that depend upon our ability to push against the solid base of the Earth: walking, running, reaching up in a cupboard for a glass, a push-up exercise, raising your hand to ask a question, and jumping.

Because of the importance of our interactions with the ground in the generation and modulation of our movements, the **ground reaction force** (GRF) could arguably be considered the most important external force acting on the body. But what is important to keep in mind is that the GRF is largely under our control via co-ordinated muscle actions. By producing a certain combination of muscle actions, we ultimately push against the ground, which pushes back against the body with an equal and opposite force. This is explained by Newton's third law of motion, which states that for every action there is an equal and opposite reaction.

When muscles contract they exert forces against external loads in the function of time and do work on the environment by using biochemical and/or elastic energy (Tihanyi, 2003). The muscles do not apply force directly to the ground or external objects however, but function by **pulling** against bones which rotate around an axis or joint. Muscles can only pull, but the system of levers within the human body allows muscle pulling forces

to exert pulling or pushing forces on external objects (Figure 9.1) (Harman, 1994).

(a)

(b)

△ **Fig. 9.1** Human body pulling and pushing objects. The function of muscles pulling against bones around joint axes produces force. The lever system in our bodies allows muscles to exert pulling or pushing forces on external objects as is evident in the diagrams where (a) shows a person pushing a car and where (b) shows a person pulling a trolley.

Force describes the interaction of an object with its environment (Enoka, 1994). The study of force and movement in biomechanics is called **kinetics** and its unit of measure is the newton (N), one newton (1 N) being the force required to produce an acceleration of 1 m/s2 on an object with a mass of 1 kg (Norris, 1999).

Forces in sport and exercise contribute to many things. Forces can:

- Get an object moving
- Stop an object from moving
- Change the direction in which the object is moving
- Change the speed at which it is moving
- Balance another force to keep an object moving at a steady speed
- Balance another force to keep an object still
- Change an object's shape.

The most common forces that affect the human body, and therefore human movement, are those produced by the musculature and internal system of ligaments and tendons, and the external forces occurring as a result of gravity, inertia and contact. To be more specific, the external forces that must be taken into consideration when analysing sports techniques are:

1. Gravitational pull
2. Normal reaction forces exerted on a body by the ground
3. Friction forces between feet and ground
4. Air resistance.

Consideration must also be given to the magnitude, direction and point of origin of the force(s) produced. Force is therefore a **vector** quantity as it is made up of both magnitude and direction (Kent, 1994). Forces can be represented diagrammatically through the use of an arrow: the **magnitude** is represented in newtons or by the length of the arrow, the **direction** of the force is represented by its direction, the **point of application** is represented by the arrowhead and the **point of origin** is represented by its origin.

Forces acting in the same line are said to be in series and are termed 'linear' (Fig. 9.2). They may also act parallel to each other or concurrent (Fig. 9.3). These are forces acting at the same point of application but at different angles. The resultant of the two forces depends on the force magnitude and the **angle of application**. Similarly, forces of equal magnitude may act in opposite directions forming a **force couple** (Chapter 11) (Norris, 1999).

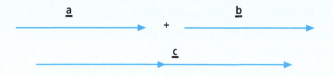

The resultant of the two forces is equal to the sum of the two forces (a + b = c)

△ **Fig. 9.2** Linear forces. The resultant of the two forces is equal to the sum of the two forces (<u>a</u> + <u>b</u> = <u>c</u>).

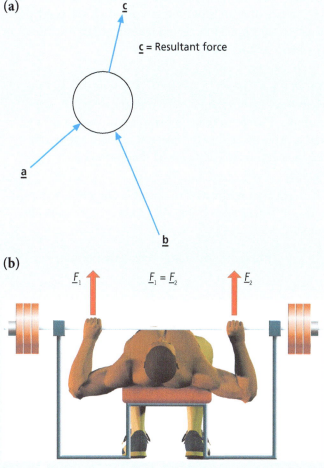

△ **Fig. 9.3** (a) Concurrent forces a and b acting on an object (b) parallel forces F_1 and F_2 acting on an object.

Newton's laws of motion

Directly applied to a practical situation

The basic principles of the relationship between force and movement were established by Newton in 1687 in his work *Philosophiae naturalis principia mathematica*.

Newton's first law of motion (law of inertia) states that a body will continue in its state of rest or uniform motion in a straight line unless acted upon by an external force. The tendency of a body to remain at rest or, if moving, to continue its motion in a straight line is described as inertia. Hence it is sometimes called the law of inertia.

A body or an object is said to be in a state of inertia (the tendency of an object to keep its current state of motion, static or dynamic) and a force must be applied to it before any change in velocity can occur. The greater the mass of a body, the more force is required to overcome its inertia. You can throw a 5 kg weight further than you can throw a 10 kg weight using the same force.

Newton's second law of motion (law of acceleration) states that a force applied to a body causes an 'acceleration' or deformation (a change in shape). The ratio of strength of the muscles involved in a sporting event such as sprinting or long jumping compared with the mass of the body parts being accelerated is of paramount importance. Newton's second law states that the acceleration of an object is directly proportional to the force causing it and is inversely proportional to the mass of the object. Or the rate of change of momentum of a body is proportional to the applied force and takes place in the direction in which the force acts.

$$F = ma$$

Therefore if $F = 0$ then $a = 0$ (constant velocity). For example, the speed that a person can throw a tennis ball is proportional to the amount of force applied by the muscles. It also depends on the inertia of the ball.

The term **momentum** of an object is often referred to in sport. It is defined as being the product of its mass and velocity.

$$\text{Momentum} = \text{mass} \times \text{velocity (kgm/s)}$$

A defender in hockey usually uses a heavier hockey stick than a forward because it allows him or her to transfer more momentum to the ball, and consequently to hit it further. Momentum can also be built up and transferred from one body part to the rest of the body, resulting in more force. For example, swinging the arms backwards and forwards before take-off transfers momentum to the rest of the body for a vertical jump. Momentum is also important in giving or receiving impact, collision, etc. It is a vector quantity.

Example

Two hockey players approach each other, just before impact player 1 moves with a velocity of 6 m/s and a mass of 90 kg. Player 2 moves with a velocity of 7 m/s and a mass of 80 kg (Figure 9.4). What is the momentum of each player before impact?

$$\text{Momentum 1} = 540 \text{ kgm/s}$$
$$\text{Momentum 2} = -560 \text{ kgm/s}$$

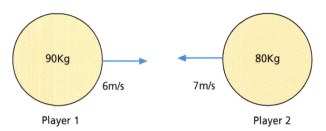

△ **Fig. 9.4** Two hockey players approach each other. Player 1 has a velocity of 6 m/s and a mass of 90 kg. Player 2 has a velocity of 7 m/s and a mass of 80 kg. Prior to impact Player 1 has a momentum of 540 kgm/s and Player 2 has a momentum of 2560 kgm/s.

The link between force and acceleration

Suppose a force (F) acts on a body of mass (m) over a time (t), and causes its velocity to change from u to v (i.e. u is initial velocity and v is final velocity). The momentum changes uniformly from $m.u$ to $m.v$ in time t. Therefore the rate of change of momentum = $(mv-mu)/t$.

By Newton's second law, the rate of change in momentum is proportional to the applied force, hence,

$$F \propto (mv-mu)/t$$

Factorizing

$$F \propto m(v-u)/t$$

And since $(v-u)/t$ = change in velocity/time = acceleration (a), therefore $F = ma$. The units are as follows: F is in newtons, M is in kg and a is in m/s^2.

Example: vertical jump kinetics

The basic mechanical principle studied when analysing a vertical jump is Newton's second law of motion:

$$\Sigma F = ma$$

where ΣF represents the summation of all forces acting on the body (i.e. the net force), m is the body mass and a is the acceleration of the body's centre of mass (C of M).

During a vertical jump, the upward push off the ground is equal to the Earth's downward attraction (weight), the net force on the body equals zero and the resulting acceleration is zero. If the ground reaction force is greater than the body weight, there is a net positive force acting on the body and the acceleration is positive. Finally, if the ground reaction force is less than bodyweight, the net force on the body is negative and the acceleration is negative.

Therefore,

If GRF = W, then $F = 0$ (no net force) and $a = 0$
If GRF > W, then $F > 0$ (net force upwards) and $a > 0$
If GRF < W, then $F < 0$ (net force downwards) and $a < 0$.

When considering changes in both the speed of the body, C of M (increase or decrease in speed) and direction of movement, positive acceleration is reflected under three conditions:

- Increase in C of M velocity as the C of M moves upwards
- Decrease in C of M velocity as the C of M moves downward
- Changing directions from moving downward to moving upward.

Similarly, negative acceleration is reflected by three conditions:

- Decrease in C of M velocity as the C of M moves upward
- Increase in C of M velocity as the C of M moves downward
- Changing directions from moving upward to moving downward.

Conservation of momentum

If the net external force applied to a system is zero, then the total momentum of the system remains constant. Conservation of momentum is typically used when a system consists of multiple objects that are colliding or separating. Objects in the system can apply force to each other. The net external force on the system as a whole must be zero.

Going back to the previous example of the two hockey players, after impact both travel together as a unit:

$M_{before} = M_1 + M_2 = 540 + (-560) = -20$ kgm/s
$M_{after} = (M_1 + M_2)(v_{after}) = (90 + 80)(v_{after}) = (170)(v_{after})$
$M_{before} = M_{after}$
$(170)(v_{after}) = -20$
$v_{after} = -20/170 = -0.12$ m/s
Therefore player 2 pushes player 1 at 0.12 m/s.

Newton's third law of motion (law of reaction)

This states that whenever force acts on one body, an equal and opposite force acts on some other body. Or, that for every **action** there is an equal and opposite **reaction**. It is important to realize that the **action** and **reaction** act on different bodies. When an object exerts a force on a second object, the second object exerts an opposite and equal force back on the first. The most common sporting illustration of this law is when an athlete pushes back against the starting blocks at the beginning of a sprint race (exerting a force on the blocks), causing the opposite and equal reaction of being pushed forward out of the blocks (Figure 9.5). During this time, force is applied to the blocks (the **action force**). At the same time the athlete will be aware of a force pushing against his or her foot (the **reaction force**). The harder the athlete pushes against the blocks the greater the reaction force. The forces are equal in magnitude and direction (see also Figure 9.6). A reaction force is always perpendicular to the surface to which it is applied (Figure 9.7).

△ **Fig. 9.5** World 400m hurdles champion Dai Green driving out of the block. This shows Newton's third law of motion in action. He is applying a force to the blocks (action); at the same time a force pushing against his foot (reaction) propels him forward.

△ **Fig. 9.6** The heptathlete Jessica Ennis applies a force (action) and as a result the shot applies a reaction force to her hand.

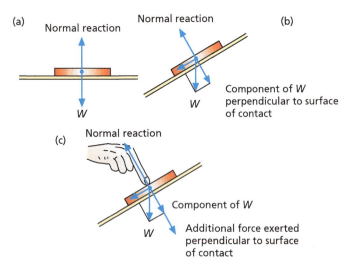

(a) Normal reaction
Normal reaction (b)
W

Normal reaction
Component of W perpendicular to surface of contact
W

(c) Normal reaction
Component of W
W
Additional force exerted perpendicular to surface of contact

△ **Fig. 9.7** Reaction forces always act perpendicular to the surface of contact. In (a) the surface is horizontal and as the reaction force acts upwards and the action or weight (W) acting downwards is equal and opposite. In (b) the surface is inclined, the reaction force is perpendicular to the surface and in this case the weight (W) acts downwards. However, a component of the weight (W) acts perpendicular to the surface of contact. (c) is the same as (b) only an additional force is added, therefore increasing the representative force vector lengths (magnitudes) and causing the force to be exerted perpendicular to the surface of contact.

Action   Reaction

△ **Fig. 9.8** Internal forces produced by a high jumper cause changes in body position.

It is possible to move one body part to cause another body part to react in the opposite direction. For example in trampolining, a half twist is achieved by swinging the arms to the right whilst rotating the rest of the body to the left. These are caused when muscles contract within the body. Muscle contractions cause changes in body shape as origins and insertions are pulled towards one another. The final body position or shape will depend on which body part moves (Figure 9.8).

Force considerations

Forces produce movement, whether internal, external or a combination of the two. However there are occa-

sions when forces acting on the human body should be reduced in order to prevent injury. For example, what are the forces acting on the head of a soccer player when he or she heads the ball?

Let us say that the initial velocity of the ball as it makes contact with the player (u) is 22 m/s, the final velocity of the ball as it leaves the player's head (v) is 17 m/s, the mass of the ball (m) is 0.5 kg and the time the ball is in contact with the player's head (t) is 0.06 s. According to Newton's second law, the force placed on the ball, $F = ma$. However, acceleration (a) can also be represented as $a = (v - u)/t$, which means that Newton's second law can be represented as $F = m(v - u)/t$.

Since u moves in the opposite direction to v it is given a negative sign.

$$F = 0.5(17 - (-22))/0.06 = 325 \text{ N}$$

This is fairly typical of the forces that occur when heading the ball during a game. Considering that players in specific positions may head the ball many times during a game, it is hardly surprising that concussion is commonplace.

An important consideration for throwing events is the amount by which the velocity of an object increases. This depends on the magnitude of the force and the time for which this force acts. Therefore, a greater range of movement will allow a force to be applied longer, which results in a greater release velocity. This principle is used in shot-putting and discus throwing. There are other occasions when reducing the magnitude of forces is appropriate. For example, stopping moving objects or stopping the human body when landing in sports such as cricket and judo respectively.

Activity 9.1

Independent learning task: Mass and weight

A man has a mass of 90 kg. What is his weight? Does his weight depend on his body position? If he was living for a period of time in a space station orbiting 450 km above the Earth, and the acceleration due to gravity is 8.8 m/s², what would his mass now be on the space station? What would his weight be?

Gravitational force

In all movements of the human body in sport, gravity acts as an external force on the body mass. The mutual attraction of two masses (gravitation) was formulated by Newton in the **law of gravitation**. The force of gravity acts through the centre of mass as a result of the Earth drawing all objects to its centre. This can be represented mathematically as $F_g = ma_g$ or $W = mg$, where F_g is the force due to gravity, often termed as the athlete's/object's weight (W), m is the mass of the athlete/object and a_g is the acceleration due to gravity, sometimes represented by the letter 'g' and is a constant (9.81 m/s^2).

As already stated in Chapter 8, the acceleration due to gravity is influenced by location on the Earth's surface because of concentrations of lighter or heavier masses beneath the surface and because the Earth is not completely spherical. Acceleration due to gravity can vary by as much as 1 per cent depending on geographical location (Harman, 1994). The 1968 Olympic Games in Mexico City was a prime example of how changes in acceleration due to gravity can have a significant effect on performance and thus on world records. Although the altitude at Mexico City (2250 m) clearly had an effect on performance, its proximity to the Equator and thus the relatively low acceleration due to gravity had even more effect. The fact that the acceleration due to gravity is approximately 0.3 per cent less than that in the UK helped many of the world records in the short duration power events. The thinner air at altitude also was also advantageous for the athletes in the power events and the sprinters – giving an extra 0.07 s over 100 m (Linthorne, 1994). The advantage of a lower acceleration due to gravity did not, however, outweigh the disadvantage of thin air in the endurance events.

The application of Newton's laws of motion

The purpose of this section is to develop an understanding of Newton's laws of motion and its applications in practice. In order to follow the exercises you will need a set of bathroom scales and access to a lift.

Start by asking yourself:

- What do bathroom scales show when you stand on them?
- Do they always indicate weight?
- Are you able to briefly change the reading indicated to show either an increased or decreased reading?

Bathroom scales are a force-measuring device that indicates the reaction force acting between your feet and the scales. The net vertical force (F_y) can be represented according to Newton's second law by:

$$F_y = ma_y$$

where m is the mass and a_y is vertical acceleration.

The upward direction is considered positive and so the equation of motion for the person on the scales now becomes

$$F_y = R - W = ma_y$$

where R is the upward reaction force and W is the weight of the person on the scales.

The reading can be used to determine the direction of the acceleration. Therefore, if:

- $R > W$ the net force (upward acceleration) is positive
- $R = W$ the net force is zero acceleration
- $R < W$ the net force (downward acceleration) is negative.

Let us take a look at the effects in three different scenarios.

1. Standing on the scales

- How hard can you push against the scales for approximately 2 s if you don't move or touch anything other than the scales?
- Can you push against the scales with a greater force than your weight? If not, why not? How big is the force that the scales pushes back against you?
- How does this relate to Newton's first and third laws?

2. Squatting on the scales

- Stand on the scales with your hands on your hips. Record your weight. Lower yourself in to a squat position. What happens to the reading as you squat?

	Is the scale reading (R) less than, equal to or greater than your weight (W)?	Acceleration direction (down, zero, up)

Squatting down
    Starting down
    Slowing
    Stopped		

Standing up
    Starting up
    Slowing
    Stopped		

△ **Table 9.1**

	Is the scale reading (R) less than, equal to or greater than your weight (W)?	Acceleration direction (down, zero, up)

Lift moving up
    Starting up
    Continuing up
    Slowing
    Stopped		

Lift moving down
    Starting down
    Continuing down
    Slowing
    Sopped		

△ **Table 9.2**

Stand up. What happens to the reading as you stand up? Complete the empty boxes in Table 9.1. Indicate whether the reading was less than (<), equal to (=) or greater than (>) your normal weight and the directions of your accelerations.

- What is the relationship between the scale reading and your acceleration? Lift your arms up to see if this relationship holds.
- How does this relate to Newton's second law?

3. Standing on the scales in a lift

- Stand on the scales in a lift as it goes up, stops, goes down and stops. What happens to the reading?
- Fill in the empty boxes in Table 9.2. Indicate whether the reading was less than (<), equal to (=) or greater than (>) your normal weight and the directions of your accelerations.
- Does the relationship between the scale reading and your acceleration still hold true for the lift? Was the net force zero at any time when moving downward? When did you feel heavier, lighter and why?

The amount by which an individual feels lighter or heavier when travelling in a lift can be determined by considering the lift's acceleration. For example, a woman of mass 60 kg travels in a lift that accelerates upwards at 0.9 m/s2. The resultant force $R - W$ acting on the woman is derived from Newton's second law.

$$R - W = ma$$

Therefore, mass = mass of woman = 60 kg

$$a = \text{upward acceleration} = 0.9 \text{ m/s2}$$
$$W = mg = 60 \times 9.81 = 588 \text{ N}$$

By rearrangement of the equations:

$$R = ma + W$$

$$R = (60 \times 0.9) + 588$$
$$R = 642 \text{ N}$$

Therefore during the acceleration phase the woman would feel heavier than normal by approximately 5.5 kg.

Enhancing force production

The stretch shortening cycle (SSC)

Human movement is made possible due to the relative contributions of eccentric (lengthening), isometric (static) and concentric (shortening) muscle actions. The elasticity of muscles and tendons has been demonstrated to affect performance in stretch shortening cycle (SSC) movements. The importance of the SSC in sprinting and jumping is typically characterised by an eccentric muscular contraction or stretch, followed immediately by a concentric muscular contraction. Using a stretch immediately prior to a concentric contraction has been shown to increase the concentric phase resulting in augmented force production, power output and a shift in the force-velocity curve to the right. The increased power output evident after a countermovement jump (CMJ) can be explained by the time to build up force, storage and re-use of elastic energy, potentiation of the contractile elements and reflex contributions. Athletes can perform various forms of strength training to improve their ability to exert power using the SSC, and understanding the mechanisms and assessment methods of the SSC may influence program design and enhance performance.

Assessing the SSC

Performance of the SSC is usually undertaken by adding a pre-stretch to a movement such as comparing a CMJ with a static jump (SJ) performance. Both of these tests have been shown to be reliable and valid tests for estimating explosive muscular power. Pre-stretch augmentation can be calculated as a percentage with % pre-stretch augmentation = [(CMJ – SJ)/SJ] × 100. Another approach is to measure the reactive strength index (RSI) (calculated as height jumped/contact time) which has been used in the practical strength and conditioning setting as a means to quantify plyometric or SSC performance.

The SSC can be classified as either fast or slow; a fast SSC is characterised by short ground contact times (<0.25s) and small angular displacements of the hips, knees and ankles such as a depth jump. Slow SSCs involve longer ground contact times (>0.25s), larger angular displacements and can be observed when performing a maximal effort CMJ. Where a training stimulus is desired to improve performance in fast SSC movements, or to assess fast SSC function, the CMJ is not an appropriate modality. Therefore, appropriate exercise selections for training and assessment of the SSC should be guided by the principle of specificity and the demands of the sport.

Friction

Friction is a force that occurs when two or more surfaces touch. The force acts in parallel to the contact surfaces and the size or magnitude of the force is dependent upon the types of surfaces involved (roughness of surfaces, μ), the applied force, the weight of the object and whether the forces are moving or stationary. Friction is a force that opposes movement; its existence allows movement to occur in accordance with Newton's third law. The friction force does not depend on the surface area.

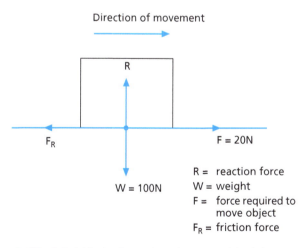

△ Fig. 9.9 A block of wood resting on a table. It has a mass of 10 kg and a force of 20 N is required to move the wood along the surface. The coefficient of friction is 0.2.

The **coefficient of friction** (μ) can be calculated using the formula:

$$\mu = F/R \text{ or } F = \mu R$$

where F is the force required to move the object and R is the reaction force.

For example, a block of wood resting on a table (Figure 9.9) has a mass of 10 kg and therefore a weight of 100 N. If a force of 20 N is required to move the block of wood along the surface, the coefficient of friction (μ) can be calculated as $F/R = 20/100 = 0.2$.

If μ = 0 then no friction exists and if μ = 1 then the object is said to be fixed. When two objects are in contact, μ will be slightly less when the objects are sliding on each other compared with when the objects are at rest but just about to start sliding. Therefore a **coefficient of sliding** or kinetic friction (μ_s) and a **coefficient of limiting or static friction** (μ_L) exist for any two objects. As $\mu_s < \mu_L$, this demonstrates that it is easier to keep an object moving than it is to initiate the movement (Figure 9.10).

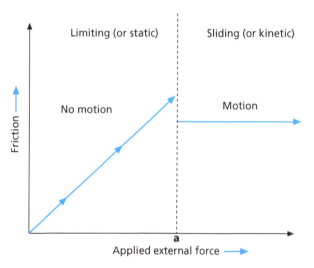

△ Fig. 9.10 A force is applied to a stationary object to initiate movement. At point a the force required to initiate movement is achieved (limiting friction). Since $\mu_s ‹ \mu_l$, the graph indicates however that less force is required to keep the object moving.

As more force is applied to a stationary object, the friction force increases. The friction will continue to increase until the instant immediately prior to the initiation of the movement. At this point the Motion ≠ 0:

$$F_{\text{friction}} = F_{\text{lim}}$$

In sport and exercise, friction forces are extremely important. Friction forces allow us to walk, accelerate, decelerate, swerve, grip and change direction. Therefore anything that changes the friction between two surfaces, for example footwear, ice, materials, mud and weather conditions, will affect the ability of the athlete to perform the skill or task optimally (Watkins, 1988). Training devices that use friction as the main type of resistance include belt- or brake pad-resisted bicycle ergometers.

Activity 9.2

Laboratory investigation: Coefficients of limiting friction

Equipment

1. Shoes, weights, various surfaces
2. Force plate and computer for data collection.

Objective

This investigation demonstrates how sliding and limiting friction change as a function of surface and weight. The frictional force encountered is dependent upon the type of surfaces involved, the applied force, the weight of the object and whether the surfaces are moving or stationary. Friction opposes motion, however its existence allows movement to occur in accordance with Newton's third law.

Points to consider

- What are some examples of manipulating the coefficient of friction or the friction force in track and field?
- How does surface area affect the coefficient of limiting friction?
- How does the reaction force affect the coefficient of limiting friction?

Exercise

1. Place a weight in one shoe and place it on the force plate.
2. Determine the total weight.
3. Pull the shoe until it begins to slide along the force plate.
4. Record the horizontal force required to get the shoe moving.
5. Repeat for different surface/shoe combinations and record your results in the table below:

Shoe/surface	Normal reaction (R)	Applied horizontal force (F)	Coefficient of limiting friction (m)

Questions

- Which shoe/surface combination gave the best grip?
- Once the shoe began to move did it require more or less force to continue to move it?
- Give an example of where a large coefficient of friction is an advantage and another where it is a disadvantage.

The free body diagram

A diagram showing in vector form all of the forces acting on an object is called a **free body diagram**. When drawing free body diagrams you should consider the following forces:

- Gravity
- Reaction/muscular forces
- Friction
- Air resistance/drag forces.

For each of the above forces, make sure you are able to do the following:

- Give a definition
- Decide whether or not it will be acting
- Decide on its origin or point of application
- Decide on its direction
- Decide on its length (magnitude)

Objects at rest are acted upon by at least two forces. One force acting will be the weight of the object and since the object is at rest it must be balanced by an equal and opposite force. In other words the force has the same magnitude but opposite direction to the weight. The resultant or net force acting must therefore be zero (Figure 9.11).

Regardless of the number of forces acting on a body or object, if an object is at rest the resultant is always zero. Likewise, if a body moves with a constant velocity, the resultant or net force is again zero. When the resultant force acting on a body is zero, the body is said to be in a **state of equilibrium**. Only when the resultant of all forces is greater than zero will the object move or change its motion from a uniform state.

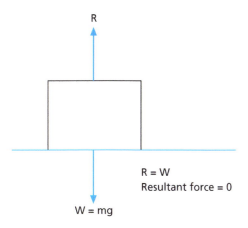

R = W
Resultant force = 0

W = mg

△ **Fig. 9.11** A resting object is acted upon by its weight (W = mg) and the equal and opposite reaction force (R) in the vertical plane. There are no horizontal forces and the resultant force is R − W = 0.

Calculating forces on objects or bodies is a simple matter if the forces are acting in series. It is simply a matter of addition or subtraction. However, when forces act at angles to each other a geometric figure called the **parallelogram of forces** is used. The two component forces form the sides of the parallelogram and the resultant or net force makes up the diagonal of the parallelogram. This principle is known as the **resolution of forces**. The idea that resultant/net forces cause acceleration is linked to Newton's first and second laws of motion, and is a fundamental property of force.

Some examples of free body diagrams showing resultant and net forces acting in various sports are given in Figures 9.12 and 9.13.

Skier

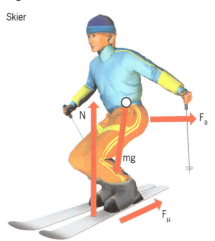

Long jumper

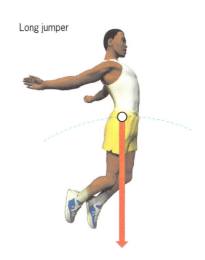

High jumper

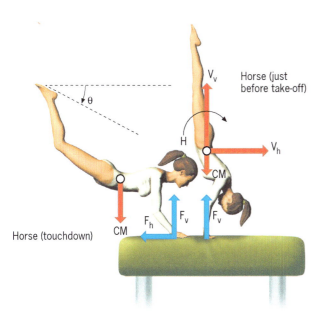

△ **Fig. 9.12** The centre of mass, force and component velocity vectors are displayed and compared at horse touchdown (TD) and takeoff (TO) during a vault.

Acceleration out
of the blocks

High jumper
at take-off

Deceleration

Sprinter at
constant
velocity

△ **Fig. 9.13** The vectors represent the nature and magnitude of the forces acting on a variety of sporting bodies. Note the changes in body positions and the corresponding changes in magnitude and direction of the force vectors (where F_R = Friction force, R = resultant force and N = normal reaction force).

Calculating forces from free body diagrams: example

A 70-kg runner has a vertical acceleration of 13.7 m/s² and a horizontal acceleration of 2.4 m/s². Draw a free body diagram and solve for the horizontal and vertical reaction forces. The athlete also had an initial vertical velocity of −0.7 m/s and an initial horizontal velocity

of 3.6 m/s. If the forces were applied over a period of 0.2 s, what are the runner's final vertical and horizontal velocities? Mass = 70 kg, a_v = 13.7 m/s², a_h = 2.4 m/s². By drawing a free body diagram (Figure 9.14) we can identify the forces acting on the athlete. Consider where the forces are acting, in what direction and their magnitude.

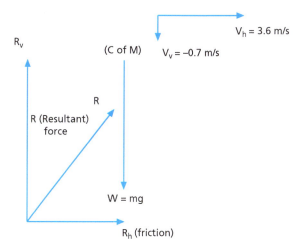

△ **Fig. 9.14** The figure illustrates magnitude and direction for the force and velocity vectors acting.

From Newton's second law:

$$F = ma$$

Therefore,

$$F_v = 70 \times 13.7 = 959 \text{ N}$$

However, we can see from the free body diagram that we have the weight of the athlete to consider. We can see that this acts downwards from the athlete's centre of mass. To calculate weight we use the formula $W = mg$:

$$W = 70 \times 9.81 = 686.7 \text{ N}$$

Therefore the total vertical reaction force = 959 + 686.7 = 1645.7 N (vertical reaction force).

The horizontal reaction force (friction force) can be calculated from:

$$F_h = 70 \times 2.4 = 168 \text{ N (friction force)}$$

We now need to find the runner's final vertical and horizontal velocities. Again we need to use the information provided and also the vertical and horizontal reaction forces calculated previously. Using the equation of motion:

$$v = u + at$$

For the vertical velocity in the y-axis

$$v_v = -0.7 + 13.7(0.2) = 2.04 \text{ m/s}$$

For the horizontal velocity in the x-axis

$$v_h = 3.6 + 0.48 = 4.08 \text{ m/s}$$

The importance of the free body diagram in the solution of mechanical problems

Students of biomechanics often have difficulty in understanding Newton's third law. For example, if a man exerts a horizontal force on an object such as pushing a box, then if the force on the box is the same as on the man's hands (action/reaction), why does it move forward? An object will only move when the resultant of all forces acting on it is greater than zero. Therefore, the student must consider **all** forces acting on the object (e.g. friction) and in order to do this it is necessary to draw a free body diagram.

Coaching application

The force vectors of the knee extensors in the free body diagram shown in Figure 9.15 indicate disproportion between the vertical and horizontal force components during the last third of the support phase. In terms of

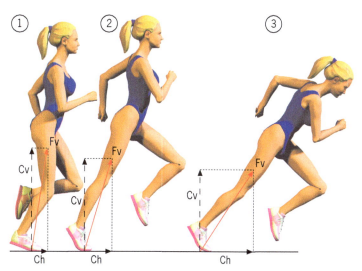

△ **Fig. 9.15** Free body diagram showing that the force vectors of the knee extension (F_v) can be split into component vectors vertical (C_v) and horizontal (C_h) components during two phases of the rear support (1) and (2) flat-out sprinting, as well as the rear support of the starting phase (3).

maximal activation of the knee extensors during this phase, you can see that an upward rather than a forward propulsion of the body occurs. Therefore it would be expected that the vastus muscles of the thigh (knee extensors) are not activating or contracting maximally, but only to a point where the optimization of the height of the flight phase guarantees stride rate maximization. During this phase, the knee extensors cannot be considered as a muscle producing forward acceleration (Wiemann and Tidow, 1995). This is disputed by Farrar and Thorland (1987), who regard the knee extensors as the key accelerators during sprinting. Figure 9.15 shows that this is only probable during the rear support of the starting phase, where the athlete is accelerating with a significant forward lean. According to Ae, Ito and Suzuki (1992) horizontal acceleration in sprinting is produced by hip extension rather than knee extension.

There is clearly some conflict here and so identification of those muscles in the hip and knee joint areas that can produce forward acceleration throughout the support phases can provide useful information. There are implications for coaching and sprint training.

Stress and strain

When forces act on objects, the force expressed per unit of area is called **mechanical stress**. Stress, then, is the force distributed over a given area and is represented mathematically by:

$$\sigma = F/A$$

where σ is stress, F is the total force applied and A is the area over which the force is applied. The SI units are the pascal (Pa) = 1 N/m^2. Pressure is stress due to a compressive force (Figure 9.16).

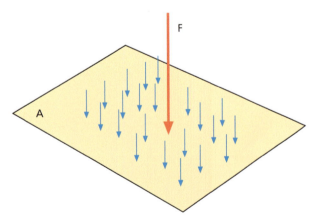

△ **Fig. 9.16** Pressure is stress due to a compressive force. The smaller arrows represent the area covered by the force, the larger arrow F represents the actual force acting over the specified area.

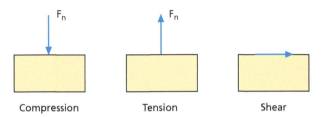

△ **Fig. 9.17** Compression, tension and shear forces are indicated in the figure.

Activity 9.3

Independent learning task: Area of contact

A kickboxer strikes another kickboxer with a force of 460 N. The contact area of the boxing glove is 0.025 m^2. What is the average stress over the area of contact? Without the glove, the contact area of the kickboxer's fist is 0.005 m^2. What would be the average stress over the area of contact if the boxer used his fist instead of the glove?

Pushing, pulling or twisting are examples of such forces. However, there are three important categories for stress that all have direct ramifications for sports injuries. These are compression, tension and shear forces (Figure 9.17).

Compression forces are pressing or squeezing forces and act perpendicular to a surface. For example the menisci of the knee or intervertebral discs when standing upright are subject to compression forces. **Tension forces** are the opposite to compression in that the forces are pulling or stretching and are directed normally (perpendicular) to a surface. For example, twisting the ankle can cause the lateral ligaments to stretch and tension stress occurs. **Shear forces** are a sliding or tearing force and are directed parallel to a surface. For example falling awkwardly at an angle may cause shearing to take place at the knee joint. Compression and tension forces or stresses occur in line with the tissue fibres. However, shearing forces occur at angles to the fibres and so potentially these are the most dangerous forces in terms of sustaining and preventing injury. The importance of acquiring correct technique at an early age and appropriate conditioning to avoid such stresses can never be over-emphasized by the responsible coach.

The relationship between load and deformation

Asymmetric loading that produces tension on one side of a body and compression on the other causes **bending**. Compressive and tensile stresses in these cases are greatest at the surface. Torsion occurs when the load or stress produces a twisting effect on a body. Shear stresses are created and are greatest at the surface (Figure 9.18).

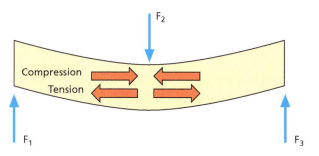

△ **Fig. 9.18** F_2 causes bending by disrupting the equilibrium of the structure. The compressive and tensile stresses are greatest at the surface.

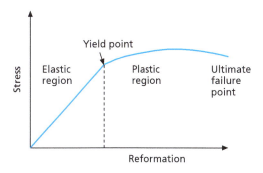

△ **Fig. 9.19** Materials behave elastically at small loads. However, loads above the yield point can create permanent plastic deformation. Rupture or fracture occurs at the ultimate failure point.

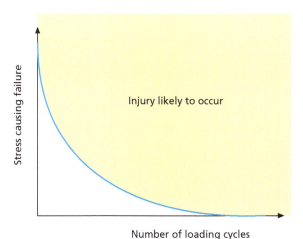

△ **Fig. 9.20** Materials such as bone, for example, can fracture. The size of the loading required to cause this decreases as the number of loading cycles increases.

However, when a stress or load is applied to a tissue, deformation occurs and this is called a **strain**. At small loads materials behave elastically, loads above the yield point, as can be seen in Figure 9.19, create permanent plastic deformation. Rupture or fracture will occur at the ultimate failure point. Different materials have their own individual curves (Norris, 1999), but the general shape is the same.

The stress–strain curve can be influenced by a number of mechanical features such as stiffness, creep, fatigue, resilience and toughness. The size of the loading required to cause a material to fail (i.e. fracture or rupture), decreases as the number of loading cycles increases (Figure 9.20). The larger the load, the fewer repetitions are required to produce failure. At the other end of the scale, however (the endurance limit), stress fractures occur in some athletes who place too much cumulative stress on a bone due to training.

Coaching application

The process of loading tissues has been shown to demonstrate the relationship between strain (the percentage of elongation or deformation) and stress (volume of loading). Each tissue has its own set degree of deformation and its own amount of stress load. Only 4 per cent of total muscular deformation (strain) is needed to initiate microinjury (Liebenson, 1996). Once microtrauma occurs in the muscle tissue, its ability to sustain more stress is reduced as a result of the rearrangement of collagen tissue (scar tissue).

The muscle's ability to return to its resting length and withstand prolonged or repetitive loading diminishes once infiltrated by scar tissue. The athlete is now vulnerable to more serious injury and may develop an intolerance to normal daily activities (Jenkins, 2003). Pain will hinder his or her ability to train and a reduction in activity levels will result in a deterioration in fitness levels, performance and motivation. If the athlete does nothing at this stage a further deterioration in the above levels will occur.

On the other hand, Hodges and Richardson (1998) have reported that certain muscles are prone to tightness and/or weakness as a result of inactivity. Poor standing or sitting postures contribute to further deconditioning and are not uncommon in today's sedentary society (Seaman, 1999, cited in Jenkins, 2003). To avoid or slow down the deconditioning process it is beneficial to apply exercises that strengthen muscles that become weak, and to stretch any muscles that become tight. For example, when seated, the hamstrings, lower back extensors and hip flexors become tight, whereas the abdominals and gluteus maximus become weak. Strengthening the muscles that have become weak and stretching those that have tightened is beneficial.

Stability and balance

What are the concepts involved in stability and balance? An object is in a state of equilibrium when it is at rest and remains at rest. For an object to remain in equilibrium, it must be stable. **Stability** is the ability of an object to resist a disruption of equilibrium, and to return to its original state if disturbed. The more stable an object, the greater is its ability to resist larger forces. **Balance** is the ability to control equilibrium. In order to remain balanced, an athlete's centre of mass must fall within its base. With regular objects this is predictable, however with the changing shape of the human body during movement the centre of mass is less predictable. In order to fully understand stability and balance, a knowledge of centre of mass and centre of pressure is essential.

Centre of mass

The centre of mass of a body is the geometric point about which every particle of the body's mass is equally distributed. Determination of the body's centre of mass is an important part of most biomechanical analyses.

The centre of mass of any object remains fixed provided that it does not change shape. Where the density of an object varies, the centre of mass is shifted towards the more weighted section. This occurs in weightlifting and powerlifting, for example, in which the masses of the athlete and the barbell become combined, effectively changing the shape of the athlete.

Activity 9.4

Laboratory investigation:

Equipment

1. Reaction board
2. Force plate
3. Computer
4. Tape measure.

Objective

This investigation outlines methods for the determination of the body's centre of mass and also shows how to solve equilibrium problems.

Concepts

Newton's law of gravitation states that each mass particle of a body has an attractive gravitational force exerted on it by the Earth. The centre of mass is the resultant of these particles and moves as the body changes position. It need not necessarily lie within the body.

Experiment 1: Centre of mass (longitudinal axis/transverse plane).

First obtain the subject's body weight in newtons. Place the reaction board on the centre of the force plate, reset the plate to zero and ask the subject to lie in a supine position on top of the board with feet aligned with the end of the board and arms by their sides. Record the force reading in newtons. Record the subject's height (H) in metres.

Taking moments about the pivot under the subject's feet, at equilibrium anticlockwise moments = clockwise moments and therefore the sum of the moments is zero.

$W \times C = R \times L$

where W is the subject's weight, C is the perpendicular distance from the subject's feet to their centre of mass, R is the net reaction force and L is the length of the reaction board in metres. Therefore

$C = (R \times L)/W$

1. Express the value obtained as a percentage of standing height. Repeat the above for three male and three female subjects.

2. Repeat the above exercise with the subject placing his or her hands overhead.

3. Tabulate your results.

Questions

● A typical value for males is approximately 56 per cent and for females the typical value is 54–55 per cent. Why is there a difference?

● What effect does the raising of the arms have on the centre of mass?

● Why was it necessary to turn the force plate to zero when just the reaction board was in place?

Experiment 2: Establish the centre of mass of the body in the sagittal axis/frontal plane and the transverse axis/sagittal plane

Repeat the above procedure with subjects standing with toes against the back of the board to determine the centre of mass in the sagittal plane in terms of distance from the front of the toes.

Repeat the experiment with subjects standing with the lateral aspect of the left foot against the end of the board to determine the centre of mass in the frontal plane in terms of distance from the left side of the foot.

A change in shape and thus a change in centre of mass is a common occurrence in human movement due to our segmental make-up. For a high jumper, for example, varying the body position can shift the centre of mass and thus affect the performance outcome (Figure 9.21). Once the human body becomes a projectile, however, the flight path of the centre of mass cannot be altered. The diagrams of the high jumper show that an athlete can move the body around the path of the centre of mass in order to optimize performance.

In the anatomical position (see Figure 8.2), the centre of mass is near the waist. In females this is 53.56 per cent of standing height and in males it is 54.57 per cent of standing height. In young children it is higher because the head is relatively large compared to the rest of the body.

a   b   c   d

e   f

g   h

i   j   k

--○-- Centre of gravity

△ **Fig. 9.21** The effect of a shift in the centre of mass due to body position on high jump performance. The centre of mass remains at the same height for each jump, but the height of the bar increases from (a) to (k) making the jumping technique in (k), the Fosbury flop, the most efficient.

217

△ **Fig. 9.22** The location of the centre of mass in irregular objects. From Hamilton and Luttgens (2002).

The centre of mass does not, however, have to lie within the physical matter of the body; think, for example, about a rugby ball or American football helmet. In humans the centre of mass will fall outside the body's physical matter in certain positions. In sport this occurs in high jumpers and pole vaulters.

The position of the centre of mass therefore varies with age, body build and sex but most importantly it varies with body position (Figure 9.22).

Segmental masses

The centre of mass or reaction board method above can show how the human body compensates for changes in body segment positioning and for the addition of external loads. It shows for example how the body adjusts when raising an arm or leg or when bending the trunk. The analysis is limited, however, to a stationary body position. The **segmental mass method** is more appropriate when analysing dynamic sport and exercise movements. This technique involves finding the centre of mass of each of the body segments through the use of a photograph, the position of these mass centres with respect to *x*- and *y*-axes and a knowledge of the individual segment weights and total body weight ratios. Identifying the centre of mass of the body segments and a knowledge of the proportionate masses of these segments enables the whole body centre of mass in any plane to be determined

by using the principle of torques (explained in Chapter 11). The sum of the torques of each of the segments will provide you with the centre of mass of the whole body with respect to the *x*- and *y*-axes.

Data has been originated through the weighing and suspension of cadaver segments. Dempster (1955) (cited in Hamilton and Luttgens, 2002) weighed eight male cadavers, then weighed the body segments and ascertained the proportion of total weight for each segment. He also located the overall centre of mass and the centre of mass for each segment. The data published by Dempster are commonly used today (Figure 9.23).

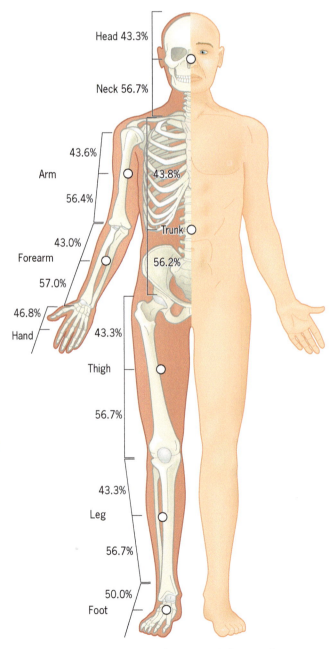

△ **Fig. 9.23** Joints/centres of mass and distance from centre of mass to joint centres. From Plagenhoef *et al.* (1983).

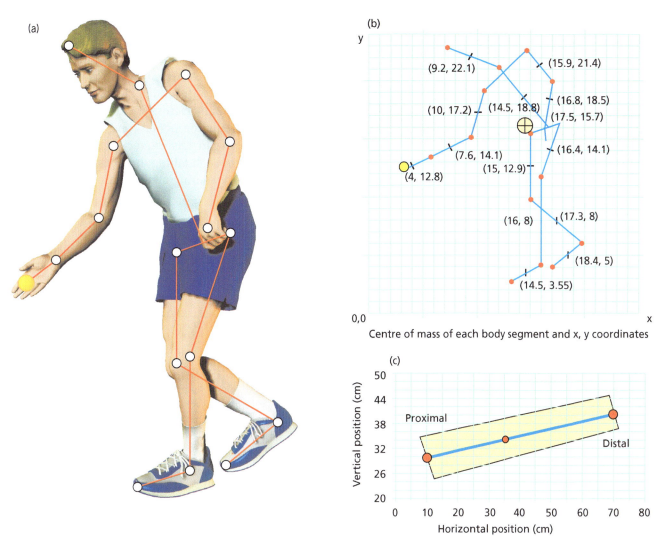

△ **Fig. 9.24** (a) Location of body segments in a handball player. (b) Centre of mass of each body segment and x, y coordinates. (c) Graphical representation of a thigh segment position. From Hamilton and Luttgens (2002).

Method for locating the centre of mass using the segmental method

Locate the individual body segments and their extremities and divide the body into the 16 segments shown in Figure 9.24(a). Estimates must be made on segment locations if hidden by other body parts. The location of the centre of mass of each segment can be found as a percentage of the distance between the segment end points using the data from Figure 9.23 (proximal and distal). For example, Figure 9.24(c) shows a graphical representation of a thigh segment position. The location of the centre of mass of the thigh can be determined from the proximal and distal point coordinates and the segment length percentage. If for the thigh the centre of mass is located at 43.3 per cent of the length from the proximal end, the specific coordinates can be determined as follows:

x position $= x_{\mathrm{proximal}} + ($ per cent length$) \times (x_{\mathrm{distal}} - x_{\mathrm{proximal}})$
$x_{\mathrm{thigh}} = 10 + (0.433)(70 - 10) = 35.98$
y position $= y_{\mathrm{proximal}} + ($ per cent length$) \times (y_{\mathrm{distal}} - y_{\mathrm{proximal}})$
$y_{\mathrm{thigh}} = 30 + (0.433)(40 - 30) = 34.33$

The centre of mass of the segment is marked by a short slash mark intersecting the segment line on the graph in Figure 9.24(b).

To determine the whole body centre of mass – an ideal point about which the torques due to body segment weights is zero – it is necessary to know the segmental centre of mass locations. To do this, you need to repeat the above procedure to calculate the x- and y-coordinates for all of the segments.

Combine these x and y segment coordinates with segmental masses to determine each segment's torque. If we go back to the above example of the thigh segment, we can see from Table 9.3 that in proportion to the total body mass, the mass of the thigh is 0.097 or 9.7 per cent of body mass. Thus, for the thigh:

$$x\text{-product} = 35.98 \times 0.097 = 3.49$$
$$y\text{-product} = 34.33 \times 0.097 = 3.33$$

Body segment	Proportion of bodyweight
1. Trunk	0.486
2. Head and neck	0.079
3. R. thigh	0.097
4. R. lower leg	0.045
5. R. foot	0.014
6. L. thigh	0.097
7. L. lower leg	0.045
8. L. foot	0.014
9. R. upper arm	0.027
10. R. lower arm	0.014
11. R. hand	0.006
12. L. upper arm	0.027
13. L. lower arm	0.014
14. L. hand	0.006

△ **Table 9.3** From Plagenhoef *et al.* (1983).

If you repeat the above procedure for each of the segments, the sum of the *x*-products represents the *x*-coordinate of the whole body centre of mass and the sum of the *y*-products is the *y*-coordinate. The values should be located and marked on to the graph (Figure 9.24(c)).

Centre of pressure

Reaction forces between the body and support surfaces are distributed over the entire contact area. The force can be summed into a single net force acting at a single point. This is called the **centre of pressure**. It is the point about which the ground reaction force is balanced. During normal standing there is often head

movement. This causes the centre of mass to move in a pendulum-like fashion, which, in turn, causes movement in the centre of pressure in the foot/ground region. This is called **postural sway** and usually takes place in the sagittal and frontal planes. Postural sway can be affected by many factors, including age, injury, obesity, fatigue or environmental conditions. There are clearly links here to performance in athletes and potential injury through falling in the elderly. Stability considerations and training methods are discussed later in this section.

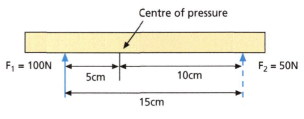

$$\Sigma T = (0.1\text{m})(50\text{N}) - (0.05\text{m})(100\text{N}) = 0$$

△ **Fig. 9.25** Force is applied to a force platform and distributed between 2 metal blocks 15 cm apart. F_1 records a force of 100 N and F_2 a force of 50 N in the sagittal plane. In equilibrium the sum of the torques (T) is 0. The centre of pressure can therefore be found at a point 5 cm from F_1 and 10 cm from F_2 in the sagittal plane.

The position of the centre of pressure of an individual standing upright can be calculated using force plates. We know that the reaction forces between the body and ground are distributed over the entire contact area. Force is distributed across two feet and the centre of pressure is located between the feet. Changes in the centre of pressure while standing on a force platform are used to evaluate control mechanisms of balance.

Stability

The **base of support** is the area enclosed by all points at which the body contacts a supporting surface. A person can move the centre of pressure to any point within the base of support, but cannot move it outside the base of support and remain stable or balanced. To remain in equilibrium, the centre of pressure must be directly below the body's centre of mass. Therefore to remain in equilibrium, the centre of mass must be within the boundaries of the base of support. This is known as **static balance** (Figure 9.26).

When a person's centre of mass is moving forward, it is important to apply a rear or posterior directed force to stop forward motion and rotation to prevent loss of

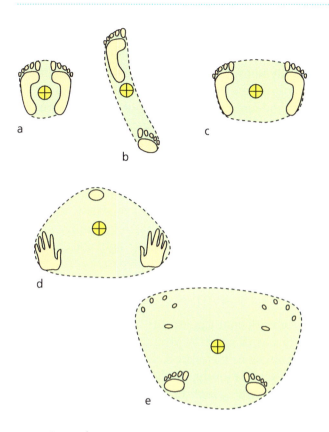

Base of support
(a), (b), (c) feet only
(d) hands and head (headstand) and (e) hands and feet

△ **Fig. 9.26** Variation in body position and stability.

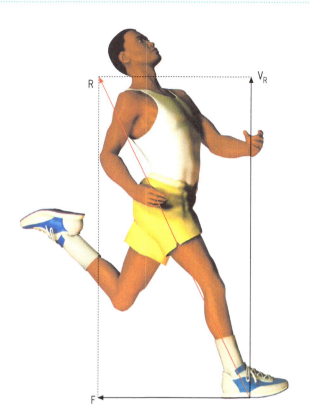

△ **Fig. 9.27(a)** Sprinter decelerating.

balance. When moving forward quickly, the centre of mass must start off more to the rear of the body since there is greater displacement during deceleration. There is a larger horizontal distance from a boundary of the base of support. As can be seen in Figure 9.27(a), by leaning backwards the centre of mass falls further behind the sprinter's lead leg and so there is a longer horizontal distance from a boundary at the base of support and balance is therefore improved.

Factors that increase stability include:

● Greater mass and moment of inertia (a larger force or torque is required to accelerate)

● Increased surface friction (friction forces are used to prevent or stop horizontal motion)

● Larger base of support (the centre of mass can travel further before crossing the boundary)

● Increased horizontal distance between the centre of mass and the base of support boundary

● Lower centre of mass (less destabilizing torque is produced and less opposing torque is needed to restore balance).

△ **Fig. 9.27(b)** Sprinter coming out of the blocks. Essentially, he is losing balance and falling, but will interrupt his fall with his first stride. This is known as 'dynamic balance'.

Stability and balance training

Good stability and balance are essential prerequisites for success in many sports. Unstable training environments have been used for many years by martial artists in particular to improve stability, balance and power. Looking around many gymnasiums and health clubs you will see an excessive use of the Swiss ball and the term 'core stability' being bandied about by the misinformed. Many of the exercises used today resemble circus acts, such as squatting with a 60 kg barbell whilst standing on a stability or Swiss ball, at the expense of the more traditional exercises and training methods. Indeed scientific studies have shown greater core muscle recruitment of the abdominals when carried out in unstable environments compared to stable environments (Vera-Garcia *et al.*, 2000).

Nevertheless, there still is much disagreement in the strength and conditioning community with regard to unstable training environments, functional training and performance enhancement. This may be due to some of the extreme methods used in an unstable environment.

The primary factor that should drive any conditioning programme is specificity; this suggests that stability and balance training in unstable training environments is no more effective than any other training method (Santana, 2002). To develop power and muscle hypertrophy high volume training, Olympic lifts and plyometrics should be used. Unstable training has a place in conditioning programmes (Santana, 2002) but should only be used to supplement other training methods and not instead of them.

Proprioceptive/balance training and rehabilitation

Balance can be defined as the ability to keep oneself in an upright position, which requires proprioceptive input. **Proprioception** is defined as the reception of stimuli within the organism. This is a neurological process. During an ankle rehabilitation programme, for example, specific balance and proprioceptive tests are used, such as a single-plane balance-board test, a clinical test for sensory interaction on balance or computerized force plate analysis that can objectively measure postural sway.

Following such tests, the next stage in the rehabilitation chain is training. Proprioceptive training can improve dynamic balance (Gilman and Newman, 1996). Ankle injuries occur more often than not in the closed chain position, where the feet remain in constant contact with the ground, and so it is essential that proprioceptive training is performed in a closed chain position. If this is too painful then open chain exercises should be used with a gradual progression through using Swiss balls, balance boards, single leg hops through to jog-to-run activities with directional changes (Hanney, 2000). There is definitely a place for unstable environment training in the pursuit of increased stability and balance following injury.

Dynamic stability

Many dynamic movements – including walking and running – are inherently unstable. They essentially consist of a series of interrupted falls and it is difficult to quantify stability during such tasks.

Static and **dynamic balance** are essential for optimum sports performance (Figure 9.7(b)). Motor skills such as kicking and jumping in contact sports like rugby and soccer are often performed under pressure. Inadequate whole body balance could result in a reduction in performance and injury. In a recent study Moosavi (2003) investigated the use of the hamstring/quadriceps strength ratio, which has traditionally been used as a predictor of hamstring injuries, as a predictor of whole body static balance in young adults. More research is required here but there are clear implications for strength and conditioning coaches in terms of athlete preparation.

Work, power and energy

Analysis of work done can identify inefficient movement patterns or wasted effort. Work and power can indicate whether muscles or forces are acting to generate or brake motion. Power relationships indicate muscles better at braking or generating motion. Energy relationships allow one to understand the relationships between applied forces, changes in speed and height.

Mechanical work done by a force (F) is the force's magnitude times the displacement of an object in the direction of the force:

$$W = F_x \times d_x + F_y \times d_y$$

where F_x and F_y are the x and y components of force and d_x and d_y are the x and y components of displacement. The SI unit of work is the joule (J) (1 J = 1 N m).

Positive work occurs when force and displacement are in the same direction. This indicates that the force is acting to speed up the object. **Negative** work occurs when force and displacement are in opposite directions. This indicates that the force is acting to slow down the object. It is determined by the directions of the vectors. In terms of muscle contractions, however, **concentric** contractions indicate positive work, **eccentric** contractions indicate negative work and **isometric** contractions indicate no work done.

Coaching application: exercising with different weights

An athlete has a 1-RM (the maximal resistance an individual can move for 1 repetition only) squat of 200 kg and trains with a squat weight of 200 kg, 150 kg and 100 kg. He has a total bodyweight of 97.5 kg but the weight above the knee joints is 90 kg. Only this body section is lifted during squatting, the feet and shanks are almost motionless. Therefore the weights that are lifted, including the body, are 290 kg, 240 kg and 190 kg and the distance that the centre of mass is raised or the difference between the highest and lowest centre of mass positions is 1 m. The 200 kg barbell is lifted once, the 150 kg barbell is lifted 10 times and the 100 kg barbell is lifted 20 times.

Using the equation above (weight multiplied by number of repeats multiplied by the distance moved), the mechanical work produced is therefore as follows:

$$290 \times 1 \times 1 = 5290 \text{ kgm}$$
$$240 \times 1 \times 10 = 2400 \text{ kgm}$$
$$190 \times 1 \times 20 = 3800 \text{ kgm}$$

The above shows that exercising frequently with a light barbell produces significantly more mechanical work than exercising less with the heaviest weight.

Power Output can be defined as the rate of work production. It can be calculated as work done in a given time:

Power = work/change in time ($P = W/\Delta t$)
Power can be generated ($P > 0$) or absorbed ($P < 0$).
The SI unit for power is the watt (W) (1 W = 1 J/s).

Instantaneous power represents the rate at which work is being performed at a specific moment. It can be calculated as force multiplied by the velocity in the direction of the force:

$$P = F_x V_x + F_y V_y$$

Because muscle force depends on the velocity at which the muscle shortens (because of the force–velocity relationship), so does its ability to generate power. In a number of sports the ability to generate power is a key element to success, and the assessment of power is important in the development of sport-specific physiological/

biomechanical profiling, the assessment of conditioning programmes over time and the evaluation of a sport's physiological demands (Kraemer *et al.*, 2003). Therefore superior ability to execute athletic movements explosively typically results in more desirable performance.

Assessing power output

When designing training programmes to increase maximal power output, it is recommended that individuals train using the load at which power output is maximised – the 'optimal' load. Previous investigations have attempted to determine at what load (i.e., a percentage of 1-RM) power output is maximised. For example, the load that elicits the greatest power output in the jump squat varies from 0 per cent of 1-RM (McBride *et al.*, 1999) to approximately 60 per cent of 1-RM (Baker *et al.*, 2001). Such large discrepancies in the optimal load reported have led to ambiguity surrounding the load–power relationship. It is therefore difficult to draw any conclusions from the literature and form specific training guidelines for developing muscular power (Cormie *et al.*, 2007).

1. Direct methods

A number of methodologies have been used for collecting and analysing power data in the literature.

Kinematic methods

Linear Position Transducer (LPT): Calculation of power through displacement data, which in most cases is collected by means of a single linear position transducer (LPT) (Alemany *et al.*, 2005; Falvo *et al.*, 2005).

Jump Squat					
Study	Methodology	Type of movements	Loading parameters	Training status of subjects	Optimal load (% 1RM)
Baker *et al.* (2001)	1-LPT+MASS	JS	25 37 50 62 75% 1RM (40 60 80 100 125kg)	32 male professional & semi-professional rugby players	55-59%
Bourque *et al.* (2003)	1-LPT	cJS	0 30 40 50 60% 1RM	8 male power (8 volleyball, 2 badminton) & 8 male endurance athletes	All 14% Power 0% Endurance 30%
Esliger *et al.* (2003)	1-LPT	cJS	30 40 50 60 70 80% 1RM	11 male & 10 female volleyball & basketball players	63%
McBride *et al.* (1999)	1-LPT+FP FP	JS CMJ	30 60 90 1RM BW BW+20kg BW+1kg	28 male nationally competitive sprinters power and Olympic	JS 30% CMJ 0%
Sieivet *et al.* (2004)	Accelerometer (similar to 1-LPT)	cJS cSS	30 40 50 60 70% 1RM	30 male rugby and basketball players	JS 60% SS 40%
Stone *et al.* (2003)	V- Scope (similar to 2-LPT)	JS cJS	10 20 30 40 50 60 70 80 90 100% 1RM	22 males (ranging from 7 weeks to 15 years training)	10%
1RM - one repetition maximum JS - jump squat c - concentric only SS - split stance JS CMJ - countermovement jump					

△ **Fig. 9.28** A summary of methodologies used for collecting and analysing power data. (From Cormie *et al.*, 2007).

- **1 x LPT:** Double differentiation of displacement data (i.e. a second order derivative of the displacement data) permits the determination of acceleration. The force (F) produced during the lift can be determined by adding the acceleration of the system (a) and the acceleration due to gravity ($a_g = 9.81 \, m/s^2$) and then multiplying the total acceleration to the mass of the system (M_{sys} = external load + body mass) [$F = M_{sys} \cdot (a + a_g)$]. In the Jump Squat (JS) and Squat (S) movements, the body and the bar move together as a unit, it is therefore assumed that the barbell velocity calculated is equivalent to the entire system. Power (P) can then be determined by multiplying force and velocity at each time point ($P = FV$).

- **1 x LPT + MASS:** Processes outlined above are used to determine the velocity of the movements evaluated. However, with this method, force (F) is equivalent to the product of the system mass (Msys) and acceleration due to gravity (F= Msys × ag), thus making force a constant throughout the movement measured. Power (P) can then be calculated as a product of this force multiplied by the velocity-time curve ($P = FV$). N.B: Methodologies using a single LPT fail to account for horizontal movements in the calculation of power, which may influence the load – power relationship.

Kinetic methods

- **Force Plate (FP):** The force platform has been widely used to calculate power within laboratory settings (Canavan and Vescovi, 2004; Linthorne, 2001). The impulse – momentum relationship is used to determine velocity and enables power to be calculated through a forward dynamics approach. However, there are two main factors associated with this method that lead to the miscalculation of power output. Firstly, this is an indirect measure (i.e. forward dynamics approach) which requires significant data manipulation, and, secondly, the inability of the FP to account for barbell movement that occurs independently of the body (e.g. when an athlete performs a power clean). As a consequence, velocity is underestimated resulting in the underrepresentation of power.

Kinematic-kinetic methods

- **1 x LPT and FP:** This method involves the combination of displacement data and vertical ground reaction forces for the calculation of power. Displacement data is collected via the single LPT whilst the FP measures the vertical force output of the system. The velocity of the system can be determined using a first order derivative of the displacement data, and power is then calculated from the product of the velocity and force data.

- **2 x LPT and FP:** To account for horizontal movement affecting the vertical displacement during multidimensional dynamic movements 2 LPTs can be used. The signals from each LPT are combined to determine vertical displacement, which are then subsequently used in combination with time to calculate velocity. Power can then be calculated by coupling this velocity with force data collected through the use of a force plate.

Although many methods for calculating power during dynamic lower body resistance exercise exist, the use of these methodologies has resulted in incomplete reporting of the load – power relationship. Standardization of data collection procedures is vital if we are to gain a clearer understanding of the load power relationship, assist coaches to more effectively monitor training (i.e. response to interventions and identification of training loads), make comparisons with other athletes (i.e. normative data) and replicate research studies.

2. Indirect methods

Using prediction equations

A number of prediction equations have been developed to estimate peak and average power from jump height. Recent research by Canavan and Vescovi (2004) has however, questioned the validity of these equations for a number of reasons. Studies validating peak power equations (Harman *et al.*, 1991; Sayers *et al.*, 1999) used separate tests to determine vertical jump height and peak power instead of pairing these values from the same jump. In some studies vertical jump was assessed using the jump and reach test (Harman *et al.* (1991) and Sayers *et al.* (1999)). This is problematic in itself as performing a jump against a wall (as is the case in this test) is likely to impede jumping technique in comparison with jumping on a force platform (Canavan and Vescovi, 2004). The study by Sayers *et al.* (1999) also included a heterogeneous sample of men and women from varied athletic backgrounds. As differences exist in jump technique and coordination between genders and between athletic/non-athletic groups (Bobbert *et al.*, 1996; Hertogh and Hue, 2002) it could be argued that a more homogenous sample was needed to fully validate these equations.

In response to these criticisms, Canavan and Vescovi (2004) compared actual peak power (PP_{actual}) measured using a force platfom with peak power estimations (PP$_{est}$) using the Harman and Sayers equations in a group of 20 recreationally trained females. In addition to generating their own peak power prediction equation, they reported that all peak power prediction equations were significantly related to PP$_{actual}$ but only the Harman et al., prediction was not significantly different to PP$_{actual}$.

Recent research by Duncan et al. (2008) has also suggested that there is a need to evaluate peak power prediction equations across sporting populations, due to the need to consider sport specific training modes and testing needs. Consequently, Watkins and Duncan (2010) compared actual peak power to peak power predicted from four widely used regression equations in a sample of elite adolescent academy footballers. The results indicated that all regression equations previously used to estimate peak power from vertical jump height overpredict PP$_{actual}$ in adolescent footballers.

The significant differences between PP$_{actual}$ and PP$_{est}$ in the current study, therefore, support research by Canavan and Vescovi (2004). Furthermore, like the current study, Canavan and Vescovi also reported that peak power equations tended to over estimate PP$_{actual}$. However, Canavan and Vescovi (2004) reported that the Harman et al. equation provided the most precise estimate of PP$_{actual}$ and Duncan et al., (2008) suggested that the Sayers-CMJ equation appeared to offer the most precise estimation of PP$_{actual}$ in their sample group of adolescent basketballers. Furthermore, they reported that peak power prediction equations tended to underpredict PP$_{actual}$ in their sample group.

This clearly contradicts the findings of the Watkins and Duncan study (2010). One explanation for this finding may be due to participant characteristics. Both the Watkins and Duncan (2010) study and the research by Canavan and Vescovi (2004) examined football players. Although of different gender groups, the similarity of these findings with those of Canavan and Vescovi and the discrepancy with Duncan et al. (2008), who used basketballers would appear to support assertions by Hertogh and Hue (2002) that there is a need to compare peak power prediction equations with actual peak power in a range of sport specific groups and that development of regression equations in more homogenous samples may be needed when dealing with sport-specific performance and testing needs. Therefore, future work would be desirable cross-validating current prediction equations and developing sport specific prediction equations specific to football players. However, the Sayers-SJ prediction appears to offer the most appropriate option for estimating peak power in football players when a direct measurement of force is not available.

Coaching application

There are many conflicting views amongst coaches with regards exercise selection and power development in athletes. Research has shown the benefits of the inclusion of the power snatch, power clean, squat snatch and squat clean in athlete training programmes, but the important question is when should these exercises be added or taken away from the meso-cycle? What about the high pull as an aid to power development?

Let us consider the squat snatch exercise. For a weightlifter to lift the weight to a level high enough so that he can drop under the bar and catch it in a deep squat position, the weight must be accelerated to a minimum velocity (another reason for coaches and sports scientists to acquire kinematic data). Let us assume that a lifter is able to move the weight with a minimum velocity of 1.83 m/s to successfully snatch a weight of 180 kg. If he can train so that he can accelerate 190 kg to the same velocity then he should be successful with that weight too. From the above formula for power, we can calculate power output. The goal of training must be to develop higher power values, and research has shown that greater power outputs (over the weight range tested) can be achieved through high pulls using a snatch-width grip for specificity.

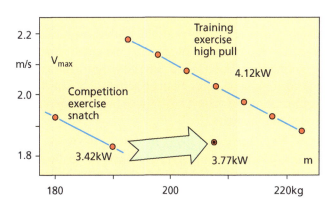

△ Fig. 9.29 The relationship between barbell mass and maximum velocity for a lifter in the snatch and high pulls. From Bartonietz (1996).

As can be seen from Figure 9.29, snatching 180–190 kg as fast as possible requires a power output of 3.42 kW, while high pulls demand more than 4 kW across the entire weight range if they are to be performed at maximum velocity. Once maximum velocity of the bar is reached it is not enough to turn over and catch the barbell. High pulls require the athlete to include an intensive shrug of the arms and shoulders and in doing so they lengthen the path of the bar by about 10–15 cm. The inclusion therefore of maximal velocity high pulls into the training programme could develop the power to snatch 190 kg because there are greater demands placed on power output.

Kinetic energy

Energy is defined as the capacity to do work. **Kinetic energy** (KE) is the energy associated with the movement of an object. Kinetic energy equals the amount of work the object can do in being brought to rest. It can be defined by:

$$Kt = \frac{1}{2}mv^2$$

where m is the mass of the object and v is the object's velocity. The SI unit of kinetic energy is the joule (J). For example, an object with a mass of 2 kg that moves at a constant velocity of 3 m/s has a kinetic energy of:

$$KE = (0.5)(2)(3)^2 = 9 \text{ J}$$

Coaching application: reversible muscle action exercises

Reversible muscle action exercises are those in which the muscle group in question is stretched immediately before shortening (stretch–shortening). An example of this is the depth jump. Here the athlete stands on a box of given height, drops off the box to the floor and immediately jumps as high as possible. With these kinds of exercises the resistance is governed by the kinetic energy of the falling body not just the mass or weight. The same magnitude of kinetic energy can be achieved by varying the combinations of velocity (height of box or dropping distance) and the mass.

According to Zatsiorsky (1995), increasing the mass will lead to a reduction in rebound velocity. Initially the moderate increase in approach velocity will result in an increase in rebound velocity. However, the rebound velocity decreases if the approach velocity is too high. The optimal approach velocity and therefore kinetic energy depends on the mass of the moving body. If training with stretch–shortening cycles it is recommended to use devices in which the mass and velocity can be changed (Zatsiorsky, 1995).

Potential energy

Potential energy (PE) is the stored energy that can be converted into work. It is associated with an object's position. Potential energy due to gravity equals the weight (W) of an object times its height (h) above a reference height:

$$PE = W \times h = mgh$$

where m is the mass of the object and g is the acceleration due to gravity (9.81 m/s^2).

For example, for a 50-kg athlete standing on a box 1 m high and preparing to jump off to perform plyometric training the potential energy just before jumping off is:

$$PE = (50)(9.81)(1) = 490.5 \text{ J}$$

Conservation of mechanical energy

When gravity is the only external force that performs work on a system, the total energy of the system remains constant.

$$KE + PE = constant$$

Projectile motion is one case when conservation of energy applies. The work produced by an external force is equal to the change in the total energy of the object acted upon.

$$W = \Delta KE + \Delta PE$$

where W is work, ΔKE is change in kinetic energy and ΔPE is change in potential energy.

Work causes changes in the mechanical energy of the system. Indeed from an applied point of view, in pole-vaulting two advantages of a flexible fibreglass pole over a rigid pole made of steel or bamboo are that a flexible pole reduces the energy dissipated in the athlete's body during the plant and that it lowers the optimum take-off angle so that the vaulter loses less kinetic energy when jumping up at take-off (Linthorne, 2000).

Example

After clearing the bar, a 75-kg high jumper has a vertical velocity of 1.7 m/s at a height of 1.25 m above the mat. Calculate the potential, kinetic and total energy.

First, let us calculate the potential energy of the high jumper at this particular point of the clearance. Using the equation for calculating potential energy:

$$PE = mgh = 75 \times 9.81 \times 1.25 = 919.69 \text{ J}$$

To calculate the kinetic energy for the high jumper we can use the following equation:

$$KE = \frac{1}{2}mv^2 = \frac{1}{2} \times 75 \ (1.7)2 = 108.38 \text{ J}$$

Therefore the total or mechanical energy:

$$(PE + KE) = 919.69 + 108.38 = 1028.07 \text{ J}$$

From here, the total height attained by the high jumper can be calculated:

$$1028.07 = 75 \times 9.81 \times h$$

By rearranging the formula:

$$h = 1028.07/735.75 = 1.4 \text{ metres}$$

Impulse and momentum

It has been previously shown that the acceleration of an object is directly proportional to the magnitude of the net force acting on it and inversely proportional to its mass:

$$F = m \times a$$

We also know that the acceleration of an object is equal to the change in velocity over a specific length of time ($a = (v_2-v_1)/t$). Therefore, the previous equation becomes:

$$F = m \times (v_2\text{-}v_1)/t$$
$$\text{or}$$
$$F \times t = m \times v_2 - m \times v_1$$

This relationship states the change in an object's velocity (v), or momentum ($m \times v$), is proportional to the impulse ($F \times t$) that is applied. Understanding the concept of impulse is extremely important as an increase in impulse can result in performance enhancement or injury, depending on the nature of the movement. Mathematically, impulse represents the integral of the force–time curve (area under the force–time curve, Figure 9.30). However, it can be estimated by multiplying the mean force by the total contact time.

Therefore, the motion of a body depends not only on the force, but also on the duration that the force is applied. When a racket, bat or foot strikes a ball, very large forces are produced. As a result large accelerations of the ball occur, but since the forces act over a very short time period it is more appropriate to represent this concept through the concept of impulse. In other words, any change in the movement state of an object depends on the applied force and the time over which it is applied. An impulse is required for any change in momentum.

Thus **impulse** is a measure related to the net effect of applying a force (F) for a time (t):

$$\text{Impulse} = F \times t$$

Impulse increases with increased force and increased duration of application. It is a vector quantity and is also equal to the change in momentum of the system.

$$I = \Delta M = M_2 - M_1 = m\,v_2 - m\,v_1$$
$$\text{Impulse} = F_{external}\,(t_2 - t_1) = mv_2 - mv_1$$

The impulse–momentum relationship comes directly from Newton's second law:

$$F_{external} = ma$$
$$F_{external} = m(v_2 - v_1)/(t_2 - t_1)$$
$$F_{external} = (t_2 - t_1) = mv_2 - mv_1$$

Impulse can be represented graphically. The area beneath a graph of force versus time between two points in time is equal to the impulse due to the force between those two times.

The impulse due to the net external force acting in a given direction on a system equals the change in the momentum of the system in that direction over the same period of time.

$$\text{Impulse}_x = (F_{external})_x(t_2 - t_1) = mv_{x2} - mv_{x1}$$
$$\text{Impulse}_y = (F_{external})_y(t_2 - t_1) = mv_{y2} - mv_{y1}$$

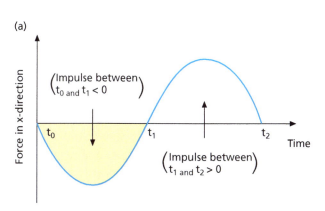
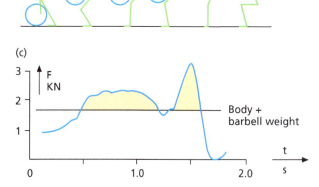

△ **Fig. 9.30** (a) Impulse as an area. (b) Vertical component of ground reaction forces versus time for the snatch lift in a decathlete. From Bartonietz (1996).

Example

Contact time is the period of time where the foot, feet or hands are in contact with the ground or external body during an activity. During the long or triple jump or during jump/power training, contact time is important to allow the body to generate force, change direction or increase velocity. When designing strength and conditioning programmes — in particular, programmes for power athletes — exercises such as squats are important to increase the forces that these specific muscle groups can generate. However, since contact times are very short in high-powered movements, the athlete needs to generate the greatest amount of force in the shortest possible time (rate of force development). Therefore, it is essential that the athlete trains these muscle groups at the appropriate velocity and range of movement. Biomechanically,

---

Activity 9.5

Independent learning tasks: Impulse-Momentum

1. Performance enhancement

1. A shot putter has come to you for advice with regards to improving her performance. After an initial assessment of her technique, you believe she should increase the time over which she applies force to the putt. She is able to apply a force 320 N over a time of 0.45 s, but you would like to increase this to 0.65 s. The mass of the putt is 5 kg. How much can she improve her final release velocity by increasing the time of force application?

2. Is it always beneficial, in terms of performance, to increase the time over which the force is applied? Why or why not? Give examples.

3. Using the impulse momentum relationship, describe technique modifications that can be made within the following sports to improve performance:

   Sprinting ; Discus ; Javelin ; High Jump.

2. Injury prevention

A gymnast comes to see you because she is experiencing knee pain when she tries to stick her dismounts. One of your assessment tools is to examine the vertical ground reaction forces that are produced during a countermovement jump. You notice that she lands very hard and suggest making modifications to her landing technique to help reduce pain.

1. In terms of the impulse-momentum theory, why might she be experiencing pain when she lands? What advice could you give to help her?

2. Why are we interested in the vertical ground reaction forces?

---

then, we are identifying and measuring ground reaction forces and the contact time. This is **impulse**.

Biomechanical data from the 1988 Seoul Olympics indicated that approach velocities in the long jump were the greatest of all and contact times ranged from 100 to 120 metres. Contact times increased to 180 metres in the triple jump, pole vault and high jump. The evidence suggested that the key to great jumping distance is to generate high force in a very short time period (Gros and Kunkel, 1990).

Impact

The term 'impact' comes from the Latin word *impingere*, meaning 'to thrust upon or drive against'. Impact is concerned with such aspects in sport as a force of contact, a collision or the striking together of two or more objects. Impact can occur vertically and horizontally, such as a falling movement or when a body is stopped by making contact with a resistive surface like a wall. In terms of receiving impact to the human body we are concerned with avoiding injury, particularly in contact sports such as rugby union, and also regaining our balance or equilibrium.

There are many sporting activities where objects collide with one another (e.g. golf club and ball, tennis racket and ball). These impacts occur over a very short period of time, involve contact forces of high magnitude, and result in rapid changes in momentum of one or both colliding objects. Also, during the brief period of impact, the two colliding objects will undergo a period of **deformation** (i.e. a change in shape) and a period of **restitution** (i.e. return to original shape). For example, consider the contact between a tennis racket and ball. The ball is in contact with the racket strings for only a few milliseconds. In the early part of the impact, the ball flattens and the strings are distorted. In the latter portion of the impact, both the ball and the racket rapidly return towards their original shapes.

The velocities of the two colliding objects following the impact depend on their velocities before impact and the nature or quality of the impact. In a perfectly elastic collision, the relative velocities of the two objects after impact (separation velocities) are the same as their relative velocities before impact (approach velocities). In an inelastic collision, the relative velocities of the colliding objects after impact are less than those before impact and some of the total energy of motion is lost (e.g. some may be transformed into heat associated with the deformation and restitution processes).

Therefore impact involves a collision in which a large force acts over a small time interval. The force acting during impact has two effects: part of the energy is absorbed and lost through deformation of the objects, and the remaining force changes the objects' direction. The total momentum of the two objects just before and just after impact are equal (Figure 9.31).

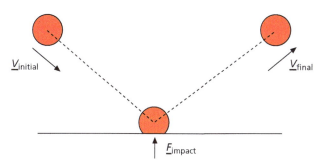

△ **Fig. 9.31** A ball approaches a fixed stationary object with velocity ($V_{initial}$). Part of the energy is absorbed and lost through deformation, the remaining force determines the object's direction and final velocity (V_{final}).

There are two different types of impact:

- **Perfectly elastic.** No energy is lost during impact. Magnitude of relative velocity between objects after impact is the same as before impact.
- **Perfectly plastic.** Objects deform and stick together. Relative velocity between the objects is zero after impact.

Most impacts fall somewhere in between these two extremes.

Avoiding impact injury

Two principles should be remembered in order to avoid injury when receiving impact:

1. The more gradual the loss of momentum (or kinetic energy) of the moving body or object, the lower the force exerted on the object or body.
2. The larger the area receiving the impact the lower the force applied per unit of surface area.

Repeated impacts, through running, for example, may lead to overuse injuries.

Maintaining and regaining equilibrium following impact

There are also two principles to remember related to maintaining and regaining equilibrium following impact:

1. Equilibrium is improved when there is a larger base of support in the direction of the force.

2. At the moment of impact, the centre of mass is centred as near as possible to the base of support. For example, a long jumper on landing will lower the hips over the feet and in so doing increase stability.

Example

A 100 kg rugby fullback runs into a goalpost while attempting to catch a high ball put up by his opposite number. If the player's initial horizontal velocity at the moment of impact was 8.9 m/s, and the impact lasts 0.2 s, what is the average force experienced by the player? We already know that the impulse/momentum relationship comes directly from Newton's second law. Since Newton's second law can be represented by the formula F = ma, we can rearrange this to F = m(v – u)/t. By substituting into the above equation we have:

$$F = 100 (0 – 8.9)/0.2 = –890/0.2 = –4450 \text{ N}$$

Therefore a force of 4450 N acts against the rugby player as he collides with the post to bring him to rest.

Coefficient of restitution

The **coefficient of restitution** (e) is an index that measures the elasticity of an impact. It ranges between 0 and 1. An e equal to 1 reflects a perfectly elastic collision, whereas an e equal to 0 reflects a perfectly plastic (or inelastic collision). The coefficient of restitution depends to a large extent on the nature of the two materials of which the colliding objects are made. It is also affected by impact velocity, the shape and size of the colliding objects, the location on the colliding objects at which the collision occurs and their temperatures.

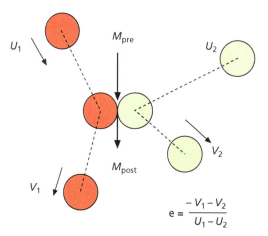

$$e = \frac{-V_1 - V_2}{U_1 - U_2}$$

△ **Fig. 9.32** e is the ratio of the relative velocities of the colliding objects before and after impact.

In quantitative terms, e is the ratio of the relative velocities of the colliding objects before and after impact:

$$e = -(v_1 - v_2)/(u_1 - u_2)$$

where v_1 and v_2 are the velocities of the two colliding objects immediately after impact and u_1 and u_2 are their velocities immediately before impact (Figure 9.32). Note that e is always a positive number. The minus sign in the equation is needed because the relative velocities before and after impact are in the opposite directions. For impacts of a ball or similar object (object 1) off a fixed surface such as the floor (object 2) the above equation can be simplified to:

$$e = -(v_1/u_1)$$

since the velocity of the floor before and after impact for all practical purposes is zero

$$(u_2 = v_2 = 0).$$

From our knowledge of uniformly accelerated motion, we know that when a ball is dropped onto a fixed surface, the velocity of the ball immediately before impact is determined by the height from which it is dropped ($v^2 = u^2 + 2as$, where $u = 0$). It is clear from the equation of uniformly accelerated motion that the velocity is proportional to the square root of the height. Similarly, the height the ball reaches after impact is proportional to the square root of the velocity of the ball immediately after impact. Therefore the equation can be simplified further for this special case when a ball is dropped onto a fixed surface:

$$e = \sqrt{(h_b/h_d)}$$

where h_b and h_d are the rebound and drop heights respectively.

Activity 9.6

Laboratory investigation:

This investigation demonstrates how the coefficient of restitution changes according to the surfaces involved and the temperature of the object.

Exercise

For each combination of ball and surface:

1. Drop the ball from 2 m onto the surface and record the rebound height. Ensure that the people responsible for measuring rebound height do not make a parallax error.

2. Record the results and calculate the coefficient of restitution.

Activity 9.6

Repeat the above experiment using three similar racket balls, one cold, one at room temperature and one warmed up.

Questions

1. Which racket ball had the highest coefficient of restitution?

2. What is another factor independent of temperature and surface that affects the coefficient of restitution? (Coefficient of restitution can be calculated using the following formula: $e = (H_b/H_d)^{1/2}$, where H_b is height rebound and H_d is height dropped.)

Further questions for consideration

1. Do aluminium bats provide any impact advantage over wooden bats? What evidence supports your answer?

2. What are the effects of ball temperature and ball inflation on collision elasticity?

Centre of percussion

A tennis racket, for example, has two sweet spots, namely the **centre of percussion** (COP) and the **vibration node**, which is located near the centre of the strings. The centre of percussion is the point where a force can be applied without causing acceleration at the centre of rotation. The vibration node differs from the centre of percussion in that you do not feel any vibration in your hand. When a ball hits the centre of percussion, your hand does not feel any force pushing against it. When a ball impacts on one of these spots the player is almost unaware that the impact has occurred since the force is relatively small. When the ball impacts away from these spots the impact can be felt in the hand/arm and can be quite painful. Some of the jarring is caused by a transferring of the shock waves from impact. Most of it, however, is caused by the player not hitting the ball at the racket's centre of percussion.

Imagine the racket is a vertical rod hanging down or pivoted from the handle by a piece of string with the centre of mass situated between the handle and the racket head (Figure 9. 33 (a)). When the racket is struck by a ball at a distance from the pivot point a reaction force occurs and the racket is pushed backwards. When the ball contacts the racket nearer the handle (or nearer the pivot), the handle moves backwards first. When the racket is struck at its centre of percussion, however, there is no reactive force. Since a tennis racket has an unevenly distributed centre of mass, the centre of percussion will depend on where the racket is gripped.

If an object is percussed at point P and this results in no reactive force at point O, then P is the centre of percussion of the object held at O.

(a)

N = vibration node
P = centre of percussion
O = acceleration pole

(b)

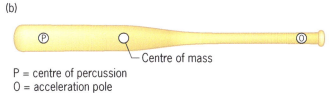

P = centre of percussion
O = acceleration pole

△ Fig. 9.33 Vibration node, centre of percussion, acceleration pole and centre of mass of a tennis racket (a) and a baseball bat (b).

The impact of a tennis ball with the strings is a very short event. It usually lasts between 4 and 8 one-thousandths of a second (0.008 s). One or two milliseconds later the shock waves of the collision reach the player's hand. Forces as a result of the shock wave are increased on the hand. The magnitude of the force depends on where

the impact occurs on the racket face, how the racket is gripped during impact and ball and racket velocity. Skilled players increase the grip force prior to impact, which can widely vary the peak forces produced using an eastern grip (Figure 9.34) (Knudson, 2001).

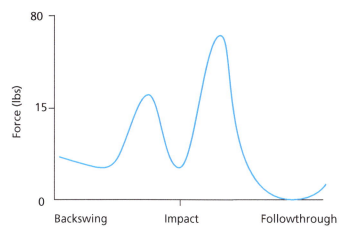

△ Fig. 9.34 Typical pattern of force applied to the base of the index finger in a tennis forehand. Adapted from Knudson (2001).

One of the most common results of repetitive impulsive and vibrational (vibrations of the racket frame after impact) loading from tennis impacts is the development of lateral elbow tendinosis or 'tennis elbow'. This is a term used to describe localized elbow pain due to overuse repetitive movements on the inside or outside of the elbow. Pain on the outside arises from irritation of the common wrist extensor attachment whereas pain on the inside arises from irritation of the common wrist flexors.

Pain indicates that rest is required and so the athlete should seek appropriate medical assistance. Knudson (2001) suggests coaches should also provide technical advice to try to reduce impulsive forces being transferred to the elbow. Biomechanically, the most important factor is to minimize eccentric or off-centre impacts. Manufacturers have produced vibration damping mechanisms that can be attached to the strings of the racket. However Li *et al.* (2003) have reported that string dampers have a negligible role in reducing vibration transferred to the arm. Modern racket design has also moved the centre of percussion nearer to the point of impact, giving benefits such as reduced fatigue and risk of injury (Bartlett, 1997). Using larger racket heads, reducing spin on ground strokes or reducing string tension, may help to avoid chronic injury in tennis players.

Summary

- This chapter has looked at the issues involved in linear kinetics and how they affect our understanding of human movement.

- Force was defined and the identification of various external forces highlighted, along with the derivation of these forces through the free body diagrams.

- Important concepts such as friction, impact and centre of mass have been addressed, including segmental masses.

- Finally, the coefficient of restitution and centre of percussion were introduced.

- Laboratory practicals and review questions have been included to develop understanding.

Review Questions

1. How much force must be applied to a 0.5-kg ball travelling horizontally at 42 m/s to bring the ball to rest in 0.25 s?

2. A 70-kg sprinter applied a horizontal force to the ground of 1300 N for 0.2 s. After the force application she has a final horizontal velocity of 9.7 m/s. What was her initial velocity?

3. Discuss the following concepts: (1) coefficient of restitution ; (2) work, power and energy ; and (3) impulse, impact and the centre of percussion.

4. Define Newton's laws of motion.

5. A weightlifter raises a 630-N barbell from 0.2 m to 1.8 m. Calculate the work done lifting the barbell. If the lift was done in 0.38 s, how much power was generated?

6. Define centre of mass and explain its importance when analysing human movement.

7. Draw free body diagrams for a long jumper in flight and a shot putter just before release.

8. How can a knowledge of the centre of percussion of tennis rackets help a coach and his athlete?

References

Ae M, Ito A, Suzuki M (1992) The men's 100 metres. *New Studies in Athletics* 7: 47–52.

Alemany, JS, Pandorf, CE, Montain, SJ, Castellani, JW, Tuckow, AP and Nindl, BC (2005) Reliability assessment of ballistic jump squat and bench throws. *Journal of Strength and Conditioning Research*, 19:33–38

Baker D, Nance S, Moore M (2001) The load that maximises the average mechanical power output during explosive bench press throws in highly trained athletes. *Journal of Strength and Conditioning Research* 15: 20–24.

Bartonietz, KE (1996) Biomechanics of the snatch: toward a higher training efficiency. *Strength and Conditioning Journal* 18: 24–31.

Bartlett R (1997) *Introduction to Sports Biomechanics*. London: Routledge.

Bobbert, M., Gerritsen, K., Litjens, M. and Van-Soest, A (1996) Why is countermovement jump height greater than squat jump height? *Med. Sci. Sports Exerc.* 28: 1402–1412.

Canavan, P.K., and J.D. Vescovi (2004) Evaluation of power prediction equations: peak vertical jumping power in women. *Med. Sci. Sports Exerc.* 36: 1589–1593.

Cormie P, McBride JM and McCaulley, GO (2007) Validation of power measurement techniques in dynamic lower body resistance exercises. *Journal of Applied Biomechanics* 23:103–118.

Duncan, M., Lyons, M and Nevill, A.M (2008) Evaluation of peak power prediction equations in male basketball players. *J. Strength Cond. Res.* 22: 1379–1381.

Enoka, RM (1994) *Neuromuscular Basis of Kinesiology* 2nd edn. Human Kinetics, Illinois.

Falvo, JM, Sciling, KB and Weiss, WL (2005) Techniques and considerations for determining isointertial upper-body power. *Journal of Sports Biomechanics.* 5:293–311

Farrar M, Thorland W (1987) Relationship between isokinetic strength and sprint times in college-age men. *Journal of Sports Medicine and Physical Fitness* 27: 368–372.

Gilman S, Newman SW (1996) *Essentials of Clinical and Neuroanatomy and Neurophysiology,* 9th edn. Philadelphia: FA Davis Company.

Gros HJ, Kunkel V (1990) Biomechanical analysis of the pole vault. In: Bruggeman GP, Glad B eds *Scientific Research Project at the Games of the XXIVth Olympiad* – Seoul 1988. Monaco: International Athletic Federation, pp. 219–260.

Hamilton N, Luttgens K (2002) *Kinesiology: Scientific Basis of Human Motion*, 10th edn. New York: McGraw-Hill.

Hanney WJ (2000) Proprioceptive training for ankle stability. *Strength and Conditioning Journal* 22: 63–68.

Harman, E, Rosenstein, M, Frykman, P, Rosenstein, R and Kraemer, W (1991) Estimation of human power output fro vertical jump. *Journal of Applied Sport Science Research*, 3:116–120

Harman E (1994) The biomechanics of resistance exercise. In: Baechle TR ed. *Essentials of Strength Training and Conditioning*. Champaign, IL: Human Kinetics.

Hertogh, C and Hue, O (2002) Jump evaluation of elite volleyball players using two methods: jump power equations and force platform. *Journal of Sports Medicine and Physical Fitness*, 42(3): 300–303

Hodges PW, Richardson CA (1998) Delayed postural contraction of transversus abdominis in low back pain associated with movement of the lower limb. *Journal of Spinal Disorders* 11: 46–56.

Jenkins JR (2003) The transversus abdominis and reconditioning the lower back. *Strength and Conditioning Journal* 25: 60–66.

Kent, J.T (1994) The Complex Bingham Distribution and Shape Analysis, *Journal of the Royal Statistical Society*, Series B, 56 285–299

Knudson D (2001) What happens at impact and why it can hurt. *TennisPro Magazine*, Sept/Oct.

Kraemer, WJ, Hakkinen, K and Triplett-McBride, NT (2003) Physiological changes with periodized resistance training in women tennis players. *Med. Sci. Sports Exerc.* 35: 157–168.

Li F-X, Fewtrell D, Jenkins M (2003) An investigation of retrofit vibration damping devices in tennis. Communications to the 12th Commonwealth International Sport Conference. *Journal of Sports Sciences* 21: 235–365.

Liebenson C (1996) *Rehabilitation of the Spine*. Philadelphia: Williams and Wilkins.

Linthorne NP (1994) The effect of wind on 100-m sprint times. *Journal of Applied Biomechanics* 10: 110–131.

Linthorne NP (2000) Energy loss in the pole vault take-off and the advantage of a flexible pole. *Sports Engineering* 3: 205–218.

Linthorne, N.P (2001) Analysis of standing vertical jumps using a force platform. *Am. J. Phys.* 69: 1198–1204.

McBride, J, Triplett-McBride, T. Davie, A. Newton, RU (1999) A Comparison of strength and power characteristics between power lifters, Olympic lifters and sprinters. *Journal of Strength and Conditioning Research* 13:58–66.

Moosavi M (2003) Relationship between lower extremity isokinetic strength and static balance among adolescent footballers. Communications to the 12th Commonwealth International Sport Conference. *Journal of Sports Sciences* 21: 235–365.

Norris CM (1999) *Sports Injuries Diagnosis and Management*, 2nd edn. Oxford: Butterworth-Heinemann.

Plagenhoef S, Evans F, Abdelnour T (1983) Anatomical data for analysing human motion. *Research Quarterly* 54: 169–178.

Tihanyi, J (2003) Determination of the Optimum Loads, Intensity and Repetitions for Individualised Strength Training. Presentation at Strength and Conditioning Symposia. Glasgow.

Santana JC (2002) Sport-specific conditioning: stability and balance training: performance training or circus acts? *Strength and Conditioning Journal* 24: 75–76.

Sayers, SP, Harackiewicz, DV, Harman, EA, Frykman, PN and MT, Rosenstein (1999) Cross-validation of three jump power equations. *Med. Sci. Sports Exerc.* 31: 572–577.

Vera-Garcia FJ, Grenier SG, Mcgill SM (2000) Abdominal muscle response during curl-ups on both stable and labile surfaces. *Physical Therapy* 80: 564–569.

Watkins, J (1996) *An Introduction to Mechanics of Human Movement*, 3rd edition, Petroc Press.

Watkins, P.H and Duncan, M (2010) A comparison of peak power prediction equations with actual peak power in elite junior footballers. *Journal of Sport Sciences* S160, 28(S1).

Wiemann K, Tidow G (1995) Relative activity of hip and knee extensors in sprinting – implications for training. *New Studies in Athletics* 10: 29–49.

Zatsiorsky VM (1995) *Science and Practice of Strength Training*. Champaign, IL: Human Kinetics.

10 Angular Kinematics

Introduction

As already discussed in Chapter 7, kinematics analyses the form, pattern or sequencing of movement with respect to time. The forces that cause this motion are not considered. Angular motion or rotation is concerned with all the points in an object or body that move in a circle about an axis of rotation. All points move through the same angle in the same time. Angular kinematics therefore involves the kinematics of particles, objects or systems undergoing angular motion.

Just as objects that move linearly are measured by the distance travelled, objects that rotate from one position to another about an axis are measured by the angular distance travelled. Angular distance is measured in degrees, and there are 360 degrees in one full revolution. In biomechanics, angular distance is measured in **radians** or degrees. A radian measures angles in the same way that degrees do but is defined differently. It is therefore just a different way of representing angular displacement. Radians are equal to the arc length of a circle divided by its radius:

Radians = (arc length)/radius

So how many radians are there in one full revolution or complete circle? We know that the formula for the circumference of a circle is $2\pi r$ and so radians for one revolution = $2\pi r/r = 2\pi$. During angular displacement a point that starts at the origin (O), completes a full circle and ends up back at its origin has gone 2π rads and not 0. The angle is cumulative unless there is a change in direction.

In short, one radian is defined as the angle formed at the centre of a circle by an arc that is the same length as the radius of the circle (Figure 10.1).

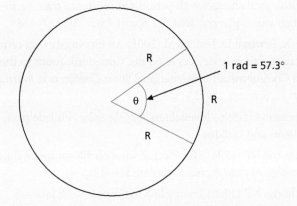

△ **Fig. 10.1** One radian showing the angle (θ) at the centre of the circle by an arc where R is the radius.

1 rev = 2π rad = 360 degrees
π rad = 180 degrees
π rad/2 = 90 degrees
π rad/3 = 60 degrees
π rad/4 = 45 degrees
1 rad = 57.3 degrees

When describing angular kinematics (Figure 10.2), **positive** angles indicate anticlockwise rotation and **negative** angles indicate clockwise rotation.

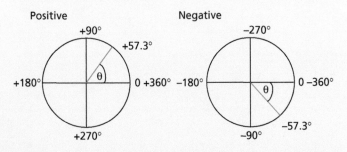

△ **Fig. 10.2** Positive angles change through an anticlockwise direction and negative angles change through a clockwise direction.

Angular displacement

Angular displacement is the change in angular position of a line segment (Figure 10.3). It is the net effect of angular motion (vector quantity). Angular displacement does not depend on the path between positions and has angular units of measurement (degrees or radians).

When standing up from a parallel squat, the knee and hip joints rotate approximately 90 degrees but in opposite directions (clockwise and anticlockwise). A diver performing a front 1.5 dive would have an angular displacement of 180 degrees (clockwise or anticlockwise).

To calculate angular displacement ($\Delta\theta$), the angular positions are subtracted. Therefore:

$$\Delta\theta = \theta_{final} - \theta_{initial}$$

For example, if the initial angular position is 0.1 rad and the final angular position is 0.9 rad, the angular displacement can be calculated using the formula:

$$\Delta\theta = 0.9 - 0.1 = 0.8 \text{ rad}$$

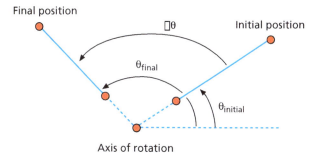

△ **Fig. 10.3** Diagram demonstrating a change in angular position or angular displacement.

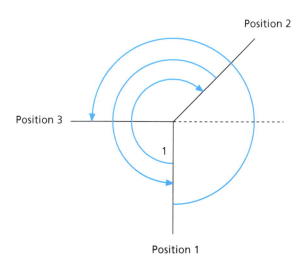

△ **Fig. 10.4** The three positions of angular displacement during the golf swing.

Taking the example of a golf swing (Figure 10.4), we can calculate the angular displacements of the phases and the overall displacement.

- Angular positions
  Position 1: $\theta_1 = 3\pi/2$
  Position 2: $\theta_2 = \pi/4$
  Position 3: $\theta_3 = 3\pi$
- Phase 1: Backswing
  $\Delta\theta = \pi/4 - 3\pi/4 = -5\pi/4$ rad
- Phase 2: Forward swing
  $\Delta\theta = 3\pi/2 - \pi/4 = 5\pi/4$ rad
- Phase 3: Follow through
  $\Delta\theta = 3\pi - 3\pi/2 = 3\pi/2$ rad
  Total angular displacement: $\Delta\theta = 3\pi - 3\pi/2 = 3\pi/2$ rad

Angular velocity

Angular velocity is the rate of change in the angular position, or how fast a body is rotating. The rotation of the arm about the shoulder joint during the delivery of a cricket ball by a medium paced bowler has a lower angular velocity than the same movement in a fast bowler.

Angular velocity = angular displacement/change in time. Its shorthand notation can be represented as:

$$\omega = \Delta\theta/\Delta t$$
$$= (\theta_2 - \theta_1)/\Delta t$$

Its standard unit is (angular units)/time = (radians/s, degrees/s).

For example, if the angular displacement ($\Delta\theta$) of an object is 0.8 rad and the time elapsed (Δt) is 0.4 s, the angular velocity can be calculated by:

$$\omega = \Delta\theta/\Delta t = 0.8/0.4 = 2 \text{ rad/s}$$

If the initial angular position of the hip joint during a movement lasting 0.4 s is 0.9 rad and its final angular position is 0.1 rad, what is the angular velocity?

$$\theta_1 = 0.9 \text{ rad}, \theta_2 = 0.1 \text{ rad}, \Delta t = 0.4 \text{ s}$$
$$\omega: = (0.1 - 0.9)/0.4 = -2 \text{ rad/s (2 rad/s clockwise)}$$

Activity 10.1

Independent learning task: Calculating angular velocity

A powerlifter performs the squat exercise and he starts from a vertical standing position (knee flexion = 0 degrees). During the first part of the downward phase, he flexes his knees 45 degrees in a time of 0.75 sec. During the next 0.5 sec, he flexes his knees another 15 degrees. What is the average knee angular velocity during each of the two periods, and for the downward phase of the lift as a whole?

Average versus instantaneous angular velocity

The formulas for angular velocity above give us the average angular velocity between an initial time (t_1) and a final time (t_2). Instantaneous angular velocity is the angular velocity at a single instant in time. Instantaneous angular velocity can be estimated using the central difference method:

$$\omega \text{ (at time } t_1) = [\theta \text{ (at } t_1 + \Delta t) - \theta \text{ (at } t1 - \Delta t)] / 2\Delta t$$

where Δt is a very small change in time. Remember to convert degrees to radians before making these calculations.

Figure 10.5 is a graphical representation of angular velocity with respect to time. You can see from the graph the difference between instantaneous and average angular velocity. This is the angular equivalent of the displacement/time graphs covered in Chapter 8.

Estimating angular velocity from angular displacement

Referring to Figure 10.6, when estimating angular velocity from angular displacement curves the following factors should be remembered:

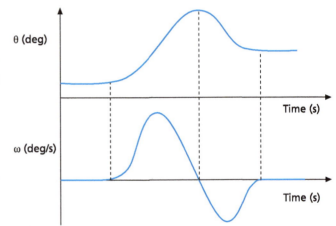

△ **Fig. 10.6** Graphical representation of instantaneous and average angular velocity with respect to time.

- The points with zero slope = points with zero velocity.
- Portions of the curve with a positive slope have positive velocity (i.e. velocity in the positive direction).

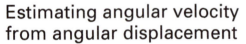
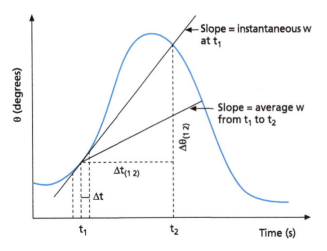

△ **Fig. 10.5** Angular velocity as a slope.

Activity 10.2

Independent learning task: Plotting angular velocity

A gymnast swings back and forth from the high bar with the angular displacement as shown in the figure below. Sketch his angular velocity.

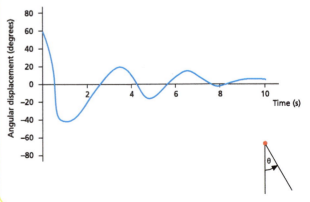

- Portions of the curve with a negative slope have negative velocity (i.e. velocity in the negative direction).

Angular acceleration

Angular acceleration ($\propto$) is expressed as the rate of change in angular velocity. For example, during a giant swing, a gymnast begins from a handstand and swings once around the bar. The gymnast accelerates clockwise on the way down, but experiences deceleration clockwise during the up phase.

Angular acceleration = (change in angular velocity)/ (elapsed time)

$$\propto = \Delta\omega/\Delta t = (\omega_{1(final)} - \omega_{0(initial)})\Delta t$$

Its standard unit of measurement is (angular units)/time² (rad/s^2).

Calculating angular acceleration

For planar motion, angular acceleration = a change in rotation speed. Taking the example given in Figure 10.7, where ω_1 = 40 degrees/s, ω_2 = 10 degrees/s and $t_1 = t_0 + 1.5$ s:

$$\alpha = (\omega_1 - \omega_0)/(t_1 - t_0) = (30 \text{ degrees/s})/1.5 \text{ s} = 20 \text{ degrees/s}^2$$

What is the relationship between angular velocity (ω) and angular acceleration (α):

1. if angular velocity and angular acceleration are in the same direction then the magnitude of angular velocity increases, and if

2. angular velocity and angular acceleration are in the opposite directions then the magnitude of angular velocity decreases (deceleration).

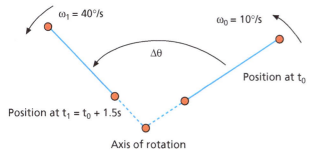

△ **Fig. 10.7** Diagrammatic representation of the above example – calculating acceleration.

Example

During a golf swing, the time taken from the end of the backswing (ω_0 = 0.0 rad/s) to impact is 0.8 s. The angular acceleration (α) is 1.5 rad/s². What is the angular velocity of the club-head at impact?

$$1.5 = (\omega_1 - 0)/0.8$$
$$\omega_1 = 1.2 \text{ rad/s}$$

Thus if angular acceleration = 0, then angular velocity does not change.

Movement patterns

When we throw an object, for example, the forces generated from the body's core transfer to the outer body segments. Each sequential segment increases in acceleration/velocity until the final or distal segment(s) at the point of release have the greatest acceleration/velocity readings. Biomechanists and coaches can quantitatively measure these angular accelerations and velocities to monitor the effectiveness of training programmes and the effects of any alterations in technique and equipment.

General movement patterns produced by an athlete include kicking, striking, pushing and pulling. They can be distinguished from sport skills in that a movement pattern is any configuration of movements in the same general spatial plane and is not specific to any sport. A sport skill, however, is specific to a particular sport and is associated with a particular mechanical purpose.

The primary mechanical purposes associated with push-like patterns include:

- To manipulate a resistance (e.g. bench pressing or squatting)
- To project an object for maximum accuracy (e.g. basketball shooting, darts).

Push-like patterns are simultaneous segmental movements where the object to be pushed does not lag behind the segments doing the pushing. The object is pushed in a rectilinear path and there is a predominance of lever movements.

In most sports involving a throwing action, the basic throwing movement is very specific to the particular sport. For example, tennis serves and javelin throwing adopt an overarm action, a tennis forehand stroke and discus throw adopt a sidearm pattern and bowling and softball pitching adopt an underarm movement pattern. These movement patterns can be divided into **open kinetic chain activities** and **closed kinetic chain activities**. Open kinetic chain activities include kicking, throwing a dart or typing whereas closed kinetic chain activities include squatting, press-ups or walking.

The kinetic link principle

This principle is based around the consideration of the body segments with regard to coordination, sequencing and timing to increase the distal segment velocity for work or athletic performance. There are several statements that can be made regarding velocity generation in an open kinetic link system. The statements relate two or more factors to a velocity increase. Examples are as follows:

- Body segments are used in a sequential fashion from the larger to the smaller segments.
- Body segments are used in a sequential fashion from the more proximal to the more distal segments.
- Body segments are used in a sequential fashion from the most stable to the most free segment.
- A more distal segment accelerates forward as its more proximal segment decelerates forward.

These mechanical principles are applicable to striking, hitting and kicking. In the sport of baseball or softball, batting is a whole body movement. Most of the power involved comes from a summation of force and a continuity of joint torque. This starts with the more powerful lower extremity joint segments and passes through the trunk to the more distal upper joint extremities (Hughes et al., 2004). Watts (1957) has reported that the hips are the most important factor involved in a successful swing, whereas Williams and Underwood (1986) believe that the hips and wrists are significant.

Performance mechanisms in various sports have been investigated using kinematic methods. The kinetic link principle involves the sequence of movements made in a multi-segment action (proximal-to-distal sequence), which is highly applicable to actions that require high end-point velocity, such as strokes in racket sports for example. With recent advances in technology, it is now possible to perform basic analyses of racket skills by obtaining data for joint angles, angular velocities and ball speeds for the tennis serve (Papadopoulis et al., 2000), backhand drive (Elliott et al., 1989a) and forehand drive (Elliott et al., 1989b). In order to analyse the tennis serve, several markers were attached to the wrist, elbow and upper arm. Using three-dimensional analysis, Van Gheluwe et al. (1987) were able to quantify the magnitudes of forearm pronation and upper arm rotation. These are considered to be important movement sequences in the production of a high-velocity racket head.

Activity 10.3

Independent learning task: Comparisons between skilled and unskilled athletes

Attend a beginners' class in badminton, golf or tennis. Observe an unskilled performer and a skilled performer in the class, and compare the segments that are used and the sequencing of the segments in a similar skill. Does the beginner use the same skill and movements? Does the beginner sequence the movements or are some of the movements used simultaneously? What differences do you notice about the speed and the accuracy of the object projected?

Similar patterns were found by Tang et al. (1995), who examined the angular kinematics of the badminton forehand smash. They reported significant wrist joint movement about the two axes of rotation. There are also implications here for specific training programmes. The stretch reflex is activated as a result of an increased supination of the forearm prior to rapid pronation (Tang et al., 1995). This in itself would increase force production and movement velocity. Coaches should not only strive to enhance correct technique in their athletes, but take advantage of the enhanced physiological and psychological opportunities that come with it.

Application: movement screening of athletes and the bilateral bodyweight squat pattern

Optimal movement can be defined as movement that occurs without pain or discomfort and involves proper joint alignment, muscle coordination and posture (Cibulka and Threkeld-Watkins, 2005). Faulty movement resulting in a faulty movement pattern has been described as a disruption in the normal balance of how muscles support and move joints (Kendall et al., 2005). The disruption may be due to the muscle being too strong or too weak, not firing at the right time or lacking the range of movement to accommodate efficient movement. When the muscle balance is disrupted, joint function will suffer which maybe detrimental to performance (Kritz et al., 2009). If the faulty pattern is associated with pain, the movement pattern may change to compensate for this discomfort or pain. If performed regularly, the change will become part of the brain's programming associated with that movement (Kendall et al., 2005). Consequently, movement quality and athletic performance maybe sacrificed (Kendall et al, 2005; Cook, 2003).

The bilateral squat is one of the most commonly used exercises in the physical preparation of athletes (Abelbeck, 2002). A bodyweight squat at or below 90° of knee flexion with proper symmetry and coordination is a good indicator of overall movement quality. Conversely, the inability to perform a bodyweight squat efficiently may imply overall stiffness throughout the body, or restricted joint mobility (Cook, 2003). The biomechanist and/or strength and conditioning coach may be qualified to screen fundamental movement competency with a basic understanding of how the foot, ankle, knee, hip and shoulder joints and the lumbar and thoracic spine function to provide efficient movement (Kritz et al., 2009). Table 10.1 illustrates what the literature considers to be the proper position (i.e. angular displacement) of each major body segment/joint during the downward and upward phases of the bilateral squat.

Downward and upward phases of the bilateral bodyweight squat, considered within the literature to be correct squatting technique					
Anatomical Region	Kinakin (2004)	Baechle (2000)	Bloomfield (1998)	Plane of Motion (POM)and Range of Movement (ROM) considerations	Summary of Squat Technique
Head	Neutral position	Neutral position	Held up	POM: Sagittal Downward gaze can increase trunk flexion by 4.5° (Donnelly *et al.*, 2006)	Maintain a neutral position, looking straight ahead
Thoracic Spine	Flat: maintain torso and shin angle	Flat: maintain torso to floor angle	Angled slight forward and held straight	POM: Sagittal Failing to maintain a slightly extended position contributes to increased compressive and shear forces of the lumbar spine	Maintain slightly extended thoracic spine position
Lumbar Spine	Flat: maintain torso to shin angle	Flat: maintain torso to floor angle	Curved slightly inward	POM: Sagittal A 2° increase in extension from neutral spine increases compressive stress within the posterior annulus by approximately 16% (Walsh *et al.*, 2007)	Maintain neutral spine
Hip Joints	Flexed: remain under shoulders	Flexed	Flexed	POM: Sagittal ROM between 0° and 135° flexion and 0° and 15° extension (Hall and Brody, 2005)	Flexed and aligned
Knee Joints	Flexed: knees over feet	Flexed: knees aligned over feet	Flexed	POM: Frontal Avoid excessive mediolateral or valgus and varus frontal plane ROM POM: Sagittal Avoid extreme anterior ROM (i.e. where the knee moves past the toes (Fry *et al.*, 2003)	Knees aligned with the feet in the sagittal plane and feet/ankles in the frontal plane
Ankles/feet	Shoulder width stance	Shoulder/remain on the floor	Shoulder width, toes point forwards	POM: Sagittal 25° ROM (15° plantar flexion, 10° dorsi flexion) (Escamilla *et al.*, 2001)	Feet remain flat, not rolling in or lifting up

△ **Table 10.1** Angular kinematics (presented qualitatively and/or quantitatively) of the bodyweight bilateral squat. (Adapted from Kritz *et al.*, 2009).

The structure in Table 10.1 can be applied to a selection of basic movements/exercises (e.g. overhead squatting, lunging) used to physically prepare athletes for competition, particularly as most sporting techniques are derived from these basic movement patterns. It should be noted that 'movement screening' is not a *diagnostic* tool; its purpose is to inform the coach of the athlete's movement pattern when there is no pathology/injury. It allows individualization of training focusing on muscle weaknesses, imbalances and asymmetries to improve the athlete's functional movement pattern. Table 10.2 illustrates the optimal viewing positions for the identification of dysfunctional movement during the bilateral bodyweight squat.

Examples of common faulty movement patterns during the bilateral bodyweight squat can be seen in Figures 10.8 (a), 10.8 (b) and 10.8 (c). Figures 10.8 (a) and 10.8 (b) illustrate medial or lateral movement of the knees when observed from the front (i.e. the frontal plane), and Figure 10.8 (c) illustrates excessive anterior motion when observed from the side (i.e. the sagittal plane). These movement patterns contribute to knee dysfunction and pain in athletes; excessive varus (lateral) and valgus (medial) frontal plane movement of the lower leg position has in part been attributed to dysfunctional rectus femoris, hip adductors/abductors and hamstring muscle groups as well as poor pelvic stability. When screening athletes, you should be familiar with the primary anatomical function of each joint and their contribution to movement efficiency.

Downward and upward phases of a bilateral bodyweight squat		
Anatomical Region	**Viewing Positions**	**Faulty Movement Pattern(s)**
Head	Side, front	- Head too far forward - Head too far back - Head movement to either side - Gaze is downwards
Thoracic Spine	Side, back	- Protracted shoulder blades (scapulae) - Over-extension of the thoracic spine
Lumbar Spine	Side	- Extension or flexion during the movement
Hip Joints	Side, front	- Medio-lateral rotation
Knee Joints	Side, front	- Alignment inside of the hip - Alignment outside of the hip - Excessive forward movement in front of the toes
Ankles/Feet	Side, front, back	- Supination or pronation of the feet, and/or heels raising off the floor during the movement

△ **Table 10.2** Optimal viewing position for identifying faulty movement patterns during a bilateral bodyweight squat

Additional biomechanical considerations when screening the bilateral bodyweight squat:

1. The movement velocity (i.e. linear and/or angular) of the athlete during the descent/ascent. Suggestion: Athlete performs the squat pattern in a controlled manner (i.e. both descent and ascent). This applies to all abilities, ages and genders.

2. The range of motion (ROM) (i.e. angular displacement) of the knee/hip joint assessed. Suggestion: Parallel squat (i.e. the top of the thighs are parallel to the floor), this applies to all abilities, ages and genders.

3. Assessment of the *quality* of the movement pattern. Suggestion: Use a video camera in both the sagittal and frontal planes with analysis software (e.g. Siliconcoach or Dartfish) and provide appropriate feedback (i.e. qualitative *and* quantitative if necessary) to your athlete(s).

4. Remediation strategy to correct faulty movement patterns: consider exercises to develop flexibility, muscle strength and control in the identified anatomical regions.

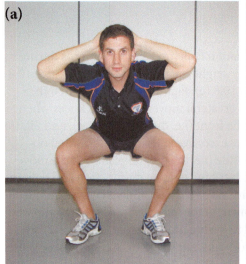  

△ **Fig. 10.8** (a) Squat movement pattern and a varus lower leg position. (b) Squat movement and a valgus lower leg position. (c) Squat movement pattern with excessive anterior motion of the knees.

Application of angular velocity and acceleration

It is important for coaches and athletes to understand the biomechanics of conditioning exercises. Athletes in strength and power sports use various power and speed/strength exercises in their training. Biomechanical and technical knowledge of both training and competition exercises can help coaches to select appropriate exercises to train for various sports. Many sports require a high power output from hip and knee extensors at the same time, and snatching from the hang position can therefore be advantageous for sports that require this. Variations in the snatch alter specific movement patterns so the coach should be concerned with developing specific performance adaptations through correct exercise selection executed with correct technique.

Example 1

Figure 10.9 shows the body positions at different points in time during the snatch and barbell path for two weightlifters from different weight categories. The figures show the body positions for the two lifters. In the **first pull** (positions 1–3) the knees and hips are extended, feet are in contact with the floor. The trunk is held almost constant to allow for an effective transition of force. Knee and hip extensor positions differ slightly between the two lifters. This is due to anthropometrical differences. The taller

athlete, Krastev, starts with a knee angle of 47 degrees to Shalamanov's 80 degrees. At the end of this phase (the first pull), the bar velocity is approximately 80 per cent of its maximum velocity.

Positions 4–6 (**second pull**) show extension of the knees, hips and ankles. When the joints are fully extended at the end of this phase the bar reaches maximum velocity.

Positions 6–8 show the **turnover and catch,** in which the lifter must move his body downwards, decelerating the downward movement of the barbell with arms locked.

Analysis of the angular displacements of elite lifters at different phases of the snatch lift should provide useful information to coaches and athletes, particularly with regard to the learning and execution of correct technique. Individual differences must be considered, however.

Example 2

Figure 10.10 shows the movement paths or trajectories of the bar and related velocity values for two weight lifters. Athlete 'S' lifted 1.5 times his bodyweight while athlete 'H' lifted 1.7 times his bodyweight. Athlete 'S' had a rapid first pull at 91 per cent of maximum velocity, whereas athlete 'H' reached 66 per cent maximum velocity.

The velocity of the bar should increase continuously. A movement coordination that results in progressive bar velocity is mechanically effective because the lifter transfers only a minimum of physical work to reach a given velocity. Too fast a first pull hinders the transition phase, resulting in a reduction in barbell velocity. There are two reasons for this: first, the highest forces can be generated at this point (see Figure 10.10) and, secondly, as a result of the force/velocity curve (parametric force/velocity

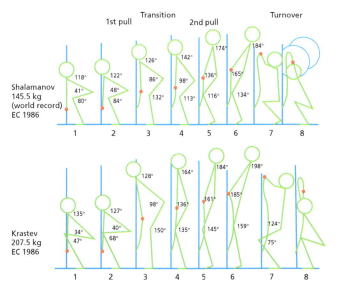

△ **Fig. 10.9** Body positions and path of the barbell during the snatch for two lifters of different weight categories. Note that the working conditions for the main muscle groups of the legs differ for each lifter as a result of differences in body dimensions. From Bartonietz (1996).

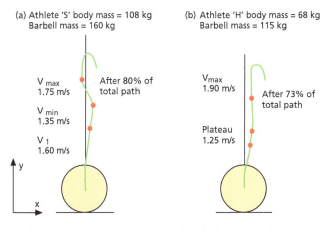

△ **Fig. 10.10** Movement paths of barbells with velocities. From Bartonietz (1996).

relationship, see Chapter 11) the force decreases as velocity increases. Therefore the barbell must approach this favoured position at a relatively low velocity to impart maximal force. These principles do not apply to those athletes who are short and where the bar is located at knee joint level in the start position before the lift (Zatsiorsky, 1995).

A slight decrease in velocity as a result of knee movement towards the bar can create higher muscular pre-tension of the quadriceps in preparation for the second pull. This is sometimes termed the double-knee bend and is indicated in positions 3 and 4.

Barbell velocity is an important factor in ensuring that training loads are applicable to the weight used. It is an indication of training intensity. When we train with varying loads and velocities we directly affect the physiological adaptations at muscular level. By selecting appropriate training loads we can determine the fibre types to be trained. Research in applied biomechanics gains insights into the internal structures of exercises. It is necessary not just to analyse movement of the barbell but also limb movements around the main joints in order to assess the work of the muscle groups used. The power of joint propulsions during the snatch lift can be calculated as the product of muscle force moment and angular velocity. Time-related changes are one way of judging the effectiveness of the movement.

Figure 10.11 shows changes in isometric force with a change in body position. The largest force produced is

about 360 kg with the barbell in a position just below the hip joint. Either side of this position there is a significant reduction in the force produced by the athlete. Consequently the force–velocity relationship will dictate a gradual but progressive pull from the floor along with many hours of technical training for the athlete and coach in order to maximize power output and training intensity.

From a biomechanical perspective, according to Bartonietz (1996) an effective variant of barbell acceleration is the relationship between velocity and time where the barbell increases in velocity gradually between the first and second pull. In other words, what causes the bar to accelerate effectively (which is what we want as athletes and coaches) is the relationship between velocity and time. Since the definition for acceleration is a change in velocity divided by a change in time (and this also applies to the angular equivalents) it can be seen that a change in one of the parameters (for example velocity) will have an overall effect on the acceleration of the barbell. Thus an increase in barbell velocity over the same given time period will have the overall effect of increasing the acceleration.

A final assessment of the lifting technique must consider the general mechanical and physiological effects as well as the lifting skills and abilities. Figure 10.12 shows the relationships between angular velocity and time for the hip and knee joint of an elite weightlifter performing the snatch lift. During the **first pull** the knees and hips are extended, feet are in contact with the floor. The trunk is

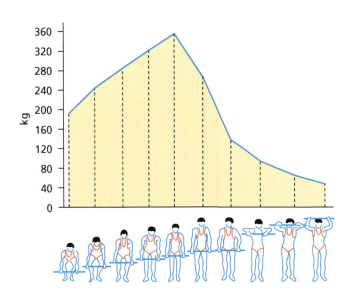

△ **Fig. 10.11** The maximal isometric force F_m applied to a bar at different body positions (at different heights of the bar). This is an example of the strength curve in a multijoint movement. From Zatsiorsky (1995).

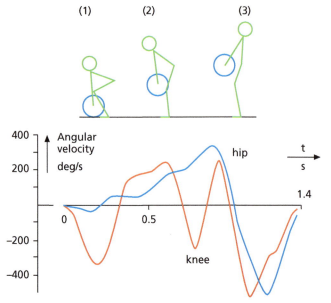

△ **Fig. 10.12** Angular velocity versus time for a squat snatch at 152.5 kg by the European Champion in 1986 at a body mass of 67.5 kg. Adapted from Bartonietz (1996).

held almost constant to allow for an effective transition of force, and accounts for the significant differences between the angular velocities of the knees and hips. The second pull occurs from the second to the third diagram and shows extension of the knees, hips and ankles. As already stated previously, when the joints are fully extended at the end of this phase the bar reaches maximum velocity. Quantitative evidence shows a rapid increase in hip and knee extension as the lifter produces a summation of force by fully extending the legs and trunk. However, a directional change in knee angular velocity halfway between diagrams 2 and 3 in (Figure 10.12) indicates that the lifter is performing a double knee bend or scoop as it is sometimes called. A final full extension of the hips and knees takes place before the lifter drops quickly to catch the bar.

Relationship between linear and angular motion

We are all aware that a tennis ball hit with a racket will travel further than a tennis ball hit with the hand, or that a golf ball hit with a driver will travel further than a golf ball hit with a pitching wedge (all other factors being the same). In each of these examples, greater force is applied to the ball because the radius of the striking implement is longer and so greater linear velocity is generated at its end-point (Figure 10.13).

In Figure 10.13 you can see that all three positions have the same angular velocity, but the linear velocity at the end of the lever is proportional to the length of the lever. Therefore the longer the radius the greater the linear velocity of a point at the end of that radius, provided angular velocity is kept constant (Hamilton and Luttgens, 2002). Although this is potentially advantageous to the athlete, it is worth remembering that longer levers require more effort and so the optimum

△ **Fig. 10.14** This is an example of the link between angular velocity and sequencing of the body segments and racket head, and the transfer to linear velocity evidenced in the ball.

length will depend on the athlete's ability to maintain the angular velocity. This relationship can be represented by the equation:

$$v = \omega r$$

Note that all equations linking linear and angular motion require the angular measures to be expressed in radians. Degrees can be converted to radians by dividing by 57.3.

Figure 10.14 shows the location of joint segments and their orientations during a tennis serve.

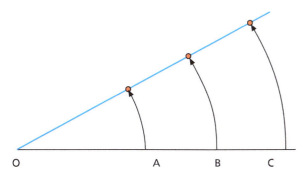

△ **Fig. 10.13** Positions A, B and C all have the same angular velocities, but point C would have the greatest linear velocity. The longer the radius the greater the linear velocity of the point at the end of the radius, provided angular velocity is kept constant.

Activity 10.4

Independent learning task: Calculating linear velocity from angular velocity

A hockey player swings her stick of radius 1.3 m at an angular velocity (V) of 345 degrees/s. Work out the linear velocity of the ball.

Activity 10.5

Laboratory investigation: Push and throw patterns and the kinetic link principle

Equipment

1. Video camera
2. Reflective markers
3. Football
4. Shot put
5. 1 × 500 W light
6. Motion analysis system
7. Mat.

Experiment 1

The aim of this experiment is to perform a kinematic analysis of a performance and to learn how to accurately collect motion data over a period of time.

Video a skilled performer kicking a football. Place reflective markers on the shoulder, hip, knee and ankle of the kicking side. Calculate the angular velocities of the trunk, thigh and calf segments. Plot a graph of this data against time.

- When did the peak angular velocity about the knee joint occur relative to the ball contact? Was this appropriate?
- Comment on the interaction of the angular velocities of the trunk, thigh and calf segments. Did the subject effectively use the kinetic link principle?

Experiment 2

Video a subject putting a shot. Place reflective markers on the shoulder, elbow and wrist of the throwing side. Calculate the angular velocities of the arm and forearm segments. Plot a graph of this data against time. (Remember the considerations for filming human movement in Chapter 8?)

- When did the peak in angular velocity of the arm and forearm segments occur relative to release? Was this appropriate?
- Compare the two graphs and comment on the changes to the movement pattern produced by the different constraints of the activities.

Summary

- This chapter introduced simple rotational or angular kinematics.
- All human movement involves angular motion or the movement of a body segment about its axis of rotation.
- The kinematics of rotation are similar to their linear counterparts in that they can be represented as vector and scalar quantities.
- Angular displacement is the change in angular position of a line segment; angular velocity is the rate of change in angular position; and angular acceleration is the rate of change in angular velocity. They are all vector quantities.
- The relationship between linear and angular velocity was discussed and calculations of displacements and velocities were introduced and practical applications provided to enhance understanding.

TIME OUT

PELVIS, TORSO AND SHOULDER KINEMATICS IN HIGH-SCHOOL BASEBALL PITCHERS

Oliver and Keeley (2010) examined the kinematics of the pelvis and torso in relation to the kinematics of the shoulder in high-school baseball pitchers. A repeated measures design was used to collect pelvis, torso and shoulder kinematics (i.e. angular displacements and angular velocities) throughout the pitching motion. Participants threw a series of maximal effort fastballs to a catcher located the regulation distance (i.e. 18.44m) from the pitching mound, and those data from the fastest pitch passing through the strike zone were analysed.

The results indicated that, for several parameters, the actions at and about the shoulder were inversely related to the actions of the pelvis and torso throughout the pitching motion. More importantly, the rate of axial torso rotation was significantly related to these shoulder parameters in a way that may help to explain the high rate of shoulder injuries in high-school pitchers. Therefore, strength training should focus on developing a strong stable core, including the gluteal musculature, in an attempt to control the rate of torso rotation during the pitch.

Review Questions

1. When kicking a soccer ball, a player had the hip and knee angles listed in the table below, measured at 0.1 s intervals. Calculate the angular velocities and accelerations at each possible frame.

Frame	Time (s)	θhip	θknee	
1	0.0	245	0	
2	0.1	200	75	
3	0.2	145	92	
4	0.3	150	145	
5	0.4	153	176	
6	0.5	155	179	(ball contact)

2. Define angular displacement, velocity and acceleration.

3. Explain the kinetic link principle and give examples from sport and exercise.

4. Observe slow-motion films of throwing, striking and kicking. Look for the application of this principle in such movements.

5. When kicking a soccer ball, a player accelerates (α) his lower leg at 4.5 rad/s^2 for 0.8 s. If the final angular velocity (Ω_f) of his leg was 1.4 rad/s, what was the initial angular velocity (Ω_i) of his leg? When throwing a javelin, an athlete has a linear velocity of 6.4 m/s, an arm length of 1.2 m and an angular velocity (Ω) at the shoulder of 2.5 rad/s. What is the linear velocity of the javelin at release? Why is it important to consider the shutter speed of a camera when collecting video for digitizing? (Hint: refer to Chapter 8.) What problem would occur if you tried to use the normal video shutter speed of 1/25 s for the analysis of walking?

References

Abelbeck KG (2002) Biomechanical model and evaluation of a linear motion squat type exercise. *Journal of Strength and Conditioning Research* 16: 516–524.

Baechle, TR, Earle, RW and Wathen, D (2000) Resistance Training. In: *Essentials of Strength Training and Conditioning*. TR Baechle and RW Earle (eds). Champaign, IL: Human Kinetics, 48.

Bloomfield, J (1998) Posture and proportionality in sport. In: *Training in Sport: Applying Sport Science*. B. Ellito (ed), New York: John Wiley and Sons, Inc, 426.

Bartonietz KE (1996) Biomechanics of the snatch: toward a higher training efficiency. *Strength and Conditioning Journal* 18: 24–31.

Cibulka MT and Threkeld-Watkins J (2005) Patellofemoral pain and asymmetrical hip rotation. *Physical Therapy* 85:1201–1207.

Cook G (2003) *Athletic Body in Balance.* Champaign, IL.

Donnelly DV, Berg WP and Fiske DM (2006) The effect of the direction of gaze on the kinematics of the squat exercise. *Journal of Strength and Conditioning Research* 20: 145–150.

Elliott BC, Marsh AP, Overheu PR (1989a) The topspin backhand drive in tennis: a biomechanical analysis. *Journal of Human Movement Studies* 16: 1–16.

Elliott BC, Marsh T, Overheu P (1989b) A biomechanical comparison of the multi-segment and single unit topspin forehand drives in tennis. *International Journal of Sports Biomechanics* 5: 350–364.

Escamilla RF, Fleisig GS, Lowry TM, Barrentine SW and Andrews JR (2001) A three-dimensional biomechanical analysis of the squat during varying stance widths. *Medicine and Science for Sports and Exercise* 33: 984–998.

Hamilton N, Luttgens K (2002) *Kinesiology, Scientific Basis of Human Motion*, 10th edn. New York: McGraw-Hill.

Hughes SS, Lyons BC, Mayo JJ (2004) Effect of grip strength and grip strengthening exercises on instantaneous bat velocity of collegiate baseball players. *Journal of Strength and Conditioning Research* 18: 298–301.

Kendall FP, McCreary EK, Provance PG, Rodgers MM and Romani WA (2005) *Muscles Testing and Function with Posture and Pain*, 5th edn., Baltimore: Lippincott Williams and Wilkins.

Kinakin, K (2004) *Optimal Muscle Testing*. Champaign: Human Kinetics, 122.

Kritz M, Cronin J and Hume P (2009) The bodyweight squat: a movement screen for the squat pattern. *Strength and Conditioning Journal*, 31: 1, 76–85.

Lauder MA and Lake JP (2008) Biomechanical comparison of unilateral and bilateral power snatch lifts. *Journal of Strength and Conditioning Research* 22: 3, 653–660.

Oliver GD and Keeley DW (2010) Pelvis and torso kinematics and their relationship to shoulder kinematics in high-school baseball pitchers. *Journal of Strength and Conditioning Research* 24: 12, 3241–3246.

Papadopoulis, C; Emmanouilidou, M and Prassas, S (2000) Kinematic analysis of the service stroke in tennis. In: Haake S, Coe AO eds *Tennis Science and Technology*. Oxford; Blackwell, pp. 383–388.

Souza AL, Shimada SD (2002) Biomechanical analysis of the knee during the power clean. *Journal of Strength and Conditioning Research* 16: 290–297.

Tang HP, Abe K, Katoh K, Ae M (1995) Three-dimensional cinematographic analysis of the badminton forehand smash: movements of the forearm and hand. In: Reilly T, Hughes M, Lees A eds *Science and Racket Sports*. London: E and FN Spon, pp. 113–120.

Van Gheluwe B, de Ruysscher I, Craenhals J (1987) Pronation and endorotation of the racket arm in a tennis serve. In: Jonsson B ed *Biomechanics X-B*. Champaign, IL: Human Kinetics, pp. 666–672.

Walsh JC, Quinlan JF, Stapleton R, Fitzpatrick DP and Mc-Cormack D (2007) Three-dimensional motion analysis of the lumbar spine during 'free squat' weight lift training. *American Journal of Sports Medicine* 35: 927–932.

Watts L (1957) Complete guide to good hitting. *Scholastic Coach* 26: 48–52.

Williams T, Underwood J (1986) *The Science of Hitting*, 3rd edn. New York: Simon and Schuster.

Zatsiorsky VM (1995) *Science and Practice of Strength Training*. Champaign, IL: Human Kinetics.

Further reading

Fry AC, Smith JC and Schilling BK (2003) Effect of knee position on hip and knee torques during the barbell squat. *Journal of Strength and Conditioning Research*, 17: 629–633.

Hall CM and Brody LT (2005) *Therapeutic Exercise: Moving Toward Function,* 2nd edn., Philadelphia: Lippincott Williams and Wilkins.

11 Angular Kinetics

Chapter Objectives

In this chapter you will learn about:

- The theory and applications of angular kinetics.
- The calculation of moment arms, torque and resultant torques.
- Resultant joint torques, anatomical torque descriptions and force couples.
- The relationships of torque/angular velocity, length–tension and their links to sport and exercise.

Introduction

Kinetics studies the relationship between the forces acting on a system and the movement of the system. When an object rotates, all the points on the object move in a circle about a single axis of rotation. All points move through the same angle in the same time. **Angular kinetics**, therefore, studies the kinetics of particles, objects or systems undergoing rotation or angular motion.

As in kinematics, comparisons can be made between linear and angular kinetics:

Linear	Angular
Mass	Moment of inertia
Force	Torque
Momentum	Angular momentum
Newton's laws	Newton's laws (angular analogues)

Inertia and moments of inertia

Resistance to angular motion (like linear motion) is dependent upon mass. However, the closer that the masses are distributed to the axis of rotation the easier it is to rotate. Therefore resistance to angular motion is dependent on both the quantity and distribution of mass. This is defined as the **moment of inertia**. Figure 11.1 shows two baseball bats. Bat A has a higher moment of inertia than bat B because the mass is distributed further away from the axis of rotation.

The moment of inertia (I), the angular form of inertia, can be calculated for a single particle as:

$$I = mr^2$$

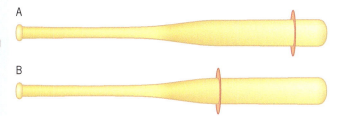

△ **Fig. 11.1** These two baseball bats have equal mass. Both have an extra mass added, but the mass on Bat A is further away from the handle end of the bat. When the bat is used, the handle end becomes the axis of rotation. Therefore, Bat A has a higher moment of inertia.

where m is the mass of the particle and r is the distance of the particle from the axis of rotation. For an object, the formula is:

$$I = \Sigma I = \Sigma mr^2$$

Its standard SI unit of measure is kg.m².

An object with high inertia is difficult to get moving, but once it starts moving it is difficult to stop. For example, if you try to push a car when it has broken down it takes a considerable amount of effort to get it moving. However, once moving at a reasonable speed you would not want to step in front of it to stop it! The amount of inertia depends on the mass of the object and is an important variable when an object travels in a straight line.

As explained above, the rotational equivalent of inertia is moment of inertia. The moment of inertia of an object depends not only on mass but also on the distribution of that mass away from the axis of rotation. Heavy objects and those in which most of the mass is located away from the axis of rotation have higher moments of inertia than those objects that are light or have most of the mass located near the centre of rotation.

It is important to recognize that rotation can occur about different axes. Each axis has its own moment of

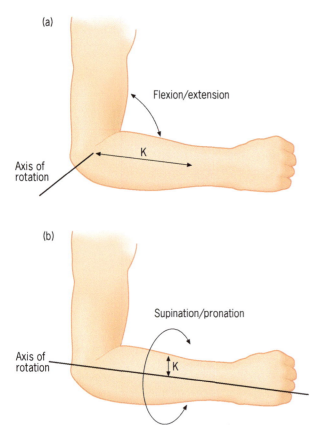

(a)

Flexion/extension

K

Axis of rotation

(b)

Supination/pronation

Axis of rotation

K

△ **Fig. 11.2** Rotation can occur about different axes and each axis has its own moment of inertia.

inertia associated with it. For example, in Figure 11.2, which shows axes of rotation in an arm in flexion/extension and supination/pronation, k indicates the mass distance (the distance the mass is away from the axis of rotation) away from the axes of rotation. The flexion/extension shows rotation about the transverse axis. The supination/ pronation shows rotation around the longitudinal axis. A larger k value will consequently increase resistance to motion.

Human movement can occur about three principal axes (frontal, transverse and longitudinal). Each principal axis has a moment of inertia associated with it. So when mass is distributed closer to the axis the moment of inertia is lower. This is evident in Figure 11.3, which shows the moments of inertia for various axes.

For a gymnast of given mass, how that mass is distributed away from the axis of rotation determines the moment of inertia at a given time. A gymnast's moment of inertia will gradually increase as she moves from a tucked to a piked and finally to a layout position when somersaulting.

Torque

During linear or translatory motion (see Chapter 8), in order for movement to occur, the object or body must be acted upon by a force (internal or external). During angular motion, in order for rotation to occur the object or system must be acted upon by **torque**. This is a vector quantity since it has both magnitude and direction. It is a measure of the extent to which a force will cause an object to rotate about an axis. A net force applied through the centre of mass of the object produces linear motion. A force whose

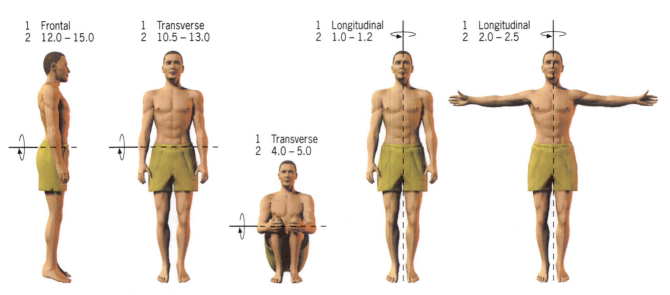

| 1 | Frontal |
| 2 | 12.0 – 15.0 |

| 1 | Transverse |
| 2 | 10.5 – 13.0 |

| 1 | Longitudinal |
| 2 | 1.0 – 1.2 |

| 1 | Longitudinal |
| 2 | 2.0 – 2.5 |

| 1 | Transverse |
| 2 | 4.0 – 5.0 |

△ **Fig. 11.3** Moment of inertia for various axes.

line of action does not pass through the centre of mass of the body on which it acts is called an **eccentric or off-centre force**. An eccentric force produces both linear and rotary movement. For pure rotation about the centre of mass, the centre of mass must remain stationary.

From Newton's first law, the net force on the object must equal zero. Therefore two forces of equal magnitude, applied in opposite directions produce pure rotation about the centre of mass. This is called a **force couple**. The magnitude of a couple is the product of one of the forces and the perpendicular distances between the forces. The larger the couple acting on an object, the greater will be the angular acceleration and therefore the speed of rotation of the object.

These terms are illustrated in Figure 11.4.

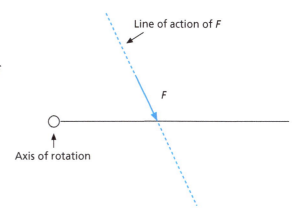

△ **Fig. 11.5** The line of action of a force is the imaginary line that extends from the force vector in both directions.

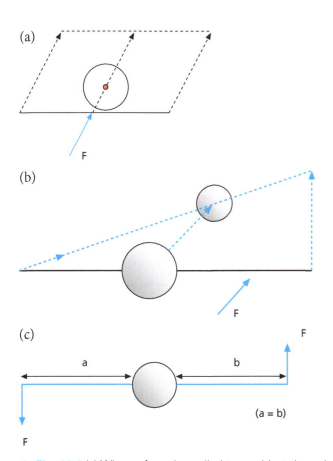

△ **Fig. 11.4** (a) When a force is applied to an object through its centre of mass, no rotation occurs. This is known as a 'centric force'. When an off-centre or eccentric force is applied to an object like a snooker ball, for example, rotation or side-spin occurs. Snooker players call this 'side'. (b) Eccentric force. (c) Force couple.

The **line of action** of a force is the imaginary line that extends from the force vector in both directions (Figure 11.5).

The use of eccentric forces to initiate rotation of the body

In many athletic activities, twists and turns are built up while the athlete is in contact with the ground. The resulting angular momentum can be achieved by (1) checking linear motion (2) transfer of momentum and (3) eccentric thrust.

Checking linear motion

This involves a change in momentum from linear to angular. It is used in many athletic activities such as vaulting and diving. For example, a gymnast vaulting over a box, after a preliminary run-up, fixes both feet momentarily at take-off, while the rest of the body rotates over and beyond the feet. This is because just prior to take-off, the line of action of the ground reaction passes behind the gymnast's centre of mass. The effect of the ground reaction force will be to simultaneously lift the gymnast's centre of mass and help create forward rotation. Forward rotation during the flight phase about the transverse axis consequently increases the chances of the gymnast landing on his or her feet. The fixing or checking of the feet at take-off is important not only because it initiates rotation but also because it causes a pre-tensioning of the muscles. The stretch–shortening cycle is induced and greater thrust at take-off is promoted.

When an object rotates freely within a plane, in other words it is not limited to rotate about a fixed axis, the rotation will take place about an axis which passes through the centre of mass of the object. Thus in the case of the gymnast the change from linear to angular momentum depends on three factors:

1. The speed of the run-up
2. The abruptness with which the feet are checked
3. The moment of inertia of the gymnast at take-off (the smaller the moment of inertia relative to the point of rotation, the faster the rotation and vice versa).

Transfer of momentum

Whether it is linear or angular, momentum can be transferred from one object to another and from part of an athlete to his or her whole body. In other words if momentum is generated in some part of the body and then that body part is checked, the momentum is not lost but is transferred to the body as a whole. Consider a gymnast lying flat on the floor. The legs are raised above the head, and from there they are swung vigorously from the hips, unjacking the body. When they are checked in the movement, angular momentum transfers to the whole body which then ends up in a sitting position.

Again, transfer of momentum can be used to initiate rotation about all axes

Eccentric thrust

When a force is applied through the centre of mass of an object, say, a block of wood, there is no rotation. However, if the force is applied out of the line of the centre of mass, there is some degree of rotation. The resulting force will be linear (upward) and angular (clockwise). The amount of angular motion will depend on the perpendicular distance of the force from the centre of mass. This is known as an eccentric or off-centre force.

The moment arm

A moment arm is the shortest perpendicular distance from a force's line of action to the axis of rotation. It is always perpendicular to the line of action and passes through the axis of rotation. Some examples are illustrated in Figure 11.6.

In order to calculate a moment arm, you need to know the distance (d) from the axis of rotation to the point at which the force is applied and the angle (θ) at which the force is applied. Trigonometry can be used to calculate the moment arm ($d^{\perp}$). For example, in Figure 11.7:

$$d^{\perp} = d\sin\theta$$

During the biceps curl exercise, in order to produce more **internal torque** (torque generated within the body as opposed to external torque which is generated outside the body) the bicep must either generate more tension or lengthen its moment arm. Appropriate weight or resistance training can increase the muscle tension-generating capacity (Durall, 2004) and consequently increase internal torque. It cannot, however, influence a muscle's maximum moment arm because this is anatomically determined, although it does change throughout the range of movement. This occurs because the line of action of the muscle moves closer to or further away from the joint axis of rotation. The magnitude of the torque produced is

(a)

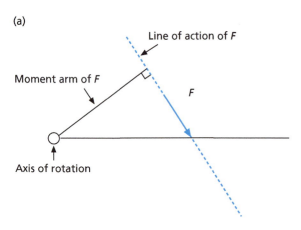

(b)

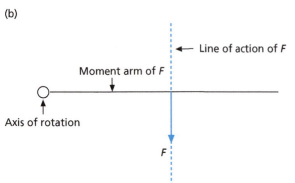

(c)

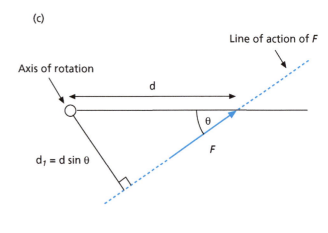

(d)

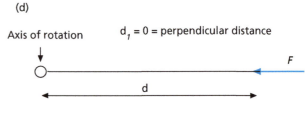

△ **Fig. 11.6** The figures represent a variety of moment arms. Each force vector is applied from a different direction but remains perpendicular to the moment arm.

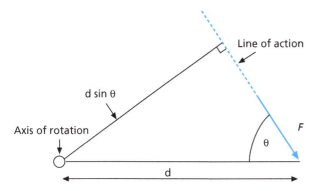

Fig. 11.7 A right-angled triangle is formed and so trigono-metry can be used to calculate the moment arm $d\sin\theta$.

dependent upon the size or magnitude of the force generated by the muscle (e.g. bicep) and the perpendicular distance between the point where the force is applied and the axis of rotation.

Torque has magnitude, direction (+ or –) and a specific axis of rotation. The size or magnitude of the torque (T) is the product of the force's magnitude multiplied by the force's moment arm ($d^\perp$).

$$T = F \times d^\perp$$

The standard unit is the newton-metre (Nm). Torque is positive when acting in an anticlockwise direction about the axis of rotation, and negative when acting in a clockwise direction about the axis of rotation.

Levers

To fully understand the musculoskeletal system in terms of human movement an understanding of levers is essential. A lever is a rigid body. When a force whose line of action does not pass through the pivot point of an object is applied, it causes rotation (Figure 11.8).

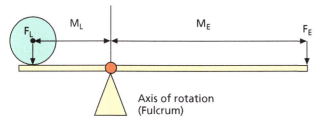

Fig. 11.8 A simple lever system: F_E is effort or applied force; F_L is resistance force or load; M_L is the moment arm of load; M_E is the moment arm of effort.

A lever system consists of:

- An axis of rotation
- A resistance force or load
- An effort force (the applied force that is used to move the load).

There are three classes of levers, known as first, second and third class levers.

First class levers

In first class levers the effort force and load force are applied on opposite sides of the axis of rotation (Figure 11.9). The effort and load forces act in the same direction. At equilibrium, anticlockwise moments = clockwise moments. Therefore,

$$F \times d^\perp{}_{load} = F \times d^\perp{}_{effort}$$

(Where $d^\perp$ = perpendicular distance from axis of rotation)

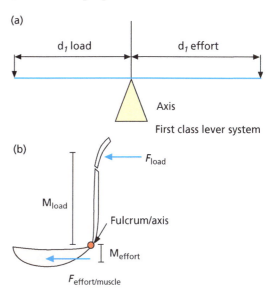

(Elbow extension against a resistance)

Fig. 11.9 Example of a first class lever system: tricep extension.

During an isometric contraction, or a constant velocity joint rotation (static equilibrium),

$$F_{effort} \times M_{effort} = F_{load} \times M_{load}$$

Since M_{effort} is significantly smaller than M_{load}, the muscular force required F_{effort} must be much greater than applied load or resistive force F_{load}. This shows how inefficient this class of lever system is. In other words a large muscular force is required to move a relatively small resistance.

Second class levers

This is a class of lever whereby the muscle and resistive forces act on the same side of the fulcrum or pivot (Figure 11.10). The muscle force acts through a moment arm longer than that of the resistive force. The effort force is applied further from the axis than the load force. The effort and load forces act in opposite directions. This class of lever is good for strength application but poor for moving loads quickly or through a large range of movement.

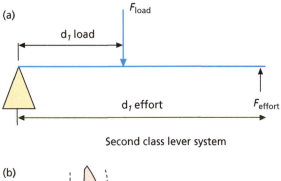
(a)

Second class lever system

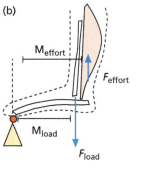
(b)

(Calf raise)

△ **Fig. 11.10** Example of a second class lever system: calf raise.

Third class levers

This is a lever system where effort force and load force are applied on the same side of the axis of rotation (Figure 11.11). However, the effort force applied is closer to the axis than the load force. The effort and load act in

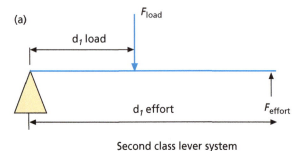
(a)

Second class lever system

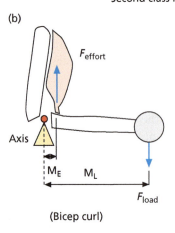
(b)

(Bicep curl)

△ **Fig. 11.11** Example of a third class lever system: bicep curl.

opposite directions. The advantage of this system is that loads can be moved quickly through large ranges of movement. The disadvantage is that muscle force or effort has to be greater than the resistive force or load and so is poor for strength generation.

Most levers in the human body (>95 per cent) are third class systems. The muscles rotate limbs or body segments about the joint (axis of rotation). As a result, internal forces acting on the body are far greater than forces exerted on external objects by the human body. This explains the high incidence of injury to soft tissues such as tendons and ligaments. These are often seen in sports such as powerlifting, where it is not uncommon for the patella tendon to become dislodged from its attachment during squatting as a result of the incredibly high muscular forces required to move the weight.

Coaching application

You must remember that although this biomechanics section is split into separate chapters and sections, this is done for simplicity, and as a result topics become categorized. However, in reality the situation is more complex, and many factors are brought to bear on each scenario. Let us look, for example, at the case of a weightlifter attempting a power clean. Throughout the lift the human body constantly changes shape and this can cause the lifter many difficulties. There are many biomechanical concepts that can be considered: stability, balance, centre of gravity, levers, torque, power and acceleration to name but a few. When an athlete lifts a bar there are two centres of mass: the weight to be lifted and the athlete. The lifter and the bar become one once he or she exerts force against the bar. There is now a combined centre of mass situated towards the heavier part of the combined body (Figure 11.12).

In order to move the weight from the floor, power is needed to overcome its inertia. Let us assume that the power required to overcome the resistance is 110 units (hypothetically). This is achieved eventually by the athlete through progressive resistance training (Figure 11.13).

However, the athlete must lift the weight as efficiently as possible and so emphasis must be placed by the coach on correct technique. The bar must therefore follow a path that has the least line of resistance. By looking at Figure 11.14 you can see that the closer the weight of the bar (weight/resistance arm) passes to the axis of rotation or fulcrum (so for the lifter we are predominantly concerned with the hip joint) the greater is the power/effort arm and so the athlete can apply greater force or more power to the bar.

In the diagram of the weightlifter shown in Figure 11.15, the bar reaches knee height, but in order to achieve

Centre of mass

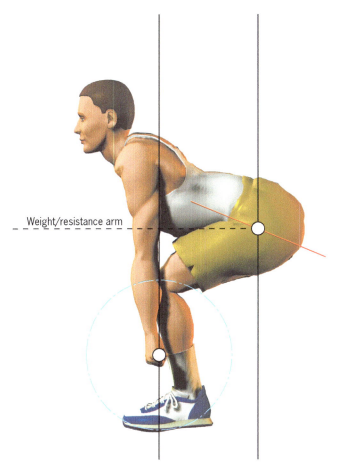

Weight/resistance arm

△ **Fig. 11.12** The hip joint is the axis of rotation. The lifter attempts to keep the moment or resistance arm as small as possible to maximize technique.

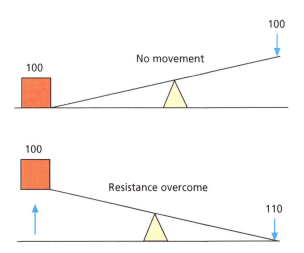

No movement

100

100

Resistance overcome

100

110

△ **Fig. 11.13** Power is needed to overcome inertia.

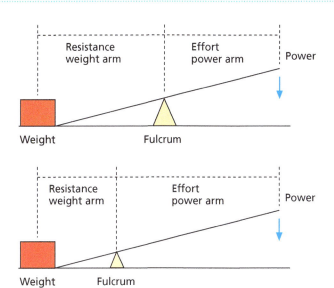

Resistance weight arm

Effort power arm

Power

Weight

Fulcrum

Resistance weight arm

Effort power arm

Power

Weight

Fulcrum

△ **Fig. 11.14** The closer the weight passes to the axis of rotation, the greater the effort arm and so the athlete can apply greater force to the bar.

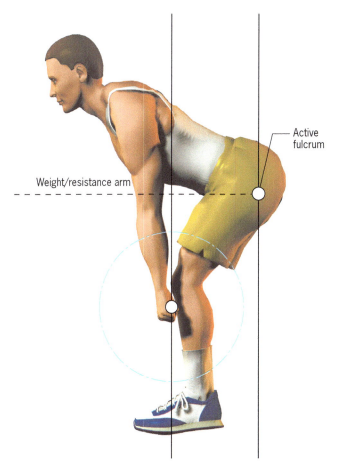

Active fulcrum

Weight/resistance arm

△ **Fig. 11.15** The bar takes the path of the least line of resistance.

maximum upward extension to enable the lifter to rack the bar on the shoulders a dynamic movement is required. This occurs by a summation of force. There is a steady build-up of the muscle groups being used in sequence, initially using the legs, closely followed by a vigorous extension at the hips, knees and ankle joints as soon as the bar passes above the knees. The bar is kept in as close as possible to the body following the least line of resistance. It also allows the athlete to apply maximum torque to the resistance. This build-up of maximum force must be smooth. A knowledge of basic biomechanics is essential for coaches teaching weight training/lifting exercises.

Lever systems and mechanical advantage

During human movement, deciding on what lever system is in operation depends on where the axis of rotation or fulcrum lies. During leg extensions and flexions, the location of the axis of rotation changes throughout the range of motion. This affects the length of the moment arm of the quadriceps and hamstrings. When performing a leg extension, the patella prevents large changes in the mechanical advantage of the quadriceps. Since the knee is not a true hinge joint the patella stops the patella tendon from getting too close to the axis of rotation. When weight training, the moment arm through which the weight acts changes as it moves through its range of movement. The change in the joint angle or angle of pull causes changes in the moment arm of the effort and changes in the moment arm of the load or resistive force. Therefore understanding the principle of mechanical advantage is of greater importance than classifying the lever.

What is mechanical advantage?

Mechanical advantage is the ratio between the moment arm through which an effort force acts and that through which a resistive or load force acts.

$$\text{Mechanical advantage} = d^{\perp}_{\text{effort}}/d^{\perp}_{\text{load}}$$

When the mechanical advantage > 1, the F_{effort} required is less than the F_{load}. This is advantageous for strength application but poor for moving loads quickly or through large ranges of movement. When the mechanical advantage <1, the F_{effort} needed is greater than the F_{load}. This is advantageous for moving loads quickly or through a large range of movement, but poor for strength application. First class levers can have a mechanical advantage greater than, equal to or less than 1; second class levers have a mechanical advantage greater than 1; and third class levers have a mechanical advantage less than 1.

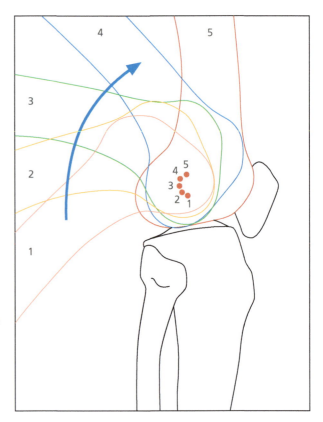

△ **Fig. 11.16** The location of the knee's axis of rotation changes continuously through the range of motion, affecting the length of the moment arm through which the quadriceps and hamstrings act. From Harman (1994).

The mechanical advantage often changes during 'real-world' activities. For example the knee joint is not a true hinge joint, and so during knee extension the axis of rotation changes throughout the movement (Figure 11.16). The patella during knee extension stops the quadriceps or patella tendon from getting too close to the axis of rotation.

Tendon insertion variation

Tendon insertion variation can affect an athlete's ability to overcome resistance. Tendons that are inserted onto the bone further away from the axis of rotation will allow the athlete to lift heavier weights as a result of the longer moment arm, producing greater torque (Figure 11.17(a)).

There is a price to pay, however, for having tendons inserted further away from the axis of rotation. The mechanical advantage gained results in a loss of maximum speed because the muscle has to contract more to make the joint move through the same range of movement. Or put another way, the same amount of muscle contraction results in a smaller angle of joint rotation (Figure 11.17(b)).

According to the force–velocity relationship (discussed later in this chapter), to produce a specific joint velocity, a muscle that has an insertion further away from the axis of rotation

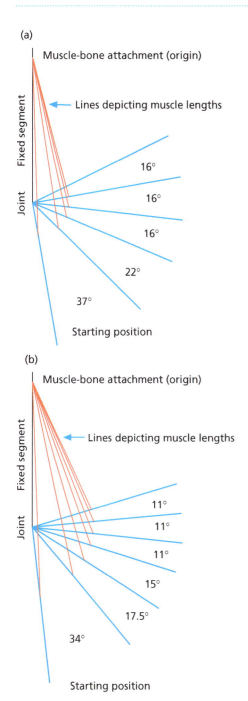

(a)

Muscle-bone attachment (origin)

Fixed segment

Lines depicting muscle lengths

Joint

16°

16°

16°

22°

37°

Starting position

(b)

Muscle-bone attachment (origin)

Fixed segment

Lines depicting muscle lengths

Joint

11°

11°

11°

15°

17.5°

34°

Starting position

△ **Fig. 11.17** The joint angle differs with equal increments of muscle shortening when the tendon is inserted (a) closer or further from the joint centre. Configuration (b) has a larger moment arm and thus greater torque for a given muscle force, but less rotation per unit of muscle contraction and thus slower movement speed. From Gowitzke and Milner (1988).

must contract at a higher velocity and so this form of tendon arrangement compromises the muscle's force capability during faster movements. It is evident that variations in anatomical structure can have implications for athletic performance. Athletes with tendon insertions further away from the axis of rotation are best suited to sports like power-lifting where the slow movement activity produces high

forces. Likewise, athletes with tendons close to the axis of rotation may be best suited to high velocity activities.

Muscle contractions

In order to prescribe effective strength and conditioning programmes an understanding of the different types of muscle contractions is essential. Muscle contraction categories traditionally begin with the prefix 'iso' meaning 'the same', and include isotonic (constant muscle tension), isometric (constant muscle length) and isokinetic (constant velocity of motion) contractions.

Isometric

In general, in an isometric contraction it is considered that the muscle length remains constant. Technically speaking, however, it is the joint angle that remains constant because there are internal movement processes that take place during muscle contraction that make it virtually impossible for the fibres to remain the same length. A contraction means to shorten, and this involves internal movement processes that shorten the muscle fibres and stretch connective tissues and tendons. It occurs when the muscular force balances the resistance and no joint movement occurs (Figure 11.18).

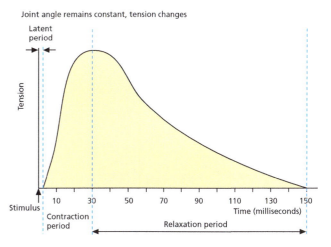

Joint angle remains constant, tension changes

△ **Fig. 11.18** Isometric contraction indicating tension changes over time.

Isotonic

The term isotonic is very misleading since it is virtually impossible for muscular tension to remain the same while joint movement occurs over an extended range. Whenever movement occurs, tension increases or decreases since acceleration or deceleration is involved and one of the stretch reflexes may be activated. The term 'dynamic' describes the type of contraction in question and should really be used instead of isotonic, which should be limited

Activity 11.1

Independent learning task: Try these sample questions

1. A 23-kg girl sits 1.5 m from the axis of rotation of a seesaw. At what distance from the axis of rotation must a 21-kg girl position herself on the other side of the axis to balance the seesaw?

2. How much force must be produced by the bicep at a perpendicular distance of 3 cm from the axis of rotation of the elbow to support a weight of 200 N at a perpendicular distance of 25 cm from the elbow?

3. A physiotherapist applies a lateral force of 80 N to the forearm at a distance of 25 cm from the axis of rotation of the elbow. The biceps brachii attaches to the radius at a 90-degree angle 3 cm from the elbow joint centre. How much force is required of the biceps (F_m) to stabilise the arm in this position? (Hint: At equilibrium, anticlockwise moments = clockwise moments, or sum of torques = 0.) What is the magnitude of the reaction force (F_r) of the humerus on the ulna? (Hint: Sum of forces = 0.)

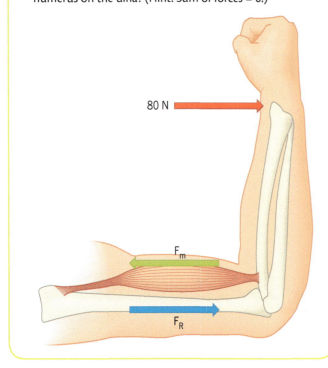

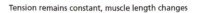

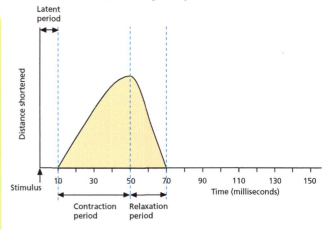

Tension remains constant, muscle length changes

△ **Fig. 11.19** Diagram represents the relationship between muscle length changes with respect to time during an isotonic contraction.

to short-range movements where muscle tension remains relatively constant (Figure 11.19).

Isokinetic

This term is used in two contexts: first as a definition of a specific muscle contraction and second as in an isokinetic testing and rehabilitation machine. The term isokinetic is often inappropriate since it is impossible to carry out a constant-velocity full range of movement muscle contraction. According to Newton's second law, a muscle that contracts from rest and then returns to that state must involve acceleration. Therefore constant angular velocity about a joint can only take place over part of that action range. It is also biomechanically impossible to design a purely isokinetic machine. Isokinetic machines therefore limit motion to approximately constant angular velocity over part of its range. They are designed this way because there is a school of thought that believes strength is best developed if the muscle tension is kept constant throughout the range of movement.

It is important to note, however, that the principle of specificity is followed when using isokinetic devices. They are not able to measure or train functional strength that is essential in many sports (e.g. explosive strength, acceleration strength). The fact that most muscular injuries occur under accelerative/decelerative and explosive conditions further highlights the limitations of isokinetic methods for testing and rehabilitation (Siff, 1993).

Concentric and eccentric muscle contractions

A **concentric contraction** concerns muscle actions that produce a force to overcome the load being acted upon. The work done during concentric contractions is referred to as positive work. Two types of concentric contraction may be identified:

1. In a dynamic concentric contraction the muscle undergoing the contraction shortens.

2. A static concentric contraction (or concentric isometric contraction) occurs if shortening is attempted but no external movement occurs.

An **eccentric contraction** refers to muscle action in which the muscle force yields to the imposed load. The work done during a concentric contraction is referred to as negative.

Two types of eccentric contraction may be identified:

● Dynamic eccentric contraction, which involves the stretching of the contracting muscle.

● Static eccentric contraction (or eccentric isometric contraction) occurs if stretching is resisted and no external movement occurs.

Both concentric and eccentric contractions are exhibited during the squat exercise. A concentric contraction occurs in the upward phase of the squat and an eccentric contraction occurs in the downward phase.

The principle of specificity distinguishes between the characteristics of the training device or machine used by the athlete and the actions produced by muscular contractions. Machines or devices may be designed so that torque or force can be controlled and remains constant throughout its range of movement. This does not mean that the joint torque or force produced by the muscles remains the same when using this equipment. This is because, as discussed earlier, torque is the vector product of force and its perpendicular distance from the line of action to its axis of rotation. So if the force or the distance changes as the joint angle changes then the torque will change also (Figure 11.20).

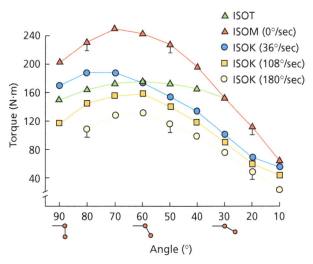

△ **Fig. 11.20** Torque and joint angle curves for knee extension. ISOT, isotonic; ISOM, isometric; ISOK, isokinetic. From Harman (1994).

For example in a training setting, fluid-resisted exercise machines are sometimes used which claim to be isokinetic in nature. They are not, however, because there is a rapid acceleration phase. Likewise they do not provide an eccentric exercise phase. When an athlete uses free weights, the agonist works concentrically when raising a weight (for example a bicep curl) and eccentrically when lowering it. When an athlete uses fluid-resisted equipment, the prime mover or agonist acts concentrically when performing the initial exercise and then the antagonist acts concentrically in order to return the weight to the start position. A lack of eccentric muscle activity means that these sorts of training devices are inappropriate for many sporting activities where eccentric muscle actions are involved (Harman, 1994). A comprehensive knowledge of muscle contraction and velocity should be obtained if the faulty design of training programmes, ineffective therapeutic treatments based on constant velocity devices and inaccurate research based on incomplete definitions are to be avoided.

Force–velocity relationship

The force–velocity relationship is an example of a relationship where movement velocity decreases as external load or torque increases. Laboratory experiments on single muscles have produced the force–velocity curve shown in Figure 11.21, which illustrates that muscle torque and joint angular velocity vary according to muscle contraction type. The relationship is not linear.

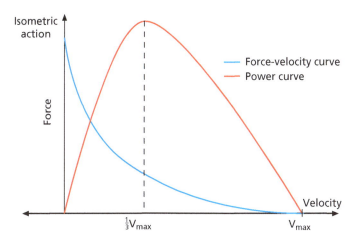

△ **Fig. 11.21** Relationship between force–velocity and power.

Figure 11.21 shows the relationship between force–velocity and power (power is the product of force and velocity). This relationship is used by sports scientists and strength and conditioning coaches to design effective strength and power training programmes. A well-designed strength–power training programme will have the effect of shifting the force–velocity curve to the right and the power curve upwards. The torque produced decreases as the movement velocity increases. The greatest forces are produced at the slowest speeds or angular velocities. When the velocity is increased to only 10 per cent of its maximum, the force is reduced by 35 per cent (Lieber, 1992). In other words, a decline in torque capabilities is at its steepest over the lower movement velocities.

One theory proposed to explain this relationship is that the union of the actin and myosin filaments is reduced at higher velocities and therefore fewer cross-bridge attachments occur (Bray *et al.*, 1986). Also as muscle shortening speed increases, the rate of cross-bridge cycling increases but the average force produced by each cross-bridge is reduced (Enoka, 1994). Mills and Robinson (2000) have suggested that maximum forces are generated through a mid-range of the joint. For example, maximum extensor torque at the hip joint might be generated through a 130–160 degree range (Enoka, 1994).

Maximal forces are produced eccentrically and so as soon as muscle begins to shorten, forces are reduced dramatically.

Length–tension relationship

The length–tension relationship states that an isolated muscle exerts its maximal force or tension when in a resting stretched position (Figure 11.22). If the muscle is overstretched or shortened then less tension can be exerted. This is explained by the sliding filament theory. Cross-bridge coupling is interfered with in the shortened position due to an overlapping of the muscle filaments. In the overstretched position the filaments are pulled

completely apart, no cross-bridges come into contact and no tension can be generated. The length of a single joint muscle is therefore linked to the joint angle, with a muscle at its shortest when the joint is fully flexed and at its longest when the joint is fully extended.

According to Bartlett (1999), tension within the muscle is at its maximum with the muscle at mid-length and when muscles are shortened and lengthened active tension is reduced. Therefore in terms of force production, it is difficult for the body to exert maximal forces when these single joint prime movers are nearing full extension or flexion. The length–tension (or force–length as it is sometimes referred to) relationship also plays an influential role in peak torque or force production following static stretching. There is much evidence to suggest that pre-exercise static stretching reduces isometric and dynamic force production (Avela *et al.*, 1999; Nelson *et al.*, 2001; Young and Elliott, 2001). As discussed in Chapter 8, there are clearly implications for strength and power athletes, as performing static stretching prior to a competition may reduce performance (Cramer *et al.*, 2004).

Mechanical and neural factors are believed to be responsible for this phenomenon, and it has been hypothesized that mechanical factors as a result of stretching include changes in the length–tension relationship of a muscle.

Centripetal and centrifugal forces

A weight rotating about an axis, such as the hammer in track and field, is engaged in linear motion. This consists of preliminary swings, followed by 3 or 4 turns and then the release of the hammer. Each individual turn contains a phase of single (one foot on the ground) and double (both feet on the ground) support (Brice *et al.*, 2011). Given the projectile nature of the hammer after release, its speed at release is the most important release parameter, and should be as large as possible (Bartonietz, 2000).

During rotation, the wire attached to the hammer's head (weight) is undergoing rotation about an axis ; acceleration to maximum speed is affected by the forces acting on the hammer and by the time interval over which these forces act. If aerodynamic forces are ignored, the forces acting are gravity and the force applied by the thrower through the cable (i.e. the cable force) to the hammer's head (Brice *et al.*, 2011). If the athlete lets go of the hammer, the cable and the head naturally fly off, experiencing linear motion. This occurs because, according to Newton's first law, an object will travel uniformly in a straight line unless acted upon by an external force. So to make the hammer change its direction an external force has to be applied that will make the head of the hammer move in a circular path.

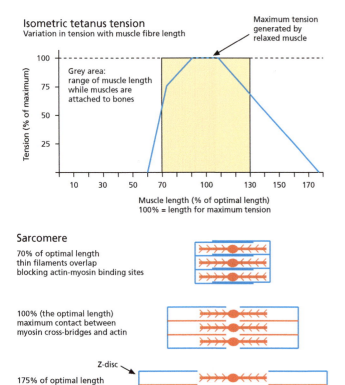

△ **Fig. 11.22** From Bastian (1994).

During rotation this force is called the **centripetal force**. It is a centre-seeking force and moves an object at right angles (tangent) to its direction of motion (Figure 11.23). The centripetal force is an external force; in hammer throwing it is applied by the athlete through the wire to the hammer head. In turn, the athlete must have a force pulling on him or her. According to Newton's third law, an object exerting a force on another will exert an equal and opposite first back onto the first body. This outward-pulling force is called the **centrifugal force**. The centrifugal force is equal and opposite to the centripetal force. The centrifugal force is therefore the reaction to the centripetal force.

The hammer head continually changes direction, and therefore even though the speed of the head remains the same, its velocity changes. In other words the weight or head of the hammer is accelerating since acceleration is defined as the rate of change in velocity divided by time taken. There are two components to this acceleration, **radial** and **tangential**. Since the radial and tangential directions are perpendicular to each other, the acceleration in the radial direction does not affect the speed in the tangential direction, but it does affect the velocity (Figure 11.24).

$$\text{Radial acceleration} = \frac{v^2}{r}$$

Tangential acceleration is what causes a skier to change speed while rounding a corner. The tangential acceleration plus the radial acceleration are equal to the direction of the acceleration vector.

Using skiing again as an example, from the formula above it is easy to understand why it is harder for a skier to make a turn with a small radius compared to a turn with a large radius. This is because the radius is in the denominator and so the smaller radius results in greater acceleration.

△ **Fig. 11.23** World Champion Tatyana Lysenko creates a centripetal force.

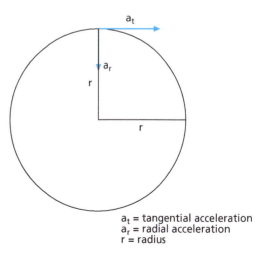

a_t = tangential acceleration
a_r = radial acceleration
r = radius

△ **Fig. 11.24** Acceleration in the radial direction does not affect the speed in the tangential direction. But it does affect the velocity.

The equation is also representative of the fact that acceleration increases as velocity increases, but reduces the further the weight or head is from the axis of rotation. When the equation for acceleration is substituted into Newton's second law we end up with the equation for the centripetal force:

$$F(\text{centripetal force}) = mv^2/r$$

This is an important equation and has many ramifications for athletic performance and training. You can see that doubling the mass will double the centripetal force and that doubling the radius decreases the force by half. However, doubling the velocity increases the centripetal force fourfold, but decreasing the velocity by half reduces the force to a quarter needed to keep the object moving along its angular path.

Other examples from sport include runners or cyclists negotiating corners. When sprinters attack the bend on a track they lean inwards. This is caused by the ground reaction force exerting a centripetal force on the athlete through the feet. If the athletes did not lean inwards they would be pulled outwards by the centrifugal force, which is equal and opposite. Centripetal forces are also important in swinging activities in gymnastics.

The above equation for centripetal force highlights the fact that athletes should be appropriately conditioned before attempting significant increases in velocity in many athletic events.

Angular analogues of Newton's laws

To fully understand the angular analogues of Newton's laws we need to first answer the question what is angular momentum? Angular momentum is a term used to describe the quantity of angular motion possessed by an athlete. It is made up of the angular momenta of the body segments of the individual. Angular momentum is directly related to the rotational speed, the centre of rotation and the distribution or configuration of the athlete's body masses. Scientifically, this would be referred to as angular velocity and moment of inertia.

Why is angular momentum important in sport and exercise? Gymnasts and divers, for example, spend much of their performance time and energy somersaulting and twisting. The laws of physics state that angular momentum is constant during airborne activities and so as the diver's body configuration changes so does the moment of inertia and angular velocity. In other words if one body part slows down another body part will speed up. This is referred to as a transfer in angular momentum (Figure 11.25).

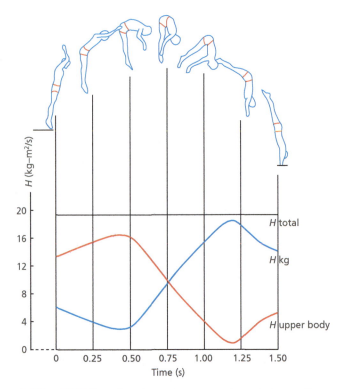

△ **Fig. 11.25** Transfer of angular momentum during a pike dive. As the diver enters the pike, the angular momentum of the legs (*Hlegs*) decreases because the legs slow down. The angular moment of the trunk and arms (*Hupper body*) then increases to maintain a constant *Htotal*. As the diver enters the water the opposite occurs: *Hupper body* decreases to give a clean entry and *Hlegs* increases to maintain *Htotal*.

The principle can be used effectively by coaches from beginner to elite level. For example, when beginner gymnasts are learning the handspring, they must first be taught to obtain sufficient angular momentum during the step or hurdle phase and the placing of the hands on the floor. This must then be followed by pushing off with the legs, and pivoting over the hands in a stretched body position. If insufficient angular momentum is generated, the gymnast will land in an unstable position or at worst on his or her back.

Angular momentum is not conserved when acted upon by external torques.

Rotational analogues of Newton's laws

1. A rotating body will continue to turn about its axis of rotation with constant angular momentum unless an external couple or eccentric force is exerted upon it:

$$\text{Linear momentum } M = mv$$

$$\text{Angular momentum } H = I\omega$$

where I is the moment of inertia and \ is the angular velocity (Figure 11.26). This is also known as the principle of conservation of angular momentum.

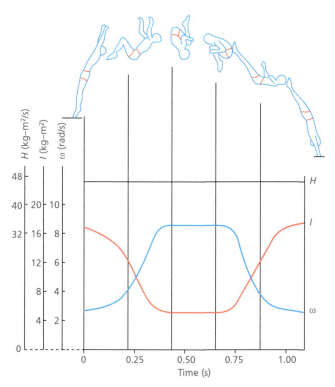

3. For every torque that is exerted by one body on another there is an equal and opposite torque exerted by the second body on the first (Figure 11.27).

△ **Fig. 11.26** Diagram represents the relationship between angular momentum, moments of inertia and angular velocity during a 1.5 back somersault dive.

2. The rate of change of angular momentum of a body is proportional to the torque causing it and the change takes place in the direction in which the torque acts

$$\Sigma T = I(\omega_f - \omega_i)/t$$
$$\Sigma T = I\alpha$$

where T is the torque, I is the moment of inertia and α is the angular acceleration.

△ **Fig. 11.27** An analogue of Newton's third law: shooting the legs out during the long-jump causes the torso to flex forward (action/reaction).

Angular momentum in long jumping

The positions of a long jumper using the hang-style technique are shown in Figure 11.28.

The following equation represents the Law of Conservation of Angular Momentum using the long jump as an example:

$$H_{total} = H_{trunk} + H_{head} + H_{arms} + H_{legs} = constant$$

△ **Fig. 11.28** The moments of inertia of the arms and legs are smaller than the total body moments of inertia. As a result, the angular velocities of the arms and legs respectively must be larger to produce angular momenta of the arms and legs large enough to accommodate the total angular momentum. Note: do remember the relationship between angular momentum, angular velocity and moments of inertia.

where H is angular momentum. At take-off the athlete wants to generate maximum horizontal and vertical velocity. Horizontal velocity is generated via the run-up, whereas maximum vertical velocity is achieved by the athlete driving into the board with his or her take-off leg. In order to generate maximum height, the athlete lowers the hips and centre of mass. As a result the ground reaction force passes just behind the jumper, the eccentric force generated produces upward vertical velocity and forward angular momentum about the transverse axis through the centre of mass. If the athlete does not alter the position or orientation of his or her body parts to each other he or she will continue to rotate forwards, landing awkwardly. To prevent the trunk and head from rotating forward, the athlete needs to rotate the arms and legs to account for H_{total}. The angular momentum of the arms and legs about the transverse axis through the centre of mass will be increased at the expense of the trunk.

Analysis of rotation

The principles that govern rotation are based on torque production and the outcomes as a result of the torque produced. Some of the general principles can be stated as follows:

- **Torque**: The torque about an axis of rotation is the product of magnitude of the force and the perpendicular distance from the line of action.

- **Summation of torques**: The resultant torque of a system is the sum of the individual torques. So if a maximum resultant force is required, all body segments must contribute to the movement. Torques can be summed simultaneously or in sequence. For example a sequential summation of force occurs when a weightlifter carries out a power clean.

- **Conservation of angular momentum**: Angular momentum cannot be altered unless acted upon by an external torque. Therefore a decrease in the moments of inertia will result in an increase in the angular velocity of the system.

- **Levers**: When the effort torque equals the resistance torque the lever will balance. Effort can be calculated if the resistance torque and the length of the effort arm are known.

- **Transfer of angular momentum**: Total angular momentum remains unchanged but can be transferred from one body segment to another. In many movements, the transfer moves from the larger to the smaller body segments.

Activity 11.2

Independent learning task: Try these sample questions

1. Just after taking off for a 2.5 back somersault dive, a diver has a moment of inertia (I) of 18 kgm^2 and an angular velocity of 7.9 rad/s. Calculate the angular momentum (H) at that moment. If the diver changes his body position from a layout to a tuck and reduces his moment of inertia to 6.4 kgm^2, what is the resulting angular velocity?

2. Calculate the angular momentum (H) of a gymnast who takes off for a back flip with an angular velocity of 8.9 rad/s and a moment of inertia of 10.5 kgm^2. If the gymnast reduces his moment of inertia to 2.9 kgm^2 what is his angular velocity?

3. An athlete rehabilitating following a knee injury performs leg extension exercises against a fixed resistance of 15 N. Calculate the amount of torque produced at the knee as a result of the leg extension machine for each of the four positions shown, given that the centre of mass of the 15 N resistance is at a distance of 0.4 m from the knee joint centre.

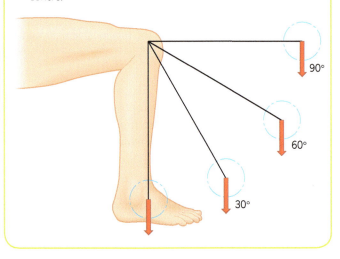

Coaching application: The squat – safety and specificity

Many strength and conditioning coaches consider the squat exercise essential for the full development of athletic potential (Fleck and Kraemar, 1997). O'Shea and Wegner (1981) stated that the full squat must be considered the cornerstone exercise because it quickly stimulates overall strength increases in both men and women. Yet controversy remains over its safety (Chandler *et al.*, 1989a). Concern has been directed primarily toward the potential for back and knee injuries. Research in the 1960s indicated that full squats (below parallel) damaged the knee joint by stretching the ligaments (Mein, 1961). Despite the fact that more recent evidence does not support this contention (Morehouse,

Activity 11.3

Laboratory practical investigation: Application of bio-mechanics to athlete training

Equipment

1. Video camera

2. Leg press machine, squat stands, Olympic bar and free weights, leg extension machine, hack squat machine, running track, isokinetic dynamometer.

Objectives

This investigation examines some biomechanical factors associated with strength and conditioning and demonstrates the use of video analysis of movement.

Basic concepts

The competitive performance of an athlete is optimally enhanced via the performance of training exercises that are performed in a manner very specific to the competitive event. Thus, where possible, exercises should be performed with the same velocity of movement, range of motion, muscle groups, movement pattern, body posture and muscle contraction as used in the target activity. Probably the most important and difficult part of developing a successful training routine for an athlete is the selection of exercises, which allow the competitive performance to be accurately simulated. Such exercise selection requires a good understanding of the biomechanics of the event and of the possible exercises.

The purpose of the investigation is to qualitatively assess the specificity of the squat, leg press, leg extension, hack squat and bounding exercises to the support phase in sprint running.

Methods

1. Taking turns using the video camera, video a subject performing a sprint run, leg press, squat, leg extension, hack squat and bounding exercises. Make sure you position the video camera to capture the relevant muscular actions of each activity. The sagittal plane view is best, so position yourself perpendicular to the athlete concerned.

2. View the videos recorded.

Group discussion questions

1. Of the five exercises undertaken, rank them in terms of their relative specificity to the support phase in sprint running. During this process consider how specific the exercises are with reference to the:

   - Velocity of movement
   - Range of motion
   - Movement pattern
   - Body posture
   - Muscle groups used
   - Type of contraction.

2. List the main disadvantages of using the leg extension exercise to enhance sprint running.

3. Given that the principle of specificity of training is so important, some people will argue that athletes should spend all their time performing the actual competitive event and should not perform any assistance training such as weight training. In light of this statement explain why specialized training like weight training is an important part of most athletic training routines.

4. How could the leg press exercise be modified to increase its specificity to the support phase of the sprint running action?

5. If a sprinter performed squats, what minimum knee angle would you recommend that he or she descend to and why?

The use of video as performed in this laboratory is an excellent means through which to demonstrate to athletes why they are performing the training exercises prescribed and also what aspect of their performance should improve as a result of it. Video cameras can also be used to qualitatively analyse competitive performance.

1970; Chandler *et al.*, 1989a), it is still accepted by some practitioners in the strength and conditioning field (Chandler *et al.*, 1989b). From a biomechanical perspective, studies have reported that the forward movement of the knees during the squat exercise is associated with an increase in shearing forces on the knees and could contribute to injury (Ariel, 1974). What is certain however is that maximal torques for the knee during a parallel squat occur in the bottom position, while maximal torques for the hips occur during the second half of the concentric phase.

A study undertaken by Fry *et al.* (2003) compared two squat conditions on joint torque. The two conditions analysed were an unrestricted squat where the knees are able to move anteriorly as far as necessary and a restricted squat where a vertical board restricts anterior knee displacement. This second technique is one which is commonly used by power lifters. The results indicated that greater torque was produced at the hips and significantly less torque produced at the knees in the restricted squat. From an injury perspective, however, although there appears to be less knee torque using a restricted squat technique, the athlete is forced to lean further forward in order to complete the lift successfully. As a result, increased shearing forces occur in the lumbar region, increasing the risk of lower back injuries.

Coaches should be aware of the transfer in joint torques when teaching correct technique. There may be a case for adopting the restricted squat technique as part of the rehabilitation process following minor knee injuries. More importantly, guidelines for exercise technique should not be based on the torque characteristics of just one joint (knee joint) at the expense of others (lower back).

Correct application of strength training that includes the squat exercise has been shown to lead to increased ligament and tendon strength and bone density (Block *et al.*, 1986), development of the large muscle groups in the lower back, hip and knee (Atha, 1981 ; O'Shea and Wegner, 1981 ; O'Shea, 1985), improved strength, speed and power of the leg and hip musculature (Stoessel *et al.*, 1991), and improved neuromuscular efficiency that aids performance in biomechanically similar movements (Palmitier *et al.*, 1991).

Activity 11.4

Laboratory practical investigation: Turntable experiments

Equipment

1. Turntable
2. Stop watches
3. Baseball bat

Methods

1. Hold a baseball bat and stand on the motionless turntable. Swing the bat overhead in large circles. Discuss what happens and explain why.

2. Repeat the above but hold the bat by the other end. What is the difference and explain in mechanical terms why this occurs?

3. Repeat the above but without holding the bat, just using the arms. What is the difference and explain in mechanical terms what occurs?

4. Flex the left arm to the horizontal position and abduct the right arm to the horizontal position. Swing them both vigorously to the left. What happens and why?

5. Have a subject stand on the turntable with his or her arms abducted to the horizontal position. Start the subject spinning at a low angular velocity. Time one revolution and calculate the angular velocity. Instruct the subject to adduct his or her arms to the sides and again time one revolution and calculate the angular velocity. Determine the change in angular velocity and estimate the ratio of moments of inertia for arms abducted and arms at the sides postures. Remember:

   $I_1 \backslash \Omega_1 = I_2 \backslash \Omega_2$ (conservation of angular momentum)

   Ratio of moments of inertia = arms abducted: arms at sides

   With the results of the above experiment in mind, explain why an ice skater will bring her arms up close to the axis of rotation of her body when performing a pirouette.

Summary

- In this chapter angular kinetics has been considered.
- The concepts and principles that surround angular kinetics have been discussed.
- Moments of inertia and torque have been defined and calculated; levers, their classifications and their link to muscle contraction and length during body segment rotation were also covered.
- The application of force–velocity and length–tension relationships to athlete training was introduced.
- Finally, the very important topic of angular momentum and the ways in which rotation is initiated and controlled in sport have been discussed. All concepts have been developed through laboratory activities and review questions.

Review Questions

1. Define the following terms:
   a. Torque
   b. Angular momentum
   c. Lever
   d. Mechanical advantage
   e. Centripetal force
   f. Centrifugal force.

2. Balance a pole across the back of a chair. Hang a 2-kg weight at each end of the pole and let one end represent the effort and the end the resistance. The effort arm and resistance arm should be equal and therefore symmetrical. Add another 2 kg on the resistance end and adjust the positioning of the pole so that it again balances. Measure the effort and resistance moment arms. What is the relationship between the following:
   a. The resistance and the resistance arm, and the effort and the effort arm?
   b. The resistance and the resistance arm given a changing resistance and a constant effort?

3. How can angular momentum be generated? Give examples from sport and exercise.

4. Explain the mechanics of muscle contraction and the length–tension relationship. What is their link with levers and sports performance?

5. Look at a video of a gymnast somersaulting. Consider ways in which the gymnast initiates rotation.

References and further reading

Atha, J (1981) Strengthening muscle. *Exercise and Sports Science Reviews* 7: 163–172.

Ariel BG (1974) Biomechanical analysis of the knee joint during deep knee bends with heavy load. In: Nelson RC, Morehouse C eds *Biomechanics* IV. Baltimore: University Park Press, pp. 44–52.

Avela J, Kyrolainen H, Komi PV (1999) Altered reflex sensitivity after repeated and prolonged passive muscle stretching. *Journal of Applied Physiology* 74: 2990–2997.

Bartlett RM (1999) *Introduction to Sports Biomechanics.* London: E and FN Spon.

Bartonietz, K (2000) Hammer throwing: problems and prospects. In V M Zatsiorsky ed. *Biomechanics in Sport: Performance Enhancement and Injury Prevention* (Vol 4) (pp 458–486). Oxford: Blackwell Science Ltd.

Block JE, Genant HK, Black D (1986) Greater vertebral bone mineral mass in exercising young men. *Journal of Medicine* 145: 39–42.

Bray JJ, Cragg PA, MacKnight ADC, Mills RG, Taylor DW (1986) *Lecture Notes on Human Physiology.* Oxford: Blackwells.

Brice SM, Ness KF and Rosemond D (2011) An analysis of the relationship between the linear hammer speed and the thrower applied forces during the hammer throw for male and female throwers. *Journal of Sports Biomechanics* 10(3):174–184.

Chandler TJ, Wilson GD, Stone MH (1989a) The effect of the squat exercise on knee stability. *Medicine and Science in Sport and Exercise* 21: 299–303.

Chandler TJ, Wilson GD, Stone MH (1989b) The squat exercise: attitudes and practices of high school football coaches. *National Strength and Conditioning Association Journal* 2: 30–34.

Cramer JT, Housh TJ, Johnson GO, Miller JM, Coburn JW, Beck TW (2004) Acute effects of static stretching on peak torque in women. *Journal of Strength and Conditioning Research* 18: 236–241.

Durall CJ (2004) Injury risk with the horizontal arm curl: biomechanical and physiological considerations. *Strength and Conditioning Journal* 4: 52–55.

Enoka RM (1994) *Neuromechanical Basis of Kinesiology*, 2nd edn. Champaign, IL: Human Kinetics.

Fleck SJ, Kraemar WJ (1997) *Designing Resistance Training Programs*, 2nd edn. Champaign, IL: Human Kinetics.

Fry AC, Chadwick Smith J, Schilling BK (2003) Effect of knee position on hip and knee torques during the barbell squat. *Journal of Strength and Conditioning Research* 17: 629–633.

Gowitzke BA, Milner M (1988) *Scientific Basis of Human Movement*, 3rd edn. Baltimore: Williams and Wilkins.

Harman E (1994) The biomechanics of resistance exercise. In: Baechle T ed. *Essential of Strength Training and Conditioning*, 2nd edn. Champaign, IL: Human Kinetics, pp. 19–50.

Lieber RL (1992) *Skeletal Muscle Structure and Function.* Baltimore: Williams and Wilkins.

Mein KY (1961) The deep squat exercise as utilised in weight training for athletes and its effect on the ligaments of the knee. *Journal of the Institute for Physical and Mental Rehabilitation* 15: 6–11.

Mills SH, Robinson PD (2000) Assessment of scrummaging performance using a pneumatically controlled individual scrum machine. In: Hong Y, Johns D eds *Proceedings of the 18th International Symposium in Sports.* Chinese University of Hong Kong.

Morehouse CA (1970) Evaluation of knee abduction and adduction: the effect of selected exercise programs on knee stability and its relationship to knee injuries in college football. Final project report. Department of Health, Education and Welfare, Pennsylvania State University.

Nelson AG, Allen JD, Cornwell A, Kokkonen J (2001) Inhibition of maximal voluntary isometric torque production by acute stretching is joint angle specific. *Research Quarterly for Exercise and Sport* 72: 68–70.

O' Shea JP (1985) The parallel squat. *National Strength and Conditioning Association Journal* 7: 4–6.

O' Shea JP, Wegner J (1981) Power weight training and the female athlete. *Physician and Sportsmedicine* 9: 109–120.

Palmitier RA, Kai-nan A, Scott SG, Chao EYS (1991) Kinetic chain exercise in knee rehabilitation. *Sportsmedicine* 11: 402–413.

Siff MC (1993) Understanding the mechanics of muscle contraction. *National Strength and Conditioning Association Journal* 15: 30–33.

Stoessel L, Stone MH, Keith R, Marple D, Johnson R (1991) Selected physiological, psychological and performance characteristics of national calibre United States women weightlifters. *Journal of Applied Sports Science Research* 5: 87–95.

Young W, Elliott S (2001) Acute effects of static stretching, maximum voluntary contractions on explosive force production and jumping performance. *Research Quarterly for Exercise and Sport* 72: 273–279.

Further reading

Hay JG (1993) *The Biomechanics of Sports Techniques*, 4th edn. Englewood Cliffs, NJ: Prentice Hall.

Maynard J, Ebben WP (2003) The effects of antagonist prefatigue on agonist torque and electromyography. *Journal of Strength and Conditioning Research* 17: 469–474.

12 Fluid Biomechanics

Chapter Objectives

In this chapter you will learn about:

- Buoyancy, centre of buoyancy and drag forces.
- Swimming and fluid mechanics.
- The Bernoulli and Magnus effects.
- The application of these principles to sporting performance.

Introduction

Fluid mechanics refers to the application of biomechanical principles relevant to the propulsion of a body (e.g. a swimmer) or object (e.g. a javelin) through water and/or air. It is important to understand how to minimize resistive forces and maximize propulsive forces in the fluid medium for efficient motion and optimal performance. However, as sport scientists/strength and conditioning coaches, you should be aware of factors that offer greater fluid resistive forces to your athletes and therefore offer different forms of overload in terms of the physical preparation of the muscles. Many sports are affected by the forces produced by water and air. These include swimming, parachuting, downhill skiing, cycling, rowing and motor racing, to name but a few. On the other hand, there are also sports where these forces do not affect performance. Boxers and gymnasts, for example, are not concerned with air resistance and the consequent effects during competition.

Water and air differ in terms of density and viscosity but they exert similar forces on a body, both acting like fluids. The forces acting on an object moving in a fluid include weight or hydrostatic pressure, buoyancy, drag and lift. So how do these forces affect athletic performance and what can an athlete and technology do to overcome them?

Buoyancy

Buoyant force is related to hydrostatic or underwater pressure and operates from Archimedes' principle. This states that a solid immersed in a fluid will be buoyed up by a force that is equal to the weight of the fluid displaced.

In other words, if the solid is weighed in the fluid it will be lighter than its true weight by the weight of the fluid displaced.

Pressure on a body immersed in fluid increases with depth. Therefore there is a greater upward thrust on a body or athlete, which keeps it afloat. This is known as a buoyant force or the force of buoyancy (Carr, 1997). The buoyant force is equal to the weight of the fluid displaced by the body. Archimedes' principle occurs because of the relationship between the density of the fluid and the density of the object. A body's density depends on the volume (in this case the space the athlete takes up in the fluid) of the body and its mass. Therefore:

$$\text{Density} = \text{(mass of athlete)}/\text{volume}$$

So, according to the above formula, an increase in the mass of the athlete will result in an increase in body density, providing the volume remains constant. Conversely, an increase in the athlete's volume will result in a reduction in body density, providing the mass remains constant. Swimmers who wear wetsuits increase their volume, but the changes in mass are virtually insignificant. Thus the buoyant force (upward thrust) is increased. You can see that body density can be intentionally altered to some degree, affecting the performance of the athlete through changes in buoyancy.

There are many sports affected by buoyancy. Water sports like swimming, sailing and kayaking for example. When a buoyant force is greater than the force of gravity pulling the athlete or object down then the athlete is pushed to the surface. A swimmer who weighs 85 kg will float if the weight of the water that he or she displaces is greater than 85 kg.

An athlete with large muscles and bones and low body fat will have an increased body density (because a cubic centimetre of muscle and bone is more dense than a cubic centimetre of fat) and will be more likely to sink. This can be overcome by swimming in salt water. The salt increases the density of the water, which increases the upward thrust on the athlete. If the displaced water weighs more than the athlete then he or she will float (Carr, 1997). Similarly, as shown in Figure 12.1, when the pull of gravity is less than the upward buoyant force the athlete floats.

△ **Fig. 12.1** If the displaced water weighs more than the swimmer then he or she will float.

The assumption that the body tissues of all humans have a constant density can be challenged, however, and breaks down when different ethnic groups are compared. Black athletes, for example, tend to have a higher bone mineral density than white athletes. This becomes evident in elite swimming competitions in which, to date, it is almost unheard of for a black athlete to make it to the finals of a major championship. The variation in body density may also apply to athletes with large muscle mass, or female athletes who lose bone mineral density during intensive training. Children also have lower bone mineral densities, less potassium and more water per fat-free mass (FFM) than adults (Lohman, 1986). Body composition plays an important role in swimming, particularly distance swimming.

Body fat not only aids buoyancy but also resists heat loss. Athletes with low body fat tend to be more at risk of hypothermia than athletes with high body-fat levels. Scuba divers, for example, wearing wetsuits at surface level experience an increase in upthrust and so float. However, when they descend, the suit is compressed and this allows the diver to descend. The diver can pump air from the tank into a buoyancy vest when making the decision to ascend or to just hang in the water. Inhaling air into the lungs and expanding the rib cage can also be used to increase buoyancy at the surface.

Temperature also affects density, the warmer the water the less dense it is.

Centre of buoyancy

The centre of buoyancy is the place where force concentrates its upward thrust on an athlete's body. The centre of buoyancy is generally not the same as the athlete's centre of mass. This is because the upper body takes up more space than the legs in the water. The upper body also contains the lungs and so weighs less than the water it displaces. As a result there is greater upthrust on the upper body compared to the legs and the floating athlete tends to adopt the position indicated in Figure 12.2(b). Salt water allows an athlete to lie higher in the water.

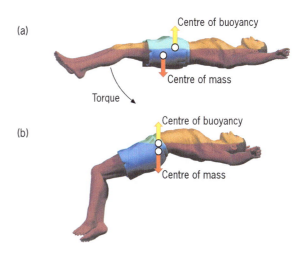

△ **Fig. 12.2** Centre of buoyancy: due to torque, the swimmer adopts a floating position where the centre of buoyancy is directly above the centre of mass.

There are rules that limit the amount of salt that can be added to swimming pools used in competition. This is because the increased upthrust due to the salt allows the athlete to adopt a more streamlined body position, leading to faster swimming times.

Drag

Resistive forces encountered by athletes and objects are as a result of collisions with the fluid molecules. Drag forces push and pull an athlete or object in all directions. The molecules are pushed aside to fill the gap behind the athlete/object. This is not done in an orderly manner, however, and turbulence is exhibited. As a result pressure differences occur between the front and the back of the athlete called 'form' or 'pressure' drag. Drag varies and depends on the type of fluid (air or water), its temperature and viscosity. Drag also depends on the size and shape of the athlete or object. The pressure at the front of the athlete or object is increased and is proportional to the speed and the density of the fluid.

(1) $P = \frac{1}{2}\rho \times v^2$

where ρ is density and v is velocity.

The form drag forces are proportional to the pressure differential multiplied by the cross-sectional area (Ap) of the body or object, and a constant for proportionality, CD, representing a streamlined shape is also considered:

(2) $F_p = \frac{1}{2}\rho \times v^2 \times CD \times A_p$

(CD is the drag coefficient whose value depends on the type of flow, on the geometry of the object and on the object–fluid relative velocity.)

Variations in these factors will cause changes in drag forces that the object will be subjected to. Velocity changes are particularly critical. As can be derived from equation (2) a doubling of the velocity will result in the quadrupling of the drag force. The relationship occurs within a defined set of circumstances and then the fluid behaviour changes.

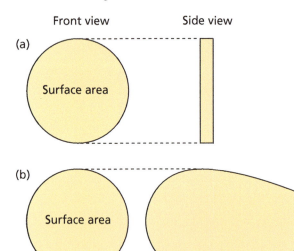

△ **Fig. 12.3** Objects (a) and (b) appear to be the same when viewed from the front but have very different shapes when viewed from the side.

Resistance met by an object moving through air or water also depends on the size and shape of its surface area. For example, Figure 12.3 shows two objects, a and b, that appear to be the same when viewed from the front but have very different forms when viewed from the side. In sports such as cycling, where high velocities are reached, the effect of air resistance becomes more significant. At lower speeds, there is a linear relationship between velocity and energy cost. At higher speeds the relationship becomes curvilinear and so the energy cost increases disproportionately. It is clear to understand why attempts are made to reduce air resistance through equipment design and positioning. There are also implications for coach and athlete in terms of nutritional requirements, particularly

during sustained endurance events like the Tour de France. Reducing air resistance will reduce cyclists' time so calorific intake will be affected in both training and competition.

Equation (2) also shows that doubling the surface area doubles the resistance force. Therefore the aim in sports like downhill skiing and long jumping is to reduce air resistance. It is especially important in swimming to reduce water resistance. The degree of streamlining available to the athlete depends on possible body positions and equipment rules. The ideal streamlining position has a 'tear drop' profile (Figure 12.4). This is displayed by pursuit cyclists' helmets, which come close to resembling this shape.

△ **Fig. 12.4** The 'tear drop' profile maximizes streamlining.

The athlete or object is also subjected to friction drag (F_f) as a result of the fluid particles moving over the surface of the body. The magnitude of the drag is dependent upon velocity of the flow in relation to the body, surface area and its characteristics.

For example, when a swimmer swims at the surface, water is pushed away and waves are formed due to differential water velocities around the swimmer. Wave length and amplitude increase due to an increase in swimming velocity. Further increases in velocity trap the swimmer in a trough, limiting greater increases. The total drag therefore that a swimmer is subjected to is a combination of form or pressure drag, surface or (skin) friction drag and wave drag (Blazevich, 2010; Toussaint *et al.*, 2000). Marinho *et al.* (2009a) used a computational fluid dynamics methodology to analyse the effects of body position on the drag coefficient (CD) during submerged gliding in swimming. Two common gliding positions were investigated: a ventral (i.e. lying face down- or prone) position with the arms extended at the front, and a ventral position with the arms placed alongside the trunk. The simulations were applied to flow velocities of between 1.6 and 2.0 m/s, which are typical of elite swimmers when gliding underwater at the start and in the turns. The gliding position with the arms extended at the front produced lower drag coefficients than with the arms placed along the trunk. The authors recommended that swimmers adopt the arms in front position rather than the arms beside the trunk position during the underwater gliding.

For each of the three sources of drag force acting on an object moving through water we can describe and give examples of that drag force, list factors that affect that drag force and give examples of how we (as sport scientists, coaches, and strength and conditioning specialists) might reduce that particular drag force.

- **Surface or (skin) friction drag:** This is the sum of the friction forces acting between the fluid molecules and the surface of the object. As a fluid molecule slides past the surface of an object, the friction between the surface and the molecule slows down the molecule. A large surface area in contact with the water will therefore increase the surface drag.

- **Form or pressure drag:** This is the equivalent to the sum of the impact forces resulting from the collisions between the fluid molecules and the object. As a fluid molecule first strikes the surface of an object moving through it, it bounces off but then the molecule strikes another fluid molecule and is pushed back toward the surface of the object. The molecule will tend to follow the curvature of the object's surface as the object moves past it. On the leading surfaces of the object, the forces exerted by the fluid molecules have components directed to the rear of the object and contribute to the form drag. On the trailing surfaces of the object, the forces exerted by the fluid molecules stay close to the surface of the object by being deflected by other fluid molecules. This occurs during laminar flow. Several factors affect the magnitude of form drag, including the velocity of the body, the magnitude of the pressure gradient between the front and rear ends of the body and the size of the surface area perpendicular to the fluid flow. Streamlining the body shape reduces the magnitude of the pressure gradient as well as the amount of turbulence created, which in turn minimizes the negative pressure that is created at the rear of the object. The nature of the boundary layer at the surface of a body moving through fluid can influence form drag by affecting the pressure gradient between the front and rear ends. The nature of the boundary layer depends on the roughness of the body's surface area and the body's velocity relative to the flow. Form or pressure drag is probably the largest resistive force in swimming. The ability to be more streamlined is closely related to the amount of buoyancy possessed by the swimmer.

- **Wave drag:** This type of drag acts at the interface of two different fluids (i.e. air and water). Bodies that are completely submerged in a fluid are not affected by wave drag. For example, when a swimmer moves a body segment along, near or across the air and water interface, a wave is created in the more dense fluid. The reaction force exerted on the swimmer is known as wave drag, the magnitude of which increases with greater up and down body movement and increased swimming speed. At swimming speeds above 3m/s, wave drag is generally the largest component of the total drag force acting on the swimmer. As a result, competitive swimmers propel themselves underwater to eliminate wave drag for a small proportion of the race where rules permit, and lane lines in pools may also be designed to dissipate moving surface water. Breaststrokers typically do this, surfacing at the end of each stroke to breathe. Any fast movements of the body such as arm recovery should be performed in the air rather than while in contact with the water.

Swimming and fluid dynamics

There is a vast difference between the swimming abilities of recreational and competitive swimmers and there are even variations at the competitive level, which makes some swimmers faster than others. Many factors contribute to these differences in performance, but one major factor is the swimmer's ability to use the fundamental principles of fluid dynamics to their advantage. Swimming is a relatively slow sport: the fastest swimmers can only reach speeds of 6.5–8 km/h (4–5 mph) and they waste considerable amounts of energy in the process. Swimmers attempt to conserve energy by minimizing water resistance, which is the motivation behind body shaving before major competitions.

There are four primary forces acting on the swimmer: buoyancy, thrust, propulsive force and drag. In the vertical plane the weight of the swimmer is offset by the **buoyancy**. People have varying natural abilities to float and so other means of overcoming their weight must be employed. This can be accomplished by the arm stroke and leg kick. Pressing down in the water causes lift that holds the swimmer higher in the water due to equal and opposite reaction forces.

A force that pushes a body forward is called **thrust**. For example an aeroplane generates thrust through its engine.

In swimming, the arm stroke produces most of the thrust. The fastest swimmers generate thrust by moving their arms and hands like a propeller to produce lift. If the arms and hands are pulled straight back, high pressure occurs on the palm of the hand and low pressure will appear on the back during the front crawl stroke. Since this force is acting in the direction of travel there is a positive effect here which is beneficial to the swimmer. This is sometimes called propulsive drag. Surface drag can be minimized by shaving body hair, the eddy currents reduce the friction by allowing water on water interaction resulting in the swimmer being able to slip, slide and glide through the water more effectively. Other methods such as wearing swimming caps also help to reduce surface drag. An additional drag force is wave drag. This is a reaction force applied to the swimmer due to the swimmer exerting forces in the water, with the corresponding result being that waves are formed.

Maximal propulsion produced by a swimmer will depend on swimming technique and muscle forces. Friction of water particles is the prerequisite for force production as a result of movement in water. Propulsion therefore moves the athlete forward; at the same time, however, the athlete encounters drag forces. At a constant velocity the propulsive force equals the drag force but as the swimmer accelerates the propulsive forces produced also increase, along with the drag forces. Eventually the swimmer achieves maximal velocity and a balance is again achieved. Therefore at constant swimming velocity the drag and propulsive forces are equal in magnitude but opposite in direction.

In the case of form drag the swimmer should attempt to minimize the cross-sectional area of the body. The front crawl implied that propulsion was achieved from the upper body by pulling the arms straight back under the body. Until recently, swimmers adopted an 'S'-shaped path by angling the hand and allowing it to move sideways during the pull phase, thus producing a force called 'lift'. This lift force in conjunction with the propulsive drag force gives the swimmer effective propulsion (Figure 12.5).

This technique was used consistently by coaches and their athletes, until Rushall *et al.* (1994) undertook a three-dimensional analysis and contradicted the theoretical underpinnings discussed. They found that front crawl propulsion was primarily attributed to drag forces based on Newton's third law and that the 'S'-shaped path failed to consider body rotation. The Rushall analysis took body roll or rotation into account and developed the concept of the 'straight-through' pull. Further studies by Riewald (2002) showed that propulsive drag forces produced by the hand stop before the hand reaches the hip and that the propulsive forces due to lift are negligible throughout the stroke.

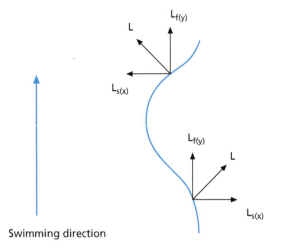

△ **Fig. 12.5** The lift force in conjunction with the propulsive drag force give the swimmer effective propulsion. L is lift force; $L_f(y)$ is the y-component of the lift force acting in a forward direction; $L_s(x)$ is the x-component of lift force acting in a sideways or lateral direction.

The stroke is now taught as an early catch with an early exit. The 'modern' front crawl swimmer or freestyler now uses the principles of equal body rotation and balance in the water, along with an early catch, exit and a straight-through pull arm stroke.

As well as providing propulsion, the leg kick has a stabilizing effect on the swimmer. The armstroke in the back and front crawl is asymmetrical. In other words when one arm is out of the water, the other is producing propulsion. This force acts to the side of the centre of gravity and causes the swimmer to twist in the water (a moment). The body becomes less streamlined and so pressure drag is increased. A good leg kick keeps the body streamlined and reduces drag, the amplitude of which should be small, since increasing the kick amplitude increases the frontal surface area and consequently form drag (Blazevich, 2010). In relative terms only a small amount of propulsion occurs from the leg kick. This is due to the fact that when the swimmer's feet enter the water, air enters the water too. This has the effect of reducing propulsion and increasing drag. In order to reduce drag and increase propulsion the swimmer should keep the feet in the water and should move their feet during the leg kick to produce lift.

Coaches at the elite level are beginning to incorporate biomechanical research into athlete training. The early catch phase can be explained by a fingers first entry, with the early exit being just above the beltline. The stroke is a straight line, not an 'S' shape. Coaches use the analogy of the swimmer pulling themselves along a step ladder just beneath the surface of the water with equal body rotation of at least 45 degrees along its long axis. The head position should be neutral on the spine as if standing on a flat surface. Correct technique avoids excessive internal rotation of the shoulder (Figure 12.6). In terms of hand position, Marinho *et al.* (2009b) analysed the hydrodynamic characteristics of a true model of a swimmer's hand with the thumb placed in different positions (i.e. thumb fully abducted, partially abducted and adducted) using numerical simulation techniques. The model was obtained through computerised topography scans. The results showed that the thumb adducted position presented slightly higher values of drag coefficient compared with the thumb abducted positions. Moreover, the fully abducted thumb position allowed for increases in lift coefficient of the hand at angles of attack (i.e. the angle between the hand and the flow direction) of 0° and 45°. This suggests that for hand models where lift force plays an important role, the abduction of the thumb may be beneficial, whereas at higher angles of attack, in which the drag force is dominant, the adduction of the thumb maybe preferable for swimmers.

Shoulder impingement and overuse injuries are common because of the number of arm strokes undertaken during training. Asymmetric body roll and a hand entry that crosses the midline of the long axis contributes to the incidence of shoulder injuries in freestyle swimming. Cole *et al.* (2002) have reported fewer shoulder injuries in recent studies as a result of biomechanical knowledge being applied correctly by coaches.

An effective conditioning programme can go a long way to reducing shoulder injuries in freestyle swimming and coaches are beginning to recognize the importance of

TIME OUT

THREE-DIMENSIONAL CFD ANALYSIS OF THE HAND AND FOREARM IN SWIMMING

Marinho *et al.* (2011) analysed the hydrodynamic characteristics of a realistic model of an elite swimmer hand/forearm using three-dimensional computational fluid dynamics (CFD) techniques. A three-dimensional domain was designed to simulate the fluid flow around a swimmer hand and forearm model in different orientations (0°, 45° and 90° for the three axes O_x, O_y and O_z). The hand/forearm model was obtained through computerized tomography scans. The drag coefficient presented higher values than the lift coefficient for all model orientations. The drag coefficient of the hand/forearm model increased with the angle of attack, with the maximum value of the force coefficient corresponding to an angle of attack of 90°. The drag coefficient obtained the highest value at an orientation of the hand plane in which the model was directly perpendicular to the direction of flow. An important contribution of the lift coefficient was observed at an angle of attack of 45°, which could have an important role in the overall propulsive force production of the hand and forearm in swimming phases, when the angle of attack is near 45°.

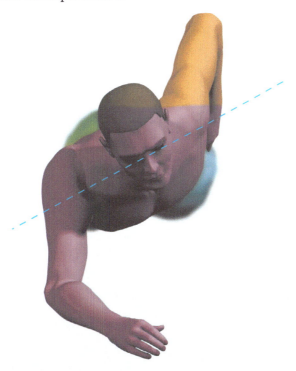

△ **Fig. 12.6** When executing correct front crawl technique, the body rotates along the longitudinal axis through 45 degrees.

strengthening the kinetic chain for injury prevention and rehabilitation (Johnson, 2003).

The Bernoulli effect

When flow lines in a fluid get close together, their velocity increases and the pressure drops. This is known as the Bernoulli effect. Figure 12.7 shows that at the constriction (the narrowest section), the flow lines get closer together, which causes the flow velocity to increase, and a low pressure reading is obtained. After this constriction the

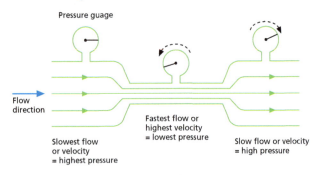

△ **Fig. 12.7** The Bernoulli effect describes the inverse relationship between flow velocity and pressure.

pressure increases again but not to a level greater than the initial value. This is due to energy loss as a result of friction.

Bernoulli's principle can be applied to many situations. For example, on an aerofoil (e.g. the wing of an airplane) (Figure 12.8) the flow lines on the upper surface have to travel further than the flow lines on the lower surface so they are accelerated so that they reach the back at the same time as the lower flow lines. As the flow rate increases the pressure decreases. Therefore a pressure differential arises between the two surfaces and the result is an upward lift force.

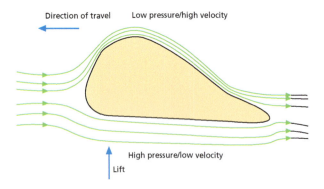

△ **Fig. 12.8** Bernoulli's principle applied to an aerofoil.

Application of the Bernoulli effect to sporting situations

Discus and javelin throwers aim to encounter minimum drag forces yet at the same time achieve maximum lift.

Ski jumpers aim to achieve the same result by maximizing lift and increasing flight time, the outcome being a greater distance jumped. Grand Prix racing cars also have streamlined aerofoils. The idea is to increase the friction force between the ground and the tyres particularly during turns. This is done by mounting an upside down aerofoil so that the 'lift' acts downwards instead of upwards, thereby pushing the car against the track surface.

The Magnus effect

Spin is an essential factor in determining the trajectory of a rapidly moving ball. Athletes who participate in sports that involve kicking, throwing or striking a ball deliberately apply spin, with the intention of deceiving opponents or overcoming obstacles (Bray and Kerwin, 2003). When a ball rotates, the fluid that is in contact with the ball (usually air) rotates with the ball. The corresponding force applied to the ball as a result of the spin is due to the Magnus effect. It is often referred to as 'lift' or, on occasions, a sideways force, although this can sometimes be misleading since the force can often be directed downwards.

The way the Magnus effect works is as follows: as the spinning ball moves through the air a boundary layer of air clings to its surface. On one side of the ball, the boundary layer of air moves in the same direction as the air passing it by. This has the effect of increasing the velocity of the air and so producing a relative reduction in pressure. On the opposite side of the ball, the boundary layer of air moves in the opposite direction to the air passing it by. Air velocity is reduced and a relative increase in pressure occurs. The pressure difference creates lift, which causes the ball to move in the direction of the pressure differential (Figure 12.9).

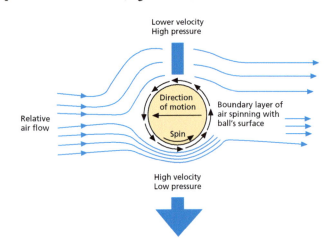

△ **Fig. 12.9** When a ball rotates, the fluid that is in contact with the ball rotates with the ball. The corresponding force as a result of the spin is known as the Magnus effect.

The wake of a non-spinning but moving ball is symmetrical and the airflow separates equally around the ball's surface. With a spinning ball, however, separation occurs earlier for points advancing and later for those receding into the flow (Bray and Kerwin, 2003). A non-symmetrical wake and a force that is perpendicular to the direction of movement is the result. Cricket balls that possess raised seams can also produce non-symmetrical airflow and a resultant force that swings the ball in flight.

The Magnus effect can be applied to many sporting situations. Topspin, backspin, sidespin can be applied in tennis; golfers also use this effect to produce hooks, draws and fades. Alaways and Hubbard (2001) have shown that the lift coefficient in baseballs is affected significantly by the rotating seam, depending on whether the ball is pitched in a two or four seam orientation. This is not the same as seam bowling in cricket, although the bowler does attempt to deliver the ball inclined at an angle to the line of flight (Mehta, 1985).

How does drag affect the flight of various balls?

The various balls used in different sports such as tennis, golf (dimpled) and cricket (seamed) all have the same round shape but have different surfaces. There are significant differences in the way that drag affects these balls. Circular objects are poorly streamlined but a knowledge of how these objects move through air can dramatically affect performance outcome. As you are already aware, the magnitude of the drag force is related to its velocity through the fluid. So let us look at a smooth-surfaced ball with a low velocity. At this velocity the flow of the air around the ball is laminar or streamlined. Because the velocity is low, the ball is affected predominantly by surface drag. The air particles are able to (or have the time to) clasp on to the boundary layer of the ball (Figure 12.10).

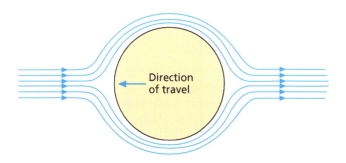

△ **Fig. 12.10** Laminar flow: when velocity is low, the ball is affected predominantly by surface drag. The air particles hug the boundary layer of the ball.

If we increase the velocity of the ball, the laminar flow starts to break down. As well as being affected by surface drag, the laminar flow now becomes distorted and a turbulent flow pattern occurs at the rear of the ball. Turbulent low pressure or a wake appears at the back of the ball relative to the high pressure at the front (Figure 12.11(a)). As the ball increases in velocity the pressure differences continue to increase and the place where the boundary layer breaks away from the ball moves from the rear to the front of the ball. As a result, an increase in turbulent low pressure or wake appears at the back of the ball and form drag dramatically increases (Figure 12.11(b)).

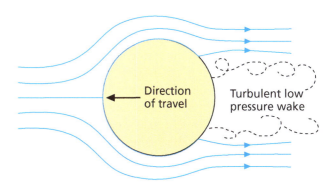

△ **Fig. 12.11 (a)** Turbulent flow pattern: the high velocity causes the boundary layer to break away to the back of the ball, causing an increase in turbulent low pressure, which significantly increases form drag.

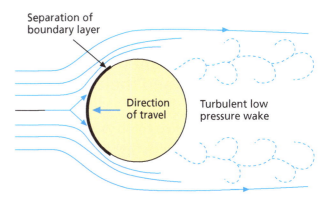

△ **Fig. 12.11 (b)** At even higher velocities the boundary layer moves to the front of the ball and pressure differences continue to increase.

For a given shape the type of flow can be characterized by an adimensional parameter called the **Reynolds number (Re)** that considers the size of the object (D), the density (ρ) and the viscosity (μ) of the fluid as well as the object–fluid relative velocity (V):

$$Re = \rho DV/\mu$$

Low values of Re indicate that the flow is laminar, but for higher values of Re a turbulent flow appears in the rear section of the ball. The **Reynolds critical number (Re$_{crit}$)**

is the point at which the flow moves from a laminar to a turbulent state. Research has shown that when Re is about 3000 or more, the flow is turbulent. In brief:

Re < 2000	Flow is laminar
Re > 3000	Flow is turbulent
2000 < Re < 3000	Flow is unstable

The relevance of Re_{crit} to sport becomes clear when we consider golf balls – and their dimples. Golf ball technology seeks to minimize spin when it is not wanted and maximize it when it is, and to design simple patterns that suit these constructions (Farrally *et al.*, 2003). Golf balls can be driven a long way. What are the factors that allow this to happen? Is it the strength and power of the golfer, the equipment, technique or aerodynamics? The answer really is a combination of the above, however a major contributing factor lies in the aerodynamics of the sphere and the effect of the Re_{crit}. The basic principles regarding the effect of drag on balls have already been covered. We saw that flow moves from a laminar to a turbulent state at Re_{crit} and that the turbulent state produces less drag because the flow remains attached to the ball longer (Figure 12.11(a)). This effect is put to use in the development of the dimpled golf ball. The dimples reduce Re_{crit} so that the flow becomes turbulent at a lower velocity than for a smooth ball or sphere, thus reducing the drag. Because the velocity of the dimpled golf ball is higher than its critical value, the drag does not change much.

What is the interaction between laminar flow, turbulent flow, separation points and their effect on motion through air?

The magnitude of the drag forces of an object moving through air is directly related to its velocity through the fluid (air). The flow of air around smooth-surfaced balls with a low velocity is laminar or streamlined. Since the velocity is low, the ball is affected predominantly by surface drag whereby the air particles are able to clasp onto the boundary layer of the ball. If the velocity of the ball is increased, the laminar flow starts to break down. As well as being affected by surface drag, the laminar flow now becomes distorted and a turbulent flow pattern occurs at the rear of the ball. Turbulent low pressure or a wake appears at the back of the ball relative to high pressure at the front. As the ball increases in velocity the pressure differences continue to increase and the place where the boundary layer breaks away from the rear to the front of the ball (separation point). As a result, an increase in turbulent low pressure or wake appears at the back of the ball and form drag dramatically increases. For a given shape, the type of flow can be characterized by the Reynolds number (Re) that considers the size of the object, the density and viscosity of the fluid as well as object–fluid relative velocity. Low values of Re indicate that the flow is laminar, but for higher values of Re a turbulent flow appears in the rear section of the ball. The Reynolds critical value (Re_{crit}) is the point at which the flow moves from a laminar to a turbulent state. Whether turbulence in the boundary layer occurs or not depends on the size of the ball and its surface roughness. In table tennis the ball is too small and smooth to be able to generate turbulent flow and take advantage of a reduced drag force. However, the dimples in a golf ball help it to go turbulent and increase the distance of a drive, although the dimples also have an important influence on flight due to the ball's spin.

Aerodynamics of the discus

The discus is significantly influenced by aerodynamic forces, with greater distances being achieved when it is thrown into a moderate headwind. This is due to aerodynamic 'lift' produced by the discus when it is in flight. If you look at a discus in side view you will see that it is symmetrical: the upper and lower surfaces have the same shape. Applying Bernoulli's principle to a discus that is released with an angle of attack, the stagnation point (the point where a flow of fluid impinges on a solid object) will move from the centreline of the discus to the lower surface. Air travelling over the upper surface has to travel faster than the air travelling on the lower surface. Pressure differences occur with the higher pressure on the lower surface producing the lift. If the angle of attack is too large the flow will separate and so there is a reduction in the amount of lift produced. This is often seen in competition when the discus is said to stall. This occurs at approximately 26 degrees for the angle of attack (Figure 12.12).

With a moderate headwind, however, distances can be increased. An increase in wind velocity increases the air speed travelling over the discus. The increased lift gives the discus a longer flight time and so distance is improved. The thrower must be more accurate to take advantage of this, therefore experienced throwers who are more consistent are more likely to benefit.

Another factor that comes into play with the discus is gyroscopic stability due to its spin. When you place your hand out of the window of a moving car, you feel it being pushed backwards due to drag force. You will also notice that, depending on how your hand is angled, your hand is pushed up or down due to lift. You will also find that it tends to be forced into a position so that your open palm faces the wind. This twisting action is due to torque. In the case of the discus, this torque would cause it to

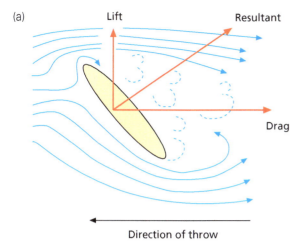

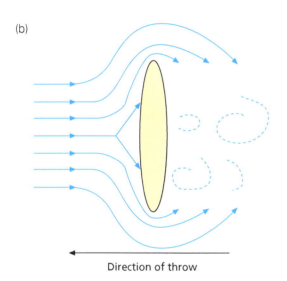

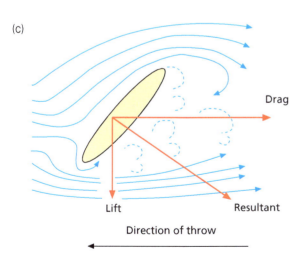

△ **Fig. 12.12** Aerodynamic forces acting on a discus. (a) Drag and force combine to produce a resultant force that pushes the discus upwards and backwards. (b) A discus thrown in this manner gets no lift. (c) A discus thrown with a negative angle of attack generates downward lift.

pitch-up and stall if it was not spinning. However, because it is spinning, the discus has angular momentum, which, as has already been discussed in the previous chapter, remains constant in time unless acted upon by an external force. The angular momentum therefore plays an important role in its stability and is proportional to its mass and angular velocity. Consequently, heavy objects such as the discus do not require as much spin for stability as lighter objects such as frisbees.

Aerodynamics of the javelin

The javelin is an aerodynamic implement that follows the laws of physics. Biomechanical research into javelin throwing has concentrated on the technique of the thrower and the aerodynamics of the implement. Throwing kinematics have been based on video or film and 2-D and 3-D motion analysis. Javelin distance depends on the release parameters and the aerodynamic forces acting during the flight phase (Viitasalo *et al.*, 2003).

The factor with the greatest influence on the distance a javelin can be thrown is the **release velocity**. This is defined as the magnitude of the velocity vector of the javelin's centre of mass relative to the ground (Bartlett and Best, 1988). All other things being equal, the greater the release velocity the greater the distance thrown.

While specific techniques are individual to the athlete, there are basic fundamentals that should be developed by all javelin throwers. These include alignment and the **angle of attack**, which can be defined as the angle between the javelin's long axis and the relative wind vector (Bartlett *et al.*, 1996). Many studies neglect the effect of wind velocity, which is termed the uncorrected angle of attack (Miller and Munro, 1983) and represents the javelin's longitudinal axis and velocity vector of the javelin relative to the ground (or the difference between the path of flight of the centre of mass and the angle of the javelin to the ground). Also important is the **angle of force** (the difference between the path of the javelin's flight at release and the path of the force exerted on the javelin by the thrower) of the javelin at release.

A javelin does not produce lift in the same way as the discus. As the air travels around the shaft of the javelin the flow separates on the upper surface. This separation would normally increase the drag force, but in this case the direction of the drag force is opposite to that of gravity and so the separation of flow increases flight time.

The modern javelin is also designed so that the centre of pressure, the point through which the aerodynamic forces (lift and drag) act, is behind the centre of mass.

The angle of attack influences both the javelin's lift, which acts perpendicular to its direction of motion, and its drag force which acts parallel to it (Figure 12.13(a)).

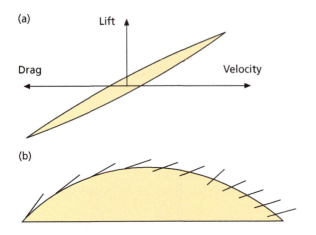

△ **Fig. 12.13** If the centre of pressure is ahead of the centre of mass of the javelin the angle of attack is increased in the early stages of flight. During the later stages of flight the pitching moment acts in the opposite direction, causing the angle of attack to decrease until the javelin hits the ground.

The location of the centre of pressure on the newer javelins varies in flight; however it remains behind the centre of mass, which causes a nose-down pitching moment. If the centre of pressure is ahead of the centre of mass, it increases the angle of attack in the early stages of flight and in the later stages the pitching moment acts in the opposite direction, causing the angle of attack to decrease until it hits the ground (Figure 12.13(b)). This was the case with the old style javelin. The redesign came into being in an attempt to make the sport safer after throws of over 100 m were recorded at some major international competitions during the 1980s. The nose-down pitching moment of the newer javelin makes sure that the javelin lands point first and so does not slide along the ground. This makes distances shorter but easier to measure.

The javelin also rotates around its longitudinal axis during flight and the speed of spin can be as high as 25 revolutions per second (Viitasalo *et al.*, 2003). As in the discus, the spin stabilizes the javelin in flight.

Oscillation of the javelin about its length occurs at a frequency of approximately 25 Hz. This oscillation has a negative effect on performance and needs to be minimized. This can be done by making sure that the delivery is in the same vertical plane as the flight path of the javelin.

Aerodynamic factors in cycling

Cyclists have to overcome wind resistance, which is tiring and inefficient. Aerodynamic efficiency can be attained by a streamlined position that allows the athlete to travel much faster with less effort. As the cyclist increases velocity, however, so does the energy expenditure. Competitive cyclists focus not only on developing physiological parameters but also on achieving greater aerodynamic efficiency. There are volumes of literature on the physiological side of cycling. The mechanical efficiency of the cyclist is attributable to frame size, saddle height, crank length, top tube length, stem length and saddle nose position in relation to the bottom bracket (Burke, 1983). In addition, the correct mechanical action of the muscles and joints, and a proper strength programme when cycling will enable the athlete to reach their full potential.

How can we reduce resistance and improve performance?

Designers have concentrated on improving equipment. Some recent designs have concentrated on shifting from round tubes to oval or tear-shaped tubes whilst maintaining a good strength-to-weight ratio. Disc wheels, while generally heavier with a higher moment of inertia (barrier to acceleration) than their spoked equivalents, produce less drag and turbulence when they spin (Kearney, 1996). The fly wheel is a compromise between spokes and solid wheels that uses flat blades instead of spokes. Cyclists also use helmets with streamlined tails that guide wind smoothly over the cyclist's back.

Drafting in cycling

Whilst improvements to equipment have improved aerodynamics, the cyclist is the greatest obstacle to performance. The human body is not very streamlined and so body positioning is essential. A lot of technology goes into battling air resistance. Using drop handle bars allows the cyclist to reduce their frontal area, which reduces the amount of resistance that they must overcome. Also tight-fitting synthetic clothing can help to reduce surface or skin friction. Relative velocity as referred to in the equation ($F = \frac{1}{2}\rho \times v^2 \times CD \times A\rho$, where ρ = density, v = relative velocity, CD = a constant for proportionality and $A\rho$ = cross-sectional area of the body) has the greatest effect on drag force and is therefore the most important variable that the athlete can control. Cyclists use the concept of drafting whereby they rely on a team mate or opponent to lead a race and they draft or 'sit-in' behind the leader. The fluid is disturbed by the person leading the race and so as a result, the relative velocity of the fluid passing the trailing cyclist is much lower, exerting smaller drag forces on this cyclist. The cyclist must ride very close to the rider in front to maximize this effect. The leading cyclist is doubly disadvantaged because the drag force is larger due to the faster relative velocity of the fluid and the cyclist(s) behind doesn't have to work as hard because the drag force is lower due to the slower relative velocity of the fluid.

Strength and conditioning for cyclists

We must not forget that it is all very well to talk about aerodynamic positions in cycling but the reality has to be linked to strength and conditioning if the body position is to be maintained for long distances (e.g. in the Tour de France). Appropriate strength and conditioning programmes must be implemented and adhered to if performances are to be maximized and injuries prevented. Small changes in body position at high velocities can have a significant effect on a race outcome. For example, Parsons (2010) has reported that maintenance of the aerodynamic position requires good overall body functional flexibility whilst maintaining a strong streamlined fixed position. In addition, the thoracic spine needs to be flexible but strong enough to hold the position, and the upper body musculature (i.e. the major pushing and pulling muscles) need to be strong and balanced with a high degree of grip strength to deal with torsion forces applied from the hips and upper body to the handle bar. The glutei should also be flexible and strong in order to deal with the anterior/posterior and lateral forces demanded from track cycling and maintaining an aerodynamic position.

All coaches and athletes should realize that a strength and conditioning programme will be of benefit to their performance. The primary muscle groups in cycling involve the hip and lower body. Analysis of hip and knee joint angles should indicate the development of a weight-training programme for hip strength specific to joint angle, for example. The development of specific programmes such as squatting with a similar foot placement to that of riding a bike along with range of movement, specific muscle contractions and velocities must also be considered. Upper body strength training should not be neglected either. Upper body strength maintains equilibrium and counteracts forces being developed against the pedals in the lower body. It can also help to offset fatigue in distance cycling. It is important for the athlete and coach to identify areas of concern at the start of the annual training cycle. For example, do the arms need to be strengthened to help in climbing or sprinting? Or are the abdominals strong enough to compensate for the hours of stretching that occurs in the lumbar region brought about by trying to maintain the most efficient aerodynamic position?

Summary

- This chapter has provided the reader with an understanding of the forces involved when a body moves through a fluid (air and water).
- The equations of fluid flow were identified to give the reader an understanding of the relationship between the variables.
- Different types of fluid flow were defined and an appreciation of how these forces affect sport and exercise movements has also been covered (aerodynamics of the discus, for example).
- Other mechanisms such as those that cause lift are also covered (Bernoulli and Magnus effects).
- Application to performance is highlighted throughout.

Review Questions

1. Explain the difference between an angle of release and angle of attack. Include a diagram with your explanation.
2. Draw diagrams indicating the Magnus effect during topspin, backspin and sidespin.
3. What factors affect the floating ability of an individual?
4. Describe what is meant by the centre of buoyancy.
5. Name and describe each of the four sources of drag force acting on an object moving through water at its surface.
6. How can a swimmer use hydrodynamic lift to produce propulsion during swimming?
7. List and explain the advantages of using a bent arm pull with a curved pathway.
8. Explain with the use of a diagram how lift can be produced on the hand moving through water.
9. Using the swerve or bend of a soccer ball as an example, provide a scientific explanation of why it occurs.
10. What is the difference between laminar and turbulent flow? Give examples from sport and exercise.
11. What aspects of mechanical design have been incorporated into the modern racing bicycle to reduce aerodynamic drag?
12. During a golf match you need to produce a hook shot to clear some trees on your approach to the green. Draw a diagram and explain the Magnus effect to achieve this goal.

References

Alaways LW, Hubbard M (2001) Experimental determination of baseball spin and lift. *Journal of Sports Sciences* 19: 349–358.

Bartlett RM, Best RJ (1988) The biomechanics of javelin throwing: a review. *Journal of Sports Sciences* 6: 1–38.

Bartlett, R.M, Muller, E, Lindinger, S, Brunner, F and Morriss, C (1996) Three-dimensional javelin release parameters for throwers of different skills. *Journal of Applied Biomechanics,* 12, 58–71

Blazevich AJ (2010) *Sports Biomechanics the Basics,* 2nd edn. A&C Black, London.

Bray K, Kerwin DG (2003) Modelling the flight of a soccer ball in a direct free kick. *Journal of Sports Sciences* 21: 75–85.

Burke ER (1983) Improved cycling performance through strength training. *National Strength and Conditioning Association Journal* June–July.

Carr G (1997) *Sports Mechanics for Coaches.* Champaign, IL: Human Kinetics.

Cole A, Johnson JN, Fredericson M (2002) Injury incidence in competitive swimmers. Paper presented at USA Sports Medicine Society and American Swim Coaches Association Meeting. Las Vegas, September.

Farrally MR, Cochran AJ, Crews DJ, Hurdzan MJ, Price RJ, Snow JT, Thomas PR (2003) Golf science research at the beginning of the twenty-first century. *Journal of Sports Sciences* 21: 753–765.

Johnson JN (2003) Swimming biomechanics and injury prevention – new stroke techniques and medical considerations. *The Physician and Sportsmedicine* 31: January.

Kearney J (1996) Training the Olympic athlete. *Scientific American* June.

Lohman TG (1986) Applicability of body composition techniques and constants for children and youths. *Exercise and Sports Science Review* 14: 325–357.

Marinho DA, Reis VM, Alves FB, Vilas-Boas JP, Macado L, Rouboa AI (2009a) Hydrodynamic drag during gliding in swimming. *Journal of Applied Biomechanics* 25: 3, 253–257.

Marinho DA, Rouboa AI, Alves FB, Vilas-Boas JP, Machado L, Reis VM and Silva AJ (2009b) Hydrodynamic analysis of different thumb positions in swimming. *Journal of Sports Science and Medicine* 58–66.

Marinho DA, Silva AJ, Reis VM, Barbosa TM, Vilas-Boas JP, Alves FB, Macahado L and Rouboa AI (2011) Three-dimensional CFD analysis of the hand and forearm in swimming. *Journal of Applied Biomechanics,* 27: 1, 74–80.

Mehta RD (1985) Aerodynamics of sports balls. *Annual Review of Fluid Mechanics* 17: 151–189.

Miller DI, Munro CF (1983) Javelin position and velocity parameters during final foot plant preceding release. *Journal of Human Movement Studies* 10: 166–177.

Parsons B (2010) Resistance training for elite-level track cyclists. *Strength and Conditioning Journal* 32: 5, 63–68.

Riewald S (2002) Biomechanical forces in swimmers. Paper presented at USA Sports Medicine Society, Colorado Springs, April.

Rushall BS, Sprigings EJ, Holt LE (1994) A re-evaluation of forces in swimming. *Journal of Swimming Research* 10: 6–30.

Toussaint HM, Hollander AP, Berg C van den, Vorontsov A (2000) Biomechanics of swimming. In: Garrett WE, Kirkendall DT eds *Exercise and Sport Science.* Philadelphia: Lippincott, Williams and Wilkins, pp. 639–660.

Viitasalo J, Mononen H, Norvapalo K (2003) Release parameters at the foul line and the official result in javelin throwing. *Sports Biomechanics Journal* 2: 15–34.

Further reading

Bown W, Mehta RD (1993) The seamy side of swing bowling. *New Scientist* August: 21–24.

13 Analysis Methods

Chapter Objectives

In this chapter you will learn about:

- A variety of techniques used for data collection, processing and analysis in sports biomechanics.
- Identifying models that contribute to qualitative analysis of human movement.
- The importance and application of qualitative and quantitative methods of analysis in sports biomechanics.
- The importance of cinematography, two- and three-dimensional analysis, force platforms, electromyography, isokinetic dynamometry, electrogoniometry, accelerometry and pressure platforms.
- The use of review questions and practical investigations to enhance learning and understanding.

Introduction

Advanced technology has allowed biomechanists to gather very accurate measurements about many of the parameters associated with human movement. High-speed filming techniques have been used extensively to examine in detail bodily movements that occur too fast for the human eye to detect. Force platforms have been installed in many elite sports training and research institutes to analyse the various forces applied during sporting events.

Data collection and analysis procedures in sport and exercise biomechanics continue to develop. Data collection methodology and the need for standardization of these procedures is required if a greater awareness of the actions involved in sport and human movement is to be achieved.

Analysis services

Quantitative versus qualitative

Biomechanists working in sports spend a great deal of time devising techniques to measure those biomechanical variables that are believed to optimize performance. These scientists might investigate the pattern of forces exerted by the foot on the starting block using force plates, the sequence of muscle activities during running using electromyography, or the three-dimensional movements of each body segment during a high jump using high-speed cinematography. These studies are examples of quantitative performance analyses. A quantitative analysis is intended for use by researchers. Coaches and teachers do not always have the available equipment to perform these analyses, and must therefore rely on any information they can obtain, visual or aural, to assess performance. This is known as qualitative analysis. Qualitative analysis requires a framework within which a skill or performance can be observed. Good qualitative analysis leads to good quantitative analysis.

The levels of analyses provided by the sport and exercise biomechanist can vary depending on the needs of the client (Hay and Reid, 1982), or the overall framework of activities undertaken.

Qualitative analysis

Qualitative analysis can be defined as being 'the systematic observation and introspective judgement of the quality of human movement for the purpose of providing the most appropriate intervention to improve performance' (Knudson and Morrison, 1996). Within undergraduate programmes, qualitative analysis can be used to illustrate the application of biomechanical principles because it is the most universal professional skill within the sport and exercise sciences. Despite a call for introductory biomechanics courses to emphasise a 'conceptual' or 'qualitative' understanding of biomechanical variables, many courses adopt quantitative research approaches. Introductory biomechanics should however emphasise a conceptual rather than a quantitative approach (Knudson, 2001).

Although quantitative analysis tools have significantly improved our understanding of the movement and the performance of elite athletes, many analytical skills faced by the coach are qualitative in nature (McPherson, 1996). According to Watkins (1987), qualitative analysis can be used in two ways:

1. It can be particularly useful when providing an individual with feedback during various stages of skill acquisition as a teacher or coach. Initially an observation is made to identify strengths and weaknesses followed by a prescription or instruction to eliminate any discrepancies.

2. In the context of performance analysis, it can be used to distinguish between individuals during competition, for example, in judging trampolining or gymnastics. Coaches are faced with a difficult task. Armed only with knowledge of a skill and their perceptual abilities, they must observe and analyse human movement and the complex movement patterns. Based on observations, coaches are required to make instantaneous decisions regarding skill or technique and then provide effective feedback to their athletes. Providing accurate feedback based on qualitative analysis requires critical observation of a motor skill performance and the subsequent identification of errors.

Skill analysis is a complex process that is made up of the equally important components of abstract task analysis and a visual dissection of the observed performance. It is the synthesis of the components that results in analytical competence (McPherson, 1996).

Biomechanical observational models

Brown (1982) developed a biomechanical model based on 19 visual evaluation techniques. The techniques are split into five areas: vantage point, movement simplification, balance and stability, movement relationships and range of movement. Hudson (1985), however, developed a model called POSSUM (purpose/observation system of studying and understanding movement). This approach is classified according to the purpose of the movement or activity. The purpose must be associated with an observable parameter of the activity. In other words, visual variables must be

selected, differentiated between skill levels, observed by the naked eye and ultimately subject to change by the athlete.

Hay and Reid (1988) have suggested that observers should identify faults in two ways. First, the coach breaks down a specific movement and then compares it to a mental image of what he or she perceives from experience to be a 'perfect model'. This is called the sequential method. The main problem with this approach is the assumption that all elite athletes use this technique and the lack of validity for determining the perfect model. McPherson (1996) supports this view and suggests that when considering technique, an athlete may use modifications of a technique and still produce outstanding performances. For example, Michael Johnson, the 1996 Olympic 200m and 400m sprint athletics champion, had an unorthodox sprinting style.

As an alternative, a mechanical method is proposed whereby a deterministic biomechanical model is initially used and a mechanical purpose or result is identified. The factors that directly influence or determine the result are then obtained. (Figure 13.1). These are the underlying factors of the performance and those that cannot be changed are not considered. During observation of the performance and identification of faults, all sensory information should be used.

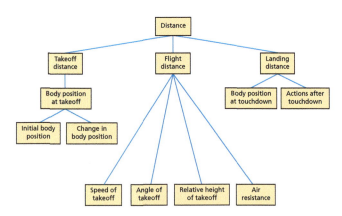

△ **Fig. 13.1** An example of a deterministic model of a long jump used in qualitative movement analysis. From Hay (1985).

Accurate error detection and correction require the ability to critically analyse and observe performance. A paradigm for skill analysis developed by McPherson (1996) based on previous biomechanical models consisted of four major phases: pre-observation, observation,

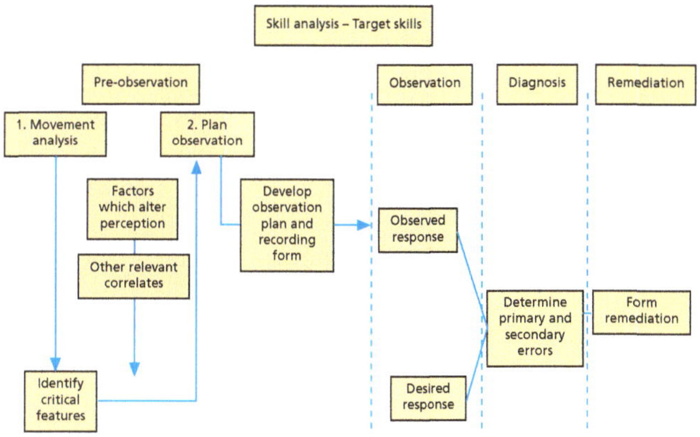

△ **Fig. 13.2** The McPherson qualitative analysis model. From McPherson (1990).

diagnosis and remediation (Figure 13.2). The pre-obser-vation phase was divided into four stages:

1. Conceptual **movement analysis** that identifies the goals or purpose of the skill, simplification of the movement, identification of the mechanical constructs and the critical features.

2. The preparation of an **observation plan and recording form**. This defines the selection of the critical features that will form the focus of the observation and the selection of the optimal viewing positions to take up throughout the observation process.

3. A consideration of the **direct and indirect constraints** so as to promote an awareness of the factors that affect the perceptual process as well as other factors that may interfere with the performance of the skill.

4. Determination of an **acceptable response range**.

During the observation phase visual sensory input is received and appropriately monitored and filtered. Diagnosis involves the identification of performance er-rors followed by a discrimination between primary and secondary errors. Following this, remediation takes place. The final stage involves the provision of feedback based on a remedy (McPherson, 1996).

The biomechanist, coach or observer should identify the mechanical principles involved in the movement. Careful observation is required and so cinematographic or video equipment should be used. A good knowledge of the correct techniques of the specific movement and an ability to communicate with the client group is expected. However, qualitative analysis is essentially descriptive and may be all that is required in some cases (Watkins, 1987).

The qualitative movement analysis in Activity 13.1 has been adapted from McGinnis (2005) and contains procedures common to existing methods. It can be used effectively by sport and exercise students, PE teachers and

Activity 13.1

Qualitative movement analysis

Pre-observation: Plan the analysis

1. What is the purpose or goal of the motor skill being observed (i.e. its desired outcome)? (e.g. bilateral squats, squat jump, bench press, gait analysis (walking)).

2. Identify the problem(s) or question(s) of interest (e.g. is a given skill being performed as effectively as possible).

3. Prior to collecting observational data identify the following:

   a. Appropriate viewing perspectives (i.e. angles, planes)

   b. Reasons for selection of viewing perspectives

   c. Appropriate viewing distances

   d. Reasons for selection of viewing distances

   e. The number of movement performances to be observed

   f. Reasons for the number of movement performances to be observed

   g. Appropriate clothing, lighting or background required to facilitate the observation

   h. Appropriate use of video recordings (i.e. necessary, use-ful, not applicable?) and rationale.

4. Complete the table below (NB: you must complete a new table for each phase identified). Break the skill down into phases and list the *critical features* of the most effective technique. This can be done by either observ-ing as many performances of the skill by elite perform-ers as possible, or, by assessing the actions or positions common to the techniques illustrated in coaching and teaching manuals. Highlight the mechanical principles that underpin each critical feature previously identified (**Hint:** consider the kinematic (i.e. position, displacement, velocity, acceleration) and kinetic (e.g. Newton's laws, forces, torque, levers, impulse-momentum, centre of mass, stability, equilibrium, energy, work, power, inertia, moments of inertia, coefficient of restitution) principles covered throughout this module and decide whether they are applicable or not).

   a. Divide the skill into phases and list the critical features within each phase.

   b. Identify the joints involved and movements occurring.

   (Continues)

sports coaches for example; and it provides a systematic way of biomechanically analysing human movement.

Quantitative analysis

Quantitative analysis can provide objective information that is relevant to the skill or sport. The information may include the measurement of such variables as velocities,

Activity 13.1 *(continued)*

	Phase of Motion	Joint Motion	Muscle Contraction	Active Muscle Group(s)	Mechanical Features (i.e. Kinematics, e.g. rapid acceleration)	Mechanical Features (i.e. Kinetics)
Shoulder						
Elbow						
Hip						
Knee						
Ankle						

Conduct the analysis

1. Carry out an observational analysis of the motor skill and highlight any errors in technique (NB: Errors in technique should be identified during or immediately after observing it).

2. Comment on the following features of the motor skill during each phase, if applicable:

   a. The position of the body or body segments at specific time intervals (i.e. usually at the start and end of a force-producing phase).

   b. The duration and range of motion of the body and its segments during specific phases of the motor skill.

   c. The velocity and acceleration of body segments during specific phases of the skill.

   The timing of body segment motions relative to each other.

Evaluate the performance

Consider the causes of errors as well as their effects by responding to the following:

1. Does the error expose the performer to the danger of injury?

2. Who is your client (i.e. elite, novice, female, child, etc.)?

3. How easy is it to correct the error?

4. Is the error a result of another error that occurred earlier in the performance?

5. How great an effect does the error have on the performance?

6. Is the error or deficiency due to poorly designed or inappropriate equipment?

Instruct the performer: Remediation

The final step is to instruct the performer by correcting errors or deficiencies identified and ranked in the previous stage:

1. Clearly communicate what the performer did incorrectly.

2. Clearly communicate what you want the performer to do.

3. Devise a strategy/strategies for the performer to correct the errors.

accelerations, linear and angular displacements, work, power or forces and torques. Data collected can be compared with those produced by previous research, followed by biomechanical profiling whereby comparisons of a movement are made with previous performances of the same or other athletes. This would require a database and a model for the movement in question.

Quantitative analysis demands rigorous methodology and experimental design. It also requires sophisticated equipment such as:

- High-speed video, cinematography, incorporating digitizing equipment and computers
- Force plates linked to analogue-to-digital converters and computers
- Electromyographs (EMG)

- Isokinetic dynamometers
- Accelerometers, electrogoniometers and pressure platforms.

Procedural issues

When conducting research with human volunteers it is essential that ethical principles are followed and comply with appropriate codes of safe practice. Many research institutions have an Ethics Committee. Issues are addressed in the British Association of Sports Sciences (BASES) Code of Conduct (www.bases.org.uk).

When working with a client group as a sport and exercise biomechanist, specific requirements such as the type of analysis undertaken should be discussed. Is the study or analysis to be carried out in the field or in a laboratory setting? Is a quantitative or qualitative approach to be used? There are also other issues such as is the athlete familiar with the testing procedure and equipment? Responses to testing such as anxiety when using EMG equipment for the first time can affect performance and results. Communication with the client group is essential beforehand. Conditions should be kept the same when testing is undertaken at regular intervals in order to improve reliability. It should not be detrimental to the athlete, however, and again the programme or intervention must be designed in partnership with the client group.

When reporting, procedures must be standardized. All information and details should be included if the study is to be repeated by other biomechanists with the same technical ability. Details should follow agreed protocols (Bartlett, 1992a). The validity, reliability and objectivity of methodology should also be evaluated along with reasons for and justification of statistical techniques used.

When studies are undertaken for coaches, athletes and other client groups, the principle of confidentiality should be followed and discussed in advance.

Equipment considerations

Image-based motion analysis

Video or cine images can be extremely beneficial when undertaking qualitative biomechanical analysis. Images are often recorded and a qualitative analysis undertaken before carrying out any quantitative studies. Measurements can then be taken from these video or cine images.

Equipment can take the form of cinematography, or video-based and automated or semi-automated systems that use video cameras interfaced to computers.

- **Cine cameras and projectors:** In practice 16-mm cameras tend to be the norm as there is a compromise here between the expensive 35-mm and the poor quality and size of the 8-mm (Bartlett *et al.*, 1992). These should have the ability to take pictures at the required frame rate with minimal optical distortion. Variable frame rates of 500 Hz are suitable for most biomechanical analyses; higher rates may be required however for some applications.

- **Video cameras and players:** Most European standard video cameras record at 25 frames per second, whereas American standard video cameras record at 30 Hz. Cameras should be electronically shuttered to allow unblurred picture pausing and good quality

for analysis. Lens quality is also important as lenses can cause image distortion. Video cameras record at a standard rate so no frame calibration is required.

- **Coordinate digitizer:** These are required in order to obtain quantitative information from film or video recordings. According to Tan *et al.* (1995), the resolution of the digitizer along with the technical abilities of the operator has a direct influence on measurement accuracy. Software and hardware are available for both cine and video digitizing. They should be routinely calibrated as part of the reliability process.

- **Automated and semi-automated motion analysis systems:** These systems require markers to be fitted to individuals being analysed. The markers can be active (e.g. LEDS with SELSPOT) or passive (e.g. retroreflective markers). The system digitizes the locations of the markers. Automatic systems collect the information produced immediately for analysis. Semi-automatic systems store images and process the data later. These systems are sensitive to incandescent light (e.g. sunlight) and therefore studies are limited to indoor analyses. Although they increase the potential for more people to be analysed, and increase rates of feedback, they are not ideal for use in competitions since markers must be attached to the athlete.

Two-dimensional analysis

In two-dimensional analysis, movements of body segments are assumed to move in a single plane (planar). Some calibration should be placed in the analysis plane and all movements should occur in that plane. In sport and exercise, movement is generally not planar and such constraints are not easily enforced. This is particularly evident in a competitive environment (Bartlett, 1997).

Three-dimensional analysis

During three-dimensional analysis (multiplanar), two or more cameras are used and synchronization is very important. Calibration is required to allow for the reconstruction of digitized points. It is recommended that experience in all that is involved in two-dimensional procedures is obtained before using three-dimensional analysis. The digitizing gives x, y and z coordinates for each stage of a movement. From here, researchers reconstruct three-dimensional pictures in the form of stick or block figures that can be viewed from any angle.

Activity 13.2

Laboratory practical: Measurement techniques in biomechanics – 2D filming

Aim

To introduce the principles and procedures for two-dimensional (2D) filming methods to record human movement on video useful for analysing technique. By the end of the session and after completion of the related follow-up activities you should be able to:

- Collect 2D kinematic data suitable for further analysis
- Demonstrate an appreciation of the benefits and limitations of 2D filming
- Demonstrate an understanding of the measurement issues for 2D filming
- Present semi-quantitative kinematic data in a manner appropriate for an athlete or coach.

Considerations

The variables of interest and the subsequent movements of the participant dictate what settings should be used on the camera. In general, considerations that influence the quality and usefulness of the video image obtained include:

Aim	Variables
Participant's movements:	Distance; speed; angle to camera
Cameras:	Focal length (i.e. zoom); exposure time (shutter speed); iris/aperture (i.e. amount of light); sampling frequency
Errors:	Camera system; parallax; perspective
Collection procedures:	Adherence to recommended guidelines (see *Handout 1* for **Activity 13.2** on http://cw.tandf.co.uk/sport/)

Equipment/preparation

1. Pre-prepare participants, by informing them for example of venue, time, what clothes to wear and what action they will perform
2. Informed consent
3. Body markers (tape measure, tape, marker pen)
4. Calibration frame and 25m tape measure.
5. Camera, tripod, video, charged batteries (and charged before returning them)
6. Paper/pen
7. Miscellaneous (e.g. timing gates)

Task

By considering the notes above (primarily: aim of filming; subject care and body markers) and the procedure for 2D filming (see *Handout 1* for **Activity 13.2** on http://cw.tandf.co.uk/sport/), and whilst considering the questions below, during this session:

1. Familiarise yourself with the digital camera
2. Prepare your participant appropriately (e.g. warm-up, body markers?). If using body markers: markers of sufficient size and contrasting colours are beneficial.
3. Ensuring appropriate collection protocols and procedures are used for each movement (see *Handout 1* for **Activity 13.2** on http://cw.tandf.co.uk/sport/), capture several trials of either bilateral and unilateral squats or countermovement jumps in both the frontal and sagittal planes.

Questions

1. Critique the images you have collected for your movement trials.
2. Define key moments in the activities you have filmed. What are the benefits of defining key moments in sport/exercise movements for the biomechanist?
3. Consider the measurement issues in (see *Handout 2* for **Activity 13.2** on http://cw.tandf.co.uk/sport/) with respect to 2D filming considerations.

Basics of data acquisition

Usually a raw input signal (before conversion to a numerical value) will be in the form of an analogue voltage whose amplitude varies continuously over time. The voltage is monitored by the hardware, which can modify it by amplification and filtering, processes called 'signal conditioning'. After signal conditioning, the analogue voltage is sampled at regular intervals. The signal is then converted from the analogue to digital form before transmission to the computer (computers need digital data). The software displays the digital data directly, which can be saved for later retrieval. The software is able to analyse the data in a variety of ways. Most of the parameters that affect data acquisition can be set by the user through

software, but the parameters must be appropriate for the signals being recorded (e.g. sampling rates, ranges and filter settings) and should not be applied blindly. You need to know the underpinning science (i.e. what you are recording (e.g. a raw EMG signal), why you are recording it and what relation it bears to real phenomena) and technique (how best to record, and what limitations or compromises are inherent in the process).

Sampling rate: This replaces the original continuous analogue signal by a series of discrete values (samples) taken at regular time intervals. The appropriate sampling rate is dependent upon the signal being measured, if it is too low information is irreversibly lost and the original signal will not be represented correctly. If it is too high there is no loss of information, but the excess data increases processing time and results in unnecessarily high data files. Recordings of periodic waveforms sampled too slowly may be misleading as well as inaccurate because of 'aliasing' (e.g. an analogy can be seen in old films whereby spoked wagon wheels appear to stop or even go backwards when their rotation rate matches the film frame speed). To prevent this, the sampling rate must be at least twice the rate of the highest expected frequency of the incoming waveform. This is known as the Nyquist frequency, the minimum rate at which digital sampling can accurately record an analogue signal. For example, if a signal has maximum frequency components of 100Hz, the sampling rate needs to be at least 200Hz to record it accurately. In most cases the highest frequency will be known, if you are unsure of the frequency range (bandwidth) of your signal, a useful rule of thumb would be to choose a sampling rate high enough to allow at least 5 to 20 samples for any recurring waves in the signal.

Filtering: Any analogue waveform can be described mathematically as the sum of a number of pure sine waves at various frequencies and amplitudes. Low frequencies characterise the slow changing elements of a waveform, and high frequencies the fast changing parts. A filter therefore removes selected frequencies from a signal, for example, low-pass filters allow low frequencies to pass and stop high frequencies. They are commonly used to help reduce random noise and give a smoother signal. A high pass filter removes any steady component of a signal as well as slow fluctuations. Filters are not perfect, for example a 200Hz low-pass analogue filter may allow frequencies of up to 150Hz untouched (its cut-off frequency), and reduce higher frequencies more and more. Filtering can change the signal to some extent, and so its use must be balanced against the distortions that it can remove, such as noise, baseline drift and aliasing to avoid vital information being lost from the signal. (ADInstruments, 2003.)

Reporting of a study

Less rigour is required for reporting qualitative compared to quantitative studies. Minimal requirements are expected, however, when reporting quantitative studies. It may be appropriate to refer to standard references to explain procedure. Studies should include the full details of experimental, reconstruction and data-processing procedures.

Force plates/platforms

Force plates can be used as an excellent teaching aid in undergraduate sport and exercise practical sessions for demonstrating and studying the production of forces during human movement. They measure contact forces between a person and their surroundings, commonly called **ground reaction forces,** in accordance with Newton's third law of motion. Examples of ground reaction force assessments include: isometrics, jumping, landing, weightlifting, running, centre of pressure/balance

What is the application of the force plate to sport and exercise?

Performance:	Maximal force, rate of force development (RFD), impulse, work, power, velocity
Technique:	Timing and point of force application
	Movement mechanics
Training Tool:	Quantitative feedback
Monitoring:	Adaptation following a training intervention
	Fatigue throughout the season
	Injury potential and rehabilitation progress

Force plates are sophisticated electronic devices, rectangular in shape, approximately 0.4 m × 0.6 m in dimension, that typically use either piezoelectric (e.g. Kistler) or strain gauge (e.g. AMTI) transducers attached at each corner. The plate measures force using these transducers, giving an electrical output that is proportional to the magnitude of the force on the plate. The deformation of each transducer is also proportional to the magnitude of the force, and produces a voltage output in proportion to the amount of deformation, and a linear relationship exists between the voltage output and force. Traditional tri-axial plates possess four force sensors embedded in the plate so that ground reaction forces can be measured in three axes: transverse axis (x-axis), sagittal (y-axis) and vertical axis (z-axis). These are more often than not a measurement between a person's foot and the ground, but other forces may be measured if the platform is appropriately positioned.

Force–time data, when combined with joint positions and segmental parameters, can be used to evaluate the effectiveness of numerous movement patterns in sport and exercise. They provide measurements for resultant kinetics only; however they frequently combine with motion analysis techniques to provide kinematic data on segmental dynamics. For example, force platform analysis of vertical jumping provides a demonstration of the kinematics and dynamics of one-dimensional motion. The height of the jump can be calculated from the flight time of the jump, by applying the impulse–momentum theorem to the force–time curve and by applying the work–energy theorem to the force–displacement curve (Linthorne, 2001). The output can be recorded in analogue format (e.g. using an oscillograph) or converted to a digital format by computer processing.

Validity of measurement depends on adequate sensitivity, low threshold, high linearity, low cross-talk and the elimination of cable interference, temperature and humidity changes. The platform must be of a suitable size for the movement under analysis and external vibrations should be excluded. When designing force platforms the purpose of the measurement, the range of forces to be measured with the required accuracy, frequency response and response time should be ascertained. Environmental conditions as well as appropriate cost and weight of the platform should also be taken into consideration.

Figure 13.3 shows a force plate designed to measure the reactions from the ground to the foot. The sum of all the reactions from the ground shown in Figure 13.3(b) is equivalent to the sum of the forces measured by the four sensors in the plate ($R_1 + R_2 + R_3 + R_4$) as shown in Figure 13.3(c). The reaction force (R) is the sum of these forces and has three components: x, y and z (R_x, R_y, R_z). The z-component is used to support the body in the vertical direction and measures the upward force of the body when jumping vertically. The y-component reflects propulsion or the breaking force, and the x-component indicates forces that propel a body laterally. Most of the commercial force platforms are based on

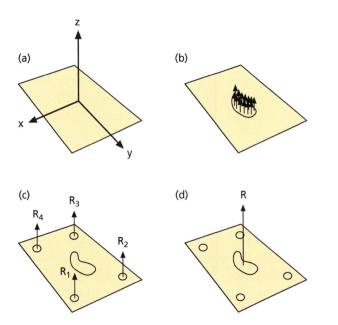

△ **Fig. 13.3** The reactions from the ground to the foot when using a force plate.

Activity 13.3

Independent learning task: Force plates

Basic concepts

Force plates are commonly used in biomechanics research to measure ground reaction forces (GRFs) during activities such as walking, running, lifting, landing, balance and standing as well as various sport and ergonomic movements. The use of portable force plates is becoming more common in strength and conditioning to measure the leg power of athletes. They can measure forces in three directions called the vertical GRF (*Fz*), the anterio-posterior (*Fy*) GRF and the medio-lateral (*Fx*) GRF (see below). Typically the vertical GRF is of most interest to applied biomechanists working within the field of strength and conditioning, or those interested in changing force capabilities of muscle. Therefore the force plate is one of the best pieces of equipment to detect these changes. You should understand as much as possible about this assessment device.

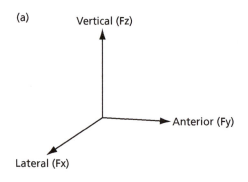

(a) Vertical (Fz) — Anterior (Fy) — Lateral (Fx)

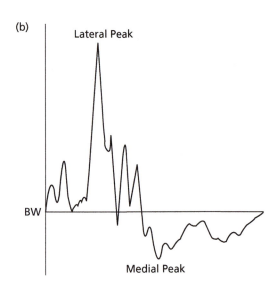

(b) Lateral Peak — BW — Medial Peak

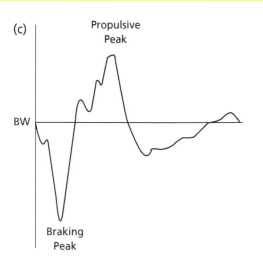

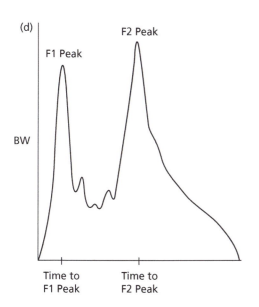

△ a diagrammatic representation of ground reaction forces (GRF) during foot contact (where BW = bodyweight, b = Fx-Medio-lateral direction, c = Fy- anterior-posterior direction, d = Fz- vertical direction)

Discussion questions

1. Which of Newton's Laws is the basis from which force platforms produce their signals?
2. What does a sampling rate of 500 Hz mean?
3. What forces (e.g. Fz) would be of interest if you were studying hopping motion?
4. What forces would you be of importance if you were interested in quantifying the standing balance/stability of a person?
5. What is impulse and how does horizontal impulse affect the velocity of an athlete?

either the piezoelectric (e.g. Kistler platform) or strain gauge (e.g. Bertec platform). Strain gauge devices have been considered to have too low a dynamic response for sports applications. These problems have been improved, although piezoelectric devices are preferred for rapid changing and short duration activities in sport and exercise biomechanics.

When mounting platforms they should be located in a solid floor, usually the lowest floor in a building. If the platform is to be used outdoors, suitable bases can be provided by specially constructed frames. Permanent outside installation is not recommended because of the danger of inadequate drainage and rusting of the frame.

When data processing, the three orthogonal components of the ground reaction force are available instantly. Accuracy of the measurements depends on careful calibration of the force platform (Bobbert and Schamhardt, 1990).

When reporting the study the principle of replicability should be applied and should include: the platform type and mounting arrangements, calibration checks, the recorder used, initiation time and duration of data collection, ranges of sensitivities chosen, clear definitions of key events in the force patterns and full details of any calculations performance.

Portable force plates

Jumping and landing tasks are commonly used to examine various parameters regarding performance and injury. The accepted standard for force measurement research involves very sensitive, heavy, strain-gauge type systems embedded immobile systems that require subjects to be tested in the laboratory (Walsh *et al.*, 2006). Newton and Kraemer (1994) have suggested that force plate analysis feedback during plyometric training should be considered to maximise training efficiency. However, use of traditional force plates may be limited to a laboratory setting and not typically used where plyometric sessions are conducted. A portable method of force measurement in a field-testing situation is therefore required. The development of a valid and reliable portable force plate (e.g. Fittech, Australia) allows testing in a variety of locations that were not previously considered practical (e.g. locations where sports team conduct their training sessions). In addition, the validity (ranged from 0.94–0.97) and reliability (ranged from 0.97–0.99 for force measurements) of the portable force plate also advocate its use in large scale injury epidemiological studies that require kinetic analysis during jumping and landing (Walsh *et al.*, 2006).

TIME OUT

AUGMENTED ECCENTRIC LOADING

Watkins and Sapstead (2010) investigated the acute effects of augmented eccentric loading (AEL) during the box squat exercise (BSE), using a loading strategy for enhancing power production, on subsequent counter-movement jump (CMJ) performance. Seven resistance-trained sport and exercise science students volunteered for this study. Participants performed maximal effort repetitions of the BSE followed by the CMJ during 3 testing sessions using the following loading conditions: Condition 1, Baseline measures; Condition 2, 50/50% BSE eccentric/concentric; Condition 3, 70/50% BSE eccentric/concentric. A portable force plate (Fittech, Australia) calculated force and power parameters at a sampling rate of 500Hz. A one-way repeated measures ANOVA revealed no significant differences ($p > 0.05$) between peak concentric force (PF), peak concentric power (PP) and peak rate of force development (PRFD) across all loading conditions, although the highest peak power was observed during the 70/50% loading condition. The authors suggest that individuals selecting augmented eccentric load thresholds for optimal concentric power production in the BSE may be most effective when loads are selected individually. More research using multiple concentric loads across populations is required to confirm if such thresholds exist.

Activity 13.4

Laboratory practical: Drop jumps and optimal drop height using a portable force plate

Equipment
- Portable force plate
- Digital weighing scale
- 3 × plyometric boxes (i.e. 30cm, 60cm and 100cm heights)

Purpose

Drop jumps are used to increase the reactive strength of the athlete. It is very instructive to test the athlete's jump performance dropping down from different heights and measuring the subsequent contact and flight times.

Instructions

1. In your groups, select a student to be the subject and instruct them in a 5-minute warm-up for the lower body.

2. Instruct the subject to perform drop jumps from approximately 0, 30, 60, and 100 cm.

3. Complete three trials of each condition and record the best performance for each condition.

4. All jumps should be completed with the hands on the hips (or holding a stick)

Record the contact time, flight time, and estimate jump height.

Drop Height	Contact	Flight	F/C ratio
0			
30			
60			
90			

Discussion questions

1. Which drop height resulted in the greatest flight to contact ratio? Explain why.

2. Discuss the significance of this drop height for testing and training.

3. Plot a graph of contact, flight and ratio against drop height.

4. What is the interaction of these three with drop height?

Electromyography

Electromyography (EMG) is a technique used for recording changes in electrical activity or the electrical potential of a muscle during contraction. It is the only method for objectively assessing when a muscle is active (Grieve, 1975). It has many applications in neurology, bioengineering, physiotherapy and sport and exercise biomechanics. Most projects in sport and exercise biomechanics that use EMG are related to those undertaken in kinesiological electromyography. This is the study of muscular function and coordination (Jonsson, 1973). Clarys and Cabri (1993) reported that EMG research in sport and exercise science includes: muscle activity and function, muscle activity during isometric voluntary contractions with increasing tension up to a relative maximum, evaluation of functional anatomical muscle activity, muscle fatigue and the relationship between EMG and muscle force.

Many factors can influence the recorded EMG signal. These are the intrinsic characteristics of the signal and the filtering features of the environment that give the signal its extrinsic characteristics (De Luca and Knaflitz,1990). The electromyographer has little control over the **intrinsic** factors. These include:

- **Physiological factors:** firing rates of motor units, the mean value and variation, muscle fibre type, conduction velocity of the muscle fibres and the characteristics of the conduction volume (shape, conductivity, Permeability of tissue and shape of the boundaries).

- **Anatomical factors:** muscle fibre diameters, the relative positions of the fibres of a motor unit, the distance of the electrodes from the muscle fibres.

The **extrinsic** factors can be modified by altering methods used to detect and record the signal. This includes the location of the electrodes with respect to the motor end plates and the orientation or position of the electrodes with respect to the muscle fibres.

Electrodes

Electrodes are an important link in the recording chain. Any conducting material may be applied to the skin to pick up myoelectric signals. Passive surface electrodes are usually silver chloride (AgCl) discs used with conducting gels. The quality of the recorded signal depends on electrode placement (Clarys and Cabri, 1993). Kramer *et al.* (1972) have advised the placing of the electrode as close to the muscle belly as possible in order to obtain EMG potentials from fusiform muscles. However, fixing the electrodes on the geometric middle of the muscle belly under contraction with its detection surface along the length of the fibres is the most reliable technique. Surface electrodes are used because of their non-invasive character. When using a bipolar configuration, the electrodes are placed parallel to the muscle fibres with an interelectrode spacing of 3 cm (Clarys and Cabri, 1993) (Figure 13.4). This minimizes cross-talk and maximizes reception.

A ground electrode for each pair of electrodes is positioned on the opposing muscle to that being tested as recommended by Okamoto *et al.* (1987). Skin preparation in accordance with the protocol of Okamoto *et al.* (1987) is also recommended. The results of any EMG investigation are often considered to be only as good as the preparation of the electrode attachment sites. Other important equipment considerations include amplifiers that are the heart of the EMG measuring system, cables and recorders. Currently, analogue-to-digital conversion and storage of the digitized signal using a microcomputer is commonly used by sport and exercise biomechanists.

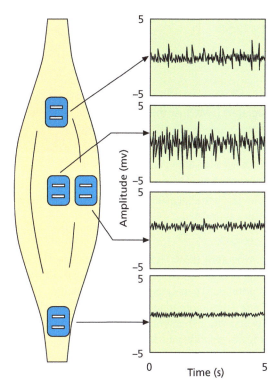

△ **Fig. 13.4** The EMG signal is affected by the location of the electrode with respect to the innervation zone (top electrode), the myotendinous junction (bottom electrode). The greatest amplitude and, therefore, preferred signal is detected in the midline of the muscle belly between the nearest innervation zone and the myotendinous junction. From De Luca (1993).

There are a number of ways of presenting EMG data: raw or unprocessed rectified (all negative values made positive) or integrated. Estimating the amplitude of raw EMG can be difficult as a result of positive and negative deviations. Also, the raw signal has an integral of zero (Bartlett, 1992b) and only a few levels can be discerned with certainty (Hof, 1984). However, monitoring the raw EMG signal before analysis can provide information as to the phasic relationships between various muscles involved in the activity.

Quantification is necessary in order that the results become comparable. However, confusion has resulted with regard to terminology, standardization, and processing, etc. and so comparisons between studies are difficult to make (Clarys and Cabri, 1993).

In reporting an EMG study, the general principle should again be that of replicability. The report should follow the recommendations of the International Society of Electrophysiology and Kinesiology (ISEK) (1980), Dainty *et al.* (1987), and Bartlett (1992b).

Activity 13.5

Laboratory practical: Electromyography

Aims

The aim of this practical is to introduce data capture of electromyography (EMG). At the end of this practical you should be able to:

- Prepare appropriate muscle sites for EMG data collection in accordance with approved guidelines
- Identify considerations in planning and performing EMG studies using the ME6000 portable EMG system (or other available system).

Overview

Electromyography (EMG) is the study of muscles through electrodes measuring the spatial and temporal electrical signals (action potentials) generated by motor units during contraction. These action potentials are generated within the muscle fibres' postsynaptic membrane as depolarization propagates along the fibres shifting ions.

Areas of research

There are many areas of research that use EMG to study phenomena. These include sport science, ergonomics, industry, medicine and rehabilitation. Within biomechanics the uses are often directed at assessing function and coordination of movement and postures in many varied populations and settings. Populations include healthy, skilled or novice participants, and settings can include during training and performance, and in the laboratory and field. In addition to research, using EMG to describe muscle activity, data analysis is progressing rapidly in attempting to quantify the signal accurately with respect to the percentage of maximal contraction and the force that is generated.

Sport and exercise biomechanics is often associated with assessing optimisation or prevention of injury through training, equipment or movement. The combination of EMG with other measures (e.g. filming) and disciplines (e.g. psychology) is often beneficial. For example, ground reaction forces will also be present in running, along with muscle activity and the mental attitude as well as the technique of the athlete will influence these forces and muscle activations.

Benefits

EMG enables the researcher to gain a confirmed insight into the activity of muscles. Some argue this is simply binary (i.e. muscle is or is not active), whilst others suggest that with appropriate methods and data analysis quantification of the magnitude of contraction is provided. There is no doubt that an appropriate method can be used to provide much more than binary information, but the researcher's views and empirical evidence often dictate their interpretation. In addition, if their interpretation is suitably reported it can enable the reader to infer their own views in order to use the research to their best advantage.

Processes within research

Based on the research question, theory, previous literature and practicalities, a number of considerations need to be made in planning an EMG study. A number of examples are presented in the table below:

Process	Example considerations
Neuromuscular Recording	Appropriate muscle selection Signal detection, volume conduction, signal amplification, input impedance, signal characteristics System and calculations (e.g. rectification, linear envelope, normalization)
Data Processing	Appropriate to data collected and research question
Qualitative/Quantitative Analysis Findings/Reporting	Relate to research question, replicable

Processes and considerations in performing EMG research

Other considerations within methods

Additional considerations mostly relate to the research question. These include which movement, which muscles and where electrodes should be located. Use of previous research can often assist in determining these. Importantly, such information is crucial to report in addition many other details that are described by Burden and Bartlett (1997).

Limitations

As with all measurement techniques there are limitations. With respect to theory, especially if attempting to quantify the signal there are varying views on the relationship between the EMG signal and the percentage of maximal contraction. These views are often that a linear or quadratic relationship exists and the use of one will be greatly different to the use of the other. It is most probable that these differences arise from methodologies, and that one or another is still to be proven. Methodological problems arise from non-standardized methods at this stage, and the skills of the researcher. Standardized methods will include equipment, procedures (e.g. electrode locations) and analysis. Skills of the researcher that are essential are knowledge of neurophysiology, anatomy and the movement in question.

Tasks

1. Prepare a participant's biceps brachii of their dominant arm and collect EMG data using your portable EMG system:

- Using a 7.5kg dumbbell, participant performs a 3 repetition biceps curl

(Continues)

- Using a 10kg dumbbell, participant performs a 3 repetition biceps curl.

N.B: Place the electrodes along the length of the muscle on the midpoint of the muscle's bellys they can be in accordance with published locations (e.g. Clarys and Cabri, 1993).

2. What benefit is there in collecting data of maximal isometric contractions?

3. What benefits and limitations are there in using this system to conduct an EMG study?

4. Based on the definition and theory of EMG, begin to consider what variables could be measured and what type of research questions these could be used to address.

Isokinetic dynamometry

This is the assessment of the dynamic function of a joint during movements at a controlled velocity. A body segment or limb is attached to the input arm of an isokinetic dynamometer and rotational movement around its axis of rotation occurs. The angular velocity of the input arm and segment is controlled by the dynamometer. Constant angular velocity conditions (zero acceleration) require the application of a resistive dynamometer moment equal to the resultant moment applied to the moment arm over the range of movement.

Muscle and joint function assessment are essential in sport and exercise, for both performance and rehabilitation purposes (Kellis and Baltzopoulos, 1995). Isokinetics is one of the safest forms of exercise and testing. Due to the fact that the resistive moment does not exceed the net applied moment, no joint or muscle overloading and therefore risk of injury occurs. Isokinetic dynamometers are equipped with various force and angular displacement transducers. They are calibrated to measure the moment applied to the input arm and its relative position.

The analogue signals are sampled through analogue-to-digital conversion boards and appropriate computer software provides for data collection and processing of the digital moments and angular displacements.

Parameters such as maximum moment, work, power and ratios of these parameters for bilateral (left–right comparisons) and reciprocal muscle groups (e.g. joint flexors-extensors) during concentric or eccentric muscle contractions are obtained.

There are several mechanical problems associated with the measurement and interpretation of isokinetic variables. During concentric tests, the system is accelerated by the active muscle groups. The calculation of the joint moment that is accelerating the system during the initial period up to the development of the constant preset angular velocity is complicated, because it requires accurate angular acceleration measurements. During eccentric tests the active mechanism of the dynamometer accelerates the input arm and the segment. The joint exerts a resistive (eccentric) moment that tends to resist the movement of the dynamometer. In some isokinetic dynamometers the body segment during a concentric contraction is allowed to accelerate up to the level of the preset angular velocity. The moment of the resistive force will be zero for a brief initial period. The resistive mechanism is then activated and required to apply a large moment to prevent further acceleration of the segment.

Figure 13.5 is a free body diagram of the lower leg segment during a concentric knee extension isokinetic test. F_q is the quadriceps muscle force acting through the patella tendon, F_h is the force in the hamstrings muscle group (antagonist), F_c is the compressive component of the joint reaction

TIME OUT

THE EFFECT OF STANCE WIDTH ON THE EMG ACTIVITY OF EIGHT SUPERFICIAL THIGH MUSCLES DURING THE BACK SQUAT WITH DIFFERENT BAR LOADS

Many strength trainers believe that varying the stance width during the back squat can target specific muscles of the thigh. Research undertaken by Paoli *et al.* (2009) measured the activation of 8 thigh muscles while performing back squats at 3 stance widths and with different bar loads. Six experienced lifters performed 3 sets of 10 repetitions of squats, each one with a different stance width, using 3 resistances: no load, 30% of 1-repetition maximum (1-RM), and 70% 1-RM. Sets were separated by 6 minutes of rest. Electromyographic (EMG) surface electrodes were placed on the vastus medialis, vastus lateralis rectus femoris, gluteus maximus, gluteus medius and adductor major. Analysis of variance (ANOVA) and Scheffe post hoc tests indicated a significant difference in EMG activity only for the gluteus maximus; in particular, there was a higher electrical activity of this muscle when back squats were performed at the maximum stance widths at 0 and 70% of 1-RM. There were no significant differences between EMG activity and the other muscle groups. The findings suggest that a wide stance squat is necessary for greater activation of the gluteus maximus during back squats.

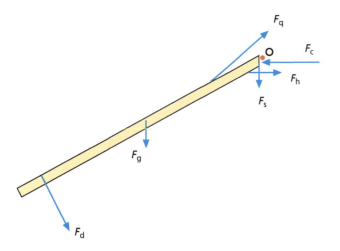

△ **Fig. 13.5** Free body diagram of the lower leg segment during a concentric knee extension isokinetic test.

force, F_g is the gravitational force and F_d is the dynamometer resistive force. The moments of these forces are calculated around the axis of rotation of the segment assumed to pass through the point O. Approximate estimation of individual muscle and ligament forces acting around joints during isokinetic testing is only possible using appropriate techniques for the distribution of the recorded moment to the individual force-producing structures (Kaufman *et al.*, 1991). It follows therefore that moment–velocity or moment–position relationships from dynamometer measurements refer to the joint in general. They must not be confused with force–velocity or force–length relationships of isolated muscles or a muscle group, even if that group is the most dominant and active during the movement.

Equipment considerations

Isokinetic dynamometers use electromechanical or hydraulic components for the control of the angular velocity of the input arm. Older models such as the Cybex II had a passive resistive mechanism and resistance was therefore developed only as a reaction to the applied joint moment. This mechanism allowed only concentric muscle action. More recently, electromechanical dynamometers with active mechanisms have been developed such as Biodex, Cybex6000, Lido and KinCom. These dynamometers are capable of driving the input arm at a preset angular velocity irrespective of the moment applied by the person being tested. Eccentric muscle actions at constant angular velocities can be assessed. The operation of the resistive mechanism and the control of the angular velocity and acceleration in these systems requires the use of appropriate microcomputers. The use of microcomputers facilitates both gravitational and inertial corrections and the provision of visual feedback of muscular effort during the test (Baltzopoulos and Brodie, 1989).

When reporting an isokinetic dynamometry study full details of the following should be included:

- Dynamometer make and type
- Details of data acquisition (sampling frequency, resolution of moment and angular position measurements, etc.)
- Calibration procedures
- Test settings (eccentric–concentric, range of movement, angular velocity, feedback type and method)
- Subject positioning, axes alignment and angle of adjoining joints (e.g. hip and ankle joint position during knee tests)
- Data processing (gravitational–inertial correction methods, data smoothing and differentiation)
- Definition and method of calculation of the different parameters used for the assessment and joint function, reporting of whether these parameters were calculated from the constant velocity (isokinetic) part of the movement – this is particularly crucial at higher angular velocities.

Activity 13.6

Independent learning task: Isokinetic dynamometry

Basic concepts

Isokinetic dynamometry involves contractions at a constant velocity, that is, the limb angular velocity is controlled at a constant value (velocity) throughout the range of movement by providing a resistive torque which is variable and equivalent to the resultant torque. However, periods of acceleration and deceleration are seen at the start and end of the range of motion (ROM). Force/torque capability therefore can be assessed at varying velocities and force–velocity curves generated. A number of other variables can also be assessed e.g. angle of peak torque, time to peak torque, etc. Traditionally, isokinetic dynamometers allowed only single-joint (uni-articular), concentric-only contractions around the knee joint. However, more recent dynamometers allow multi-joint (bi- or tri-articular) actions at a number of joints in both eccentric and concentric contractions. Isometric contractions and passive stiffness tests can also be conducted. An example of an isokinetic dynamometer printout is depicted in the figure below.

An isokinetic dynamometer can be used for comparing injured to non-injured limbs and thereafter making conclusions with regards to limb symmetry and agonist-antagonist ratios. For example, hamstring-hamstring, quadricep-quadricep and hamstring-quadricep ratios are often compared for diagnostic and prognostic purposes. Some hamstring-quadricep ratios (peak torque) for various populations described in various research articles have been summarized in Table 13.1.

(Continues)

Activity 13.6 (continued)

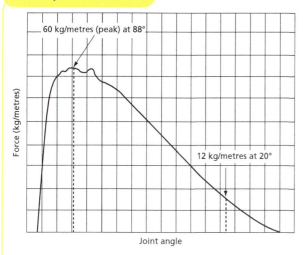

Example of a Cybex isokinetic dynamometer printout

Researcher(s)	Participants	Angular Velocity (deg/sec)	Ratio (%)
Alexander (1989)	23 Elite sprinters	30 230	63.0 78.3
Cronin & Hansen (2005)	26 Professional rugby league players	60 300	56.9 67.5
Dowson et al. (1998)	8 Rugby players	60 150 240	63.9 66.0 70.5
	8 Track sprinters	60 150 240	60.3 66.0 70.4
	8 Sportsmen	60 150 240	59.6 65.2 64.4
Ellenbecker and Roetert (1995)	87 elite junior tennis players Male Male Female	 180 300 300	 60.0 62.5 67.5
Housh et al. (1984)	16 Adolescent throwers	180	70.5
	11 Adolescent jumpers	180	76.7
	12 Adolescent middle distance runners	180	81.0
	23 Adolescent sprinters	180	71.0

Table 13.1 Peak torque/hamstring: quadricep ratios for various populations measured by an isokinetic dynamometer

Discussion questions

1. Comment on any trends you observed in Table 13.1.

2. Calculate the ratios between legs (quadriceps and hamstrings) and between agonist-antagonist pairing using the following data:

Left Leg: Peak quadricep force = 1348 N, Peak hamstring force = 531 N

Right leg: Peak quadricep force = 1406 N, Peak hamstring force = 706 N

3. Interpret your findings and suggest a course of action.

4. Table 13.2 below shows the forces applied to the lever arm (force applied at the 0.5m mark on the lever arm) of a Cybex dynamometer during an isokinetic leg extension at 90°/s. Assume that the lever arm velocity is constant throughout the range of motion (i.e. 0–54°)

(Continues)

Force (N)	Time (s)	Angle (°)	ΔAngle (°)	Torque (N.m)	Work (J)
50	0–0.1	0–9			
120	0.1–0.2	9–18			
500	0.2–0.3	18–27			
600	0.3–0.4	27–36			
300	0.4–0.5	36–45			
20	0.5–0.6	45–54			

△ **Table 13.2** Forces applied to the lever arm during an isokinetic leg extension at 90°/sec

Complete Table 13.2 and answer the following:

a. What is the peak torque?

b. What is the time to and angle of peak torque?

c. What work is done to the Cybex lever arm (assume force on the lever arm acts for each 0.1s, e.g. 50 N acts on the lever arm for the first 0.1s).

Other motion analysis techniques

As technology has developed, other measurement techniques have become increasingly popular in sport and exercise biomechanics.

Electrogoniometry

Electrogoniometers measure angular position and displacement with respect to a reference, normally the longitudinal axis of a limb. An electrogoniometer generally has two 'arms', at the junction of which is an angular position sensor such as a potentiometer or digital decoder. Any relative movement of the arms and therefore rotation of the sensor results in a change of voltage that is directly proportional to the rotation. The analogue signal can be recorded to provide an angular displacement–time history. Normally the signal is converted to digital form, providing an instantaneous readout of the angle between the arms. Single and double differentiation of the data will produce angular velocities and accelerations.

The kinematics of hinge joints can be assessed using uniaxial electrogoniometers; biaxial and triaxial joints require biaxial and triaxial goniometers. Multiaxial goniometers are both more expensive and difficult to set up (Chao, 1980). Electrogoniometry is an appropriate technique when angular data only are needed. Examples of this include standardizing the initial knee angle in a vertical jump, or where subjects remain almost stationary relative to recording equipment as in treadmill running (Sheeran, 1980). Despite having to be attached to the subject, they have the advantage of being relatively cheap to buy and can be used on-line to provide immediate feedback (Yeadon and Challis, 2008).

TIME OUT

ISOKINETIC LEG STRENGTH PROFILING OF ELITE MALE BASKETBALL PLAYERS

Bradic *et al.* (2009) determined the isokinetic leg strength of elite male basketball players and evaluated the positional differences in their absolute and relative leg strength. Forty-three elite male basketball players (15 guards, 14 forwards and 14 centres) performed maximal isokinetic concentric knee extension and flexion efforts (60°/s and 180°/s) and ankle plantar flexion and dorsiflexion efforts (30°/s and 60°/s). Significant ($p < 0.01$) positional differences in absolute leg muscle strength were observed, with centres having significantly ($p < 0.05$) greater peak torques compared with guards in all of the tested muscle groups and at all angular velocities. Moreover, centres also possessed significantly ($p < 0.05$) stronger plantar flexors at 30°/s and dorsiflexors at 60°/s compared with forwards, and forwards possessed significantly ($p < 0.05$) stronger knee extensors, plantar flexors and dorsiflexors compared with guards. The results suggest that the positional differences in quadriceps and hamstring muscle strength of elite male basketball players are the result of respective differences in body size, whereas factors other than body size are responsible for the positional differences in ankle plantar flexor and dorsiflexor strength. These results together with the isokinetic leg strength data can be useful for coaches and therapists when designing and evaluating position-specific strength training programmes of elite male basketball players.

Accelerometry

Accelerometers detect accelerations by quantifying the forces that are applied (Yeadon and Challis, 2008), and normally consist of a mass attached to a fixed base, acceleration of which is detected and measured by the tension or compression of a distortion-sensitive element. Direct acceleration measurement using accelerometers in principle provides greater accuracy than that derived from double differentiation of displacement data (Ladin and Wu, 1991).

Four types of accelerometer exist, all of which produce an electrical output. **Strain gauge accelerometers** work by a change in resistance in strain-sensitive wires owing to an acceleration. The output is proportional to their resistance. **Piezoresistive accelerometers** work in a similar way to strain gauges except that piezoresistive elements are used, the resistance of which is proportional to stress. **Piezoelectric accelerometers** work by causing stress on a compressed piezoelectric material when accelerated. **Inductive accelerometers** use changes in the coupling in magnetic coils caused by an acceleration, resulting in a change in electrical output.

The first three types are small and light (1–2 g) and have a high natural frequency (2000–5000 Hz) combined with good frequency response characteristics. Inductive accelerometers are slightly heavier, with considerably lower natural frequencies, which makes them unsuitable for many biomechanical applications. Piezoelectric accelerometers are often the most appropriate for biomechanical studies because of their excellent range (0.01–10 000 times gravitational acceleration) (Nigg, 1994a).

Segmental acceleration data can be confounded by various factors, including differences between soft tissue and bone accelerations and incorrect location of the accelerometer. There are problems associated with mounting the accelerometers on the human body. For the measurement of bone acceleration with a skin-mounted accelerometer there should be minimal underlying tissue, in addition, there can be significant noise added to the signal due to soft tissue movement (Yeadon and Challis, 2008). The measured acceleration is that at the accelerometer-to-mounting junction, which may not correspond with the point of interest. Accelerometers are normally mounted directly onto the skin, which may well move relative to underlying tissue. Tightness of attachment also changes both the stiffness and damping which influence the measured acceleration (Nigg, 1994a). The drawback of three-dimensional accelerometers is their expense, which is not proportional to their axiality. Accelerometers used in biomechanical analysis are small and unobtrusive. However, accelerometry is the most difficult of all the measuring techniques used in sport and exercise biomechanics to use appropriately (Nigg, 1994a).

When reporting an accelerometry study the normal condition of replicability should be adhered to, with the following information included as a minimal requirement:

- Full details of the equipment used
- Full details of experimental procedures, including the method of mounting the accelerometers and their sampling frequency
- Full details of data processing.

Pressure measurement

Pressure measurement in sport and exercise biomechanics has become increasingly popular. The principles are identical to those for force platforms (i.e. it involves the distortion of a force-sensitive device), except that many sensors of known area are used. This produces a measure of pressure over the contact surface. While the measurement of reaction forces by force platforms is often useful in sports biomechanics, information about the local forces acting between the measuring device and parts of the contacting body may be of importance (Nigg, 1994a).

Current systems measure pressure either between the foot and ground (pressure platforms such as the Dynamic Pedobarograph and Musgrave Footprint) or between the foot and shoe (pressure insoles such as Tekscan and Pedar).

Pressure platforms are similar in appearance to force platforms. Their size is variable and they are normally placed directly onto the floor. It is important that their height is small to minimize kinematic and hence kinetic changes. Despite requiring the direct connection of online computer and appropriate circuitry such as an analogue-to-digital converter, the smaller platforms are portable. When the subject being studied is required to move across a platform, small platforms may be unsuitable because of targeting, which may alter natural technique. They are equally unsuitable when the subject's feet fail to remain on the platform and repeated stress (contact phases) are required. Other limitations include sensitivity of these systems to temperature, humidity, non-linear response, susceptibility to mechanical breakage, the need for trailing wires and the fact that the matrix insole may re-distribute pressure (Yeadon and Challis, 2008).

When reporting a pressure measurement study the normal conditions of replicability should be observed. The following should be included as a minimum requirement:

- Full equipment details
- Full details of experimental procedures including the sampling frequency and method of mounting (platform) or attachment (insole)
- Full details of data processing.

Summary

- In this chapter, many of the methods used to analyse and record data in sports biomechanics have been introduced.

- Biomechanical researchers have proposed models to apply the principles of mechanics to qualitative analysis of human movement.

- Many authors have made contributions to this area, however only a selected number were reviewed.

- The reader is introduced to qualitative models and quantitative techniques and their appropriateness is discussed.

- Quantitative techniques used to measure specific biomechanical parameters include force plates, EMG, cinematography, accelerometry and isokinetic dynamometers to name but a few.

- Some advantages, limitations and reporting methods have been covered.

Review Questions

1. What is the result – the measure of the outcome of the performance – for:

    a. A standing long jump

    b. A handspring vault in gymnastics

    c. A block in volleyball

    d. A giant slalom in skiing

    e. A javelin throw

    f. A ski jump

    g. A rugby place kick

    h. A Basketball free throw.

2. The result in a triple jump is the distance from the front edge of the takeoff board to the nearest mark that the performer makes in the landing pit. How might this distance be divided for the purposes of qualitative analysis?

3. Select a skill in which you are proficient or one in which you wish to become proficient.

    a. List the primary biomechanical principles that apply to the skill.

    b. List the fundamental components and subcomponents of the skill.

4. Complete the following qualitative model for running.

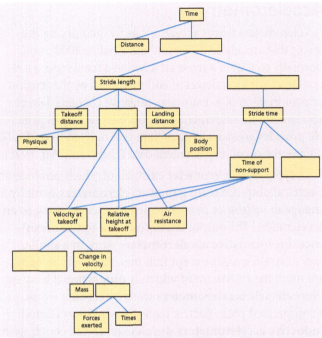

5. Why do you think cinematography and video analysis are important aids to the coach and his or her athlete?

6. List any benefits and limitations of two-dimensional and three-dimensional recording to analyse technique and performance.

7. What are the advantages and disadvantages of the two main types of force plate/platforms?

8. What are the recommendations for EMG electrode placement according to Clarys and Cabri (1993)?

9. Describe the uses of an isokinetic dynamometer in human movement analysis.

10. Describe the limitations on the use of accelerometry in sports biomechanics.

References

ADInstruments (2003) *Basics of data acquisition: Application Note*. ADInstruments, October.

Alexander MJL (1989) The relationship between muscle strength and sprint kinematics in elite sprinters. *Canadian Journal of Sports Sciences* 14: 148–157.

Baltzopoulos V, Brodie DA (1989) The development of a computer system for real time display and analysis of isokinetic data. *Clinical Biomechanics* 4: 118–120.

Bartlett RM (1992a) *Biomechanical Analysis of Performance in Sport*. Leeds: British Association of Sports Sciences.

Bartlett RM (1992b) Electromyography. In: Bartlett RM ed. *Biomechanical Analysis of Performance in Sport*. Leeds: British Association of Sports Sciences, pp. 28–37.

Bartlett RM (1997) *An Introduction to Sports Biomechanics*. London: E & FN Spon.

Bartlett RM, Challis JH, Yeadon MR (1992) Cine/video analysis. In: Bartlett RM ed. *Biomechanical Analysis of Performance in Sport*. Leeds: British Association of Sports Sciences, pp. 8–23.

Bobbert MF, Schamhardt HC (1990) Accuracy of determining the point of force application with piezoelectric force plates. *Journal of Biomechanics* 23: 705–710.

Bradic A, Bradic J, Pasalic, E and Markovic, G (2009) Isokinetic leg strength profile of elite male basketball players. *Journal of Strength and Conditioning Research* 23(4): 1332–1337.

Brown EW (1982) Visual evaluation techniques for skill analysis. *Journal of Physical Education, Recreation and Dance* 53: 21–26.

Chao EYS (1980) Justification of a triaxial goniometer for the measurement of joint rotation. *Journal of Biomechanics* 13: 989–1006.

Clarys JP, Cabri J (1993) Electromyography and the study of sports movements: a review. *Journal of Sports Sciences* 11: 379–448.

Cronin JB and Hansen KT (2005) Strength and power predictors of sports speed. *Journal of Strength and Conditioning Research* 19(2): 349–357.

Dainty DA, Gagnon M, Lagasse PP, Norman RW, Robertson G, Sprigings E (1987) Recommended procedures. In: Dainty DA, Norman RW eds. *Standardising Biomechanical Testing in Sport*. Champaign, IL: Human Kinetics, pp. 73–100.

De Luca CJ (1993) The use of surface electromyography in biomechanics. Delsys Wartenweiler Memorial Lecture. International Society of Biomechanics. 5 July.

De Luca CJ, Knaflitz M (1990) *Surface Electromyography: What's New?* Boston, MA: Neuromuscular Research Center.

Dowson MN, Nevill ME, Lakomy HK, Nevill AM and Hazeldine RJ (1998) Modelling the relationship between isokinetic muscle strength and sprint running performance. *Journal of Sports Sciences* 16: 257–265.

Ellenbecker TS and Roetert EP (1995) Concentric isokinetic quadriceps and hamstring strength in elite junior tennis players. *Isokinetic Exercise Science* 5: 3–6.

Grieve DW (1975) Electromyography. In: Grieve DW, Miller DI, Mitchelson D, Paul JP, Smith AJ eds. *Techniques for the Analysis of Human Movement*. London: Lepus Books, pp. 109–149.

Hay JG, Reid JG (1982) *The Anatomical and Mechanical Bases of Human Movement*. Englewood Cliffs, NJ: Prentice-Hall.

Hay JG, Reid JG (1988) *Anatomy, Mechanics and Human Motion*, 2nd edn. Englewood Cliffs, NJ: Prentice-Hall.

Hof AL (1984) EMG and muscle force: an introduction. *Human Movement Science* 3: 119–153.

Housh TJ, Thorland WG, Tharp GD, Johnson GO and Cisar CJ (1984) Isokinetic leg flexion and extension strength of elite adolescent female track and field athletes. *Research Quarterly for Exercise and Sport* 55: 347–350.

Hudson J (1985) POSSUM: purpose/observation system for studying and understanding movement. Paper presented at the AAHPERD National Convention, Atlanta, GA.

International Society of Electrophysiological Kinesiology (ISEK) (1980) *Units, Terms and Standards in the Reporting of EMG Research*. USA: ISEK.

Jonsson B (1973) Electromyographic kinesiology: aims and fields of use. In: Desmedt J ed. *New Developments in EMG and Clinical Neurophysiology*. Basel: Karger, pp. 498–501.

Kaufman KR, An K, Litchy WJ, Chao EYS (1991) Physiological prediction of muscle forces-11. Application of isokinetic exercise. *Neuroscience* 40: 793–804.

Kellis E, Baltzopoulos V (1995) Isokinetic eccentric exercise: a review. *Sports Medicine* 19: 202–222.

Knudson, D (2001) *Application of Biomechanics in Qualitative Analysis*. Biomechanics Symposis, University of San Francisco.

Knudson D, Morrison C (1996) An integrated qualitative analysis of overarm throwing. *Journal of Physical Education, Recreation and Dance* 67: 31–36.

Kramer H, Kuchler G, Brauer D (1972) Investigations of the potential distribution of activated skeletal muscles in man by means of surface electrodes. *Electromyography* 12: 19–27.

Ladin Z, Wu G (1991) Combining position and acceleration measurements for joint estimation. *Journal of Biomechanics* 24: 1173–1187.

Linthorne NP (2001) Analysis of standing vertical jumps using a force platform. *American Journal of Physics* 69: 1198–1204.

McGinnis, P.M (2005) *Biomechanics of Sport and Exercise,* 2nd edn. Human Kinetics.

McPherson M (1990) A systematic approach to skill analysis. *Sports* 11: 2.

McPherson MN (1996) Qualitative and quantitative analysis in sports. *The American Journal of Sports Medicine* 24: 6.

Newton, R and Kraemer, W (1994) Developing explosive muscular power: Implications for a mixed methods training strategy. *Journal of Strength and Conditioning Research*, 16(5), 20–31

Nigg BM (1994a) Acceleration. In: Nigg BM, Herzog W eds. *Biomechanics of the Musculo-Skeletal System*. Chichester: Wiley, pp. 237–253.

Okamoto T, Tsutsumi H, Gotto Y, Andrew PD (1987) A simple procedure to attenuate artefacts in surface electrodes by painlessly lowering skin impedance. *Electromyography and Clinical Neurophysiology* 27: 173–176.

Paoli A, Marcolin G and Petrone N (2009) The effect of stance width on the electromyographical activity of eight superficial thigh muscles during back squat with different bar loads, *Journal of Strength and Conditioning Research* 23(1): 246–250.

Sheeran TJ (1980) Electrogoniometric analysis of the knee and ankle of competitive swimmers. *Journal of Human Movement Studies* 6: 227–235.

Tan J, Kerwin DG, Yeadon MR (1995) Evaluation of the APEX digitisation system. In: Watkins J ed. *Proceedings of the Sports Biomechanics Section of BASES*. Leeds: British Association of Sports Sciences, pp. 5–8.

Watkins J (1987) Qualitative movement analysis. *British Journal of Physical Education* 4: 177–179.

Watkins, P.H and Sapstead, G (2010) Augmented eccentric loading and force and power production during the countermovement jump. *Journal of Sports Sciences* (abstract), 28(S1), S160.

Walsh MS, Ford KR, Bangen, KJ, Myer GD and Hewett TE (2006) The validation of a portable force plate for measuring force–time data during jumping and landing tasks. *Journal of Strength and Conditioning Research* 20(4): 730–734.

Yeadon, M.R and Challis, J.H (2008) The future of performance-related sports biomechanics research. *Journal of Sports Sciences* 12: 3–32.

Further reading

Baltzopoulos V (1992) Isokinetic dynamometry. In: Bartlett RM ed. *Biomechanical Analysis of Performance in Sport*. Leeds: British Association of Sports Sciences.

Basmajian JV, De Luca CJ (1985) *Muscles Alive: Their Functions Revealed by Electromyography*. Baltimore, MD: Williams and Wilkins.

Hay JG (1985) A system for the qualitative analysis of motor skill. In: Wood GA ed. *Collected Papers on Sports Biomechanics*. Nedlands, Western Australia: University of Western Australia Press.

Kreighbaum E, Barthels KM (1990) *Biomechanics: A qualitative Approach for Studying Human Movement*, 3rd edn. New York: Macmillan.

Nigg BM (1994b) Force. In: Nigg BM, Herzog W eds. *Biomechanics of the Human Musculoskeletal System*. Chichester: John Wiley.

III Psychology

14 Psychology of Sport

Chapter Objectives

In this chapter you will learn about the contribution that sport psychology has made to understanding:

- Performance anxiety
- Aggression in sport
- Audience effects
- Groups and teams
- Leadership
- Individual differences
- Mental imagery
- Motivation whilst participating in sport.

Introduction

Psychologists like definitions. This is because much of what we deal with can seem rather elusive. Questions like: 'Where is your mind?' 'What colour is it?' 'Where does it go when you are asleep?' are enough to upset anyone. In keeping with this approach we should ask the obvious question, What is **sport psychology**? If we can answer that then at least we can agree on what we are talking about. But here is the worry: sport psychology does not really exist. Before you become concerned, let me explain.

Psychology has been around for well over a hundred years and there is much we now know, and even more we do not know, about the mind and behaviour of human beings. Sport psychology takes some of the theories and applies them to what human beings do and think when they play sport and take exercise. I do not think there is a theory of sport psychology that is independent of 'mainstream psychology', so sport psychology is a form of applied psychology – it takes theories of psychology and applies them to sport. There is a good

reason for this and that is because sport is about human life – sport psychology is just a special version of it.

So if psychology is the scientific study of the mind and behaviour then sport psychology is the scientific study of the mind and behaviour of those participating in sport. What sport psychology does is to take what we know about psychology and apply it to sport.

Some interesting questions arise from the study of sport psychology:

- Why are performances often inconsistent?
- Why do some players rise to the occasion whereas others do not?
- Why do some people love sport whilst others hate it?
- Why is one player more successful than another despite being the same on paper?

And the best bit is that, armed with some of the answers to these questions, sport psychology can help to solve some of these problems and help to improve performance.

Performance anxiety

Most people who play sport experience apprehension before competition. A racing cyclist may be asking questions like 'Did I tighten up that wheel properly?', 'Are my water bottles full?', 'Can I beat my opponent?', 'If I crash will it hurt?' and so on. In practice, the cyclist knows that the answers to these questions are yes and once they get riding they generally stop worrying. These kinds of experiences are common in sport. As Hackforth and Spielberger (1989) points out, the idea that sports performance makes the individual 'nervous' or 'anxious' is hardly new. It is clearly an obvious thing for a sport psychologist to study, though. This led Martens (1977) to ask questions like 'What makes an athlete nervous?' and 'Why do some rise to the occasion whilst others buckle?' (Figure 14.1.)

Hackforth and Spielberger (1989) sees the problem thus:

Stressor → perception and appraisal of threat → anxiety

So there is something that is the source of the threat. We perceive this danger and this makes us anxious. In our case the:

Stressor = the sport + the competition
+ the fitness + the occasion

In other words the sum of the nature of the sport, how tough the competition is, how fit we are, and how big the occasion is. And the:

Threat = past experience + potential failure
+ all the things that can go wrong

Performance anxiety is always future orientated. That is, we are concerned about a potentially harmful event which is yet to happen; harm in this case being physical or psychological. These concerns are mediated by complex mental processes. The central issue is to do with how you manage your 'nerves'. If you can deal with them then they are not really a problem, but if you cannot then your performance will suffer and you will stop enjoying your sport.

Along with the anxiety (and indeed a fundamental part of it) comes a whole range of physical signs and symptoms due to the activation of the parasympathetic and sympathetic nervous systems. In the case of sports performance, anxious athletes experience the following symptoms:

- a feeling of being under scrutiny and likely to fail
- palpitations and rapid respiration
- sweaty palms
- dry mouth
- stammering

△ **Fig. 14.1** Anxiety before a sporting event is common and often needs addressing.

- inability to think clearly
- incontinence
- vomiting.

In some areas of the literature (and among some coaches and athletes) this experience is known as **choking**. This is a rather woolly term which many speak of but few have defined in a very helpful way. The experience of choking seems usually to be to do with a pattern of behaviour: things start to go wrong, performance continues to deteriorate, and the competitor seems unable to regain control. As well as losing control (or not being able to regain it), you experience some of the physical signs and symptoms listed above. However, whether we call it choking or performance anxiety, the key component seems to be a loss of attentional capacity. Instead of concentrating on the physical task in hand – like hitting a ball, you become distracted by your own worries and fears of failure. This stops you applying yourself to physical task and leads to decisional errors, poor timing and judgement and a general catastrophic failure. For example, Asafa Powell was at the 11th IAAF World Championships where he was running against Tyson Gay. Powell at the time was the fastest man in the world over 100 metres but had a history of not rising to the big occasion. On this occasion he said, 'I stumbled from the blocks, felt him coming on, then I started to panic and that slowed me down'.

Applying psychology theories and ideas to sport is often useful and there is a great deal of research on stress and anxiety in sport. A number of studies have been performed in specific sports, including football, basketball, badminton and athletics (e.g. Hanin, 1989; Hošek and Man 1980; Sanderson *et al.*, 1989).

Whether or not an athlete will experience this anxiety seems to be related to practice, physical activity, perceived or experienced success or failure, level of competition, situation factors and the skill and experience of the athlete. Hollingsworth (1975) reported that athletes feel less anxious in practice than in competition and also showed that more practice leads to lower levels of anxiety in competition (see also Gill, 1986). Milillo (1975) found high anxiety levels just prior to competition for marathon runners, tennis players and archers. He also demonstrated a positive relationship between the amount of strenuous motor activity required for a particular sport and the level of anxiety. Anxiety is highest for marathon runners and lowest for archers (tennis players are somewhere in the middle). Noyes (1971) found greater levels of anxiety in those who experience failure, as did Scanlan and Passer (1978) and Martens (1977). On the other hand, in women's badminton Sanderson and Ashton (1981) demonstrated a reduction in anxiety following success.

The problem is that these results only imply a correlational rather than a causal relationship, i.e. it was not shown that too little or too much arousal and stress impaired performance. This begs the question: does increased arousal just go with what happens, or is it a side effect? What is arousal? The Yerkes–Dodson law says that arousal is beneficial but only up to a point. As arousal increases so does performance. There is a point which is optimum (known as the **optimal arousal point (OAP)**): go below this and performance is not what it could be; go above it and performance drops off (Figure 14.2).

Clearly it would be useful to identify what makes one go beyond the OAP. Wilson (1994) asserts that the variables that contribute to this are trait anxiety, task mastery and situational stress. **Trait anxiety** refers to how anxious you

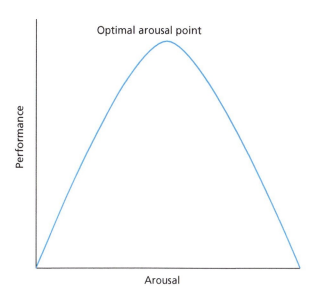

△ **Fig. 14.2** Optimal arousal point.

are as a person; **task mastery** is about how good you are at the task; and **situational stress** refers to the difficulty of the situation in which you are performing the task. If you are a nervous person, you are not very good at your given sport and there is a big crowd watching you, then Wilson's prediction is that you would become so aroused that your performance would suffer. In fact he would go further than this and say that rather than your performance dropping off in a nice curve (like the Yerkes–Dodson law) it would rapidly plummet (Figure 14.3). Hardy and Parfitt (1991) call this the **catastrophe theory** because the rate and amount of performance impairment is great and catastrophic.

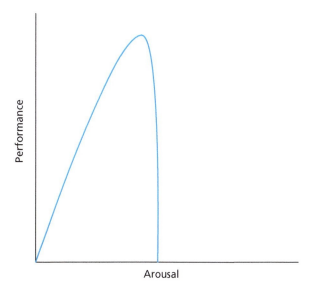

△ **Fig. 14.3** Catastrophe theory predicts excess arousal will lead to a collapse in performance for some individuals.

Of course if you are less nervous or have practised a great deal or have much experience of performing in front of a crowd then you'll be fine and your performance will not be affected so much.

However, there is an alternative view: (Apter, 1982) that says it is the *experience* of arousal that is important and not the *amount*. We can illustrate this with the scenario described in **Box 14.1**.

Box 14.1

Lion and sex

Imagine you are waiting at a bus stop and a lion walks out from behind the bushes. My guess is your heart rate would go up (let's say to 180 beats per minute because you are very scared). You wish the bus (any bus!) would come, the local zoo keeper would appear or the lion would remember a prior engagement. In short, you are scared.

On another occasion, I come round to where you live, let myself into your house and find you having sex. It is okay to be having sex with this person, so don't worry. Because I'm an experienced sport psychologist I can reach over and take your pulse without interfering with the proceedings and your heart rate is 180 beats per minute (and you are having a great time). You would not mind if this carried on and you could even live with having a higher heart rate.

Can you see that in both situations you have the same amount of arousal, or at least heart rate, but in the case of the lion it is an aversive and horrible experience and in the case of the sex quite the opposite. So it is not necessarily the amount of arousal that is the issue but rather the experience of the arousal. More or less arousal is not better or worse – it is the interpretation of the arousal that is the issue.

Furthermore, Hackforth and Spielberger (1989) point out that the inverted U does not include feelings of anxiety and is better at describing than explaining how anxiety affects performance. This, combined with the findings of Apter (1982), implies that the relationship between arousal and performance is more complicated than the Yerkes–Dodson law would suggest.

If you experience these often horrible signs and symptoms it seems only natural to seek help to alleviate them. This is often the job of a sport psychologist, who would usually start by trying to ascertain if these anxieties are rational. In other words, have you got grounds to worry? Are you properly prepared for competition or are you out of your depth? If you are well prepared, do you have the skills for the sport? If not, can you go away and learn them? If you have these problems, then the anxiety is justified and useful and you can have a long hard think about what you can do to fix yourself. Perhaps you could take on more/different training, compete at a lower level, try to decide if this sport is really for you and so on.

If, on the other hand, there is no obvious reason for you to be so anxious then there is a need to 'cure' the problem.

Interventions to 'cure' anxiety
Drug treatment

In non-sporting activities anxiolytics and tranquillizers (not to mention alcohol) are sometimes used to reduce anxiety, but because they impede general brain processes they tend to impair performance. Of course this is neither desirable nor permissible in sport. It is said by some (e.g. Wilson, 1994) that even coffee can have deleterious effects on performance because, whilst the performer feels alert, the audience regards them as jittery. Beta-blockers are more selective in the way in which they affect the nervous system and they are allowed in some sports (for example, for some players in snooker). They act on the visceral parts of the nervous system (the parts controlled by adrenaline) without directly affecting the higher mental processes. However, beta-blockers do have harmful side effects (e.g. nausea, tiredness, depression and sleep disturbance) and they are banned in most sports, so perhaps psychological interventions are better.

Psychological treatments

Methods originally developed to treat phobias are often used to counter anxiety in sportsmen and women. For example, psychologists often use different types of **de-conditioning**. The basic premise is that because the fear has been learnt you should be able to unlearn it. One technique used in de-conditioning is **flooding**. This is an old (and some would say barbaric) way of dealing with phobias which can be (and sometimes is) used in sport. You have probably seen this kind of thing on TV. One takes someone who is scared of spiders, lock the person up in a room full of the little blighters and wait for them to stop screaming. After a while they realize that nothing much happens and with a bit more time they unlearn their fear. You can probably see how you might apply this to sport. It seems to me that this has the potential to be a very traumatic 'cure' and not one that most people would readily sign up for. But it has the advantage of being quick and that is probably why it is still sometimes used.

A much gentler, although much slower intervention, is that of **systematic desensitization**. Let's say that an athlete always feels very anxious just before a sprint race and this anxiety lingers for the first few metres of the race. This inevitably affects the athlete's performance. The first step in controlling this anxiety is to learn **progressive muscular relaxation (PMR)** as described in **Box 14.2**.

Box 14.2

Progressive muscle relaxation

The aim of progressive muscle relaxation is (obviously) to become more relaxed. It is also to learn to control your anxieties.

First get yourself into a comfortable position and start thinking about your feet. Scrunch up the muscles in your feet and then relax them – take about 10 seconds to scrunch them up and then the same to relax. Do this several times. Once your feet are relaxed, do the same thing with your calf muscles. Then with your thigh muscles. The idea is that you work up and around your body, right up to your head, until you feel totally relaxed. Once you have done this a few times you will find that you can do it easily.

For it to be a truly useful technique you need to practise progressive muscle relaxation in a variety of settings so that you can 'turn it on and use it' whenever you need to.

At first, even when the athlete simply thinks about the period of time just before the race, it is enough to make them feel anxious. The use of PMR will enable the sprinter to gain 'control' over these fears. Once the feelings of anxiety subside then the athlete can think about the start of the race again – and this time the feelings are not so strong and the PMR quickly sees them off. After several repeats the athlete can think about this period of time without feeling anxious at all. Eventually the athlete can think through the whole race without feeling troubled – then they move to looking at videos, then actually going to the start of a race (without competing) and so on until they are fine again.

The problem with de-conditioning is that it is time-consuming and it does not always work.

It seems surprising how many athletes do not learn even basic psychological skills to help them with their sport, and whilst athletes are very self-aware when it come to their bodies, their personal psychological life is often a mystery to them. One of the keys skills of dealing with performance anxiety is realizing when one feels anxious, and people often need help to do this. A good way to do this is to use **biofeedback**. This is a form of self-monitoring which enables one both to identify when one is feeling anxious and to monitor how successful one is at being at achieving relaxation. In a laboratory, various equipment can be used and typical techniques available are muscle feedback (EMG), thermal feedback, electro-dermal feedback (e.g. GSR), cardiovascular feedback (ECG) and electroencephalogram feedback (EEG). Most of these are not suitable for use 'in the field' or in the stadium, however, and most of the equipment associated with these techniques is really expensive. So in practice we use heart rate monitors. Whilst these are not always the best tool for the job – because not everyone's heart rate goes up when they feel anxious – they are cheap, mostly effective and many athletes are accustomed to using them. An athlete about to compete in a race can keep an eye on their heart rate monitor, notice when their heart rate seems to be increasing and then use something like progressive muscle relaxation to reduce it. The athlete can then follow their progress by watching their heart rate monitor.

Another useful technique often used by athletes is **meditation**. This usually involves the adoption of a yoga-like position and concentrating on and repeating a mantra, which is a non-threatening set word or saying. An example of a technique suggested by Benson (1976) is shown in **Box 14.3**. One problem with meditation is that not everyone can do it. Some people are also put off by the idea that it is all a bit 'hippy trippy'. Having said that, many well-known athletes swear by it.

Box 14.3

Meditation: Six steps to relaxation as recommended by Benson (1976)

1. Get into a relaxed position and close your eyes.

2. Concentrate on your breathing and try to slow it down. Breathe out for longer than you breathe in and try to make the breathing out as pleasurable as possible.

3. Try to feel the position you are in and feel each part of you being supported by the chair or bed so that you can relax your muscles further.

4. Search your body for signs of tension; focus on this and as you breathe out try to relax it away.

5. Keep breathing through your nose and begin to say or think the word 'one' to yourself. Keep doing this for five to ten minutes. If you get distracted then go back and continue repeating it. After a few sessions you might want to go back and change the word (e.g. to 'calm').

6. When you are ready to end, open your eyes and sit up slowly. Take one or two deep, slow breaths.

A specialized technique for reducing performance anxiety is **hypnotic suggestion**. Hypnotism can work really well for many people and is a well-respected and long-established discipline. It is a shame that stage hypnotists have made many people believe either that hypnotism is a party trick used to humiliate or that it is somehow fake. For many people their fears are not conscious: although they know they are scared they often do not know why. Hypnotism can often uncover these feelings and there are many techniques that can be used to place a suggestion in the unconscious that would be useful for those who find themselves fearful. Anyone interested in pursuing this option further should find a professional registered hypnotist.

A common way to address performance anxiety is to use **cognitive orientation techniques.** Central to this is the idea that there is a limit to the amount of attention one has to allocate during a performance (Wolverton and Salmon, 1991). An athlete divides their attention between themselves, the crowd and the activity. Cognitive therapies train the athlete to concentrate on the sporting activity at hand (often called 'getting focused') since during a competition they cannot change themselves or the audience. Many athletes seem very possessed by the notion that their performance has to be flawless, and this tends to make them feel anxious. Cognitive therapies aim to get athletes orientated to the idea that not all mistakes are a disaster and that many top players make mistakes. They just need to make fewer mistakes than their opponents and to forget about them when they happen.

Of course one of the worst things about performance anxiety is feeling anxious about feeling anxious. Inevitably this makes it more likely to happen. The last technique to consider is **stress inoculation**, which exists to deal with exactly this problem. In many ways this is fundamental to all performance anxiety-reducing techniques. It grew out of work that was trying to persuade cardiac patients that they need to do more exercise. Of course when you exert yourself – even if only a little – your heart rate increases. This often scares cardiac patients, who are understandably reluctant to do anything which (as they see it) puts strain on their heart. One of my colleagues (a physiologist and a very competent runner) was amazed to learn that some people associate high heart rate with anxiety. He just thinks that a high heart rate goes with exertion. This surprised me because I always thought that one of the horrible things about running fast is that it makes your heart feel like it is trying to bash its way out of your chest, which is a feeling I always associate with being scared (I'm no runner!).

In essence, stress inoculation is about learning to recognize anxiety symptoms for what they are (i.e. to some

extent normal) and to not worry about them too much (Meichenbaum, 1985). One tries to internalize the idea that it is pretty normal to feel a bit nervous before a race, for example, and it is not necessarily inevitable that these 'nerves' will develop into a full-blown panic attack. As pointed out earlier this is a fundamental point; it underlies most of the other interventions and is often used with various relaxation techniques.

Aggression in sport

There has long been a lively debate about aggression in sport. The arguments seem to focus on whether there is too much or not enough aggression in sport, or whether sport would be better with or without aggression. This is still a controversial topic today (Kerr, 1999a & 1999b, 2002; Tenenbaum *et al.*, 1997, 2000). I like setting my students an essay entitled 'Aggression in sport is inevitable, discuss' because there are just so many different ways they can answer it – and it is good fun to mark!

At the start of this chapter we asserted that psychologists are always keen to define things so that they know what they are talking about. In the area of aggression in sport this concern could not be more important. There are many different definitions and it is important to distinguish, for example, between instrumental aggression, reactive aggression, hostile aggression and assertiveness.

- **Hostile aggression** is the intent to harm. For example you foul the player because you want to hurt them.
- **Instrumental aggression** is the intent to harm to gain an advantage. For example, you foul the player so you can stop a run at goal.

Clearly both of these are outside the rules of sport and there are rules and punishments that go with committing these misdemeanours. It is interesting to think of boxing, which is a sport founded on instrumental aggression. To this extent boxing is an example of a special case: instrumental aggression is the norm, but other types of aggression – especially those that are outside of the rules – are not acceptable (e.g. headbutting, hitting below the belt, hitting with elbows, biting your opponent's ear, etc.).

- **Reactive aggression** is a reaction to the behaviour (verbal or physical) of others (Moyer, 1976). An opponent says or does something you don't like and you react to it in a way that intends harm.
- **Assertiveness** is where goal-centred behaviour is pursued but there is no intent to harm and this behaviour is within the rules.

The problem with these definitions is that trainers, commentators and the media tend to use these four synonymously and they are probably wrong to do so. Clearly assertiveness is the most desirable.

Why is aggression such an issue in sport?

One reason for the presence of aggression as an issue in sport is the common belief (e.g. Parsons, 1951; Lorenz, 1963) that aggression is an instinct, so it is likely to be a natural part of many human activities. As we pointed out earlier, sport is just a special version of life, so the instinct of aggression is bound to be present, especially in view of the fact that sport is meant to be a socially acceptable means of releasing aggressive energy. Sport is also meant to be a cathartic experience – that is, it reduces aggression in the participant or spectator, either directly by taking part in the sport or vicariously by watching it.

There is not much evidence in the literature though that watching sport reduces aggression in the watcher. Turner (1970) observed an increase in aggression in spectators after watching basketball or American football but a decrease after watching wrestling. Kingsmore (1970) reported similar results. However, Zillman *et al.* (1972), who performed a meta-analysis of this evidence, concluded that on the whole aggression breeds aggression, especially in sport.

There are two fundamental problems with sport which bring about aggression, **reinforcement** and **frustration**:

1. **Aggression** is **reinforced** or rewarded in many ways. For example it may be rewarded with a favourable outcome such as scoring a goal or winning the match, gaining approval by the coach, by other team members or by the supporters, all of whom may praise the player for being a good, aggressive player. The opposition may also unwittingly reinforce the behaviour by being wary, scared or deferential. The golden rule of behaviourism is, of course, that if the behaviour is rewarded then the likelihood that it will be repeated is increased. If this interplay is noticed by others it can be vicariously learned and is then often adopted by them. This leads to concerns about crowd behaviour or the behaviour of children who witness aggression.

It is perhaps paradoxical that the more physical the sport is supposed to be, the less it seems that aggression spills over into the crowd. For example, rugby and American football are undoubtedly physical sports but they are characterized by far less crowd trouble than certain other less physical sports. According to Voigt

(1982) this is due to the norms and values associated with the game. Ice hockey seems to be some kind of a special case though (Smith, 1975). Whilst aggression and violence is common in ice hockey there seem to be strict, largely unwritten rules about what is and is not acceptable. For example, skaters never chop with

△ **Fig. 14.4** Ice hockey often spills into aggression.

their sticks or kick with their skates, but they do fight one another and, given the chance, the crowd seems disposed to join in (Figure 14.4).

2. **Frustration** leads to aggression. If goal-orientated behaviour is blocked, especially by intentional, arbitrary, unfair and non-legitimate behaviour, it almost inevitably leads to aggression (e.g. Zimbardo, 1990). Many competitive sports (especially team sports) are organized so that the ability to score points is blocked by the opposing team. So the goal (e.g. scoring the point) is frustrated by one's opponent. Thus,

Blocking → frustration → aggression

It is not very surprising, then, that aggression is often a feature of these types of sports.

Of course not every sport has aggression as a feature and even if (like football) aggression is far from unknown it does not happen in every game. So what are the mediators of aggression in sport? Here are four contenders:

1. **Arousal and excitement** – Geen and O'Neil (1969) point out that:

Increased excitement → arousal → increased aggression

So the more excited and aroused you are the more likely you are to become aggressive. Also the more excited you get the more frustrating blocking becomes and this of course leads to a greater likelihood of aggression.

2. **Fitness** – Zillmann (1979) points out that

Increased fitness → decreased arousal → lower aggression

So the fitter you are the less likely you are to be aggressive, and vice versa.

3. **Performance** – Both Silva (1980) and Ryan (1970) have demonstrated that

Increased aggression → decreased performance

This is because individuals become more interested in acts of aggression than in the task they are there to perform. Also, in a sort of backwards version of Geen and O'Neil's (1969) formula:

Increased aggression → increased arousal → decreased performance

So a decrease in aggression should lead to an increase or improvement in performance.

4. **Type of sport** – Voigt (1982) points out that aggression is not a feature of all sports. For example individual field sports, volley ball, drag racing and weightlifting are not normally associated with aggression, whereas soccer, rugby and ice hockey often are. This is usually due to the nature of the sport and the likelihood of coming into contact with other opposing competitors.

There are also things that happen in a game that make aggression more or less likely to occur. Volkamer (1971), who studied soccer, noticed that losers commit more fouls (probably due to frustration), that visitors commit more fouls than the home side (may get goaded by the crowd – see section on social facilitation), that the closer the game the lower the level of aggression and the fewer fouls are committed (players become more cautious and don't want to give away a free kick or a penalty), that in a game where there are more goals there is less aggression (the opposite is also true), and, finally, that lower ranked teams commit more fouls.

Social facilitation

A very straightforward and simple sounding idea is studied here: that performance is affected by the presence of an audience. In fact, this idea has been a thorn in the side of social psychology for the best part of a hundred years.

The first experiments were carried out by Triplett in 1897. These were the first sports psychology studies and, some say, the first social psychology experiments.

A keen follower of cycling, Triplett noticed that the records were set when there was an audience and when there were fellow cyclists to compete against. He called this 'release of competitive energy' by the very wonderful name of **dynamogenesis**. The study of how audiences affect performance or **social facilitation**, as it became known, continued for many years and the literature became a right old mess. There was evidence to show that audiences increased, impaired and made no difference to performance! Then in 1965 Zajonc appeared in the literature. His theory was published in *Science* (a prestigious journal) and his theory was very much respected. It stated that:

1. Audiences increase arousal.
2. This arousal inhibits learning new responses.
3. This arousal facilitates the performance of well-rehearsed responses.

In other words, when we are being watched it might be harder to learn new things but we can 'show off' the things which we can do well in front of an audience.

Zajonc used the **Hull–Spence drive model** to explain his observations. This states that the capacity for a particular stimulus to elicit a particular response is governed by the existing potential for that response, which is the multiplicative product of habit strength and drive level. Thus an increase in drive level will tend to enhance the level of emission of dominant responses (i.e. those with high habit strengths). This idea is based on the assumption that well-learned responses are dominant, that new responses to be learned are subordinate and that audiences act to raise the subject's general drive level.

This is best explained by an example. If you are used to driving in England on the left-hand side of the road in a right-hand drive car, then hiring a car when abroad is often a bit of a nightmare. You have to get used to driving on the other side of the road in a car that has the gear stick and inside mirror in the 'wrong' place. After a while you get used to this and all is well until there is some kind of emergency where you have to react quickly. This is a high arousal situation. All too often you find yourself changing gear with the window winder and looking at the screen pillar for the mirror! So the dominant response is 'emitted' or takes over (this is the one you have done a 'million' times before, i.e. gear lever on left, mirror on left) and the new response (mirror and gear lever on right) is subordinate to it.

Zajonc stated that the presence of an audience will cause increased emissions of dominant responses (such as the performance of dominant tasks) and the impairment of subordinate tasks (such as the learning of novel tasks). He further stated that his theory explained the various inconsistencies in the pre-1965 literature. At the time this was a very popular notion but, increasingly, the theory received criticism.

In 1982 Glaser published some searing criticisms of Zajonc, pointing out seven problems with the theory:

1. It is hard to demonstrate that audiences arouse drive.

2. It is hard to replicate Zajonc's findings. If you design an experiment where subjects perform a novel task, then according to Zajonc their performances should be worse in front of an audience than unobserved (they should be better, though, if the task is well learnt). It seems quite hard to get these experiments to 'work'.

3. It is hard to find a physiological correlate of arousal and social facilitation. If audiences increase arousal then there should be some kind of physiological measure we can use as a marker for this, for example heart rate. The trouble is this just does not seem to work.

4. The notion of mere presence is more complicated than would at first appear. I think this is a less convincing criticism. I think that all Zajonc was saying is that just having someone there is enough to produce an effect. But Glaser is right to point out that more people, people you know well, people with high status, people who stare or smell may affect different people differently.

5. The Hull–Spence drive model is generally considered to be problematic in itself and this is true although we do have the benefit of hindsight – which was something that Zajonc did not have.

6. Zajonc's theory does not in fact explain the pre-1965 literature. This is also true although it does explain quite a lot of it.

7. The 'seductive' appeal of his theory inhibited the reporting, publication and citation of studies failing to provide support for it. This may be true although this is a common problem when people criticize a very popular theory.

There are other criticisms of social facilitation in general. For example the nature of the 'alone' condition in social facilitation experiments. If one asks a person to perform a task whilst alone they probably think that their performance is being monitored in some way (otherwise why ask them to perform the experiment?). It has been found that if you ask subjects in an alone condition if they feel they were monitored they usually say 'Yes', and if you ask them 'How?' they start talking about hidden cameras and spyholes and the like. In other words, in the absence of an obvious monitor they make one up – they infer monitoring (Griffin and Kent, 1998; Griffin, 2001). This makes the alone condition a poor control and begins to give us an idea of why some of the results of some of the experiments seem, at best, a bit odd.

Since the publication of Zajonc's theory there have been several other theories of social facilitation. They are far less ambitious and deal with more specific instances of audiences affecting performance. Although they do not provide us with one overarching meta-theory of social facilitation, these newer contributions tend to be easier to apply to sport.

Alternative theories of social facilitation

Evaluation apprehension (Geen and Gange, 1983)

This theory states that audience effects are due to the socially learned expectation of the evaluation of others.

- **Relevance to sport** Evaluation apprehension explains how audiences, peers, competition and big events affect performance. Because players know that every 90 minutes of a football match is evaluated both during and for many hours after the match, they feel troubled and this interferes with their performance. This is especially true if a player makes a mistake.

Distraction conflict (Sanders, 1981; Thibaut and Kelly, 1986)

The presence of others is arousing because they are a distraction. Distraction has drive-like properties in that one is 'driven' to resolve the 'distraction'; usually by paying attention to it thus leaving less attention for the sport.

- **Relevance to sport** Distraction conflict explains how a hostile crowd or a big event can affect performance. Athletes are 'put off' by the behaviour of the crowd.

Attentional overload (Manstead and Semin, 1980)

The presence of others serves to increase attention on a simple task but leads to attentional overload for a complex or difficult task. So if one believes that we have a finite amount of attention. A demanding task exceeds an attentional capacity, and our performance suffers.

- **Relevance to sport** Attentional overload explains 'rising to the occasion' and 'failing under pressure'.

Self-attention (Duval and Wicklund, 1972; Carver and Sheier, 1982, 1988)

Self-awareness increases arousal, and self-awareness can be increased by the presence of an audience. Both these sets of researchers carried out really neat experiments which demonstrated that this was true. They both used a mirror. We all know that mirrors increase our self-awareness (that is what they are for). They devised an experiment in which a task was performed either alone, in front of an audience, or in front of a mirror. Given that mirrors increase self-awareness, if the effects of an audience on performance are the same as those from a mirror then it is probably fair to say that an audience increases self-awareness. That was just what they found.

- **Relevance to sport** Self-attention seems related to confidence and performance anxiety.

Self-presentation (Baumeister, 1984)

Audience effects are caused by the 'performer's' concern to make a favourable impression. This seems very true. For example I cannot believe that lipstick makes you run faster, but during the Olympics many women athletes wear lipstick in competition. They know they will be seen by many and I guess they like to feel that they look good. If they feel good they will probably run well (if it was just the lipstick then I guess the men would wear it too!). So if we are in front of an audience we want to look good, not bad.

- **Relevance to sport** Most sports performers are trying to do well – obviously.

Social inhibition (Berger *et al.,* 1981, 1982)

The presence of others inhibits overt practice.

- **Relevance to sport** Social inhibition is relevant if the above happens, but in recent years it has become less common. It is quite usual to see sportsmen and women overtly practising their sport beforehand.

Social impact theory (Latané, 1981)

This is a useful theory because it looks at audience size. The idea is that the bigger audience means more evaluation. If you are part of a team then that impact is shared out in the team so the impact on each individual is smaller.

- **Relevance to sport** Social impact explains the differences in the effect of an audience on team and individual sports and the impact of the size of the audience on performers.

Groups and teams

Let us define what we mean by a team:

- 'A collection of individuals who have relations to one another that make them interdependent to some degree.' (Cartwright and Zander, 1968).
- 'A team must have a shared sense of purpose, structured patterns of interaction, interpersonal attraction, personal interdependence, and a collective identity.' (Miller, 1992).

What happens in a group or team? According to Baron and Byrne (1989) the answer to this can be broken down into three parts: roles, norms and cohesiveness. A **role** is determined by which tasks are performed by each member. So, in a family it might be father, gardener, fixer of broken toys, lover and so on. In a team it might be captain, defender, playmaker, etc. These roles are performed in the context of the **norm** for that group. These norms are tied in some way to rules about what is and is not okay. So in a family it might be to do with being respectful to one another, not playing on Dad's motorcycles, not kicking your ball at Mom's greenhouse and so on. In a team it might be to do with not rubbishing others' efforts, listening to the coach, not being rude to the mascot, etc. What determines whether or not these roles and norms 'work' is **cohesiveness**. This is a force that causes group members to want to stay and can encompass things like how attracted one is to other members of the group and how much it would cost to leave. In high-cohesive groups there is an increase in the desire of members to socialize with one another and this often seems to go with an improvement in task performance – that is, the team does well and becomes more successful. In a low-cohesive group the group just cannot get on with one another and the tendency is for members to leave – usually leading to task performance being reduced.

There are a number of factors that determine the way in which a group performs a task. According to Steiner (1972) it is a balance between task demands (i.e. the resources that are required to complete the task) and human resources (i.e. all the relevant knowledge, skills, abilities and tools that members of the group possess).

So this boils down to something like: Given that we are playing hockey on Saturday what do we need to win? and have we got the players to make this way of winning work? The potential productivity of the group is a function of the type and number of the requisite skills that the group possesses. The more complete this collection the more potential the group has for success.

I expect that, like me, you have sat with your friends and tried to work out the 'dream team' for a particular sport, taking all the best players from any point in history and trying to construct a completely unbeatable team. It is good fun but it probably would not work because getting this dream team to all pull together is not a trivial matter.

We have an equation to try to understand this a bit better:

$$\text{Actual productivity} = \text{Potential productivity} - \text{Process losses}$$

These **process losses** 'get in the way', are many and varied and may be due to intra- and inter-group processes, leadership, communication, consensus and democracy, the presence of others, etc.

Let's pick **consensus** out of this list and look at it more closely. Consensus is about getting almost everyone to agree a strategy, for example. But the problem is that reaching it takes time, especially if the 'correct' answer is not apparent to everyone (Thomas and Fink, 1961). It also ignores minority opinion (Maier and Solem, 1952). This is a great shame because, by definition, minority opinion brings new and interesting ideas. Consensus is often a bit safe and conservative and consequently rather grey and stagnant. So, in terms of process loss, consensus can be slow (i.e. inefficient) and loses some of the richness of potential.

Another good example of process losses is seen in the Ringelmann effect (1913). This says that the bigger the group the less each individual achieves. In its simplest form this idea is best understood by a tug-of-war competition. The argument is that if you pull as part of a team you do not pull as hard as if you are pulling alone. For example, an average man pulls with a force of 63 kg but two men only pull with a force of 118 kg (losing 8 kg), three with 168 kg (losing 29 kg) and eight men lose 256 kg. Ringelmann accounts for this loss in terms of faulty process and coordination losses. A more psychological argument is put forward by Stroebe and Frey (1982) who posit the notion that we are also seeing motivational losses – that is, team members let the others do the work. This is easy to see in this example because an individual input is hard to identify but each individual shares in the group product. It is also known as social loafing (Latané *et al.*, 1979) (Figure 14.5).

△ **Fig. 14.5** Tug of war is a great example of the Ringelmann effect.

Zaccaro (1984) points out that by manipulating the nature of the task by making it more or less pleasant, then social loafing could be decreased or increased. The Ringelmann effect can be made to disappear if team members believe that their individual performance can be evaluated.

Earlier in this section we suggested that cohesive teams 'got on' better and tended to socialize more. There has been quite a lot of work in this area. An early piece of research was performed by Yaffe (1975), who studied two football teams. He asked members of two teams from Budapest 'Who is your best friend in the team?' and 'Who can carry the game?' (i.e. Who gets the spirits up?). He turned these into **sociograms**, which are a way of plotting affiliations between individuals. You get a line for each instance of affiliation. The resulting **sociograms** are shown in Figure 14.6. Remember, the more lines there are, the more affiliations there are between team members. Ferencvaros, who were doing well in the league, have many positive relationships and these seem shared amongst most of the team. MTK have several solitary players, which implies a lack of cohesiveness and a poor level of functioning. This team was struggling to escape Division One relegation.

(a) Ferencvaros

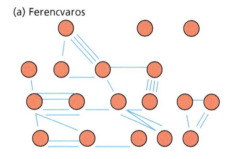

(b) MTK

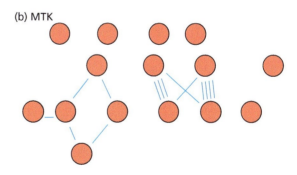

△ **Fig. 14.6** Sociograms of these two teams. More lines equal more affiliation.

As well as affiliation, cohesiveness and functioning were measured by recording who passed the ball and to whom. Players who were 'friends' passed between one another the most and much less to those with whom they were less friendly, did not like or did not know. This was despite the fact that on innumerable occasions it would have been better to pass to these other people.

There seem to be two consequences of this. The first seems obvious – the promotion of friendships within a team is a good idea. The second is to use this knowledge as a training opportunity. If one explains the principle and shows video evidence of this phenomenon to the team it should be possible to change patterns of play, improve cohesiveness and increase team effectiveness.

There are other differences between successful and unsuccessful teams. Veit (1968) shows that teams who are the most effective can agree on who is the best player in a particular situation and have an effective playmaker whom the team can all identify. In contrast, ineffective teams cannot agree who is the best and cannot agree on or cannot identify who is the effective playmaker.

To make things worse, Essing (1970) and Veit (1968) point out that 'bad' teams also show numerous and frequent changes in the team whereas 'good' teams show few and irregular changes in the team. Essing (1970) also points out that team stability is conducive to the development of mutual knowledge and the anticipation of the actions of other team members, and cites a strong positive correlation of 0.62 between stability and success.

I think all of this is pretty interesting. A colleague, Kirsten Spencer, and I have looked at similar issues in the women's national hockey league and have come to similar conclusions about affinity and success. However, not everyone would agree that this approach is very useful. The trouble with it is that it does not address the degree of interaction or the meaning that the interaction might have for the player. One solution is to ask the player who he or she would most like to interact with during a match; for example with someone who is fast/ slow, strong/weak, leader or follower. The nature of the preferred team members can be much more precisely determined and thus preferences and performances can be compared. Also, if we can identify the interactions occurring in the game, it enables a check to see if the coach's instructions about interactions actually happen on the field. Of course, the same technique can be used to assess the opposition, which enables them to be played more effectively the next time. These measures can be classified into activities done to or by the player who are successful or unsuccessful.

Leadership

And once again here comes the definition:

> The behavioural process of influencing individuals and groups toward set goals (Barrow, 1977).

There is a big difference between a leader and a manager. A **manager** takes care of such things as scheduling, budgeting and organizing, whereas a **leader** provides vision and is more concerned with the direction of an organization, including its goals and objectives, usually having a view of how great the organization or team can be. There are also differences in how leaders are chosen. There are those who are appointed, usually by some authority, to a leadership position (e.g. health club manager, coach, head athletic trainer), and there are those who emerge from a group and take charge (e.g. captain of an intramural team, exercise class, student leader). So our two categories are: **appointed** or **emergent** leaders.

The function of a leader is to ensure that the group meets its goals and objectives whilst making sure that the group needs are satisfied. As ever in psychology there is more than one approach to studying leadership, and in sport these divide into the trait approach, the behavioural approach and the interactional approach.

The **trait approach** asks the question: What personality characteristics are common to great leaders? The opposite view of this is that leaders have a variety of personality characteristics and there is no particular set of personality traits that make a leader successful.

The **behavioural approach** asks what are the universal behaviours (not traits) of effective leaders? If you can identify these behaviours then you might be able to learn or teach them. People who make a living out of teaching leaders often take this view, especially those who teach coaching. Typical ideas are that effective coaches focus on the positive while providing clear feedback and technical instruction, that facilitating positive coaching behaviours (frequent use of reinforcement and mistake-contingent encouragement) assures greater enjoyment, higher self-esteem, and lower drop-out rates in young athletes.

The Coaching Behaviour Assessment System (CBAS) has two categories of behaviour: reinforcement and spontaneous behaviours. Common examples of **reinforcement behaviours** are:

- Mistake-contingent encouragement ('Yes that was a mistake but it was a good effort and you were nearly there')
- Mistake-contingent technical instruction ('Yes that was a mistake. Don't do it like that next time, try this instead')
- Punishment (pretty obvious I think)
- Punitive technical instruction ('If you do it like that next time then you'll owe me for a new racket')
- Ignoring mistakes (obvious and very powerful)
- Keeping control (obvious and very important in many groups – especially when there is danger).

Examples of **spontaneous behaviours** are reasonably straightforward and include:

- General technical instruction
- General encouragement
- Organization
- General communication.

On the basis of 25 years' of research, Smith and Smoll (2002) provide eight guidelines for coaching young athletes:

1. Do provide reinforcement immediately after positive behaviours and reinforce effort as much as results.

2. Do give encouragement and corrective instruction immediately after mistakes. Emphasize what the athlete did well, not what the person did poorly.

3. Don't punish after athletes make a mistake. Fear of failure is reduced if you work to reduce fear of punishment.

4. Don't give corrective feedback in a hostile, demeaning or harsh manner, as this is likely to increase frustration and build resentment.

5. Do maintain order by establishing clear expectations. Use positive reinforcement to strengthen the correct behaviours rather than punishment of incorrect behaviours.

6. Don't get into the position of having to constantly nag or threaten athletes to prevent chaos.

7. Do use encouragement selectively so that it is meaningful. Encourage effort but don't demand results.

8. Do provide technical instruction in a clear, concise manner and demonstrate how to perform the skill whenever possible.

The **interactional approach** takes the view that both person and situation factors must be jointly considered to understand effective leadership; the implication of this is that no one set of characteristics ensures successful leaders (but characteristics are important). Another important principle is that effective leader styles or behaviours fit the specific situation. What follows from this is that leadership styles can be changed. Two types of leadership styles – relationship- and task-oriented leaders – are compared. A **relationship-oriented leader** focuses on developing and maintaining good interpersonal relationships; a **task-oriented leader** focuses on setting goals and getting the job done. The effectiveness of an individual's leadership style stems from its 'matching' the situation. Task-oriented leaders are effective in very favourable or unfavourable situations. Relationship-oriented leaders are effective in moderately favourable situations.

How this might work was explored by Cooper and Payne (1972), who looked at effective leadership among soccer players. They identified three characteristics for leadership: **task orientation** – how interested the player is in getting the job done; **self-orientation** – how much the individual requires personal rewards regardless of the job; and **interaction orientation** – the extent to which 'keeping everyone happy and harmonious' is important even if it interferes with the task at hand.

Cooper and Payne found that team success was positively correlated with high task orientation scores in coaches and trainers. The most successful teams had more players who were high on self-orientation (but low on the other two) and attacking players were more self-oriented than defenders.

Individual differences

Of course not all people are alike but to what extent are the differences between them important and worth studying in the context of sport? In mainstream psychology the study of individual differences concerns itself with intelligence tests and personality. In sport psychology it is mostly to do with personality. The idea that there might be sporting or non-sporting types of people is a very seductive sounding one as the notion that some kinds of people excel at one sport over another sounds quite plausible. As ever in psychology though, not everyone agrees that this is much of an idea. The two sides roughly divide up between those who think that personality is an important part of individual and team sporting success – we sometimes call these the **credulous group** – and those who believe that personality traits make little or no useful impact on the study of sport – the **sceptical group**.

To make things worse, there are many approaches to the study of individual characteristics, but broadly we can put them into two categories: trait theory and state theory. So you can be credulous or sceptical about both or either.

Trait theory states that people display enduring personality characteristics, like idleness, sportiness, bookishness, being cheerful or miserable and so on. Cattell (1965) thought that personality could be categorized into 16 personality factors (16PF) and Eysenck (1967) thought that people could be categorized as **introverted** or **extroverted**. He devised a measurement of this known as the EPI (Eysenck Personality Inventory). Eysenck thought that extroverts were thrill-seekers, likely to be attracted to dangerous or extreme sports. But of course not all people experience this in the same way. Many participants in so-called high-risk sports do not perceive them to be as dangerous as the spectators do. If they thought they were going to get killed or injured they probably would not compete. For example, I race motorcycles and if I thought I was going to get hurt I would not do it.

Today there are more sophisticated ways of measuring personality (for example, the NEO Personality Inventory, Costa and McCrae, 1985). The fundamental message is the same: personality is essentially stable and centred around traits, but the newer idea has been added that it only becomes 'set' by the age of 30 years. Personality is unlikely to change unless you have great misfortune (or equally good fortune). This 30 years-of-age deadline is interesting for us; in many sports the magic 30 spells the beginning of the end of a sporting career, so maybe these traits are not quite as stable in sportsmen and women simply because they tend to be younger.

△ **Fig. 14.7** A base jumper may well be an example of an extreme personality type.

State theory is where we categorize people by the behaviour they are exhibiting at the time. It acknowledges that an individual may spend more time in one particular state than others but it is the actual behaviour at the time that is important. So one might be assertive and 'in charge' on the field but quite the reverse at home (cf. Rogers, 1961).

The advantages of trait theory are predictability and plausibility. We tend to behave as though traits are 'true'. We say things like 'He's always been like that' or 'I bet he would like an Abba CD for Christmas – he's always liked them' (!). The real problem with trait theory is that if a person behaves out of character then the trait is left high and dry. Few people would expect even the most cheerful person to be really happy if they had just had the bad news of the death of a near relative. Few people are truly always the same – they have good and bad days. If individuals were always that close

to their trait then life would be much more predictable (and boring).

The disadvantages of trait theory are that it does not allow for moods, it does not allow for different reactions, it does not allow for ageing, it does not allow for self-understanding (i.e. self-cognition), and it does not allow for the effect of the situation. Also, whilst it might explain one's choice of sport (but see below) it does not explain success.

The advantages of state theory are that it can make accurate and sensitive descriptions of behaviour, but on the other hand it makes predictions difficult.

If we apply personality and personality theory to sport we uncover some quite interesting observations. Hemery (1986) reported that 89 per cent of the 63 international athletes he interviewed said that they were shy and introverted before they became successful at sport. It makes me wonder about the concept of trait (although the modern theories do point out that up to 30 years old personality is still fairly uncrystallized). It also makes one wonder if playing sport positively changed their personalities (if you think being shy is a bad thing).

Cooper (1969) said that athletes were more self-confident and socially outgoing than non-athletes. Does sport do this for you or do people like this take up and enjoy sport?

Whilst we are asking ourselves questions, here are a few more:

- Do all sports people have the same personality?
- Is it sport that 'makes' you like this or are these the sort of people sport attracts?
- Do different sports attract different personalities?
- Is there a type of person for each sport?
- Are all people within a sport the same?

I cannot help thinking that the answer to all these is probably no. If I think of my own sporting colleagues I think the only things we have in common is a love of the sport. It seems to me we are not all the same at all. I know it is easy to say 'rugby types' or 'people into football', but these sports attract people from many different backgrounds, cultures and nationalities – how could they possibly be the same? True, they have their sport in common and we can all sit round telling lies about our sport, but having things in common with one another is not the same as all being the same.

Other studies have produced mixed results. Gabler (1976) (and others) came to the conclusion that participation in sport does not affect personality (by 'personality' they mean personality traits), Folkins and Sime (1981) demonstrated that self-concept changes following training, and

Sack (1975) found little difference in terms of personality between athletes and non-athletes.

Yet another good question goes like this: Does all this conflict arise due to the methodology used to investigate these ideas? Most of these theories and ideas are not based on either very rigorous experimentation or long-term studies. It would be quite hard to control and perform experiments that gave an unequivocal answer to these questions – and I guess that is why no one has really resolved these issues.

One possible way to address these problems is to evoke an interactionist approach. Bowers (1973), for example, tries to take into account personal factors, the situation the individual is in and the interaction between the two. When the situational factors are very strong these will dominate the interaction and will tend to affect behaviour more than personal factors. For example a normally very placid sports person may become uncharacteristically animated when they win a prestigious competition. When situational factors are not strong then personality is more likely to affect behaviour. This approach is very useful for describing behaviour but very poor at predicting it.

Imagery and mental rehearsal

What is fascinating about this area of research is that we know that mental rehearsal works but we are less sure why that is. As ever, let us start with a definition: mental rehearsal is mentally practising a particular sporting activity. For example, before throwing a javelin the athlete would imagine (something like): picking up the javelin, running forward, moving their arm backwards, throwing their arm forwards, letting go, stepping forward, stopping, watching the flight of the javelin, seeing it hit the ground, turning round and walking away. In other words they mentally rehearse what they actually do when they throw the javelin.

According to Jowdy *et al.* (1989) and Orlick and Partington (1988) elite performers use imagery extensively and successfully. Orlick and Partington (1988) report that 99 per cent of their sample used it. Murphy (1994) showed that 90 per cent of American Olympic athletes used it, as did 94 per cent of their coaches. Hall *et al.* (1996) and Mahoney *et al.* (1987) reported that elite athletes were more proficient than non-elite athletes at imagery. In the literature (see Murphy 1990, 1994; Murphy and Jowdy, 1994) there is a distinction made between imagery and mental rehearsal:

- **Imagery** is defined as a symbolic sensory experience that may occur in any sensory mode and is seen as a

mental process (Murphy and Jowdy, 1992) or a mode of thought (Heil, 1985).

- **Mental rehearsal** is defined by Hardy *et al.* (1996) as the employment of imagery to mentally practise an act – so it is seen as a technique rather than just a mental process.

Feltz and Landers (1983) and Weinberg (1981) suggest three robust findings:

1. Mental rehearsal is better than no task at all.
2. Mental rehearsal combined with physical practice is better than either alone.
3. The effects of mental rehearsal are greater for cognitive than for motor tasks.

Two other findings are that mental practice is more efficacious if accompanied by physical stimulation of the target acts, e.g. practice swinging a golf club (Meacci and Price, 1985; Ross, 1985) and mental rehearsal in 'real time' is as efficacious as in 'slow time' (Andre and Means, 1986) – research in this area seems to be 'thin' though. This makes one (me!) wonder if speeding mentally speeding up the swing has an effect on the outcome.

According to Murphy and Jowdy (1992), mediators of this effect are imagery ability, imagery perspective and imagery outcome.

How able is the individual to form vivid images? If they can produce them, can they control them? This **imagery ability** has been shown to influence the effects of mental rehearsal on performance (Start and Richardson, 1964; Housner, 1984; Goss *et al.*, 1986) and helps to distinguish between elite and non-elite or successful and less-successful sports performers (Meyers *et al.*, 1969; Highlen and Bennett, 1983; Orlick and Partington, 1988). Orlick and Partington (1988) also showed that elite performers have often had to learn to be good at imagery (i.e. learn to control it).

This begs two questions: (a) How do they learn to control it? (there is very little research on this) and (b) Can you learn to have vivid imagery? (also very little research on this).

Imagery perspective is to do with mentally rehearsing, either seeing oneself perform a task (rather like watching a video) – this is called external imagery – or feeling oneself performing a task – known as internal imagery. Mahoney and Avener (1977) distinguished between internal and external imagery:

- **External imagery**: An individual perceives themselves from the perspective of an external observer.
- **Internal imagery**: Real life phenomenology where the person imagines being in their body experiencing the sensations that might be expected in the actual situation.

Some researchers used to think that it mattered which was best, but nowadays this is not so clear. Some researchers support the idea that internal imagery is best because it allies the perceptual and kinaesthetic experience of performance (e.g. Corbin, 1972; Lane, 1980; Suinn, 1983; Vealey, 1986) and, according to some researchers (e.g. Mahoney and Avener, 1977; Rotello *et al.*, 1980; Mahoney *et al.*, 1987), that is what successful athletes do. But Meyers *et al.* (1969) failed to replicate this.

Those researchers who have adopted an experimental approach have failed to find a difference between the two (e.g. Epstein, 1980; Mumford and Hall, 1985).

No surprise then that Mumford and Hull (1985) and Murphy (1994) assert that the differential evidence is somewhat equivocal! The question as to which is best might be redundant anyway, since many sports performers use a combination of both (Jowdy *et al.*, 1989; Orlick and Partington, 1988).

White and Hardy (1995) assert that the difference between internal and external mental practice is often not clearly defined by researchers (e.g. Mahoney and Avener, 1977). They further conclude that there is little evidence to support the idea that 'kinaesthetic imagery is easier to perform in conjunction with internal visual imagery as opposed to external visual imagery'.

The third mediator, **imagery outcome**, is about mentally rehearsing the end of (for example) a race. Powell (1973) and Woolfolk *et al.* (1985) showed that mental rehearsal that involved negative outcomes degraded performance. In other words if you mentally rehearse losing then you will do so! (Do not do it!) Why this should be has evoked much speculation and little evidence. Murphy (1994) suggests that negative imagery negatively affects confidence and motivation. This might be relevant to stage fright too. Presumably the opposite should also be true. So, if you imagine successful outcomes then you should do better. Using the Sports Imagery Questionnaire, Hall *et al.* (1991), Moritz *et al.* (1996) and Vadocz *et al.* (in press) have provided some evidence for this.

We said at the start of this section that we know more about getting mental rehearsal to work than we do about why it works, but there some theories.

Psychoneuromuscular theory

This posits the idea that during mental rehearsal the body runs the performance programme but with the gain turned down (as it were) Jacobson, 1930. Thus during mental rehearsal EMG activity that mirrors the EMG activity that occurs when the task is actually performed should be recorded, but at a lower level of intensity. Unfortunately the studies that have attempted to demonstrate this putative link have either lacked experimental controls (e.g. Jacobson, 1930; Suinn, 1976) or have failed to demonstrate the effect (e.g. Hale, 1982).

Symbolic learning theory

This suggests that mental rehearsal allows the performer to practise the cognitive aspects of the task – things like task strategies, spatial and temporal sequences (e.g. Sackett, 1934; Schmidt, 1975). If this is true then mental rehearsal should be important for new skill acquisition (e.g. Fitts, 1962; Schmidt, 1975), although Richardson (1969) holds the opposite view, i.e. that mental rehearsal is most useful when the performer is familiar with the task (the 'I can't use imagery until I know what to image' argument).

There is published evidence to support the symbolic learning theory (e.g. Wrisberg and Ragsdale, 1979; Minas, 1980). Furthermore, Johnson (1982) compared the two theories and found evidence to support symbolic learning theory but not psychoneuromuscular activity.

Bio-informational theory

Proposed by Murphy (1990) and Murphy and Jowdy (1992) who think that symbolic learning theory is too simplistic, this is taken from Laing's (1977, 1979, 1984) clinical work. The idea is that an image is a 'functionally organised set of propositions stored by the brain' and that imagery is divided into propositions about stimulus and about response and that there is a relevant physiological component to both. I might be missing the point here but this seems more of a description than an explanation.

All of which seems to leave us with – we know it works but we don't know why.

Motivation

If there was ever a topic in sports psychology that everyone thinks they understand it is motivation. Coaches are particularly taken with this idea. They seem to think that their job would be much easier if they could only motivate their athlete to do more of what they say. If only athletes did what they were told then they would win more often. This is somewhat worrying because it is not clear that it is their job to make people do something they don't want to do. On the other hand it is not always clear to the athlete why they are doing what the coach tells them to do and they often need a bit of convincing that 'it will work'. Of course the other problem is we are all a bit lazy (when was the last time you changed channel on your TV without using the remote?) and we could all (as my mother used to say) do with 'a bit of livening up' from time to time.

It is also true that in some situations it is important to maintain motivation or you will not finish what you started; in some sports, such as mountain climbing, if you do not continue then you die. Commonsense definitions of 'motivation' tend to be diverse and vague. When people talk of motivation they say things like 'she's highly motivated' (psychologists would call this an intrinsic motivation – see below), or 'I need something to motivate me' (extrinsic motivation) or 'I wanted it too much and was over-motivated' (a behavioural explanation). Of course definitions are important (no news here then) but coaches who refer to 'motivation' stand a good chance of being misinterpreted by their athletes. It is therefore essential to explain exactly what we mean when we use the term 'motivation'.

One definition of motivation is that it 'is the direction and intensity of one's effort' (Sage, 1977). Here **direction of effort** refers to whether an individual seeks out, approaches or is attracted to certain situations and **intensity of effort** refers to how much effort a person puts into a given situation.

Both direction and intensity are often considered separately but in reality they are related (i.e. those who are highly directed tend also to participate more intensely; those who are less directed tend to participate less intensely).

As implied earlier, an interesting way of thinking about motivation is to divide it up into intrinsic and extrinsic (Deci, 1975):

- **Intrinsic motivation** is where one performs a task because just doing it is 'good' in some way – that is, just doing it makes us feel good.
- **Extrinsic motivation** is performing a task because there is some kind of reward at the end of it.

By now you will have realized that I like riding motorcycles. When I throw my leg over one and hit the starter I am having a great time, I just love doing it. There is an intrinsic motivation for me that comes from riding motorcycles. However, in the winter I need to go to the gym to help keep fit for next season. I do not really like it all that much and I get bored quickly but I do it because in the spring I will be fitter than I might otherwise be, I will not have put on weight and I will be well set-up for the season. I am extrinsically rewarded for this behaviour but I do not much enjoy it. This does not really matter though because I will have achieved my goal. I probably would hardly go to the gym if I did not race but for as long as I can get on one I will ride a motorcycle.

Box 14.4

Understanding motivation

John was a retired schoolteacher and on Wednesday afternoons he liked to watch his favourite TV programme. During the summer he had his windows open but the local children played noisily in the street outside. Normally he wouldn't mind but after a lifetime dedicated to children he thought he deserved one afternoon a week where he could watch his programme in peace and quiet. So he went out to the children and offered them all 50p to make as much noise as they could on a Wednesday afternoon. He offered the noisiest boy an extra pound if he could get more of his friends to join in, explaining that now he was retired he missed hearing children and as he was older his hearing wasn't so good. The next week lots of children turned up and made a terrible racket outside John's window. After a while he went out, congratulated the noisy boy on his organization and paid up. He reminded them to come again the following week. They turned up as arranged, made the same row and John went out and paid them. The noisy boy said 'Same time next week?'. John told him okay, but as he was a pensioner and as money was a bit tight, he could only afford to pay half the rate. The boy looked a bit doubtful but agreed. The following week John paid up but told the children that he would in future only be able to pay 10p per person and no extra for the organizer. The noisy boy became really angry and shouted 'We won't be coming back, it's not worth playing here for such a small amount of money!'

△ **Fig. 14.8** A mountaineer, their direction and intensity of effort is up.

At some level many people think that intrinsic motivation is better than extrinsic motivation, if only because if you really want to do something that you really enjoy then you will be both successful and easy to work with. There is also something faintly unsporting about extrinsic motivation (e.g. the idea that 'you are only doing it for the money').

Lots of sports commentators comment on this, especially when it comes to footballers. The arguments seem to focus on not being 'hungry' for success because they already earn fabulous amounts of money. Some commentators attribute England's demise and Greece's success in Euro 2004 as being to do with intrinsic versus extrinsic motivation to win.

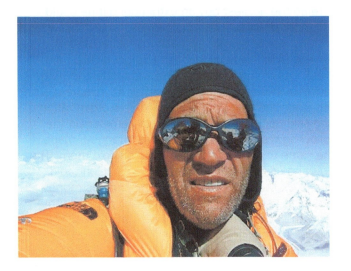

△ **Fig. 14.9** Marshall Ulrich on the summit of Mount Everest. He has won the Badwater Ultra-marathon four times, climbed the highest mountain in each continent, and has run across America in record time.

Figure 14.10 shows Marshall Ulrich. I think from his list of achievements it is clear that he is a highly motivated individual. He did, after all, run across America in 2009. In recent years I have crewed for him in the Badwater Ultra-marathon. This is a 136-mile foot race through Death Valley in July, where it is common for temperatures to reach 50°C. He often asks me if I will run it with him. He says it is easy, all I have to do is want to do it badly enough. So the question is 'How bad do you want it?' In the case of the Badwater for me the answer is 'not very much!' (crewing for it is quite bad enough!).

This idea that success in sport is related to how much you want to win is a very common and seductive one. The 'need to achieve', or achievement motivation, has been and still is studied extensively by many psychologists. The first work was performed by McClelland (1961) and Atkinson (1964). The idea is that when faced with a challenge we either embrace (that is approach) or reject (that is avoid) it.

Our need to embrace or achieve (known as N_{Ach}, i.e. the need to achieve) will depend on two things: our motive to achieve success (M_s) and our motive to avoid failure (M_{af}). So our need to achieve will be a function of the difference between our motive to achieve and our fear of failure. So when we elect to 'have a go' we will risk the failure to achieve the success and when we baulk we will not risk the failure to gain the success. When we try to decide what we are going to do (go for it or go to the pub) we also make a judgement about the probability of success (P_s) and the incentive value of that success (I_s). Finally, we work out what the extrinsic reward will be if we are successful (M_{ext}). This gives us a formula:

$$N_{Ach} = (M_s - M_{af})(P_s \times I_s) + M_{ext}$$

So we are motivated to 'have a go' (or to answer our question of how bad we want it) on the basis of a calculation that we make about what the costs and reward will be.

Thus if you want success, do not fear failure; see how you might be able to do it. Think doing it is a good thing and see some kind of extrinsic reward, then you will be highly motivated. But if you are not very bothered about the success, think you will fail, do not believe you are up to it and do not think the extrinsic reward is worth the effort then you will be unmotivated. There is a middle ground where you are (as it were) motivated for some bits and not for others.

Summary
Performance anxiety

- Performance anxiety is so common it is pretty much seen as normal, but it becomes a problem when it interferes with sporting performance.

- It is future orientated and produces many physiological symptoms.

- It happens in all sports but is most prevalent in those sports with high motor activity.

- It is often described by the Yerkes–Dodson law but this idea is thought to be rather descriptive and some think faulty.

- Catastrophy theory is thought by some to describe it better while others point out that the experience of arousal is very important.

- We can 'fix' anxiety by a number of behavioural and cognitive methods and develop strategies for understanding and coping with our anxieties.

Aggression in sport

- Defining aggression is quite easy but often commentators, pundits and coaches use the term incorrectly, which leads to a general confusion about the concept.
- Aggression is common in sport because it is common in life. Sport is meant to be a socially acceptable means of expressing aggression and many promote its cathartic effect.
- Mostly though, researchers believe that witnessing sporting aggression tends to make people more not less aggressive. This is not true for all sports.
- All too often aggression is reinforced in sport and this in turn tends to teach people to be more aggressive.
- Aggression does not occur in all sports and even in those where it is common psychologists have identified a number of mediating factors that make it more, or less, likely to happen.

Social facilitation

- The idea that audiences affect performance has been around for a long time.
- Trying to figure out how it works has been a thorn in the side of psychology for many years.
- Zajonc had an elegant theory which said that audiences increased drive or arousal, that this arousal made performing novel tasks in front of an audience difficult but we can often perform well-rehearsed tasks better.
- Glaser and others pointed out several problems with this theory.
- Slightly more modern theories of social facilitation try to be less overarching and can be applied more directly to sport.
- These more focused ideas can enable us to understand more of what goes on when an athlete competes in front of a crowd.

Groups and teams

- The way in which a team interacts is to do with roles, norms and cohesion.
- Whilst we may be able to identify what kind of team we need to win, having it and making it work are two different things.
- Process losses 'get in the way' of the team reaching its maximum potential.
- The Ringelmann effect is one example of where individuals do not try as hard as they can.

- The more cohesive a team the better it plays and that cohesion is often related to more group socializing off the field.

Leadership

- There is a difference between a coach and a manager.
- The trait approach says that managers are born, not made.
- The behavioural approach says that individuals can learn to manage and can learn to manage better too.
- The interactionist approach says that the secret to leadership is to get the right kind of manager in the right place.

Individual differences

- 'Credulous' psychologists believe that personality is an important part of sporting prowess.
- 'Sceptical' psychologists do not.
- Personality as a subject divides up into state and trait theorists.
- Traits are permanent, stable parts of personality whereas states are not.
- Traits make predictions about the future easier than states, which are better at describing what goes on in a particular situation.
- The inflexible nature of traits means that they are often 'wrong' because people can be moody and inconsistent, for example.
- State theory is better at describing behaviour than it is at predicting it.

Imagery and mental rehearsal

- We know that many athletes use mental rehearsal as part of their training and competition.
- Not everyone has the ability to perform mental rehearsal but most people seem to be able to improve performance using it, but it is most useful if you can already perform the task (i.e. you won't learn how to ski by mentally rehearsing it but if you can already ski you can improve).
- There are three different kinds: internal where one imagines how doing it feels; external where one imagines seeing oneself doing it; and outcome imagery where one imagines successfully finishing the competition.
- There are three main theories of why it works: psycho-neuromuscular theory, symbolic learning theory and bio-informational theory.

- There is equivocal evidence for all of these theories and we are left with the conclusion that we know it works but we don't know why.

Motivation

- We can divide motivation into intrinsic (doing it for its sake) and extrinsic (doing it for a reward).
- We can also think of the amount of motivation we have as a function of the relationship between our need to achieve and our fear of failure.

Review Questions

Performance anxiety

1. What is the nature of performance anxiety?
2. Describe two theories of how it 'works'.
3. To what extent is sport anxiety a good thing?
4. What are the ways that we can use to 'cure' it?

Aggression in sport

1. Give four definitions of aggression which are relevant to sport.
2. To what extent is aggression inevitable in sport?
3. In what ways is aggression reinforced in sport?
4. Why is aggression not seen in all sports or even in all competitions?

Social facilitation

1. What was Zajonc's theory of social facilitation?
2. List and explain the problems with Zajonc's theory.
3. Describe and explain the seven 'alternative' theories of social facilitation.

Groups and teams

1. What makes a team work?
2. What prevents a team from working?
3. How do we get a team to work better?

Leadership

1. Are leaders born or made?
2. What makes a good leader?
3. Can you make someone a better leader?

Individual differences

1. Explain the difference between a credulous and sceptic view of personality.
2. Explain what a trait is.

3. What are the problems with trait theory?
4. What is state theory and what are the good and bad things about it?

Imagery and mental rehearsal

1. Describe the different types of mental imagery.
2. To what extent can you learn to 'do' mental imagery?
3. Explain the three theories of why mental rehearsal 'works'.

Motivation

1. What is the difference between intrinsic and extrinsic motivation?
2. Explain achievement motivation.

References

Andre JC, Means JR (1986) Rate of imagery in mental practice: an experimental investigation. *Journal of Sports Psychology* 8: 124–128.

Apter MJ (1982) *The Experience of Motivation*. London, Academic Press.

Atkinson JW (1964) *An Introduction to Motivation*. London, Van Nostrand.

Baron RA, Byrne D (1989) *Social Psychology*. London, Allyn and Bacon.

Barrow (1977) The variables of leadership: a review and conceptual framework. *Academy of Management review* 2: 2, 231–251.

Benson H (1976) *The Relaxation Response*. New York, Avon.

Berger SM, Carli LL, Garcia R, Brady JJ (1982) Audience effects in anticipatory learning: a comparison of drive and practice inhibition analyses. *Journal of Personality and Social Psychology* 42: 478–486.

Berger SM, Hampton KL, Carli LL, Grandmason PS, Sadow JS, Donath CH, Herschlang LR (1981) Audience induced inhibition of overt practice during learning. *Journal of Personality and Social Psychology* 40: 479–491.

Bowers KS (1973) Situationalism in psychology: an analysis and critique. *Psychological Review* 80: 307–336.

Cartwright D, Zander A (1968) *Group Dynamics: Research and Theory*. New York, Harper Row.

Carver CS, Scheier MF (1982) Outcome expectancy, locus of attribution for expectancy and self-directed attention as determinants of evaluations, and performance. *Journal of Experimental Social Psychology* 18: 184–200.

Carver CS, Scheier MF (1988) A controlled perspective on anxiety. *Anxiety Research* 1: 17–22.

Cattell RB (1965) *The Scientific Analysis of Personality.* Baltimore, Penguin.

Cooper L (1969) Athletics Activity and personality: a review of the literature. *Research Quarterly* 40: 17–22.

Cooper R, Payne R (1972) Personality orientations and performance in soccer teams. *British Journal of Sociology and Clinical Psychology* 11(1): 2–9.

Corbin CB (1972) *Mental Practice.* New York, Academic Press.

Costa PT, Jr McCrae RR (1985) *The NEO Personality Inventory.* Odessa, Fl, Psychological Assessment Resources.

Deci EL (1975) *Intrinsic Motivation.* New York, Plenum Press.

Duval S, Wicklund RA (1972) *A Theory of Objective Self Awareness.* New York, Academic Press.

Epstein ML (1980) The relationship of mental imagery and mental practice to performance of a motor task. *The Journal of Sports Psychology* 2: 211–220.

Essing W (1970) *Team Line Up and Team Achievement in European football.* Chicago, Athletic Institute.

Eysenck HJ (1967) *The Biological Basis of Personality.* Springfield, C C Thomas.

Feltz DL, Landers DM (1983) The effects of mental practice on motor skill learning and performance: a meta-analysis. *The Journal of Sports Psychology* 5: 25–27.

Folkins CH, Sime WE (1981) Physical fitness training and mental health. *The American Psychologist* 36: 373–389.

Gabler H (1976) Entwicklung von Personlickeitsmerkmalen bei hochleistungssportlern (Development of personality traits in top-level sport performers). *Sportwissenschaft* 6: 247–276.

Geen RG, Gange JJ (1983) *Social Facilitation: Drive Theory and Beyond.* London, Wiley.

Geen RG, O'Neil EC (1969) Activation of cue-elicited aggression by general arousal. *Journal of Personality and Social Psychology* 11: 289–292.

Gill DL (1986) *Psychological Dynamics of Sport.* Champaign, Il, Human Kinetics.

Glaser AN (1982) Drive theory and social facilitation: critical re-appraisal. *British Journal of Social Psychology* 21: 265–282.

Goss S, Hall C, Buckolz E, Fishburn G (1986) Imagery ability and the acquisition and retention of movements. *Memory and Cognition* 14: 469–477.

Griffin M (2001) The phenomenology of the alone condition: more evidence for the role of aloneness in social facilitation. *The Journal of Psychology* 135(1): 125–127.

Griffin M, Kent MV (1998) The role of aloneness in social facilitation. *The Journal of Social Psychology* 138(5): 667–669.

Hackforth D, Spielberger CD (1989) *Anxiety in Sports.* New York, Hemisphere.

Hale BD (1982) The effects of internal and external imagery on muscular and ocular concomitants. *Journal of Sports Psychology* 4: 379–387.

Hall, Craig R, Rogers, Wendy M, Barr, Kathryn A (1990) The use of imagery by athletes in selected sports. *The Sport Psychologist,* Vol 4(1): 1–10.

Hall CR, Rogers WM, Buckholtz E (1991) The effect of an imagery training programme on imagery ability, imagery use and figure skating performance. *The Journal of Applied Sports Psychology* 3: 109–125.

Hanin YL (1989) *Anxiety in Sports - An International Perspective.* New York, Hemisphere.

Hardy L, Parfitt CG (1991) A catastrophe model of anxiety and performance. *British Journal of Psychology* 82: 163–178.

Hardy L, Wyatt S (1986) *Immediate Effects of Imagery upon Skilled Motor Performance.* New Zealand, Human Performance Associates.

Hardy L, Jones G, Gould D (1996) *Understanding Psychological Preparation for Sport.* Chichester, Wiley.

Heil J (1985) The role of imagery in sport: as a training tool and as a mode of thought. Paper presented at The World Congress in Sports Psychology, Copenhagen.

Hemery D (1986) *The Pursuit of Sport and Excellence.* London, Collins.

Highlen PS, Bennett BB (1983) Elite divers, wrestlers: a comparison between open and closed skilled athletes. *Journal of Sports Psychology* 5: 349–390.

Hollingsworth B (1975) Effects of performance goals and anxiety on learning a gross motor task. *Research Quarterly* 46: 162–168.

Hošek V, Man F (1980) Theoretical foundations of the psychological education of coaches and teachers in the field of physical education. SO: *Psychologie-v-Ekonomicke-Praxi.* Vol 15(1): 26–36.

Housner LD (1984) The role of visual imagery in recall of modelled motoric stimuli. *The Journal of Sports Psychology* 6: 148–158.

Jacobson E (1930) Electrical measurement of neuromuscular states during mental activity. *American Journal of Physiology* 94: 22–34.

Johnson P (1982) The functional equivalents of imagery and movement. *Quarterly Journal of Experimental Psychology* 34A: 349–365.

Jowdy DP, Murphy SM, Durtschi S (1989) *An Assessment of the Use of Imagery by Elite Athletes: Athlete, and Coach Psychological Perspectives.* Colorado Springs, Co, United States Olympic Committee.

Kerr JH (1999a) The role of aggression and violence in sport: a rejoinder to the ISSP position stand. *The Sports Psychologist* 13: 83–88.

Kerr JH (1999b) *Experiencing Sport*. Chichester, Wiley.

Kerr JH (2002) Issues in aggression and violence in sport: the ISSP position stand revisited. *The Sports Psychologist* 16: 68–78.

Kingsmore JM (1970) *The effects of a professional wrestling and a professional basketball contest upon the aggressive tendencies of spectators*. Chicago, Il, Proceedings of the Second International Congress of Sports Psychology.

Laing PJ (1977) Imagery in therapy: an information-processing analysis of fear. *Behaviour Therapy* 8: 862–886.

Laing PJ (1979) A bio-informational theory of emotional imagery. *Psychophysiology* 17: 495–512.

Lane JF (1980) *Improving Athletic Performance through Visuo-motor Behaviour Rehearsal*. Minneapolis, Burgess.

Lang PJ (1984) *Cognition in Emotion: Concept and Action*. New York, Cambridge University Press.

Latané B (1981) Psychology of social impact. *American Psychologist* 36: 343–356.

Latané B, Williams K, Harkins S (1979) Many hands make light work: the causes and consequences of social loafing. *The Journal of Personality and Social Psychology* 37: 823–832.

Lorenz K (1963) *On Aggression*. New York, Harcourt, Brace and World.

Mahoney MJ, Avener M (1977) Psychology of the elite athlete: an exploratory study. *Cognitive Therapy and Research* 1: 135–141.

Mahoney MJ, Gabriel TJ, Perkins TS (1987) Psychological skills and Exceptional Athletic Performance. *The Sports Psychologist* 1: 181–199.

Maier NRF, Solem AR (1952) The contribution of a discussion leader to the quality of group thinking. *Human Relations* 5: 277–288.

Manstead ASP, Semin GR (1980) Social facilitation effects: mere enhancement of dominant responses? *The British Journal of Social and Clinical Psychology* 18: 191–202.

Martens R (1977) *Sport Competition Anxiety Test*. Champaign, Il, Human Kinetics.

McClelland DC (1961) *The Achieving Society*. New York, Free Press.

Meacci WG, Price EE (1985) Acquisition and retention of golf putting skill through the relaxation, visualisation and body rehearsal intervention. *The Research Quarterly for Exercise and Sport* 56: 176–179.

Meichenbaum D (1985) *Stress Inoculation Training*. New York, Pergamon.

Meyers AW, Cooke CJ, Cullen J, Liles L (1969) Psychological aspects of athletic competitors: a replication across sports. *Cognitive Therapy and Research* 3: 361–366.

Milillo MD (1975) A study of trait anxiety, state anxiety, defense mechanisms, and personality in three individual sport groups. *Dissertation-Abstracts-International* 36(6b): 3058.

Miller B (1992) *Team Athletics*. Melbourne, Blackwell.

Minas SC (1980) Mental practice of a complex perceptual-motor skill. *Journal of Human Studies* 4: 102–107.

Moritz SE, Hall CR, Martin K (1996) What are confident athletes imaging?: an examination of image content. *The Sports Psychologist* 10, 171–179.

Moyer KE (1976) *The Psychobiology of Aggression*. New York, Harper Row.

Mumford P, Hall C (1985) The effects of internal and external imagery on performing figures of figure skating. *The Canadian Journal of Applied Sports Sciences* 10: 171–177.

Murphy SM (1990) Models of imagery in sports psychology: A review. *The Journal of Mental Imagery* 14: 153–172.

Murphy SM (1994) Imagery interventions in sport. *Medicine and Science in Sports and Exercise* 26: 486–492.

Murphy SM, Jowdy DP (1992) *Imagery and Mental Practice*. Champaign, Il, Human Kinetics.

Noyes RC (1971) The effects of success and failure in physical performance upon state anxiety and bodily concern of college students varying in anxiety proneness. *Dissertation-Abstracts-International* 31(9a): 4529.

Orlick T, Partington J (1988) Mental links to excellence. *The Sports Psychologist* 2: 105–130.

Parsons T (1951) *The Social System*. New York, The Free Press of Glen Coe.

Powell GE (1973) Negative and positive mental practice in motor skill acquisition. *Perceptual Motor Skills* 37: 312.

Richardson A (1969) *Mental Imagery*. New York, Springer.

Ringelmann M (1913) Recherches sur les moteurs animés: travail de l'homme. *Annales de l'Institut National Argonomique*, 2e srie(tom 12): 1–40.

Rogers C (1961) *On Becoming a Person*. Boston, Houghton Mifflin.

Ross SL (1985) The effectiveness of mental practice in improving the performance of college trombonists. *The Journal of Research in Music Education* 33: 221–230.

Rotello RJ, Gansneder B, Ojala D, Billing J (1980) Cognitions and coping stratergies of elite skiers: an exploratory study of developing athletes. *The Journal of Sports Psychology* 4: 350–354.

Ryan ED (1970) The cathartic effect of vigorous motor activity on aggressive behavior. *Research Quarterly* 41(4): 542–541.

Sack HG (1975) *Sportliche Betatigung und Personlichkeit (Sport Participation and Personality)* Ahrensburg, Czwalina.

Sackett RS (1934) The influences of symbolic rehearsal upon the retention of a maze habit. *Journal of General Psychology* 10: 376–395.

Sage G (1977) *Introduction to Motor Behavior: A Neuropyschological Approach.* Reading, Ma, Adderson Wesley.

Sanders GS (1981) Driven by distraction: an integrative review of social facilitation theory and research. *Journal of Experimental and Social Psychology* 17: 227–251.

Sanderson FH, Ashton MK (1981) Analysis of anxiety levels before and after badminton competition. *International Journal of Sport Psychology* 12(1): 23–27.

Sanderson WC, Rapee RM, Barlow DH (1989) The influence of an illusion of control on panic attacks induced via inhalation of 5.5% carbon dioxide enriched air. *Archives of General Psychiatry* 46: 157–162.

Scanlan TK, Passer MW (1978) Factors related to competitive stress among young, male youth competitors. *Medicine and Science in Sport* 10: 103–108.

Schmidt RA (1975) A schema theory of discreet motor learning. *Psychological Bulletin* 76: 92–104.

Silva JM (1980) *Understanding Aggressive Behavior and Its Effects upon Athletic Performance.* Ithaca, Movement.

Smith MD (1975) The legitimating of violence: hockey players perception of their preference group's sanction for assault. *The Canadian Review of Sociological Anthropology* 12: 72–80.

Smith RE, Smoll FL (2002) *Youth Sports as a Behavior Setting for Psychosocial Interventions.* Washington, DC, US: American Psychological Association.

Start KB, Richardson A (1964) Imagery and mental practice. *British Journal of Educational Psychology* 34: 280–284.

Steiner ID (1972) *Group Processes and Productivity.* New York, Academic Press.

Stroebe W, Frey BS (1982) Self-interest and collective action: The economics and psychology of public goods. *British Journal of Social Psychology* 21: 121–137.

Suinn RM (1976) *Visual Motor Behaviour for Adaptive Behaviour.* New York, Holt.

Suinn RM (1983) *Imagery and Sports.* New York, Wiley.

Tenenbaum G, Sacks DN, Miller JW, Golden AS, Doolin N (2000) Aggression and violence in sport: a reply to Kerr's rejoinder. *The Sports Psychologist* 14: 315–326.

Tenenbaum G, Stewart E, Singer RN, Duder J (1997) Aggression and violence in sport: an ISSP position stand. *The Sports Psychologist* 11: 1–7.

Thibaut JW, Kelley HH (1986) *Social Psychology of Groups.* New Brunswick, Translation Books.

Thomas EJ, Fink CF (1961) Models of group problem solving. *Journal of Abnormal and Social Psychology* 63: 53–63.

Triplett N (1897) The dynamogenic factors in pace making and competition. *The American Journal of Psychology* 9: 507–533.

Turner ET (1970) *The Effect of Viewing College Football, Basketball and Wrestling on the Elicited Aggressive Response of Male Spectators.* Chicago, Il, Proceedings of The Second International Congress of Sport Psychology.

Vealey RS (1986) Conceptualisation of sport confidence and competitive orientation: preliminary investigation and instrument development. *Journal of Sports Psychology* 8: 221–246.

Veit H (1968) *Interpersonal relations and the effectiveness of ball game tennis.* In Proceedings of the Third International Congress of Sport Psychology, Madrid.

Voigt HF (1982) *Die Struktur von Sportdisziplinen als Indikator fur Kommunikationsprobleme und Konflikte (The structures of particular sports as an indicator for communication problems and conflicts)* Schorndorf, Verlag Karl Hofmann.

Volkamer M (1971) Zur Aggresivitat in konkurrenzorientierten sozialen Systemene. Eine Untersuchungan Fussballpunktspelen (Aggression in rivalry-oriented social systems. An investigation into soccer) *Sportwissenschaft* 1: 33–64.

Weinberg RS (1981) The relationship between mental preparation strategies and motor performance: a review and critique. *Quest* 33: 195–213.

White A, Hardy L (1995) The use of different imagery perspectives on learning and performance of different motor skills. *British Journal of Psychology* 86: 169–180.

Wilson GD (1994) *Psychology for Performing Artists.* London, Jessica Kingsley.

Wolverton DT, Salmon P (1991) *Attention Allocation and Motivation in Music Performance Anxiety.* Amsterdam, Swets and Zeitlinger.

Woolfolk RL, Parrish W, Murphy SM (1985) The effects of positive and negative imagery on motor skill performance. *Cognitive Therapy and Research* 9: 335–341.

Wrisberg CA, Ragsdale MR (1979) Cognitive demand and practice level: factors in the mental practice of motor skills. *The Journal of Human Movement Studies* 5: 201–208.

Yaffe M (1975) *Some Variables Affecting Team Success in Soccer.* London, Lepus.

Zaccaro SJ (1984) Social loafing: the role of task attractiveness. *Personality and Social Psychology Bulletin* 10(1): 99–106.

Zajonc RB (1965) Social facilitation. *Science* 149: 269–274.

Zillman D (1979) *Hostility and Aggression.* Hillsdale, NJ, Earlbaum.

Zillman D, Katcher AH, Milavsky B (1972) Excitation transfer from physical exercise to subsequent aggressive behavior. *Journal of Experimental Social Psychology* 8: 247–259.

Zimbardo PG (1990) *Shyness: What It Is, What To Do About It.* Reading, Ma, Adderson-Wesley.

Further reading

Fitts DM (1962) *Skill Training.* Pittsburgh, University of Pittsburgh Press.

Hardy L, Callow N (1999) Efficacy of external and internal visual imagery perspectives for the enhancement of performance on tasks in which form is important. *Journal of Sport & Exercise Psychology*, Vol 21(2): 95–112.

Jones G, Hardy L (1990) *Stress and Performance in Sport.* Chichester, Wiley.

Vadocz EA, Hall CR, & Moritz SE (1997). The relationship between competitive anxiety and imagery use. *Journal of Applied Sport Psychology* 9, 241–253.

15 Exercise and Psychological Health

Chapter Objectives

In this chapter you will learn about:

- Abnormality
- Body image
- Self-esteem
- Depression
- Self-efficacy
- Injury and recovery from injury
- The end of a sporting career.

Introduction

There is a great deal of concern about the relationship between physical activity, mental health and well-being. There are increasing numbers of people who are diagnosed as suffering from depression and increased concerns that some aspects of modern life are detrimental to well-being. In this chapter we shall consider the evidence that some of these concerns are justified. We will also examine the evidence that physical activity can positively affect mental health. Finally, we will look at the effects that stopping or reducing physical activity has on mental health and well-being, especially when sports persons come to the end of their sporting careers.

Abnormality and illness

Before we can start to think about the impact of sport and exercise on mental health we need to decide what mental health is. One way to do this is to consider a lack of mental health or what some might call 'illness' (see below for a critique of this approach). We might also try to decide what we mean by being mentally 'abnormal'. This is nothing like as easy as it seems. Clearly some things are not at all 'normal', but sometimes even extremes of behaviour might not be seen by some people as all that abnormal.

As ever, we need to define abnormal and although there are many ways of doing this, none of them are very satisfactory.

Statistical infrequency

Under this definition we would say that the rarer something is, the more abnormal it is. So if we think of a normal distribution (**Box 15.1**) abnormality would lie at the ends of the curve. This is a real problem because by its nature many very wonderful things are very rare. Indeed, elite athletes would be categorized here just because they are very rare and unusual. Do we really want to call very clever people, very gifted or talented people, very tall people, and very beautiful people abnormal? This seems very unsatisfactory.

Violation of norms

Can we consider something abnormal if it 'violates social norms' or makes observers feel threatened or anxious? There are many problems with this, not the least of which is that social norms are far from universal, so the questions that follow here are: Whose norms are we considering? How easily is an individual threatened? Are you abnormally easily threatened? Put another way, it is much harder to be this kind of abnormal in a tolerant society. We do not usually call criminals abnormal but by the nature of their activities they violate norms.

Personal distress

The idea here is that something is abnormal if it creates distress. While distress often goes with abnormality, many forms of abnormality produce no distress on the part of the person displaying it (indeed that very lack of distress contributes to the idea of them being abnormal). For example,

Normal distribution

What most researchers, statisticians and (if they were animate) statistics want to see is a normally distributed population. This is based on the idea that in any population there will be lots of 'typical' people (these are known as 'normal' in mathematics and means just that, i.e. a term used in mathematics) and a few 'atypical' ones.

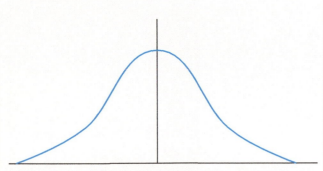

If we think about height, I suppose that most adults would be between 1.5 and 1.8 metres tall. These are what I am calling the 'typical' ones (normal in maths). The further we go above or below these heights the fewer people there are, and these are what I am calling (although not in any pejorative, that is, negative sense) 'atypical'. In other words it is more unusual to be 2.1 metres or 1.2 metres tall than, say 1.7 metres. Because we want to be able to say most about most people (i.e. in this case the ones between 1.5 and 1.8 metres), we want to be sure that our population is 'normally distributed'. That is, it has most people in the middle and a few either side.

in some psychiatric patients we see a symptom called incongruity of effect. That is, giving the wrong emotional response (e.g. laughing when they should be crying). In other circumstances it is normal to be distressed (e.g. bereavement), so in this case becoming distressed may be an example of normality.

Disability or dysfunction

Disability or dysfunction are terms used when an individual is impaired in some important aspect of life, for example through blindness, having a learning disability or being a wheelchair user. There is probably a distinction to be made between disabling and disabled. For example, someone who is phobic about flying has a disabling problem but is not necessarily disabled. Clearly they may be prevented from working if they need to travel by plane to do their job. Do we really want to go to the next stage and call disabled people abnormal? I suspect not. Some might

consider being a transvestite dysfunctional. But many transvestites only cross-dress in private and do not cross-dress for work. They hold down a job, form relationships and seem by many 'normal' markers to be getting on with life quite successfully. It is hard to see where the dysfunction lies. I do not feel inclined to get involved in the 'are they abnormal' debate.

Unexpectedness

This is where a reaction or piece of behaviour is outside what we might usually expect. It may be out of all proportion to the situation, for example if you are absolutely beside yourself because you are £5 overdrawn at the bank, your behaviour would be unexpected. The opposite might also be true, for example being unconcerned that you are 30 kg overweight would also be unexpected. I think that these kinds of things are often surprising (because they are unexpected) but are they really abnormal? Equally, there is often a context to the behaviour – for example, some people almost never react in an overtly emotional manner, so getting a bit excited is really remarkable. Others may be quite excitable people so, for them, being emotional is nothing special.

Normality

One way to address the problem might be to define normal instead. This can often be just as hard, as an absence of abnormality will not do. I think we might say things like 'that behaviour is not surprising given the circumstances'. But there are lots of 'normal' behaviours that we might have trouble explaining to someone from another planet. Take, for example, smoking, drinking until we cannot stand up, Morris dancing (Figure 15.1), playing sport or rowing across the Atlantic.

△ **Fig. 15.1** Is this 'normal' behaviour?

How common is mental 'illness'?

It is worth just taking the time to find out some statistics about mental health problems. Research shows that:

- Mental health problems are more common in women although there are some very interesting debates about this.

- Three out of five visits to the GP are mental health related.

- One out of five people will have a diagnosable mental health problem.

- Two out of five people will have some kind of mental health problem that does not qualify for a mental health diagnosis (e.g. insomnia).

Do you find these statistics alarming? Does this tell us anything about our attitudes to mental health?

Our definition of mental 'illness' tends to be subsumed by a medical model. This says that if you have an illness or disease that can be cured by some medical intervention (i.e. surgery, drugs or therapy just like a physical illness) it is there to be got rid of. Many would argue that this approach has not been completely successful (although it is very useful). Many people are not 'ill' in the conventional sense, but they are troubled and often in a way that is not very susceptible to medical treatments. By this I mean they have not got psychiatric symptoms – they don't hear voices or feel suicidal – but they perhaps feel very anxious, sad and tense, have trouble sleeping and/or concentrating. They still get up, go to work and have some sort of a 'normal' life but they feel unhappy and troubled. The last thing these people may need are drugs but they might need help, for example, counselling or psychotherapy.

Body image

Perceptions of our physical bodies are part of self-concept, and they form an integral part of overall self-worth. The evaluation of one's size, weight or other aspects of the body that determine the manner in which the body is viewed are the essential components of the physical aspect of body image. I was in my local newsagents the other day and I did my usual thing of casting a psychologist's eye over the magazines. I am always amazed at how many magazines concern themselves (if only obliquely) with body image (Fig. 15.2). There seems to be endless reams of advice and comment on various aspects of body image. This reflects (or maybe causes) the extent to which people seem to be concerned about body image, which may also

△ **Fig. 15.2** The media constantly concern themselves with issues of body image.

be related to something of an 'epidemic' of eating disorders (Thompson, 1990: Leit, Gray and Pope, 2002).

Did you know that one of the top ten reasons for taking out a personal loan (include a car, a holiday, home improvements) is cosmetic surgery?

For many people participation in sport and exercise may be associated with body dissatisfaction, and athletes in particular face tremendous pressure to maintain an ideal body. This pressure may be partly responsible for sustaining the cyclical repetitious nature of eating disorders which is seen in some athletes. A good example of this is what has become known as the **female athlete triad:** disordered eating, amenorrhoea and osteoporosis. According to many psychologists this is the physical manifestation of a pathological adherence to exercise, and is often linked with an inappropriate diet (Scully *et al.* 1998).

Body image assessment

Body image assessment techniques were initially developed to help psychologists understand body image disturbances. The measurement procedures have mainly focused on two aspects of body image: a perceptual component and a subjective component.

- The **perceptual component** is also known as size perception accuracy, and is measured by subjects matching the width of the distance between two points to their own estimation of their body size or a particular body site. It has also been measured using schematic figures of different body sizes, where individuals are asked to choose the body size they think reflects their own.

- The **subjective component** refers to the degree of satisfaction or dissatisfaction felt about the body's appearance and function. This has been measured by comparing actual and ideal body sizes. Many questionnaire measures, such as Secord and Jourard's Body Cathexis Scale (1953) (Figure 15.3) also assess subjective representations of physical appearance, whereby respondents rate the degree of satisfaction they feel about various body parts (Thompson, 1990).

Although research into body image initially focused on eating-disordered populations, it has now progressed to the general population. This research has shown that many people seem unhappy with their bodies (known as a normative discontentment) and this is even true for parts of the population that one might have thought would have less to worry about than the rest of us. In a study of 148 fitness instructors, for example, Nardini (1998) found 64 per cent perceived an ideal body as one that was thinner than their current body. Despite having lower than average body fat levels, the instructors were as dissatisfied with their bodies as other people who were the same age as them.

Given how few people seem satisfied with how they look one might wonder where we get these 'standards of bodily perfection' from. According to Willis and Campbell (1992) they are imposed on people through the media, who communicate messages about how the body should ideally look. The extent to which the body matches the cultural ideal clearly has a huge role to play in determining the degree of body satisfaction that is experienced. Although research is unclear about how this affects the way men perceive their bodies, it seems that the number of men who suffer from problems in this area is increasing (Davis, 1997). Evidence for this is the increase in the numbers of men diagnosed with anorexia or **muscle dysmorphia** (the 'body builder's illness'). This latter is a condition in which men, despite being very muscled in appearance, see themselves as 'puny' looking. One of my students referred to this as a kind of backwards anorexia.

Body image and physical activity

Several studies have found that physically active men and women evaluate their physical appearance more positively and are significantly more satisfied with various parts of

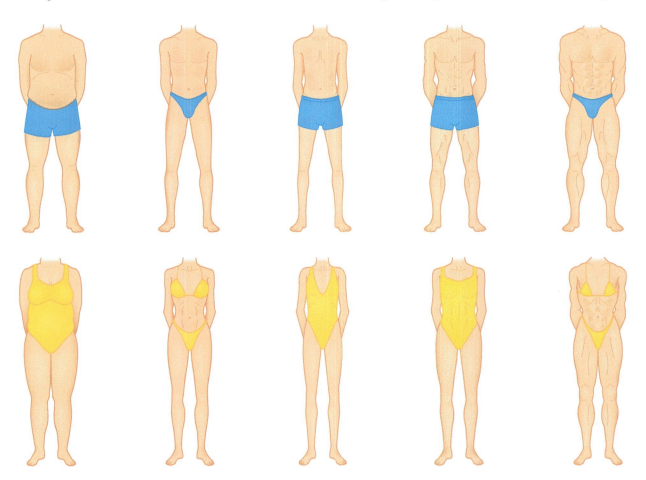

△ **Fig. 15.3** Body Cathexis Scale. Respondents were asked to rank their preferred body type in order and to mark the body type they most resembled

their bodies than those who are not physically active (Davis, 1997; Guinn et al., 1997; Lowland, 1998, 1999). Bartlewski et al. (1996) found that women who enrolled in an aerobic exercise class showed a decrease in social physique anxiety (a bodily self-consciousness resulting from perceived evaluations by other people) throughout the duration of their attendance but this did not change significantly for non-exercising control subjects.

Lox et al. (1995) investigated the effect of exercise on subjective well-being in HIV-1-infected men over a period of 12 weeks. They were assigned to either an aerobic exercise training group, a resistance weight-training group or a control group. Both exercise groups demonstrated improvements in perceived physical ability and subjective physical appearance, while the control group exhibited decreases in these two measures. The aerobic exercise group displayed greater increases in perceived physical appearance than the weight-training group.

Marsh (1998) examined physical appearance self-concept among elite and non-elite athletes using a measure called the Physical Self-Description Questionnaire (PSDQ). There were large between-group effects favouring elite athletes and significant group-by-gender interactions for total physical self-concept. Elite athletes scored more favourably than non-elite athletes across all the PSDQ subscales except health concept, while there were no significant group-by-gender interactions for appearance, body fat or global physical self.

Individuals who have physical disabilities often elicit a stigmatized response from other people, as they fall outside the range of what is considered to be 'normal' (Figure 15.4). They are assumed to be different from conventional standards of body build and attractiveness. Taub et al. (1999) explored the possible alteration of the image of a disabled body through involvement in sport and physical activity. Male subjects with a variety of disabilities, including paraplegia, quadriplegia and cerebral palsy, were included in the study. The majority perceived participation in physical activity as a positive experience, and they believed they had an enhanced bodily appearance as a result. They also considered sport and physical activity to be compensatory to stereotypical perceptions about the appearance of a disabled body. For most participants the important thing was participating in the experience itself rather than the type or intensity level of the physical activity.

Dekel et al. (1996) also examined self-esteem and body image in adolescents with postural deformities who were diagnosed as having structural and non-structural adolescent idiopathic scoliosis (AIS). Individuals who engaged in physical activity perceived their bodies more positively than those who did not.

△ **Fig. 15.4** Wheelchair basketball players challenging the stereotypical perception of the disabled body.

Not all studies have found a positive relationship between body image perception and physical activity. Marsh et al. (1995) did not find any significant differences in physical appearance self-concept between athletes and non-athletes, even though they did find differences in self-esteem. Baldwin and Courneya (1997) found a significant correlation between exercise participation and global self-esteem in women who had been treated for breast cancer, but physical acceptance was not correlated significantly with exercise participation. Davis et al. (1993) investigated physical appearance with particular regard to men. Although they hypothesized that appearance anxiety would be inversely related to physical activity participation, this association was only weak. Upper body esteem accounted for nearly half the variance in appearance anxiety, suggesting that male body dissatisfaction is most notable for the chest and waist.

The relationship between body image and self-esteem

For most people how they look is important to them. From time to time we all tend to show off our best bits and hide the parts we are less comfortable with. By the age of 11 children have begun to rate themselves on particular aspects of their appearance, and have already formed an opinion about whether or not they are attractive. Page and Fox (1997) noticed how certain features become increasingly important as well as consistently relating to self-esteem. Their work indicated that the perception of appearance is the strongest correlate of self-esteem for both boys and girls compared with other areas of life.

The body is also of great importance to the self at times of life other than childhood. This is evident from people's willingness to take part in procedures that can be

 (a)

 (b)

△ **Fig. 15.5** (a)&(b) Ideals of body perfection.

unhealthy or expensive, such as cosmetic surgery and the use of sun beds. Also the use of steroids by body builders and the adoption of bad eating habits in young girls in order to look slim (Figure 15.5 (a) and (b)) can be viewed as self-presentation strategies (Fox, 1997).

Secord and Jourard (1953) speculated that if someone's status and security were dependent on their attractiveness, and they did not consider themselves to be attractive, they would exhibit a loss in self-esteem. They found inter-correlations between body image scores and self-concept scores of students for males and females both. This suggests that individuals have a moderate tendency to cathect (invest emotional energy in) their body and self in the same direction and to the same degree (Figure 15.6).

Guinn *et al.* (1997) found a negative relationship between self-esteem and body fatness in female adolescents. Body image exerted a stronger influence over subjects' self-esteem scores than exercise.

The role of gender

It is commonly assumed that women worry about their appearance more than men. There is indeed a substantial amount of information to support this, but I have a sneaking feeling that this is slowly changing. In both athletes and non-athletes, men have been found to have a significantly higher self-esteem ($p < 0.01$) and physical appearance self-concept ($p < 0.01$) than women (Marsh *et al.* 1995). Marsh (1998) found that group effect (elite/non-elite athletes) was substantially larger than gender effect for total physical self-concept and most scales on the PSDQ, with the exception of appearance, body fat and global physical scales. For these scales the male gender

effect was substantially larger than the group effect. Loland (1998, 1999) discovered that both inactive and active women were more concerned with appearance and weight and were less satisfied with weight and most parts of their bodies than their male counterparts.

Davis and Katzman (1998) found that Asian women reported significantly more body dissatisfaction than Asian men. Gender patterns mirror those reported in samples from white populations with respect to body image. In

△ **Fig. 15.6** A female body builder. In this example, her body is likely to be important to her sense of self.

contrast, Secord and Jourard (1953) did not find any significant differences between means of scores on the Body Cathexis Scale for the two genders, although women did cathect their bodies more highly than men, indicating a poorer perception of body image.

In an exploratory study of motives for exercising and body image satisfaction, it was discovered that women who experienced the most body dissatisfaction exercised for appearance and weight control. Women also exercised for appearance-related reasons more than men (Smith *et al.,* 1998).

Self-esteem, sport and exercise

Most but not all mental health problems seem related to reduced self-esteem either as a consequence or a cause of the illness. In other words, poor self-esteem causes mental illness or having a mental illness causes poor self-esteem. So:

- If we could fix self-esteem we could perhaps fix the underlying problem.
- If we could fix self-esteem we could use that as a marker to help us to see if we are helping the underlying problem.
- If we could fix (i.e. improve) self-esteem then that would be a good thing in itself.

What is self-esteem?

According to Campbell (1984) self-esteem is an awareness of good possessed by self. It is sometimes seen as a self-rating of how well the self is doing. This worth is dictated by both the individual and the culture in which he or she operates. So it is both a personal thing based on the things that you value most, and a societal thing, based on the things which those around you value the most. The personal is the most important and many things may contribute to it.

The symptoms of low self-esteem are depression, anxiety, neuroses, suicidal thoughts or behaviour, a sense of hopelessness, lack of assertiveness, and low perceived personal control. Clearly the consequences of these are likely to be negative.

We are assuming here that high/good self-esteem is important. I think this is because the alternatives are undesirable. It is seen as a key indicator of emotional stability, and adjustment to life demands is seen as one of the strongest predictors of subjective well-being (Diener, 1984). High self-esteem is valued and is associated with healthy behaviours and it is therefore important to consider it in relation to health. Also low self-esteem is related to mental illness. It seems to be one concept that the general public seems to

understand and it is even part of the National Curriculum in the UK educational system.

A considerable body of evidence suggests that one way of improving self-esteem is to do aerobic exercise, such as aerobics, running and cycling, which get your heart rate up and make you breathless. Fox (2000) cites many studies that show this positive change in children (Schempp *et al.,* 1983), in adults (Brown *et al.,* 1995) and in those with special needs (in this case people with learning disabilities (Mactavish and Searle, 1992). It would seem that children and adolescents benefit from this kind of exercise. It is very good for those with low self-esteem and is very powerful if it also encourages mastery and self-development (Marsh and Peart, 1988; Calfas and Taylor, 1994; French Story and Perry, 1995). It should in theory be good for older adults also, but there is not a great deal of evidence for this.

Characteristics of effective exercise

Numerous researchers are now working in this area and we are beginning to understand more about it. Fox (2000) states that cardiovascular activity produces a significant change in self-esteem in about 50 per cent of those that try it. However, there is growing evidence that resistance training is good for improving body image whereas swimming, flexibility training, martial arts and expressive dance produce non-significant results.

Exercise frequency, intensity and programme duration all might reasonably be expected to make a difference to the effect but there is not much evidence for this as not much research has been done in this area.

Environment

Pretty *et al.* (2004) have demonstrated through their 'green exercise' research that it is possible to improve mood, increase self-esteem and reduce blood pressure by exercising in front of pleasant rural images, whereas exercising in front of no pictures, rural unpleasant, urban pleasant or rural unpleasant pictures does not produce this effect.

Why does it work?

There has been much conjecture about the correlation between self-esteem and exercise, including ideas about some undetermined psycho-physiological mechanism, improvements in fitness or weight loss, autonomy and personal control, an increased sense of belonging and significance, but no one seems to have marshalled enough evidence to arrive at a definitive answer.

Depression

Depression is a relatively common mental illness (although some might argue it is not really an illness). I am now going to give you a quite old-fashioned way to look at this. More modern theories of depression are based on a continuum, but I want to consider it as a two-type 'illness'. This is because if you are new to this area it makes it easier to understand. The two types are:

- **Reactive depression**, which occurs because of some (usually negative) event. For example the loss of a significant other, some perceived loss or disappointment or failure. This can vary greatly from mild to severe. Mild is much more common than severe.

- **Endogenous depression**, which occurs for entirely biological reasons. Again it can vary considerably but endogenous depression is often more severe.

Of the two, reactive depression is the most common – many instances of endogenous depression are really reactive, it is just that the cause is complicated or buried or not clear or not obvious.

Many people believe that we understand the biology of depression quite well (see below) but just why these perceived negative events set off this 'biology' is less clear. In endogenous depression I guess we could look for a biological mechanism (or mechanism failure) to explain what happened.

In one sense we are also quite good at fixing the biology of depression (by the use of drugs – also see below) but the psychological component is a tougher nut to crack.

Signs and symptoms of depression

The signs of depression include feelings of unhappiness, misery, worthlessness and poor self-esteem. The person may feel unattractive, unloved and incompetent and experience a loss of interest in their usual pursuits, in work, in sex, in things in general. They may lose interest in personal hygiene and appearance. People with depression speak slowly, sit alone, look sad, sigh, moan and complain. They can only see the 'black' side of things and 'can't be bothered' with trying to fix their 'problems' as 'it's all too difficult'. They lack concentration, have a loss or increase in appetite and have little energy. They often suffer a disturbed sleep pattern, feel anxious and suffer from psychomotor retardation (are so depressed they can hardly move). They think about death a lot and have suicidal thoughts; this may lead them to attempt suicide.

None of this sounds very much fun at all. I've worked with many depressed people over the years and depression is a terrible thing to have to suffer.

It is interesting to look at who gets depression; we call this aetiology. According to Kessler *et al.* (1994) depression occurs in about 17 per cent of people (in the case of mild depression it is probably more); Murray and Lopez (1997) estimate the figure at 11 per cent of the world and project this to increase to 15 per cent by 2020 – if this were true this would move it from the fourth to the leading disease burden. Depression comprises a major proportion of so-called mental illness, is 2–3 times more common in women than men, is more common in lower socio-economic classes, is more common among Jewish men than in other men (Klerman, 1988) and has been increasing over the last 50 years. It tends to be a recurrent illness.

Depression is often linked with other psychological and physical illness. When I was thinking about this topic I used the key word depression in an internet search and got over 4 million hits compared with under 2.5 million for happiness!

The biology of depression

Many psychiatrists believe that depression is essentially a problem related to neurotransmitters – chemical substances that transmit impulses from neuron to neuron. The most important neurotransmitters in this context are norepinephrine and serotonin. Serotonin is implicated in non-bipolar depression, low levels seeming to produce depression. This is because when levels drop, transmission across the synapses of the neurological system becomes weaker or less effective, which results in the signs and symptoms of depression (Figure 15.7). The hard bit is deciding why the death of a near relative, for example, should bring about this decrease in serotonin levels.

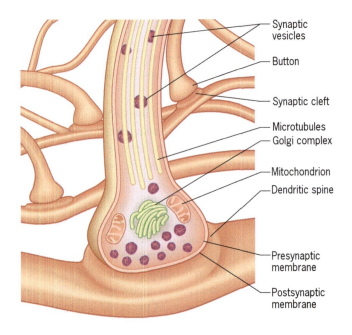

Synaptic vesicles

Button

Synaptic cleft

Microtubules

Golgi complex

Mitochondrion

Dendritic spine

Presynaptic membrane

Postsynaptic membrane

△ **Fig. 15.7** A synapse

It is important to point out that many psychologists are not convinced that this is the 'be all and end all' of depression and it must be said that actually measuring serotonin in working synapses is no trivial matter.

'Fixing' depression without using sport or physical activity

A combination of drug and psychological therapies is often used to try to 'fix' depression. Many believe that we know a great deal about the psychopharmacology of depression, and using drugs as an intervention is common and can be very effective. There are various groups of drugs commonly prescribed, such as tricyclics, monoamine oxidase inhibitors (MOAI) and Prozac (which is a selective serotonin uptake inhibitor). Some people find effective relief from their symptoms by just taking the drugs, whereas others find that the drugs at least get them in a better frame of mind so that they can begin to address their 'problems'. Psychological therapies such as counselling, psychotherapy, etc. are also used both with and without drugs. As this is very often quite complicated it can take time for things to get resolved. Of course depression will sometimes fix itself (as it were).

Using sport, exercise and physical activity to reduce depression

On the face of it there seems much evidence to support the idea that physical activity has a positive effect on depression. Overall this seems to be the case but there are some doubts and some holes in the research. The take-home message is that there is good evidence to support the idea but more studies need to be done.

Defining clinical depression

One sticking point is the problem of defining clinical depression. This is interesting. Many people say they have **clinical** depression (emphasizing the clinical bit) to make sure that people understand that they are more than just fed up. In many ways you are clinically depressed if you go and see your doctor and they evoke an intervention, i.e. give you time off work, prescribe drugs, tell you to go and see a counsellor or therapist or something.

The definition issue is also interesting because many of the studies cited as providing evidence that exercise beats depression were carried out on people who were not diagnosed as clinically depressed, or at least it is not clear that this was the case.

There are objective ways to categorize clinical depression. Classification tools include:

- The Beck Depression Inventory (BDI) (Beck 1967) – a score of 16 or above indicates depression
- The DSM-IV (the Diagnostic and Statistical Manual of Mental Disorders)
- The ICD-10 (International Classification of Diseases)
- The RCD (Research Diagnostic Criteria) (Spitzer *et al.*, 1978)

And I suppose we should pay most attention to studies that use these kinds of measures, for example the studies by Martinsen *et al.* (1985), Mutrie (1988) and Veale *et al.* (1992).

Epidemiological evidence of depression

Epidemiological evidence is evidence gained from examining populations. Morgan (1994) compared fitness levels in hospitalized psychiatric patients with those of non-hospitalized controls. They found that fitness levels were lower in the psychiatric patients than in the controls. Patients who had short (up to 61 days) hospital stays had higher levels of muscular endurance on admission than those who stayed longer (at least one year) even though they had similar initial levels of depression.

This leaves us with an interesting set of questions. Does lack of exercise cause depression, does depression cause lack of exercise, and will increasing fitness speed up recovery?

There are three important studies in this area (they all made statistical adjustments for confounding variables such as age and socio-economic background; see **Box 15.2**):

- **Farmer *et al.* (1998)** (who studied 1497 people over an 8-year period) showed that women who did little or no activity were twice as likely to develop depression as those who engaged in 'much' or 'moderate' activity. This was not true for men, but for those men who were depressed at baseline, inactivity was a strong predictor of depression at the 8-year follow-up.
- **Camacho *et al.* (1991)** performed a study spanning 18 years (baseline, 9 years later and a further 9 years later) and found a relationship between inactivity and depression. The relative risk of depression was greater for both genders in those with low activity levels, but this could be changed by manipulating activity (for example, reducing depression by increasing activity).

- **Paffenbarger *et al.* (1994)** found that men aged 23–27 who engaged in 3 hours or more of sporting activity at baseline had a 27 per cent reduction in risk of developing depression at follow-up compared with those who did an hour or less. How hard they exercised was also important, as those who expended 2500 kcal or more per week were 28 per cent less at risk of developing clinically recognizable depression than those expending less than 1000 kcal/week. And those who expended between 1000 and 2499 kcal/week had a 17 per cent risk reduction compared with those in the least active group.

One way of trying to draw research together and make sense of it is to perform a meta-analysis, which puts all the different bits of evidence together and reanalyses the area as a whole. There have been several of these: McDonald and Hodgdon (1991) and North *et al.* (1990) show that exercise does have an antidepressant effect. Calfas and Taylor (1994) also had a similar finding with healthy and at-risk adolescents – although the number of studies involved was small. Dishman (1995) thought that there was too much dissimilarity between studies to bother trying to do a meta-analysis. However it is important to point out that neither North *et al.* (1990) nor McDonald and Hodgdon (1991) used clinically depressed subjects. Craft and Landers (1998) performed a meta-analysis of only clinically depressed subjects in 30 studies and showed a difference in levels of depression following exercise in moderate to severe depression.

Self-efficacy

The term 'self-efficacy' was coined and studied by Bandura (1997), who defined perceived self-efficacy as being concerned with judgements of personal capability. This differs from self-esteem, which is concerned with judgements of self-worth. According to Bandura there is no fixed relationship between beliefs about one's capabilities and whether or not one likes oneself. Individuals can see themselves (for example) as useless at racket sports, without any loss of self-esteem whatsoever. This is because they do not invest their self-worth in that activity. I play a lot of squash but I'm pretty useless at it. Any half-competent squash player can make me look like an idiot. Except of course they can't really beat me because I only play squash to try to keep fit. Even though I'm rubbish I find it more fun than going to the gym. I don't really care if I crawl off the court a defeated, sweaty wreck because I did 45 minutes of exercise. I simply do not measure my self-worth by how good I am at squash. Of course, with racing motorcycles it's a very different story. . .!

Others may regard themselves as highly efficacious in an activity but take no pride in performing it well. So, I am great at telling off students but I do not enjoy doing it at all.

According to Bandura, people need much more than self-esteem to do well in given pursuits. Many achievers are hard on themselves because they adopt standards that are not easily fulfilled, whereas others enjoy high self-esteem because they do not demand much of themselves, or they derive self-esteem from areas other than personal accomplishments.

Consequently self-liking does not necessarily lead to high performance (in fact self-loathing may lead to high performance). Often high performance is due to what Bandura calls 'toilsome self-disciplined effort'. People need firm confidence in their efficacy to mount and sustain the effort required to succeed. Thus self-efficacy is a good predictor of success whereas self-esteem is not.

Of course the two can be aligned, so we do tend to pursue activities that give us an enhanced sense of self-worth and that we are confident we can pursue but I hope you can see that the two terms efficacy and esteem are not synonymous.

Relationship of self-efficacy to performance in athletes

There is some evidence for the relationship between self-efficacy and various types of athletic performances including tennis (Barling and Abel, 1983), gymnastics (Feltz and Albrecht, 1986; Feltz, 1988), diving (Martin and Gill, 1991), basketball (McAuley and Gill, 1983) and distance running (Morelli and Martin, 1982). High self-efficacy is accompanied by low pre-competition stress and high athletic performance.

In laboratory studies in which efficacy has been lowered or raised this has been found to affect motor performance in both competitive conditions (Weinberg *et al.*, 1979) and non-competitive ones (Gould and Weiss, 1981; McAuley, 1985).

There is the important issue of prior performance as a predictor of performance. Whilst there is some evidence that success breeds self-efficacy there is also other evidence that prior performance can be an inflated predictor of performance – that is, lead the athlete to expect to do better than they should. This is usually related to some interfering or extrinsic factor that has not been taken into account, for example presence of others, being in competition, weather, different emotional or physical state, etc. If, however, one 'dials' these variables into self-efficacy then predictions can become more accurate. So 'How capable am I of hitting that target in this wind?' is likely to bring about a more accurate prediction than 'I can always hit this target so I'll be able to now' (even though it is windy).

Efficacy beliefs in transcendental attainments

This is to do with believing that one can do something that no one else has done before, that is, realizing a potential that one thought was hitherto unobtainable. It is about not doing something because you think it is not possible. Very many people thought the 4 minute mile was unattainable so they did not beat it. Roger Bannister regarded this record as a surmountable barrier. Once this time had been achieved (Figure 15.8(a)), however, soon even high school students were beating it, and Kip Kaino bettered it more than 50 times.

In other words, once these extraordinary achievements become 'doable' they become commonplace. It is almost as though athletes do not beat records because they do not think the records can be beaten. Bob Beamon's long jump record was an example (Figure 15.8 (b)). This record stood for 23 years before Mike Powell (and nearly Carl Lewis) beat the record.

For many athletes, defending titles is harder than winning them (Gould *et al.*, 1993). They become pursued and in order to maintain their position they must defer valued life pursuits outside of their sport. Once a champion, the athlete has to work just as hard to stay in the same place, and they make big demands on themselves, demanding flawless performances to try to live up to the image of being a champion. Anyone with a shaky sense of self-efficacy is likely to be overwhelmed by these pressures (Bandura and Cervone, 1986). Of course most people who succeed heighten their perceived efficacy and inspiration and raise the intensity of their motivation, but others doubt their efficacy to repeat the feat and lower their aspirations and slacken their efforts. Gould and Weiss (1981) suggest that these people are the most vulnerable to failure.

Self-efficacy as a determinant of exercise adoption and adherence

Most of the work on self-efficacy and exercise adherence has been on structured aerobic exercise. The idea is to try to figure out the factors that foster the adoption of exercise regimens and enable lasting adherence to them. It is also about trying to get individuals to abandon their sedentary lifestyles.

It is no surprise to find that weak self-efficacy leads to high drop-out rates from exercise activities (Desharnais *et al.*,

△ **Fig. 15.8** (a) Roger Bannister did not believe that a sub-4 minute mile was impossible, although it had never been done. (b) Bob Beamons' long jump record stood for 25 years.

1986), especially if participants have inflated or unrealistic expectations about speed of benefit. The best efficacy predictor of exercise adherence is not 'Can you do the exercise?' but, given a range of impediments (social, personal and situational), 'How likely are you to make yourself go?' (Sallis *et al.*, 1988). So if we ask the question 'How likely are you to go training in November in the wet and cold?' we get an answer that better fits with actual behaviour. People who distrust this self-regulatory efficacy are poor exercise adherers – they do not go, they do not exercise at the intensity or duration that bring health benefits, and they are quick to drop out (McAulay, 1992, 1993).

In older people the stronger the perceived physical efficacy prior to the exercise programme the greater the physical (and mental) benefits. Whilst this physical functioning may decrease over time, those with high self-efficacy maintain their exercise habit and cardiovascular capacity (McAuley *et al.*, 1993).

Bandura and Cervone (1986) also point out that convenience is a major player in predicting adherence, even if efficacy is high; thus compliance for home exercisers is high, especially for those with high self-efficacy.

Injury and recovery from injury

As sport participation increases so does sport injury.

- In 2003 sport injuries cost Australians $1.5 Billion
- In Germany the cost is US $2.5 Million
- In the United States it is $100 Million in roller-skating alone.

The tradition is to blame much of this on poor fitness, the improper use of equipment, poor coaching, environmental factors and/or the aggressive nature of some sports. Nideffer (1989) points out that despite improvements in all these areas the numbers of such injuries continue to rise. According to Ryde (1965), psychological factors contribute 30 per cent to vulnerability to sport injuries.

So are some people more susceptible to injury? Is personality a factor? There have been studies that have looked at anxiety, attention, locus of control, risk acceptance and self-concept. As we saw in the last chapter, individual differences can be a knotty area for sport psychology so you will not be surprised to learn that research in this area has produced a raft of seemingly contradictory findings.

Ogilvie and Tutko (1966) suggested that injuries to athletes are related to certain aspects of personality, such as masochistic tendencies, determination to punish others, escape from competition, fear of success, need for sympathy, wish to avoid training, attempt to evade the pressures of

demanding parents and, for male athletes, their resolve to prove their manhood. There is little empirical evidence for these ideas (although that may be to do with how difficult it is to turn these ideas into an appropriate methodology). But when studies use standard personality tests such as the 16 Personality Factor Questionnaire (16PF) and the California Psychological Inventory (CPI) things do not seem much clearer. For example, Valliant (1981) found that runners who had been injured in preparation for a race came out as less tough-minded and less forthright on the 16PF. Unfortunately there are some methodological problems with this study though; there were not many subjects and the response rate was poor. Other problems are to do with causality. Did they get their score because they were injured? Or were they more likely to be injured because of their personality? These are pretty 'classic' concerns about correlations! (see **Box 15.3**).

Box 15.3

Correlations

Correlations are a way of testing how well things go together.

- In a **positive** correlation as one variable increases so does the other (e.g. the further you run the more you collect for charity).
- In a **negative** correlation one variable increases while the other decreases (e.g. the more papers you have to carry, the slower you walk).

There is an idea abroad – especially in psychology – that there is no such thing as no correlation between two variables. I'm reminded of the idea that brushing my teeth in Colchester might be correlated with a cat falling off a wall in Hong Kong. Just because we cannot see a connection does not mean there isn't one! Equally, thinking that there is (or ought to be) a connection is no guarantee that there is a correlation.

It is important to understand that correlations are not about cause – one thing does not cause another, they just seem to go together.

There have been several other studies that have examined the relationship between personality and injury both from the retrospective and projective point of view. In football, Taimela *et al.* (1990) showed that even though previously injured players scored differently on the 16PF to non-injured players, this did not predict injury in the following 12 months. Jackson *et al.* (1978) noticed that American football players who scored high on tender-mindedness at the start of the season were more likely to sustain injuries in subsequent weeks. No significant findings of this nature were reported by Brown (1971) (grid iron players) or by Abadie (1976), Govern and Koppenhaver (1965) or Kraus and Gullen (1969).

So the take-home message here is that personality does not seem to predict injury. Or maybe we have not found a methodology that enables us to test any relationship.

Are you more susceptible to injury if you are anxious?

There is a considerable literature on this question and many studies relate stress and muscle tension. For example Nideffer (1983) points out that stress and anxiety can lead to muscle tension, which can in turn lead to reduced muscle coordination and flexibility and thus more susceptibility to injury.

Do you recover slower and cope less well if you are stressed?

To investigate this question we first have to look at what we mean by stress. According to Folkman and Lazerus (1985) stress is best viewed as a transactional process between the organism and the environment. Individuals experience stress when they appraise a potential threat as likely to exceed the resources available to combat it.

There exists a quite muddled literature to address the issue of whether or not ability to cope with stress is related to injury. Blackwell and McCullagh (1990) found no relationship between coping ability and injury in 105 college American football players (but there were some relationships between life events and injury), whereas Hanson *et al.* (1992) showed the opposite – that is, coping did predict injury (or at least poor coping predicted injury!).

Smith *et al.* (1990) investigated 451 high school competitors in basketball, wrestling and gymnastics, and showed that although there were no correlations with coping skills, 22 per cent of the variance in injury was accounted for by those with low social support and psychological skills. Summers *et al.* (1993) have shown that injury is related to coping skills in four different sports, and Kirby and McCleod (1992) found that injury could be predicted from coping variables in NBL basketball players. So again, the take-home message here is that coping variables are important in this area but are not the whole story.

The big contribution to this area comes from 'life events'. The idea here is that stressful life events cost the individual and can make them more susceptible to illness and injury. The major players in the life events area are Holmes and Rahe (1967 and for ever after). The principle is that the more life events the greater the susceptibility to illness. This idea is pretty much accepted. There are clearly individual differences in the amount of impact life events have on a person but the rank order of magnitude shows a clear pattern. Holmes and Rahe have produced a list of these events and they give a score to them according to severity (or at least how likely they are to make you sick) (**Box 15.4**). As you can see, events such as the unexpected death of a near relative come high on the list.

Box 15.4

Life events

The following table lists the 19 life events with the most impact according to the Social Re-adjustment Rating Scale.

Rank	Life event	Mean value
1	Death of spouse	100
2	Divorce	73
3	Marital separation	65
4	Jail term	63
5	Death of close family member	63
6	Personal injury or illness	53
7	Marriage	50
8	Fired at work	47
9	Marital reconciliation	45
10	Retirement	45
11	Change in health of a family member	44
12	Pregnancy	40
13	Sex difficulties	39
14	Gain of a new family member	39
15	Business readjustment	39
16	Change in financial state	38
17	Death of a close friend	37
18	Change to a different line of work	36
19	Change in number of arguments with spouse	35

It is important to realize that positive and negative life events can both be stressful; getting promoted might in its own way be just as stressful as being made redundant.

Life events have been linked to the onset (and poor recovery from) tuberculosis, multiple sclerosis, cancer, psychiatric disorders and (obviously) depression, and although the theory has its critics because much evidence for it is retrospective (e.g. Millar *et al.*, 1990) it is an idea that is highly regarded.

The best regarded measure of life events is the **Social Re-adjustment Rating Scale (SRRS)**. Using this, Holmes (1970) showed that American football players were more likely to become injured if they scored high on the SRRS than if they scored low. This success led to development of the Social and Athletic Re-Adjustment Rating Scale (SARRS), which is more appropriate to an athletic population. In 1975 Bramwell *et al.* showed that players who suffered major injuries had scored significantly higher on the SARRS pre-season assessment, and that 75 per cent of the high SARRS scorers were injured compared with less than a third of the low scorers. Cryan and Alles (1983) looked at 151 footballers and found that 68 per cent of SARRS high scorers were injured.

Getting over sports injuries

There is not a great deal of sport psychology research on this area. This is partly because it is hard to collect data because of the enormous variation in injuries, sports, personalities, backgrounds, etc. The statistical analysis is hard and to some extent injured players are 'invisible' because they do not turn up to sporting events (because they are injured). Much of the evidence is based on single case studies, small numbers or is anecdotal.

The injured player has all the physical problems that go with their injury but they also suffer psychologically. Whilst frustration and boredom are the most obvious contenders, loss of self-esteem and disengagement from their usual markers of the self are also clearly important. Kolt and Kirby (1994) showed that previously injured gymnasts were more likely to report fatigue and anxiety.

An important part of recovery is rehabilitation adherence and clearly those who do not 'play the game' often have a slower or a more uncomfortable recovery. Duda *et al.* (1989) measured adherence on the basis of appointments kept, exercises completed and intensity of effort. More adherence was found in those who reported more belief in the efficacy of the treatment, had more social support and in those who were more self-motivated. A criticism of this investigation was that it was with only mildly injured players (i.e. those who were out for three weeks). Fisher *et al.* (1988) looked at more extensively injured players (out for six weeks) and

found that 'adherers' were more self-motivated, tolerated pain better and worked harder to repair their injuries.

Both of these studies leave one (OK, me!) wondering about the non-adherers and whether this is a rich area for psychological intervention.

Ievleva and Orlick (1991) have listed a number of factors important in recovery from illness and injury. These include attitude and outlook, stress control, social support, goal-setting, positive self-talk, mental imagery and, finally, belief in the effectiveness of the treatment. They looked at 32 athletes with ankle or knee injuries. Whilst some of their findings are 'difficult' from a statistical point of view, they found a relationship between rate of recovery and the injured athlete's use of goal-setting, positive self-talk and healing imagery strategies. This fits with earlier work by Weiss and Troxel (1986).

The end of a sporting career

There is much theorizing in this area, much of which uses social gerontology (the study of ageing). Rosenberg (1981) identifies six of the theories:

- **Activity theory** (Havinghurst and Albrecht, 1953; Burgess, 1960) examines the roles of self-concept and life satisfaction. Ideally the retiring athlete should aim to maintain the same amount of activity post-retirement as they had pre-retirement. They therefore need to fill the gaps left when a sporting career ends with something else that has the equivalent amount of activity. This will lead to positive self-concept and life satisfaction. Of course many people who retire want to do a bit less activity and do a bit more of putting their feet up.

- **Disengagement theory** (Cumming *et al.*, 1960, Cumming, 1961) is a sort of 'yin and yang' theory in as much as it posits that the old need to move over so the young can come up.

Neither of these theories is helpful when making predictions about how people cope.

- **Subculture theory** (Rose, 1965) looks at the social aspects of retirement. Many of us have made some very good friends and some very useful contacts whilst participating in our given sport. When you retire it can mean that you lose your friends. A large part of my sport is bench racing – that is, sitting on a bench (usually in someone's draughty workshop), telling lies and talking rubbish about motorcycle racing. Giving up the sport might mean giving this up – which would be a great shame. This is a kind of subculture and the theory is that people survive well or better if they either remain in their subculture or attach themselves to a new one. This is especially relevant to athletes.

- **Continuity or consolidation theory** (Atchley, 1981) suggests that once we retire we will need a new challenge – one where we can be just as successful as we were pre-retirement. Not doing this brings about a sense of failure.

- **Social breakdown theory** (Kuypers and Bengston, 1973) asserts that losing one's role brings susceptibility to external labelling (e.g. being called a 'has been'). Orlick (1990) calls this the 'hero to zero' experience. If this social labelling is negative there is a tendency for the individual to withdraw. This theory is clearly important to athletes. The remedy is clear but hard to deal with. The retiring athlete needs a new challenge and one that builds on their 'former glories'. Obvious examples are coaching and charity work.

- **Exchange theory** (Dowd, 1975) is based on the old but useful social exchange theory (**Box 15.5**). This is to do with minimizing costs and maximizing return. One reason for giving up a career might be because the costs of competing (e.g. training, injury, pain) outweigh the rewards (i.e. success). Training really hard only to be beaten by a younger athlete can be really soul-destroying.

Box 15.5

Social exchange theory

Let's suppose it is a Friday night and there are three people I could spend the evening with – Jim, Bill and Michael. There are good things and bad things about all three people. Jim is good fun and really makes me laugh but he is always broke, so if I want to spend time with him I'm paying. Bill is a mine of funny stories but he lives miles away, so if I see him I'll have to drive so then I can't have a drink. Michael is really influential at work and could get me that promotion, but he is very boring.

Social exchange theory says that I make the choice of who I spend time with by balancing the costs and rewards of each encounter. So to get the reward of the good thing I incur the cost of the bad thing.

Of course it is not as simple as that because if I choose to see Bill, not only do I incur the cost of the travel, but I also incur the cost of forgoing the rewards of seeing the other two (I also don't incur their costs either, of course). All of which is fine unless Bill wants to see Rachel and not me...

So the idea is that when I make a choice I incur the costs but gain the rewards of that encounter. The theory is based on the (economic) notion that I want to maximize rewards and minimize costs.

Social death

Social death is when you are treated as though you are dead even though physiologically and mentally you are very much alive. It comes from Thanatology – the study of death and dying. Clearly this death is only an analogy and is nothing like real death but the idea is that the dynamics are the same. Except that of course you are alive, so you realize what is happening! It's a bit like going to your own funeral – I know everyone has to go to their own funeral, but you know what I mean!

From a theoretical point of view the two important components of social death are awareness contexts and stages of dying.

Awareness contexts

Glaser and Strauss (1965) propose that there are four levels of awareness:

- **Closed awareness** – the athlete is unaware that there are moves afoot for them to be dropped. Other members of the team or squad may see it coming but don't tell. This is because failure is rarely discussed. The athlete often is shocked and surprised when 'it' happens.

- **Suspicion awareness** – the athlete suspects something is up because they perceive a change in things like verbal and non-verbal communication, from coaches and administrators.

- **Mutual pretence** – a make-believe world, where everyone knows but no one says that no matter how well the athlete trains and performs they are coming to the end of their career.

- **Open awareness** – where (finally) things are out in the open.

Often athletes move through these four stages but sometimes they get hit or caught by some of them.

Stages of dying

Kubler-Ross (1969) draws parallels between the reactions and coping mechanisms of the dying and those of athletes coping with social death. For example:

- **Denial** (not me)
- **Anger** (why me?)
- **Bargaining** (I'll do anything to stay in the game)
- **Depression** (now life is not worth living)
- **Acceptance or resignation** (OK, I give in).

In the case of social death, of course, the individual continues to live, so recovery from death is possible (rare in real death unless you have particular religious beliefs).

Rosenberg and Lerch (1982) think that the social death analogy is particularly useful for forced retirement. However, there are some critics of the idea. Essentially the issues are to do with stereotyping athletes' reactions and an objection to the belief that retirement is the same as death. It seems to me that the theory is helpful for those who do not cope with it, but does not deal with those that do – but then I guess they don't have the same problem!

Transition models

Transition models are derived from counselling and counselling theory and usually define transition as 'discontinuity in a person's life space' (Crook and Robertson, 1991). This refers to a situation where a person knows that things are about to change and they need to manage or deal with the change.

There is nothing easy about studying and researching transition, many variables 'slosh around' in it and it results in different degrees of loss and stress (Brammer and Abrego, 1981). According to Schlossberg's (1981, 1984) model there are three interacting factors that affect transition:

1. **The characteristics of the transition** – the trigger, timing, source and duration of the transition, the role changes involved, the current level of stress, and the individual's previous experience of transition.

2. **The characteristics of the individual** – their personality, socio-economic status, age, stage of life, state of health and coping skills.

3. **The characteristics of the environment** – the extent of their social support networks, both inside and out of the sport and the nature and quality of options for the athlete outside the sport.

All three characteristics are seen as potential assets and liabilities, depending on the individual's appraisal of the transition, self and personal environment, giving one a multidimensional approach to understanding the factors relating to successful adjustment (Crook and Robertson, 1991).

Career transition research

One of the issues which is often researched is the notion of **athletic identity**, that is, the extent to which the individual defines themselves as an athlete. For some people that is how they identify themselves, for other people athletics is one of the things that they do. The former type is referred to as **excusive athletic identity** and most researchers think that this form of self-reference makes transition hard.

Two studies are relevant to this. Werthner and Orlick (1986) showed that former Canadian Olympians who had pre-arranged alternative career options and commitments

adjusted to the transition better than those who made no plans. Kleiber et al. (1987) showed that college athletes whose careers were terminated early due to injury reported lower life satisfaction than their peers whose careers were not terminated early. Although I do not think these two studies exactly support the theory, they do give us reason to pursue the idea.

Werthner and Orlick (1986) studied the transition of 28 Canadian Olympians. They identified seven major factors that seemed to determine the nature of the transition:

1. **A new focus** – Having options or alternatives is important.

2. **A sense of accomplishment** – It helps if the athlete feels as though he or she has had a successful career.

3. **Coaches** – Poor coaches or coaches with whom there were conflicts make transition harder.

4. **Injuries and health problems** –These produce poor transition, especially if the injury/illness terminates the career.

5. **Politics or sport association problems** – Athletes who 'leave under a cloud' or feel unsupported by their association or blame the association for the end of their career have 'bad' transitional experiences.

6. **Finances** – If they have left with poor financial stability or stopped because they could not afford to continue training then athletes often suffer in transition.

7. **Social support** – Good social support makes transition easier; poor social support makes it harder.

Summary
Abnormality and illness

- Defining abnormality is very difficult. Even though at some level we 'kind of know' what we mean, the five definitions discussed in this section are insufficient and defining normality is just as tough.

- Mental 'illness' is probably more common than we feel comfortable with and as an idea it is wrapped up with a medical model which sees mental 'troublement' as something that can and should be fixed.

Body image

- Body image is an important part of many people's lives and it seems as though lots of us are discontented with the way we look.

- It is the reason why many people participate in sport and exercise and this participation often seems to reduce this discontent. This can lead to an improvement in self-esteem.

- It is common to think that women are more susceptible to this relationship than men but there is some evidence to show that this may be changing.

Self-esteem, sport and exercise

- Self-esteem is an important idea and one which many seem to understand and value.
- There is a growing body of evidence that demonstrates that increasing exercise seems to improve self-esteem but it is not clear why this is.

Depression

- Depression is the most common mental 'illness' and blights many lives.
- Whilst we are good at describing depression, and know something of the biological aspects of it, we are less clear about the psychological mechanism that triggers it.
- There is quite a lot of evidence to suggest that depressed people suffer less with depression if they take exercise and that for some, expending more than 2500 kcal in an exercise session is especially useful.

Self-efficacy

- Self-efficacy is conceptually different from self-esteem and is often an important predictor of both sporting success and exercise adherence.
- It is about making future-orientated judgements about how likely it is that we can gain a particular goal.
- This is especially salient when it comes to trying to break records, or just do some exercise.

Injury and recovery from injury

- Many factors seem to predict whether or not we become injured and some of them seem to have very little to do with the sport itself (i.e. they are often psychological or social and are to do with the other things that are going on in your life that are not to do with sport).
- Whilst it is clear that there are many psychological factors that mediate the recovery from injury, the scientific evidence for them is thin on the ground.
- We are beginning to learn more about this area through the use of case studies.

The end of a sporting career

- There is no doubt that whilst the end of a sporting career must happen it is no less painful for that.

- There are several theories that try to explain and describe this process. These break down into those that see the end of a career as a type of social death and those which concern themselves with the nature of the transition from athlete to non-athlete.
- Several theories are quite useful for developing strategies to enable athletes to cope with this change.
- It is clear that the essence of these is to be active in planning for that inevitable day.

Review Questions

Abnormality and illness

1. List five ways of defining abnormality and say what is wrong with them.
2. How rare is mental illness?
3. What is mental illness? (This is far too difficult a question – don't try to answer it!)

Body image

1. Which groups of people are often susceptible to harming themselves by being over-concerned with body image?
2. How do we measure body image?
3. To what extent can we improve body image with physical exercise?
4. How related are body image and self-esteem?
5. Are there any gender differences in body image?

Self-esteem, sport and exercise

1. What kind of exercise improves self-esteem the most?

Depression

1. What are the signs and symptoms of depression?
2. What causes depression?
3. How successful are we at treating depression with exercise?

Self-efficacy

1. In what way is self-efficacy different from self-esteem?
2. What is a transcendental achievement and how do we attain one?
3. In what way is self-efficacy important to exercise adherence?

Injury and recovery from injury

1. Describe why sports people get injured without considering the nature of the sport they play.
2. Write a psychological strategy to enable an injured athlete to recover.

The end of a sporting career

1. Devise a strategy based on the theories described in this section to help an athlete come to terms with the end of a sporting career.

References

Abadie DA (1976) Comparison of personalities of non-injured and injured female athletes in inter-collegiate competition. *Dissertation Abstracts* 15(2): 82.

Atchley RC (1981) T*he Social Forces in Later Life*. Belmont, Ca, Wadsworth.

Baldwin MK, Corneya KS (1997) Exercise and self-esteem in breast cancer survivors: an application of the exercise and self-esteem model. *The Journal of Sport and Exercise Psychology* 19: 347–358.

Bandura A (1997) *Self-efficacy: The Exercise of Control*. New York, WH Freeman.

Bandura A, Cervone D (1986) Differential engagement of self-reactive influences in cognitive motivation. *Organizational Behavior and Human Decision Processes* 38: 92–113.

Barling J, Abel M (1983) Self-efficacy beliefs and tennis performance. *Cognitive Therapy and Research* 7: 265–272.

Beck AT (1967) *Depression: Clinical, Experimental, and Theoretical Aspects*. New York, Harper Row.

Blackwell B, McCullagh P (1990) The relationship of athletic injury to life stress, competitive anxiety and coping resources. *Athletic Training* 25: 23–27.

Brammer LM, Abrego PJ (1981) Intervention strategies for coping with transitions. *The Counselling Psychologist* 9(2): 19–36.

Bramwell ST, Masuda M, Wagner NN, Holmes TH (1975) Psychosocial factors in athletic injuries: development and application of the social and readjustment rating scale (SARRS) *Journal of Human Stress* 1: 6–20.

Brown RB (1971) Personality characteristics related to injury in football. *The Research Quarterly* 42: 133–138.

Brown DR, Wang Y, Ward A, Ebbeling CB, Fortlage L, Puleo E, Benson H, Rippe JM (1995) Chronic psychological effects of exercise and exercise plus cognitive strategies. *Medicine and Science in Sports and Exercise* 27: 765–775.

Burgess E (1960) *Aging in Western Societies*. Chicago, University of Chicago Press.

Calfas KJ, Taylor C (1994) Effects of physical activity on psychological variables in adolescents. *Paediatric Exercise Science* 6: 406–423.

Camacho TC, Roberts RE, Lazarus NB, Kaplan GA, Cohen RD (1991) Physical activity and depression: evidence from the Alameda county study. *American Journal of Epidemiology* 134: 220–231.

Campbell RN (1984) *The New Science: Self-esteem Psychology*. Lanham, MD, University Press of America.

Craft LL, Landers DM (1998) The effect of exercise on clinical depression and depression resulting from mental illness: a meta-analysis. *Journal of Sport and Exercise Psychology* Vol 20(4): 339–357.

Crook JM, Robertson SE (1991) Transitions out of elite sport. *International Journal of Sports Psychology* 22: 115–127.

Cryan P, Alles W (1983) The relationship between stress and college football injuries. *Sports Medicine* 23: 52–58.

Cumming E, Henry WE (1961) *Growing Old and The Process of Disengagement*. New York, Basic Books.

Cumming E, Dean LR, Newell DS, McCaffrey I (1960) Disengagement: a tentative theory of aging. *Sociometry* 13: 23.

Davis C (1997) Body image, exercise, and eating behaviors. *The Physical Self: From Motivation to Well-being*. KR Fox. Champaign, Il, Human Kinetics: 143–174.

Davis C, Brewer A, Ratusny D (1993) Behavioral frequency and psychological commitment: necessary concepts in the study of excessive exercising. *Journal of Behavioral Medicine* 16: 611–628.

Davis C, Katzman MA (1998) Chinese men and women in the United States and Hong Kong: body and self-esteem ratings as a prelude to diet and exercise. *The International Journal of Eating Disorders* 23(1): 99–102.

Dekel Y, Tenenbaum G, Kudar K (1996) An exploratory study on the relationship between postural deformities and body image and self-esteem in adolescents: the mediating role of physical activity. *The International Journal of Sports Psychology* 27(2): 183–196.

Desharnais R, Bouillon J, Godin G (1986) Self-efficacy and outcome expectations as determinants of exercise adherence. *Psychological Reports* 59: 1155–1159.

Diener E (1984) Subjective well-being. *Psychological Bulletin* 95: 542–575.

Dishman RK (1995) Physical activity and public health: mental health. *Quest* 47: 362–385.

Dowd JJ (1975) Aging as exchange: a preface to theory. *Journal of Gerontology* 30: 584–594.

Duda JL, Smart AE, Tappe MK (1989) Predictors of adherence in the rehabilitation of athletic injuries: an application of personal investment theory. *Journal of Sport and Exercise Psychology* 11: 367–381.

Farmer M, Locke B, Moscicki E, Dannenberg A, Larson D, Radloff L (1988) Physical activity and depressive symptoms: the NHANES I epidemiologic follow-up study. *American Journal of Epidemiology* 128: 1340–1351.

Feltz DL (1988) Self-confidence and sports performance. *Exercise and Sport Sciences Reviews* 16: 423–457.

Feltz DL, Albrecht RR (1986) The influence of self-efficacy on the approach/avoidance of a high-avoidance motor task. In: Humphrey JH, VanderVelden L, *Psychology and Sociology of Sport*. New York, AMS Press.

Fisher AC, Damm MA, Wuest DA (1988) Adherence to sports injury rehabilitation programs. *The Physician and Sports Medicine* 16: 47–51.

Folkman S, Lazarus RS (1985) If it changes it must be processed: study of emotion and coping during three stages of a college examination. *Journal of Personality and Social Psychology* 48: 150–170.

Fox KR (1997) The physical self-perception profile manual. In: Fox KR *The Physical Self: From Motivation to Well-being*. Champaign, Il, Human Kinetics: 111–140.

Fox KR (2000) *The Effects of Exercise on Self Perceptions and Self-esteem*. London, Routledge.

French SA, Story M, Perry CL (1995) Self-esteem and obesity in children and adolescents: a literature review. *Obesity Research* 3: 479–490.

Glaser B, Strauss A (1965) *Awareness of Dying*. New York, Aldine.

Gould D, Weiss M (1981) Affect of model similarity and model self-talk on self-efficacy on muscular endurance. *Journal of Sports Psychology* 3: 17–29.

Gould D, Jackson SA, Finch LM (1993) Life at the top: the experiences of US national champion figure skaters. *The Sport Psychologist* 7: 354–374.

Govern JW, Koppenhaver R (1965) Attempts to predict athletic injuries. *Medical Times* 93: 421–422.

Guinn B, Semper T, Jorgensen L (1997) Mexican American female adolescent self-esteem: the effect of body image, exercise behavior and body fatness. *The Hispanic Journal of Behavioral Science* 19(4): 517–526.

Hanson SJ, McCullagh P, Tonyman P (1992) The relationship of personality characteristics, life stress, and coping resources to athletic injury. *Journal of Sport and Exercise Psychology* 14: 262–272.

Havinghurst RJ, Albrecht R (1953) *Older People*. New York, Longmans Green.

Holmes TH (1970) Psychological screening in football injuries. Paper presented at a workshop on athletic injuries. National Research Council, Washington DC.

Holmes TH, Rahe RH (1967) The social readjustment rating scale (SRRS). *The Journal of Pyschosomatic Research* 6: 213–217.

Ievleva L, Orlick T (1991) Mental links to enhanced healing: an exploratory study. *The Sports Psychologist* 5: 25–40.

Jackson DW, Jarrett H, Bailey D, Kausek J, Swansen J, Powell JW (1978) Injury prediction in the young athlete: a preliminary report. *American Journal of Sports Medicine* 6: 6–14.

Kessler RC, McGonagle KA, Zhao S, Nelson CB, Hughes M, *et al.* (1994) Lifetime and twelve month prevalence rates of DSM-III-R psychiatric disorders in the United States: results from the co-morbidity survey. *Archives of General Psychiatry* 51: 8–19.

Kirkby RJ, McCleod S (1992) The psychological antecedents of injury in National level basketball players. Paper presented at the 27th Annual conference of the Australian Psychological Society (Armidale).

Kleiber D, Greendorfer S, Blinde E, Samdahl D (1987) Quality of exit from university sports and life satisfaction in early adulthood. *Sociology of Sport Journal* 4: 28–36.

Klerman GL (1988) The current age of youthful melancholia. *British Journal of Psychiatry* 152: 4–14.

Kolt GS, Kirkby RJ (1994) Injury, anxiety and mood in competitive gymnasts. *Perceptual and Motor Skills* 78: 955–962.

Kraus JF, Gullen WH (1969) An epidemiological investigation of predictor variables associated with intramural touch football injury. *American Journal of Public Health* 59: 2144–2156.

Kubler-Ross E (1969) *On Death and Dying*. New York, Macmillan.

Kuypers JA, Bengston VL (1973) Social breakdown and comptence: a model of normal aging. *Human Development* 16: 181–220.

Leit RA, Gray JJ, and Pope HG (2002) The media's representation of the ideal male body: a cause for body dysmorphia? *International Journal of Eating Disorders* 31(3), 334–8.

Loland NW (1998) Body image and physical activity: a survey among Norwegian men and women. *International Journal of Sports Psychology* 29(4): 339–365.

Loland NW (1999) *Body image and physical activity*. Norwegian University of Sport and Physical Education.

Lox CL, McAuley E, Tucker RS (1995) Exercise as an intervention for enhancing subjective well-being in an HIV-1 population. *Journal of Sport and Exercise Psychology* 17: 345–362.

Mactavish JB, Searle MS (1992) Older individuals with mental retardation and the effect of a physical activity intervention on selected social psychological variables. *Therapeutic Recreational Journal* 26: 38–47.

Marsh HW (1998) Age and gender effects in physical self-concepts for adolescent elite athletes and non-athletes: a multi-cohort – multi-occasion design. *Journal of Sport and Exercise Psychology* 20: 237–259.

Marsh HW, Peart ND (1988) Competitive and co-operative physical fitness training programs for girls: effects on fitness and multidimensional self-concepts. *Journal of Sport and Exercise Psychology* 10: 390–407.

Marsh HW, Perry C, Horsely C, Roche L (1995) Multi-dimensional self-concepts of elite athletes: how do they differ from the general population? *The Journal of Sport and Exercise Psychology* 17: 70–83.

Martin JJ, Gill DL (1991) The relationships among competitive orientation, sport-confidence, self-efficacy, anxiety, and performance. *Journal of Sport and Exercise Psychology* 13: 149–159.

Martinsen EW, Medhus A, Sandvik L (1985) Effects of aerobic exercise on depression: a controlled trial. *British Medical Journal* 291: 100.

McAuley E (1985) Modelling and self-efficacy: a test of Bandura's model. *Journal of Sport Psychology* 7: 283–295.

McAuley E (1992) Understanding exercise behavior: a self-efficacy perspective. *Motivation in Sport and Exercise.* Roberts GC, Champaign, Il, Human Kinetics: 107–127.

McAuley E (1993) Self-efficacy, physical activity, and aging. *Activity and Aging: Staying Involved in Later Life.* Roberts GC, Newbury Park, Califtom, Sage.

McAuley E, Gill DL (1983) Reliability and validity of the self-efficacy scale in a competitive sport setting. *Journal of Sport Psychology* 5: 410–418.

McAuley E, Lox C, Duncan TE (1993) Long-term maintenance of exercise, self-efficacy, and physiological change in older adults. *Journal of Gerontology: Psychological Sciences* 48: 218–224.

McDonald DG, Hodgdon JA (1991) *Psychological Effects of Aerobic Fitness Training: Research and Theory.* New York, Springer-Verlag.

Millar TW, Vaughn MP, Millar JM (1990) Clinical issues and treatment strategies in stress orientated athletes. *Sports Medicine* 9: 370–379.

Morelli EA, Martin J (1982) *Self-efficacy and athletic performance of 800m runners.* Manuscript, Simon Fraser University, Vancouver, VC.

Morgan WP (1994) Physical activity, fitness and depression. In: Bouchard C, Shepherd, RJ, Stephens T, *Physical activity, fitness and health.* Champaign, Il, Human Kinetics.

Murray, CJ and Lopez AD (1997) Alternative projections of mortality and disability by 1990–2020: global burden of disease study. *Lancet* 349: 1498–1504.

Mutrie N (1988) Exercise as a treatment for moderate depression in the UK National Health Service. *Sport, Health, Psychology and Exercise symposium Proceedings.* London, The Sports Council and Health Education Authority.

Nardini M (1998) *Body image, disordered eating and obligatory exercise among women fitness instructors*, Indiana University.

Nideffer RM (1983) The injured athlete: psychological factors in treatment. *Orthopedic Clinics of North America* 14: 373–385.

Nideffer RM (1989) Psychological aspects of sports injuries: issues in prevention and treatment. *International Journal of Sports Psychology* 20: 241–255.

North TC, McCullagh P, Tran ZV (1990) The effect of exercise on depression. *Exercise and Sport Sciences Reviews* 18: 379–415.

Ogilvie BC, Tutko TA (1966) *Problem Athletes and How to Handle Them.* London, Pelham.

Orlick T (1990) *In Pursuit of Excellence: How to win in Sport and Life through Mental Training.* Kingswood, Human Kinetics.

Paffenbarger RS, Lee IM, Leung R (1994) Physical activity and personal characteristics associated with depression and suicide in American college men. *Acta Psychiatrica Scandinavia* 89: 16–22.

Page A, Fox KR (1997) Adolescents' weight management and the physical self. *The Physical Self From Motivation to Wellbeing.* Fox KR, Champaign, Il, Human Kinetics.

Pretty J, Sellens M, Griffin M (1993) Is nature good for you? *Ecos* 24: 3–4.

Rose A (1965) *The Subculture of Aging: A Framework in Social Gerontology.* Philadelphia, MS, Davis FA, Company.

Rosenberg E (1981) *Gerontological Theory and Athletic Retirement.* West Point, NY, Leisure Press.

Rosenberg E, Lerch SH (1982) *Athletic Retirement as Social Death: Concepts and perspectives.* Fort Worth, Tx, Texas Christian University Press.

Ryde D (1965) The role of the physician in sport injury prevention. *Journal of Sports Medicine* 5: 152–155.

Sallis JF, Pinski RB, Grossman RM, Patterson TL, Nader PR (1988) The development of self-efficacy scales for health-related diet and exercise behaviors. *Health Education Research: Theory and Practice* 3: 283–292.

Schempp PG, Cheffers JTF, Zaichowsky LD (1983) Influence of decision-making on attitudes, creativity, motor skill and self-concept in elementary children. *Research Quarterly for Exercise and Sport* 54: 183–189.

Schlossberg NK (1981) A model for analysing human adaptation to transition. *The Counselling Psychologist* 9(2): 2–18.

Schlossberg NK (1984) *Counselling in Transition.* New York, Springer.

Scully D, Kremer J, Meade MM, Graham R, Dudgeon K (1998) Physical exercise and psychological well-being: a critical review. *British Journal of Sports Medicine* 32(2): 111–120.

Secord PF, Jourard SM (1953) The appraisal of body cathexis: body cathexis and the self. *Journal of Consulting Psychology* 17: 343–347.

Smith EL, Smith KA, Gillingan C (1990) Exercise, fitness osteoarthritis, and osteoporosis. In: Bouchard C, Shepherd RJ, Stephens T, Sutton JR, McPherson BD, *Exercise, Fitness and Health*. Champaign Il, Human Kinetics: 517–528.

Smith BL, Handley P, Eldredge DA (1998) Sex differences in exercise motivation and body image satisfaction among college students. *Perceptual and Motor Skills* 86(2): 723–732.

Spitzer RL, Endicott J, Robins E (1978) Research diagnostic criteria. *Archives of General Psychiatry* 35: 773–782.

Summers JJ, Fawkner H, McMurray N (1993) Predisposition to athletic injury: The role of psychosocial factors. Paper presented at the Eighth World Congress of Sport Psychology (Lisbon)

Taimela S, Osterman I, Kujala U, Lehto M, Korhonen T, Alaranta H (1990) Motor ability and personality with reference to soccer injuries. *Journal of Sports Medicine and Physical Fitness* 30: 194–201.

Taub DE, Blinde EM, Greer KR (1999) Stigma management through participation in sport and physical activity: experiences of male college students with physical disabilities. *Human Relations* 52(11): 1469–1484.

Thompson JK (1990) *Body Image Disturbance: Assessment and Treatment*. Oxford, Pergamon Press.

Valliant PM (1981) Personality and injury in competitive runners. *Perceptual and Motor Skills* 53: 251–253.

Veale D, Le Fevre K, Pantelis C, de Souza V, Mann A, Sargeant A (1992) Aerobic exercise in the adjunctive treatment of depression: a randomised controlled trial. *Journal of the Royal Society of Medicine* 85: 541–544.

Weinberg RS, Gould D, Jackson A (1979) Expectations and performance: an empirical test of Bandura's Self-efficacy theory. *Journal of Sports Psychology* 1: 320–333.

Weiss MR, Troxel RK (1986) Psychology of the injured athlete. *Athletic Trainer* 21(2): 104–109.

Werthner P, Orlick T (1986) Retirement experiences of successful Olympic athletes. *International Journal of Sports Psychology* 17: 337–363.

Willis JD, Campbell LF (1992) *Exercise Psychology*. Champaign, Il, Human Kinetics.

Further reading

Bartlewski PP, Van Raalte JL, Brewer BW (1996) Effects of aerobic exercise on the social physique anxiety and body esteem of female college students. *Women in Sport and Physical Activity Journal* 5: 49–62.

Weinberg RS, Gould D, Yekelson D, Jackson A (1981) The effect of pre-existing and manipulated self-efficacy on a competitive muscular endurance task. *The Journal of Sports Psychology* 4: 345–354.

Weinberg R (1986) Relationship between self-efficacy and cognitive strategies in enhancing endurance performance. *The International Journal of Sports Psychology* 17: 280–293.

16 The Psychology of Motor Learning and Performance

Chapter Objectives

In this chapter you will learn about:

- Skills
- Learning
- Performance
- Abilities
- Motor skill development.

Introduction

In this chapter we are going to be considering **skills** and **abilities**. These are not the same thing at all, and are the cause of confusion for many people. I hope that by the end of the chapter you will have this sorted out. If you are interested in coaching and training people in sport you should find information about how we learn very interesting and useful. I also hope to convince you that there is no such thing as a 'learning curve'. My wife will tell you how cross it makes me when someone on the television says 'Well David, this has been a real steep learning curve for us'. The reason for this is simple: you cannot measure learning. I'll explain why later.

When we look at skills and abilities we will consider the role of individual differences in competence at sport. We will also look at how skills develop in children and show that there are some gender differences in these.

Skills

This section has two parts: one is an introduction to the idea of skilled performance and the other is a description of motor performance. It includes an introduction to the concept of skill and a discussion of the various features of this definition.

A critical feature of human existence is our capacity to perform skills. Human skills can take many forms, from those that emphasize the control and coordination of our large muscle groups (for forceful activities) to those in which the smallest muscle groups must be tuned precisely (for intricate activities). Because skills make up such

a large part of human life, understanding their determinants and the factors that affect their performance provides many applications to numerous aspects of life, including improving performance in sports.

Skill definition

In humans, skills are many and varied and this makes them difficult to define in a way that applies to all cases. However, Guthrie (1952) provided a definition that encapsulates most of the critical features of a skill: 'a skill is the ability to bring about some end result with maximum certainty and minimum outlay of energy, or of time and energy'. Such a definition sounds pretty obvious and is one of those 'do we really need a psychologist for this' moments. Well maybe we don't, but when you look a little closer at this old definition, one realizes how important and how clever it is.

Performing skills implies some desired **environmental goal**; in other words, the person is trying to achieve something like score a goal or jump a hurdle. Although skills include movements, they are usually thought of as different from movements. Movements do not necessarily have any particular environmental goals. To be skilled implies meeting this performance goal or end result with maximum certainty and not just by luck. I've hit the gold many times in archery, I've even performed a cracking serve in squash but usually no one is more surprised than me that I have. When I play sport I want to come up with something stunning but if I do I usually have no idea how I did it. You can call yourself skilled when you can do 'it' just about every time.

For many skills the minimization and thus **conservation of the energy** required for performance is critical. Some car racers can drive like a thing possessed for one lap but to be a winner one has to do it lap after lap throughout the race. The use of minimal energy requires organizing action with physiological as well as psychological or mental energy costs at their lowest. Minimizing energy costs is beneficial because expending little attention allows cognitive processes to be available for other features of the activity, such as strategies (in games) or expression (in performance).

Another important feature of many skills is to achieve goals in **minimum time**. Many sports have this as the only competition goal. However, minimizing time can interact with the other skill features mentioned. Speeding up performance often results in sloppy movements that have less certainty in terms of achieving the environmental goal. Also, increased speed generates movements for which the energy costs are sometimes higher. This shows how an understanding of skills involves optimizing and balancing several skill aspects that are important to different extents in different settings.

To summarize, skills generally involve achieving some well-defined environmental goal by:

- Maximizing the achievement certainty
- Minimizing the physical and mental energy costs of performance
- Minimizing the time used.

What are the three main features of a skill?

The performance of the skilled individual may appear simple (and very often that is how we know they are skilful), however, the goals of the skill were realized through a complex combination of interrelating mental and motor processes (Figure 16.1).

There are three elements critical to almost any skilled performance:

- **Perception** – perceiving the relevant environmental features
- **Decision-making** – deciding what to do and where and when to do it
- **Activity production** – producing organized muscular activity to generate movements.

Therefore, perceptual processes lead to decisions about what to do, how to do it, and when to do it.

Skills typically depend on the quality of movement generated as a result of those decisions. Even if the situation is correctly perceived and the response decisions are appro-

△ **Fig. 16.1** Dougie Lampkin – an excellent example of the three main skill features in action.

priate, performance will not be effective in meeting the environmental goal if actions are executed poorly.

It is interesting, but unfortunate, that each of these skill components seems to be recognized and studied in isolation from the others and by separate schools of psychologists. A major problem for the study of skills, therefore, is the fact that several components of skill are studied by widely different groups of scientists, generally with little overlap and communication among them.

The major processes underlying actions are:

- **Sensory or perceptual processes** – studied in cognitive psychology and psychophysics
- **Decision-making processes** – studied in cognitive and experimental psychology
- **Motor-control or movement-producing processes** – studied in the neurosciences, physical education and physiology.

Skill classifications

All of the processes mentioned are present in almost all motor skills. However, this is not to say that all skills are fundamentally the same. In fact, the principles of human performance and learning depend to some extent on the kind of movement skill to be performed. In general, classifications are beneficial as they enable us to both better understand as well as apply theoretical knowledge.

Therefore, with relation to motor skills, classifications help us to understand how the many varieties of skills help us to apply information about how we can control and learn motor skills and how better to instruct motor skills. Classifying motor skills into general categories is usually based on determining what components or elements of a skill are common or similar to components of another skill. There are several skill classification systems that help to organize motor skills. Each classification system tends to have its own area or context for use.

It should be pointed out that most of the methods of classification use a dichotomy. That is, skills in each classification are presented as being in one of two categories. This is clearly an artificial situation since not many skills are wholly in one classification, therefore, in this method of classification a continuum or dimension idea is used. This means that skills can be classified as being more closely allied with one category than with the other without having to totally fit into that category exclusively.

We shall consider four classification systems of skill:

- Open–closed system
- Discrete, continuous and serial skill classification
- Two-dimensional taxonomy
- The motor and cognitive skill system.

The open–closed skill classification system

One way to classify motor skills concerns the extent to which the environment is stable and predictable throughout the performance of the skill. In this classification skills are categorized by the stability of the environment.

The open–closed classification system for motor skills was originally presented by Poulton (1957) for use in the industrial setting. However, Gentile (1972) expanded it to make it applicable to the instruction of sports.

According to this classification system, **open skills** form one end of the dimension, with **closed skills** forming the other end. Poulton defined an open skill as 'One for which the environment is variable and unpredictable during the action'. Gentile saw open skills being performed in a temporarily and/or spatially changing environment. For these tasks to be performed the performer must act upon the object according to the action of the object. An example of an open skill would be carrying the ball against a defensive team in football, where it is difficult to predict the future moves of the opponent (hence future responses to the opponent) very effectively.

A closed skill, on the other hand, is: 'one for which the environment is stable and predictable'. Poulton classified a skill as closed if the environment is stable, in other words is predictable. Gentile considered that on this end of the continuum the object waits to be acted upon by the performer. Therefore the performer is not required to begin action until they are ready to do so (Figure 16.2).

△ **Fig. 16.2** An archer uses closed skills.

These open and closed designations actually only mark the end points of a spectrum with skills lying between having varying degrees of environmental predictability or variability.

Another set of terms has been used interchangeably with the open–closed skills categories. The term **self-paced skill** is synonymous with the closed skill category. This is because the performer's pace of when and how to initiate the required action is determined by the performer. The other extreme is the **forced-paced skill**, which is synonymous with the open skill. Here the initiation of action is determined by an external source, the stimulus. The performer is forced into action.

Gentile (2000) made an interesting extension of the open–closed classification system that is especially helpful in further identifying where a motor skill belongs on the open–closed continuum. They suggested that rather than merely considering a single dimension, four categories could be identified:

- Two types of variations from one response attempt to the next – change or no change
- Two types of environmental conditions during execution of the movement –stationary and in motion.

These can be presented in a 2 × 2 diagram as shown in Figure 16.3.

In keeping with our continuum approach, it is interesting to arrange Gentile's 2 × 2 diagram in a continuum

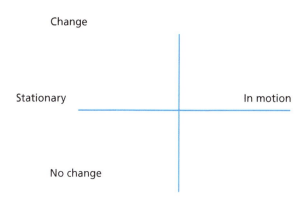

Change

Stationary   In motion

No change

△ **Fig. 16.3** 2x2 diagram of Gentile *et al.*'s extension to the open–closed classification of skills. A skill is classified according to how it fits the intersections of any two of these types of movement conditions.

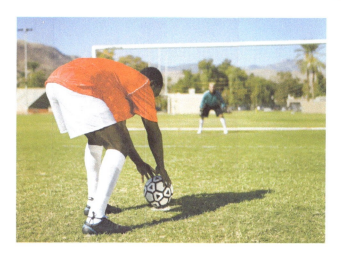

△ **Fig. 16.4** Taking a penalty is a discrete skill.

configuration. The no change/stationary conditions category becomes the extremity on the closed end of the continuum. The change/in motion category is the extremity for the open end of the continuum.

An advantage of presenting this four-point continuum can be seen when a teacher needs to determine how to modify an open skill, such as hitting a baseball thrown by a pitcher. We can use this idea to help us teach this skill and to determine progressively more difficult steps.

The discrete, continuous and serial skill classification system

A second scheme for classifying skills concerns the extent to which the movement is an on-going stream of behaviour, as opposed to a brief, well-defined action. In this classification skills are categorized by the beginning and end points of the movement. At one end of this dimension is a discrete skill; at the other end is a continuous skill. Between the polar ends of the discrete and continuous dimension is the serial skill.

A **discrete skill** usually has an easily defined beginning and end, often with a very brief duration of movement (such as throwing or kicking). If there are clearly defined beginning and end points, then the skill is categorized as a discrete motor skill (Figure 16.4).

A **continuous skill** has no particular beginning or end, with the behaviour flowing on for many minutes. Therefore, if a skill has rather arbitrary beginning and end points, the category label is continuous. The performer or some external force determines the beginning and end point of the skill rather than the skill itself. Sport skills such as swimming or running can be considered continuous in that the beginning and end points of the task are

determined by the performer and not specified by the task itself (Figure 16.5).

△ **Fig. 16.5** Running is a serial skill.

A **serial skill** is often thought of as a group of discrete skills strung together to make up a new, more complicated skilled action. Here serial implies that the order of the elements is usually critical for the successful performance. Therefore, discrete skills can be put together in a series. When this occurs we consider the skill to be a serial motor skill. Many sports skills can be considered serial skills; for example, performing an ice dance routine consists of a specific series of movements that must be performed in a specific order for the proper execution of the skill (Figure 16.6).

Discrete and continuous skills can be quite different, requiring different processes for performance. This dictates that they be taught and coached somewhat

△ **Fig. 16.6** Ice skating is an example of a continuous skill.

differently as a result. Serial skills differ from discrete skills in that the movement durations are somewhat longer, yet each movement retains a discrete beginning and end.

The use of this classification system has been especially prevalent in the motor skills research literature.

Two-dimensional taxonomy

This is another classification system developed by Gentile (2000). It deals with:

1. The environmental context
2. The function of the action.

The idea is that actions take part in an environmental context and the interaction between the two is what leads to a much more complex range of categories. Gentile uses this to generate 16 skill categories as shown in Table 16.1.

- **Stationary regulatory** conditions are when the conditions are controlled (e.g. the run up and sand trap in a long jump).
- **In motion regulatory** conditions are when the conditions are changeable (e.g. in football or basketball).
- **Inter-trial variability** is to do with the extent to which things are the same every time. Clearly in football they seldom are, but even in sports like archery even the wind can produce some variability.

- **Body orientation** is about changing position (i.e. moving, as would a goalkeeper when trying to save a penalty) or maintaining position (i.e. not moving, as would an archer prior to taking a shot).
- **Object manipulation** is about changing the position of an object (i.e. moving an object, for example hitting a golf ball) or maintaining the position of an object (i.e. not moving an object, for example holding a ball still prior to a serve).

Examples of **stationary environmental context** are things like picking up a cup, walking up a flight of stairs, hitting a ball from a tee, throwing a dart at a target. That is, it does not move unless you move it.

Examples of **in motion spatial environmental context** are stepping onto an escalator, standing in a moving bus, hitting a pitched ball, catching a batted ball – that is, it is already moving and you catch it up.

In the taxonomy table (Table 16.1) the complexity of the skill moves from top left to bottom right.

The motor and cognitive skill classification system

It is sometimes useful to consider a fourth dimension labelled 'motor and cognitive' skills. In this classification skills are categorized by the components of the skill. At one end of this dimension is a motor skill; at the other end is a cognitive skill.

With a **motor skill** the primary determinant of success is the quality of the movement itself, where perception and subsequent decisions about which movement to make are nearly absent. On the other hand with a **cognitive skill** the nature of the movement is not particularly important but the decisions about which movement to make are critical. In short, a cognitive skill mainly involves selecting what to do, whereas a motor skill mainly involves how to do it.

This dimension, like the others, is really a continuum because there is no completely cognitive skill, nor a completely motor skill. Almost every skill, no matter how cognitive it might seem, requires at least some motor output, and every motor skill requires some preceding decision-making.

Most real-world skills fall somewhere between the polar ends and are complex combinations of decision-making and movement production. Even though most sport skills are weighted quite heavily toward movement, there are strong cognitive and decision-making components.

Recognizing this combination of perceptual and cognitive factors with motor control factors has produced several

Environmental context	Action function			
	Body orientation: stability and no object manipulation	Body orientation: stability and object manipulation	Body orientation: transport and no object manipulation	Body orientation: transport and object manipulation
Stationary regulatory conditions and no inter-trial variability	1 Body stability No object manipulated Stationary regulatory conditions No inter-trial variability	2 Body stability Object manipulated Stationary regulatory conditions No inter-trial variability	3 Body transport No object manipulated Stationary regulatory conditions No inter-trial variability	4 Body stability Object manipulated Stationary regulatory conditions No inter-trial variability
Stationary regulatory conditions and inter-trial variability	5 Body stability No object manipulated Stationary regulatory conditions Inter-trial variability	6 Body stability Object manipulated Stationary regulatory conditions Inter-trial variability	7 Body transport No object manipulated Stationary regulatory conditions Inter-trial variability	8 Body transport Object manipulated Stationary regulatory conditions Inter-trial variability
In motion regulatory conditions and no inter-trial variability	9 Body stability No object manipulated Regulatory conditions in motion No inter-trial variability	10 Body stability Object manipulated Regulatory conditions in motion No inter-trial variability	11 Body transport No object manipulated Regulatory conditions in motion No inter-trial variability	12 Body transport Object manipulated Regulatory conditions in motion No inter-trial variability
In motion regulatory conditions and inter-trial variability	13 Body stability No object manipulated Regulatory conditions in motion Inter-trial variability	14 Body stability Object manipulated Regulatory conditions in motion Inter-trial variability	15 Body transport No object manipulated Regulatory conditions in motion Inter-trial variability	16 Body transport Object manipulated Regulatory conditions in motion Inter-trial variability

△ **Table 16.1** Skill categories defined by Gentile (2000).

additional labels for skills, such as 'perceptual-motor skills' or 'psychomotor skills'.

Learning

The ability to learn is essential to human beings. For all kinds of human performance, whether cognitive, verbal, interpersonal or social, learning occurs. Learning seems to be almost continuous, happening all of the time. In this section, we take a more restricted view of learning, focusing on situations involving practice in sport.

Practice can be defined as a relatively deliberate attempt to improve performance of a particular skill or action

Learning is generally defined as: 'a change in the capability of the individual to perform a skill that must be inferred from a relatively permanent improvement in performance as a result of practice or experience' (Magill, 2004). Therefore it is a set of processes associated with practice or experience leading to relatively permanent changes in the adaptability for skilled performance.

There are several important aspects to this definition of skill learning. According to this definition, learning is due to the effect of practice or experience. But there are many factors in addition to learning that improve the capability for skilled performance, such as the maturation level of the individual, the anxiety-provoking potential of the situation, etc. An example of this would be that as children grow or mature, their capabilities increase. However, as we pointed out before, this can be defined as maturation and not learning. These growth factors are not evidence of learning, as they are not related to practice.

The definition also states that learning is not directly observable, only its products are, and the changes in the processes underlying performance are generally not directly observable either. However, the products of the learning process (i.e. the performance) are directly observable. Therefore, the existence of learning is usually **inferred** from the changes in performance, with evidence about learning processes gained by examining various performance tests.

As defined, learning is indicated only by relatively permanent changes in skill. For an indicated change in skilled performance to be regarded as due to learning, it must be relatively permanent. This is because the changed capability for the performance is then an integral part of the person. Therefore, it is important to understand those practice variables that affect performance in a relatively permanent way. Many different factors affect the momentary level of skill performance, some of which are only temporary and transient in nature. For example, skills can be affected by drugs, sleep loss, moods or stress (or a com-

bination of them all). Most of these variables alter skill only momentarily, with the effects soon disappearing.

Learning versus performance effects of practice

Of vital importance, not only for the world of experimentation but also for the practical world, is the ability to understand the difference between performance and learning – learning versus performance effects. In simple terms, performance is observable behaviour, whereas learning is an internal phenomenon, which therefore cannot be directly observed. It can only be inferred indirectly.

Practice can have two different kinds of influences on performance: one that is relatively permanent and due to learning and another that is only temporary and therefore transient.

Permanent effects (learning)

One product of learning is the establishment of a relatively permanent performance capability, or learning. Therefore practice produces a permanent change in the person's performance that allows the individual to perform a particular action with certainty in the future, and this should endure over many days or even many years.

Temporary effects of practice

Many practice variations have important temporary effects (either positive or negative) as well as the aforementioned permanent ones. These transient effects are a result of simply practising, that is, making the movements necessary to perform the task over and over again.

Other influences

An instructor can generate positive influences by providing praise or instruction. Some effects are positive and contribute to increased performance levels, whereas others are negative and degrade performance somewhat.

Positive effects

- Various kinds of instruction or encouragement elevate performance and are temporary due to a motivating or energizing effect.
- Providing the learner with information about how they have progressed in the task can have an elevating effect on performance.
- Giving guidance in the form of physical help or verbal directions during practice can also alter performance.
- Various mood states can likewise elevate performance temporarily, as can various drugs.

Negative effects

- Practice effects can be negative, degrading performance temporarily. For example, sometimes practice generates physical fatigue; this can depress performance when it is compared with rested conditions.
- Lethargic performances can result if practice is boring or the performers become discouraged at their lack of progress (this is almost opposite to the energizing effect).

The measurement of motor learning

Given that changes in performance due to learning must be relatively permanent, special methods have been developed for measuring learning, Essentially these methods separate relatively permanent changes from temporary changes. In sport, measuring learning and evaluating progress is critical. However, it is not easy to measure a quality that is not directly observable. To make an accurate inference is often very difficult in any situation. In sport there are three general methods for measuring learning:

- Performance curves
- Retention tests
- Transfer tests.

Performance curves

One of the most useful methods of assessing learning is to keep a performance record over a period of time while a new skill is being practised. This record can then be transformed into a **performance curve** (PC) (Figure 16.7). A PC plots the progress made by a person or by groups of people over a specified amount of time. The measures obtained over this period of time can be represented graphically, providing a graph illustrating the performance changes that have taken place. However, this record

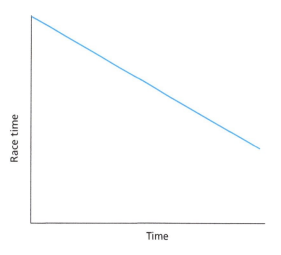
△ **Fig. 16.8** Performance curves.

should consist of a series of measures in performance that you have decided reflects the skill.

PCs can slope upward or downward, depending on whether the measured data increases or decreases with practice and experience. For example, if you are measuring speed of performance, like a hurdler over 100 metres, then the curve would slope down as the hurdler took less time to complete the distance (Figure 16.8). But if you were measuring the hurdler's completed hurdles then the curve should slope up as the errors or the hurdles knocked down decreased with skill.

A typical feature of PCs is that in the beginning there are rapid changes which become more gradual later (Figure 16.11). This general shape of PCs is a fundamental principle of practice. Snoddy (1926) called it the **law of practice**.

Although PCs are the most common way to evaluate learning progress during practice and are useful for charting a learner's progress, they do have certain limitations

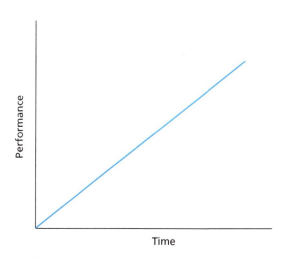
△ **Fig. 16.7** Performance curve.

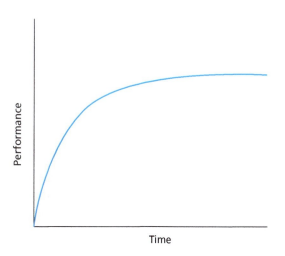
△ **Fig. 16.9** The law of practice.

and several potential difficulties that should be noticed when trying to interpret results from them:

- PCs are plots of performance against trials, they do not necessarily indicate anything about progress in the relatively permanent capability for performance. Therefore they do not show learning *per se*.
- PCs may mask between subject effects.

Note that PCs are not learning curves – these do not exist (**Box 16.1**).

Box 16.1

Learning curve

Learning curve is a commonly used term but the problem is that they do not and cannot exist. This is because we cannot see, measure or otherwise quantify learning. If you want a headache try to think about how you could measure learning to throw a ball better. Certainly you can measure the performance – that is, you could measure how far someone can throw a ball, teach them to throw better and then measure how much further they can now throw it. Can you see that you have not measured learning? All you have measured is how far you have thrown the ball.

You might well infer that now that their behaviour has changed, they have learned something they did not know before. But you cannot measure the learning independent of the behaviour. The opposite is nearly true. I can say that they can throw the ball further without worrying about whether learning has taken place.

Some of you may be thinking that you have learnt something if you can remember how to do it. I know what you mean, but memory and learning are not the same thing. I can 'go rusty' at racing over the winter without having unlearnt how to do it.

When people talk about learning curves the implication is that they have plotted their learning on a graph against time and they believe that they have learned a great deal in a short period of time (the steep learning curve part). I hope you can now see that even if someone believes they have learned much in a short period of time there cannot be a graph that shows this. What they could say is that they can show a real steep 'performance curve', but that implies that they have got better quickly (of course they might have not learned much) or – and I think this is the tricky bit – they have learned lots but have yet to put it into practice. Learning about skiing from a book seldom translates into anything other than injury on the slopes.

One of the main reasons for using performance curves is that they tend to average out the discrepant performances of different learners. By averaging a large group of people together, performance changes in the mythical average subject can be seen and it is hoped inferences can be made about general proficiency changes. The drawback is that this averaging process hides any differences between individuals, the individual differences. Because of this, the averaging method gives the impression that all subjects learn in the same way and that learning is a gradual and continual process.

Retention tests

Another method of inferring learning from performance is to give retention tests. These types of tests are commonly used in education to determine how much you know or have retained from your practice. They are called examinations and are a favourite of students everywhere. An inference is made about how much you have learned on the basis of your test performance.

The most useful way of giving a retention test in the area of motor skill is to administer an appropriate test of skill on the very first day of practice, then to administer the same test again after a period of practice. The difference between these two measures of performance should indicate learning. But to be sure that the performance increase was dependent on learning you should administer the test one final time after practice has ceased, just to see if it is practice that increased the performance and not anything else.

Transfer tests

Transfer tests are sometimes used to measure learning. These involve creating a test situation where the individual must use the skill they have been practising but in a new situation. Again the pre-practice and post-practice test method is used to establish the amount of learning that has taken place. Learning transfer tests usually involve changing the context in which the skill is performed. The performance context can be anything related to the condition of the subjects when they perform (i.e. internal or external – being stressed or tired, on a grass pitch or a clay pitch).

This method of measuring learning is particularly useful with skills that have to be performed under a variety of conditions. Therefore open skills are ones that benefit greatly from this method of measurement.

Stages of learning

As practice continues, certain changes take place in the learner. These changes can be seen in terms of what the learner attends to during the performance. They are evident in the components of the individual's skilled performance, especially in their cognitive processes and their motor movement.

There are several distinct phases or stages that can be identified in all individuals during the process of learning a skill and various approaches have been developed to describe these. The most commonly accepted approach is Fitts and Posner's (1967) three-stage model, often referred to as the **Classic Stages of Learning Model**. Fitts and Posner suggested that the stages are:

- The verbal/cognitive stage
- The motor (or associative) stage
- The autonomous stage.

The verbal/cognitive stage

In the first stage the task is completely new to the learner, making the learner's first problem one of verbalization and cognition. The dominant questions asked by the learner refer to goal identification and performance evaluation (e.g. what to do, what not to do, when to do it).

Verbal and cognitive abilities dominate at this stage, so instructions, demonstrations and other verbal information from the instructor are particularly useful. Breaking down actions and putting the skill into discrete compartments for the learner may also be useful. As learners will probably have to talk themselves through each action, this activity demands a lot of attention and cognitive energy, thus preventing the processing of simultaneous activities, such as overall game strategies.

Performance at this stage is characterized by:

- A large number of errors, with the errors tending to be large
- Jerky and poorly timed movements
- Great variability and little consistency
- Rapid and large gains in proficiency.

The motor (or associative) stage

When most of the cognitive problems have been resolved, the performer enters the motor stage. This stage generally lasts some time longer than the previous one, perhaps for several weeks or months with many sports or even longer if the learner is having difficulty. As many of the basic fundamentals or mechanisms of the skill have to some extent been learned, the learner can now focus on organizing more effective movement patterns to produce the action. Concentration is given to refining the skill.

Several factors change markedly during the motor stage. Enhanced movement efficiency reduces motor energy costs and self-talk becomes less important for the performer, therefore indicating that less cognitive energy is needed. The learner may start to identify environmental regularities to serve as effective cues for timing. Anticipation develops, making movements smoother and less rushed. In addition, learners begin to monitor their own feedback and detect their errors. The ability to detect errors is now there, even though this ability is not perfect it is still apparent at this stage.

Performance at this stage is characterized by:

- Fewer, less drastic errors
- Greater consistency in performance
- More stable and controlled movements
- Rapid improvement in performance.

The autonomous stage

Only after much practice does the learner gradually enter the autonomous stage, involving the development of automatic actions that do not take cognitive or motor energy. This stage brings increased autonomaticity in the sensory analysis of environmental patterns. The decreased attention demand frees the individual to perform higher-order cognitive activities, such as decisions about game strategy. Self-talk about the actual muscular performance is almost lacking, with performance often suffering if self-analysis happens. However, self-talk continues in terms of higher-order strategic aspects, self-confidence increases and the capability to detect one's own errors becomes more developed. Programming longer movement sequences means triggering fewer programmes during the given time, decreasing the loss on attention-demanding response initiation processes.

Here the learner is now able to not only detect their own errors but also to make the proper adjustments to correct them. However, learning is far from over. There are often gains like improved autonomaticity, further reduced physical and mental effort in skill production, improved style and form and many other factors not so directly related to actual physical performance.

The autonomous stage is the result of tremendous practice, but it allows performers to produce a response without having to concentrate on the entire movement.

Performance at this stage is characterized by:

- Few errors and these are not drastic
- Small variability in performance
- Well-developed, stable and controlled movements
- Slow performance improvements as the learner is already very capable when the stage begins.

Individual differences in skilled performance

In the previous sections we focused on the average performance in skill and the general way in which people learn skills. This assumes that all individuals are the same. However, it is obvious this is not the case. Some people seem to perform better at some tasks when first performed than others. This section will take a more differential approach, and examine the factors that make individuals perform skills differently to each other.

Differential psychology is generally concerned with identifying and measuring individual abilities or traits. This is clearly important for coaching and training since it is really useful to be able to tailor 'universal' principles to the needs of the individual.

In an earlier chapter we looked at how individual differences are defined (Chapter 14). In this area of sport psychology the **trait theory** is a common approach. Just to remind you, trait theory says that individual differences are stable, enduring differences among people. In the case of motor skills this is usually in terms of their performance in some task. Individual differences in skill performance are always:

- stable from one attempt to another attempt,
- endure over time (persist), and
- are not necessarily indicated by skill differences on one single trial.

Abilities are thought to be factors that explain skill differences among people. Therefore, it is assumed that skills involved in complex motor activities can be described in terms of the abilities that underlie their performance. Fleishman (1972) defined ability as 'the capacity of the individual that is related to the performance of a variety of tasks'. It is also defined as an 'inherited, relatively enduring stable trait of the individual that underlies or supports various kinds of motor and cognitive activities or skills.

Academics generally define abilities as being genetically determined and largely unmodified by practice or experience. They can be perhaps thought of as the basic equipment you are born with to perform various real-world tasks. This is very much a trait approach.

Ability versus skill

The lay-person often uses the terms 'ability' and 'skill' interchangeably. This is an incorrect use. It is important to know the distinction between them. We define a **skill** as one's proficiency at a particular task such as shooting at the goal in football. Skills are:

- Easily modified by practice
- Countless in number
- Represent the particular capability to perform a particular activity.

Abilities are:

- Genetically defined
- Essentially unmodified by practice or experience
- Trait like.

It might be best to think of abilities as factors that limit performance. It has been suggested that the appropriate abilities for a task limit the level of performance that a particular individual can eventually attain. For example, a colour-blind individual will have difficulty performing a task or being proficient in a task that relies on colour identification. And at 5 foot 7 inches (1.70 m) tall I was unlikely to make it as a professional basketball player (not to mention that there are no wheels in basketball).

However, it should be kept in mind that all novices generally perform poorly, and this does not necessarily mean that all novices do not possess the proper abilities for that particular task. Remember that most tasks can be performed more effectively with practice. It would be foolish to judge someone's abilities in a task when they have not moved out of the motor stage of learning. This is because several factors can change because of practice to improve performance.

As pointed out previously, beginners have to invest large quantities of cognitive energy in a task at first. For example, they have to decide what to do, remember what comes after what and try to figure out the instructions, rules, task scoring, etc. With a little practice, the performer learns the intellectual or cognitive parts of the task, and the cognitive abilities are replaced by more motor-orientated abilities, perhaps related to limb movement. Hence the abilities needed in the earlier stages of learning a task or in the cognitive stage are very different from those needed in the motor stage, where the manual, physical qualities are more important.

The differences between ability and skill are summarized in Table 16.2.

Abilities	Skills
Are inherited traits	Are developed with practice
Stable and enduring	Modified with practice
Perhaps 50 or more in number	Infinite in number
Underlie many skills	Depend on several abilities

△ **Table 16.2** Important differences between abilities and skills.

Motor ability

As mentioned previously, differential psychology is generally concerned with identifying and measuring individual abilities or traits. The motor and cognitive abilities identified have been varied and are considerably different from each other. Fleishman (1972) was one of the first to develop a taxonomy or classification system of human perceptual motor abilities. He tried to identify as many abilities as possible in the fewest ability categories, which relate to performing the widest variety of tasks.

Here is a list of abilities he identified:

- Multi-limb coordination
- Control precision
- Response orientation
- Speed of arm movement
- Rate control
- Finger dexterity
- Arm–hand steadiness
- Aiming
- Reaction time
- Manual dexterity
- Wrist, finger speed.

Fleishman assumed that particular tasks may use or rely on a specific subset of abilities for performance. For example, playing snooker may need arm–hand steadiness and visual acuity, whereas playing netball may need multi-limb coordination. From this perspective people who are particularly good at a specific skill are highly proficient in the abilities relating to that task.

The structure of motor abilities

There have been several approaches to the structure of motor ability. One of the more traditional views suggests that the structure of human abilities consists of one single factor: A general factor or g-factor. In cognitive or intellectual functioning a g-factor has been associated with intelligence. In motor research this single factor has been termed **general motor ability.**

The general motor ability notion assumes that a single inherited motor ability underlies all movement or sports tasks and that strong general motor abilities make people effective at nearly any motor task attempt.

Support for this notion has come from several proponents of this view who assume that a single factor or ability accounts for all skilled performance in the individual. McCloy (1934) developed the General Motor Capacity Test as one of his general motor ability tests. He believed

that motor capacity consists of a person's inherited potentialities for general motor performance. Brace (1927) also made attempts to provide tests that would measure general motor ability. These consisted of tests that measured whole body actions, which would presumably measure the general capability to perform athletic tasks.

In recent times the general-motor-ability hypothesis has been heavily criticized and is no longer widely accepted. This is because there has been little research which has supported it and correlations among different skill performances are generally low. It has been shown that even skill performances that appear quite similar in structure usually correlate poorly. Also, there has been little support for the predictive value of the general motor ability. Individuals who have good general motor ability are not necessarily found to be proficient at all motor skills.

I think it is easy to see why the myth that some people are just 'good at sport' came about. It may be true that some people seem to be better at sport in general than others who are generally not so adept. Put another way, some people seem to be useless at every sport they try. I think that this is a bit of a self-fulfilling prophecy, though. Once you have convinced yourself (or have been convinced) that you are good or bad at sport you either pursue it and get better or avoid it and get worse (**Box 16.2**).

Box 16.2

Self-fulfilling prophecy

The idea of the self-fulfilling prophecy comes from Irving Goffman (1968), who was the first to crystallise a notion that I'm sure many had thought about before. It is the idea that if you treat someone like an idiot they'll behave like an idiot. Think like a failure, be a failure. Hate statistics be no good at statistics. Goffman's point is that we make judgements about ourselves or allow others to make judgements about ourselves. We can either move away from them so that it becomes clear that the judgement was wrong or come to believe the judgement to be true and act accordingly.

For Goffman, fulfilling the prophecy (especially if it is negative) is no evidence that the prophecy was correct. He believes that we can make it come true (or that society or institutions can make it come true) but it is not inevitable. So finding out we are poor at football and cricket or netball and hockey (because that is what your school plays) does not mean that you are rubbish at sport and had better learn to use a computer. It might just mean that you have yet to play tennis, or go bowling or learn to ride a horse.

Few people are really good at one sport, let alone several (Figure 16.10). There are many skills which transfer across sports, for example being very fast at running is useful for the 100 metres and the long jump. Good eye–hand coordination is useful for all racket sports but there have been few, if any, players who have been world champions at the same time in, for example, squash, tennis and badminton. Many people are 'half handy' at a diversity of sports but they are not competing at a very high level. They probably have some skills that are useful for a variety of sports (like being able to run fast) which means they 'get on okay' with a number of sports but they do not excel at them. I expect you are thinking that this is a lot of old rubbish and you can think of lots of players who are good at lots of sports. Of course there are always exceptions but these are so far at the ends of an already not very normal, normal distribution that I wonder if we can say much more than 'we are dealing with some very special people here'.

△ **Fig. 16.10** Mike Hailwood was one of very few sportmen to excel at an international level at more than one sport, in his case motorcycling and motor racing.

A similar concept to that of general capacity to perform is that of a generalized ability to learn new skills. Brace (1927) called this **motor educability**, a term that was then popularized by McCloy (1934). Both authors used this term to refer to the ease with which an individual learns new motor skills. This again was thought to represent some general ability to learn new skills. However, like the concept of general motor ability, this has been heavily criticized and many of the criticisms of the 'g-factor' apply here.

Motor skill development

Motor development does not stop in infancy and early childhood. As children develop, they continue to acquire new skills that permit more sophisticated and complex motor feats. One of the arenas in which motor skills are

most important is in children's sports. Between the ages of 6 and 18, both girls and boys participate extensively in a wide range of sports. Participation in sports not only provides opportunities for refinement of motor skills but also provides a continuing opportunity to learn new social skills. Participation in sports is determined by a variety of social and cognitive factors and not just motor skills alone.

As in many domains of development, cognitive, social and physical aspects often operate together. Children's motor skills are no exception. The interplay is nicely illustrated by the curious finding that at 12 there is a sharp increase in the drop-out rate of children from organized sports programmes. In the modern parlance this seems to be the age for some children when it stops being 'cool' to play sport.

Phases of motor development

The development of a child's motor behaviour follows a sequential process, starting with very simple reflexes and terminating with very complex motor skills. Generally motor behaviour will move from reflexes to the learning of postural movements to transport locomotor movements and finally to manipulative movements. It should be pointed out that these behaviours increase simultaneously with the development of motor control, which develops in a cephalocaudal direction (head to feet) and also in a proximodistal direction (midline to extremity) but also with the development of perceptual and cognitive skills.

Using the four stages of development as a guideline:

- infancy (0–2 years),
- early childhood (2–5 years),
- late childhood (5–10 years), and
- adolescence (10–18 years)

one could roughly say that from infancy to adolescence children will move from reflexive responses to specialized skills.

Development of motor skills in infants (0–2 years)

During infancy, children move from reflexive responses to rudimentary movement abilities. These rudimentary behaviours, which include sitting, crawling, standing and walking, essentially form the foundation for the development of other fundamental abilities.

Reflexes

The first movements that can be elicited from newborn infants are involuntary responses that we call reflexes. Although reflexes are primitive motor responses, they are important for the early survival of the child. The most obvious of the survival reflexes are sucking and rooting.

Reflexes are not only important for the survival of infants but they also serve as an indicator of the maturity and soundness of the infant nervous system. Even though reflexes demonstrate whether there is normal or abnormal development they have no useful predictive value for later motor and cognitive skills. The early survival reflexes reflect the relatively undeveloped nervous system. As the nervous system matures, reflexes change from those solely under spinal control to those under brainstem control and others under midbrain control. Finally, as the highest nervous system centre matures, many of these transient reflexes are gradually inhibited. Voluntary responses then take over.

Development of voluntary movement

As the infant's nervous system matures, many of the reflexes give way to voluntary motor responses. These responses, termed **rudimentary**, are activities such as rolling over, sitting, crawling and finally walking. A number of detailed studies have been conducted on the normal development of these early motor skills (Gessell and Ilg, 1940). From these we know that although there are individual differences and hence variation in the age at which these skills are acquired, they always progress in a sequential manner and are therefore predictable. For example, rolling over will always occur before sitting and sitting will occur before crawling and creeping. The remarkable achievement of the development of posture and locomotion in infants was plotted by Shirley (1933) (Table 16.3).

Months	Movement
0	Foetal posture
1	Chin up
2	Chest up
3	Reach and miss
4	Sit up with support
5	Sit on lap
6	Sit in high chair can grasp
7	Sit alone
8	Stand with help
9	Stand holding furniture
10	Crawl
11	Walk when led
12	Pull to stand by furniture
13	Climb stairs
14	Walk alone
15	Stand alone

△ **Table 16.3** Sequence of motor development in locomotion.

One of the important implications of these motor achievements is the increasing degree of independence that children gain. They can explore their environment more freely. You can always tell when a baby has started to get around because when you go round to visit all breakable objects have migrated more than a metre off the floor! As my friend Ted says, the trouble with children is you spend hours teaching them to walk and then spend the rest of their childhood telling them to sit down.

The invariant order or acquisition of voluntary motor skills is important to the tracking of development. Enough research has been done in this area so that we can predict when the ability and skill should show itself. When it does not this enables us to spot a potential problem. For example, if a child cannot sit on one's lap by seven months we begin to wonder why not. It may, of course, be nothing to worry about but it may also be an indicator of a developmental problem, an illness or disability or even neglect.

There are variations in the age at which walking begins. Some of these differences seem to be attributable to experimental or cultural factors and some are based on biologically determined individual differences. For example, it has been shown that children raised in different European cities show variations in the average age of onset of walking. French children tend to walk slightly earlier. Even more striking are the wide differences in individual children. Although neither gender nor social class can account for such variations, it is possible that nutritional and environmental factors may contribute to these differences. The available data indicate that although the general limits to motor development may be set by a maturational pattern, the timing of the emergence of motor skills may be either enhanced or slowed by particular environmental factors.

Cross-cultural evidence indicates that variations in opportunities for practice can have an effect on motor skills. Goldberg (1972) demonstrated that Zambian infants, like many African infants, show early development of motor skills. These children are carried everywhere in a sling until they can sit, then they are allowed to sit for considerable amounts of time, thus having a lot of time to practise emerging motor skills. In comparison, Mexican infants, who are carried very tightly and even have their faces covered for the first three months, show a lag in motor development (Brazelton, 1972). According to Brazelton the mother's aim is a quiet infant, and so she tries to anticipate the infant's needs.

Development of manipulative skills

Although an infant sees an object at a very young age, reaching movements toward that object do not take place until about four months. Initially these movements are slow and awkward. But by the sixth month the infant is capable of accurate hand movements and can grasp objects. (This is not to say that an infant will not grasp an object before the age of four months, but in younger infants it will usually be an involuntary response.) Grasping moves from crude palm grabbing to pincer type grasps utilizing the thumb and forefinger, making hair and glasses a favourite target, which illustrates proximodistal development.

Since grasping appears to be such a difficult skill to develop, it seems that releasing should be easy. However, this is not the case. Most of us have undoubtedly observed a frustrated infant attempt to release a toy without success, often leading to tears! This occurs because the child has not reached a level of maturation able to consciously relax the musculature of the hand so that the object will be released. By the age of 18 months most children have mastered the skill of voluntary release.

Early childhood and the development of general fundamental skills (2–5 years)

During early childhood children develop the ability to run, jump, balance, catch and throw (Figure 16.11). These are commonly labelled general fundamental skills, since they are characterized by being common to all children and are necessary for ordinary survival. There will, however, be large individual differences in a child's ability to perform these fundamental skills.

After two years of age the child begins to gain proficiency in locomotion such that running, jumping, hopping, galloping and skipping become part of their locomotor repertoire.

△ **Fig. 16.11** A child learning to run is at a very important developmental stage.

With respect to individual differences it should be pointed out that the order in which rudimentary and fundamental skills normally develop will be the same for all children – only the rate at which these skills develop will vary.

Some time after a child becomes proficient in walking, running develops. The initial pattern resembles a hurried walk. This is not a genuine run, since the child does not have sufficient balance or leg strength to allow both feet to leave the ground momentarily. Some time between the ages of two and three years the child will demonstrate true running. By the age of four to five years a smooth coordinated running pattern will be exhibited. For this age-level researchers are typically not so much concerned with performance time in running as they are with the pattern of running.

Late childhood and the development of more specific motor skills (5–10 years)

During late childhood more specific movement skills will appear in the child's repertoire of movement abilities. The early general fundamental skills become further refined and hence will appear as more fluid and automatic. More emphasis is placed on form, accuracy and adaptability and the child can now apply these skills to sports performance. One noticeable feature of children as they move from early to late childhood is the increase in the rate of physical growth. Although growth still occurs, it is not so dramatic as it was during the first 5 years. Motor skills continue to improve, however, particularly the fundamental running, jumping and ball skills, Also during the period of late childhood children begin to experience and perform different motor activities because they are exposed to different daily interactions and environmental demands.

Motor skills thus become more specialized. The acquisition of these specialized skills and the further development of fundamental skills allow the child to participate in games and sports that call for utilization and hence development of unique motor skills. For example, children now convert their running and jumping skills, their throwing and catching to basketball playing. Sports require specific motor abilities.

Adolescence and the development of specialized motor skills

As the adolescent's specific abilities develop they can be called specialized. This process evolves slowly from late childhood through late adolescence and really depends on the amount of practice an individual has with specific skills. Adolescence is a very trying period physically, psychologically and socially for girls and boys. Sometime between the ages of 10 and 18 years (earlier for girls) significant changes take place within their bodies and they perceive themselves as neither children nor adults. Motor skills learned during childhood continue to improve for most boys through adolescence, but this is not the case for girls. For some reason, girls peak at about age 14 then either level off or decline in their performance of motor skills. Gender differences therefore become more apparent during adolescence.

Adolescence also brings the learning of more specialized motor skills because of a variety of increased novel experiences, many of them in games and sports. Those children turned on by these experiences continue to improve in these skills. Other children do not enjoy these activities and as a result their performance measures do not improve. This is probably what has happened to many girls, hence their premature peaking in motor skill performance.

Gross to fine development

It could also be said that skills develop not only on a continuum from rudimentary to specialized, but also along the continuum from gross to fine. Gross motor skills incorporate large muscles, usually several muscle groups of the body. Examples of gross motor skills include running, jumping, balancing and throwing. Fine motor skills involve small muscles and limited activities of the body extremities. Examples of fine motor skills are threading a needle, typing and hand writing.

Physical activity, sport and the implications for the development of motor skills

Our present family-rearing practices and educational practices tend to give the impression that motor skills, if left alone, will somehow develop by themselves. Motor skill learning and development should not be left to chance. Do we leave cognitive skills to chance?

Zaichkowsky (1974) argues that motor skill learning is essential for all children for a variety of reasons. As pointed out earlier, motor skill development is intricately related to cognitive development and affective or social-psychological development, and certainly these domains of behaviour do not develop in a vacuum independent of each other.

The work of Piaget (1972) gives powerful evidence supporting the need for motor exploration in cognitive development. Positive childhood experiences with a variety of motor activities will also serve to enhance self-concept. Feeling good about one's ability in motor skills will do a

lot to encourage further participation in physical activity and sport, thereby increasing the chances of developing an individual who is physically and psychologically fit.

Motor skill development is essential in its own right for simply surviving. If fundamental motor skills such as walking, running and balancing were not adequately developed, we would have a great deal of trouble avoiding being hit by cars while crossing a street.

The neonate progresses from uncontrolled and non-adaptive reflexive and spontaneous behaviours to controlled accurate and adaptive behaviours. These later behaviours must include skills of the type seen in ball games, gymnastics and the shaping of physical surrounding, but also the subtle movements used in the advanced communication of thoughts and emotions.

The progressive refinement in our capability to mobilize and control our motor system almost certainly continues over the first 20 years, and there are changes throughout life in the way we employ this system which we are mastering, to both adapt and control our environment. There is also development of progressively more abstract knowledge of our environment and our relation to it. We build knowledge or skills on the basis of active interactions with our environment.

The term 'behaviour' is linked to motor, whereas the term 'skill' is reserved for perception, cognition and meta-cognition. This use of terms is meant to highlight the important fact that a motor skill cannot exist in isolation from perceptive and cognitive skills. One generalization springs from this view of development. That is, the basis of behaviour becomes broader and broader as development progresses, and the older the individual the greater the number of perceptual and cognitive factors that can potentially affect what, how and when he or she does it. By the time a child or an adult can plan and control behaviour that is adaptive to specific sets of circumstances the movement he or she generates is controlled from a broad base of perceptual and cognitive skills, which is not to say that the capability to generate a reaction is lost.

Gender differences

As mentioned earlier, abilities are thought to explain the differences in skill performance and they are thought to be genetically determined. One identifiable genetic difference which is thought to explain skill differences is gender. The relative importance of the genetic difference between the gender in the performance of motor skill is perhaps best seen when taking a developmental perspective, as this may uncover whether males/females are born with different abilities for skill performance.

Infancy

During the infant stage of development few significant gender differences in motor development exist, even though we know that in most cases girls have a nervous system that is more mature than boys.

Early childhood

During early childhood it appears that the socialization process has begun to make an impact on motor skill acquisition. At this time we see emerging gender differences in hopping, skipping and galloping.

Girls are more proficient at hopping and skipping – however, they do this in their play more often than boys. It also appears that girls perform better than boys in tests of static balance, although studies are somewhat equivocal here.

Boys excel in galloping, a task they practise more than girls. Boys tend to perform more skilfully in tests of throwing in both form and distance.

Late childhood

The late childhood years show larger gender differences in a variety of tasks. The performance of boys exceeds that of girls in numerous tests of running speed, jumping ability, throwing ability and some tests of balance.

Girls are more flexible on numerous measures. They also appear to do better in tasks requiring fine motor coordination.

Boys consistently do better on tests of strength, but differences are slight compared with what they will be during adolescence.

Adolescence

Adolescence sees even greater gender differences than those observed during childhood years. A consistent pattern that emerges in the research literature shows boys improving in performance on most motor skills through adolescence, whereas girls peak at about age 14 then either level off or decrease in performance.

Girls are generally thought to do better at tasks that require flexibility and fine motor coordination (Keogh, 1981; Zaichkowsky, 1974).

Boys are generally thought to do better on strength-type gross motor tasks.

The literature on sex differences generally offers four different variables to explain the differences in performance in boys and girls. It is important to point out that these are far from uncontested. The four are:

- **Body size** – According to Newsholme *et al.* (1994) women are generally smaller than men

- **Anatomical structure** – According to Newsholme *et al.* (1994) the wider pelvis of a woman decreases mechanical efficiency in running by increasing the length of the thigh bone to bring the knees closer together. For the same body weight women possess about 10 per cent more fat than men, which increases the weight to be carried. In women the Achilles tendon (important in the elastic recoil of running) is shorter.

- **Physiological functioning** – According to Newsholme *et al.* (1994) cardiac output in women is about 10 per cent less than in a man with the same body size – due to a smaller heart size. For the same volume of blood women have about 10 per cent less haemoglobin and about 20 per cent fewer blood cells than men.

- **Social and cultural factors** – Many people would say that the previous three points are not as important as they sound. For example, Cashmore (2000) points out that women are 'able to hold their own in virtually every sporting match up in which raw physical strength is not the sole determining factor'. Many researchers (for example Hargreaves, 1986; Cashmore 2000) think that the reason boys and girls are different is that they are brought up differently, are encouraged to do different things, and win the praise of their parents by behaving in gender-specific ways. For example girls are praised for being pretty whilst boys are encouraged to be tough.

● Summary

Skills

- Knowledge about skills has come from a variety of scientific disciplines and can be applied to many settings, such as sport, physical education, industry and physical therapy.

- A skill is usually defined as the capability to bring about an end result with maximum certainty and minimum time and energy.

- Motor skills have been defined as skills in which physical movement is required to accomplish the goal of the task.

- In skilled performance many different components are involved. These include perceptual or sensory processes, decision-making and movement output.

- Skills may be classified along numerous dimensions, such as: open and closed, discrete, continuous and serial, motor and cognitive. These classifications are important because the principles of skill and their learning often differ for different categories.

- It is important to remember that these are just descriptive theories – they really explain very little, and often much of what we deal with fits poorly in to these categories, the usual solution to this is to add ever more categories until 'it' (the classification system) 'works'.

Learning

- Learning produces an acquired capability for skilled performance.

- It results from practice or experience, is not directly observable, is inferred from performance changes and produces relatively permanent, not transitory changes.

- Performance curves are plots of individual or average performance against practice trials. They can either increase or decrease with practice, depending on the particular way the task is scored.

- The law of practice suggests that improvements are rapid at first and much slower, later. This is a nearly universal principle of practice.

Individual differences in skilled performance

- Individual differences in skilled performance are always stable from one attempt to another attempt,

endure over time (persist) and are not necessarily indicated by skill differences on one single trial.

- Skills are easily modified by practice, countless in number, and represent the particular capability to perform a particular activity.

- Abilities are genetically defined, essentially unmodified by practice or experience and trait-like.

Motor ability

- Whilst the idea that there is some kind of general motor ability to learn is a seductive one there is little evidence to support it.

Motor skill development

- Motor skill development always follows an invariant order or pattern, so we literally cannot run before we walk.

- Cognitive, social and physical aspects often operate together.

- This development is in a cephalocaudal direction (head to feet) and also in a proximodistal direction (midline to extremity).

- We are usually born with reflexes which are then further developed. This leads to voluntary movement and increased independence.

- There are some cross-culture differences in the rates of change but not in the order.

- We develop manipulative skills which develop into more specialized skills.

Gender differences

- Gender differences have been demonstrated in the performance of motor skills.

- These can be seen through the stages of motor development.

- In pre-pubertal children the physical differences are small but these differences seem to increase after puberty.

- Some of these differences can be explained by physical differences but some researchers feel that these are not as important as social and cultural factors, which tend to treat the genders differently to the detriment of girls' sporting achievements.

Review Questions

Skills

1. What are the four main skills classifications?
2. What are the main criticisms of these classifications?

Learning

1. Why is there no such thing as a learning curve?
2. Explain what the different kinds of performance curve describe.
3. What is the difference between practice and learning?
4. What are the key components of the definition of learning?
5. What are the effects of learning?
6. Describe the methods we can use to infer that learning has taken place.
7. Describe the three phases of learning.

Individual differences in skilled performance

1. What is the difference between an ability and a skill?
2. Describe what is meant by an ability.

Motor ability

1. What are the major criticisms of general motor ability or general motor ability to learn?

Motor skill development

1. What are the key components of motor skill development?

Gender differences

1. What are the physical and performance differences between boys and girls?
2. What are the four main reasons for these differences?

References

Brace DK (1927) *Measuring Motor Ability.* New York, Barnes AS, and Co.

Brazelton TB (1972) Implications of infant development among the Mayan Indians of Mexico. *Human Development* 15(2): 90–111.

Cashmore E (2000) *Making Sense of Sports.* London, Routledge.

Fitts PM, Posner MI (1967) *Human Performance.* Belmont, CA, Brooks/Cole.

Fleishman EA (1972) On the relationship between abilities, learning, and human performance. *The American Psychologist* 27: 1017–1032.

Gentile AM (1972) A working model of skill acquisition with application to teaching. *Quest* 17: 3–23.

Gentile AM (2000) Skill acquisition: action, movement, and neuromotor processses. In Carr JH, Shepherd RB *Movement Science: Foundations for Physical Therapy,* 2nd edn., Rockville MD, Aspen: 111–187.

Gesell A, Ilg FL (1940) *The First Five Years of Life; A Guide to The Study of the Preschool Child.* New York, Harper.

Goffman E (1968) *Stigma: Notes on the Management of Spoilt Identity.* Harmondsworth, Penguin.

Goldberg DP (1972) *The Detection of Psychiatric Illness by Questionnaire.* London, Oxford University Press.

Guthrie ER (1952) *The Psychology of Learning.* New York, Harper and Row.

Hargreaves J (1986) *Sport, Power and Culture.* Oxford, Polity Press.

Keogh J (1981) Motor development. *Perspectives on The Academic Discipline of Physical Education.* Brooks GA. Champaign, Il, Human Kinetics.

Magill, RA (2004) *Motor Learning and Control.* New York, McGraw-Hill.

McCloy CH (1934) The measurement of general motor capacity and general motor ability. *Research Quarterly* (supplement) 5: 46–61.

Newsholme E, Leech T, Duester G (1994) *Keep on Running.* Chichester, Wiley.

Piaget J (1972) Intellectual evolution from adolescence to adulthood. *Human Development* 15: 1–21.

Poulton EC (1957) On prediction of skilled movements. *Psychological Bulletin* 54: 467–478.

Shirley MM (1933) *The First Two Years: A Study of Twenty-Five Babies.* Minneapolis, University of Minnesota Press.

Snoddy GS (1926) Learning and stability: a psychophysical analysis of a case of motor learning with clinical applications. *Journal of Applied Psychology* 10: 1–36.

Zaichkowsky LD (1974) The development of motor sequencing ability. *Journal of Motor Behavior* 6: 230–235.

Further reading

Marshall EL (1937) A comparison of four current methods of estimating physical status. *Child-Development* 8: 89–92.

Zaichkowsky LD, Fuchs CZ (1986) *The Psychology of Motor Behavior.* Ithaca, New York, Mouvement.

Index

Entries for illustrations and figures are in italics, e.g. *8*. Entries for tables are identified with a lower-case t after the page reference, e.g. 351t.

3-C (4-C) models, body composition 125
abdominal muscles *41*, *46*
abilities vs. skills 356–8
abnormality, definition of 325–6
acceleration 181–7, *194*, 195, 204–8, 213–14
   angular acceleration 237, 241–3
acceleration-time graphs *186*
accelerometry 136, 295
acetabulum 40
acetaldehyde 85
acetyl-CoA carboxylase 20, *101*, 102
achievement motivation 318
Achilles tendon (calcaneal tendon) *41*, 44
acid-base homeostasis 14–15, 17, 18, 19, 102–8
   oxyhaemoglobin association/dissociation curve 73–4
acini (pancreatic cells) 81
acromial process 39, *40*, *47*
actin 34–5, 95, 96, 258
action potentials (muscle) 91–2, 93–5
   cardiac muscle 67–9
activity theory 338
adductor muscles 41–2, 43
adenosine diphosphate (ADP) 99–100, 106
adenosine triphosphate (ATP) 20–1, 35–6, 95, 99–110
adipocytes *30*
adipose tissue 29, 31, 37, 121–6, 137, 152, 267
adolescents 359–61, 362
   mental health 329–30
Adolph, E A 139
ADP (adenosine diphosphate) 99–100, 106
adrenal glands *60*, 62
adrenaline 62
adrenergic neurons 58
Ae, M 214
aerodynamics 272–7
afferent nerves 52, 55, 59–60
ageing
   and body fat 121, 122
   and cardiovascular disease 98
   diabetes 63
   and energy requirements 137
   loss of bone mass 32
   macular degeneration 158
   muscles 113
   and retirement 338–40
aggression 306–8
agonist muscles 39

agranulocytes 77
air displacement plethysmography 124
air resistance 195, 267–8, 272–7
airborne phase (running) 189
airways (respiratory system) 70–1
Albrecht, R 338
alcohol abuse 150, 153–6
alcohol, metabolism of 79–80, 84–5, 127, 128
Alewaeters, K 166
alpha-actinins (ACTN) 96–7
alveoli 71
American football 307, 336–7
amino acids 7–8, 16–18, 107–9
ammonia 85, 99–100
AMP deaminase 20
amphetamines 158
amyl nitrates 159
anabolism 21, 99
anaemia, treatment of 65
anaerobic testing 193
anatomy 25–48. *see also* biomechanics
   anatomical position 27, *28*, 173
   gender differences 362
anconeus *47*, *48*
angiotensin-converting enzyme (ACE) 87
angle of attack 274, 275, 276
angular kinematics 234–44
angular kinetics 247–64
angular motion, definition 172, 173
animal experiments 22
anions 13
ankle joint 43, *45*, 187, 294
annulus fibrosus *56*
antagonistic muscle action 39
anterior cruciate ligament (ACL) 42, *43*
anterior superior iliac spine *40*
anthropometric measurements 25–6, 31, 32, 119–25
anthropometry, definition 166
anti-inflammatory drugs 54
antibodies 78–9
antidiuretic hormone (ADH) 61, 86, 87
antioxidants 152–3
anus *80*, 82
anxiety (performance anxiety) 302–6, 337–8
appendicular skeleton 39–40
appendix 75
Apter, MJ 304
Arabatzi, F 169
archaea 5, 9
Archimedes' principle 266
areolar connective tissue 32
arginine *16*, 18
Ariel, BG 263
Aristotle 166
arousal 303–4, 307–8, 309–10
Ashton, MK 303
associative stage, learning 355
asthma 71–2

asymmetric loading 214–15
Atchley, RC 339
athletes
   identity 340
   training diets 118–19, 139
athletics. *see also* high-intensity exercise
   doping 65, 72, 75
   field events 193, 199, 258–9, 275–6
   inter-species comparisons 9
   world records 4, 131t, 207, 335
Atkinson, J W 318
atoms 13–14
ATP (adenosine triphosphate) 20–1, 35–6, 95, 99–110
atria (heart) 63–4, *65*, 67–8
atrial natriuretic peptide (ANP) 86
atrioventricular node (heart) 67–8
Atwater chambers 134
audiences 307, 308–10
auditory system 60
autocrine actions, definition 60
autonomic nervous system 55–8
autonomous stage, learning 355
Avener, M 315–16
axes of rotation (anatomical reference axes) 174

babies, developmental stages 358–60
bacteria 5, 9
balance, definition 216, 222
balance (equilibrium) 60
Baldwin, MK 329
ball and socket joints 39, 40–2, 187
Baltzopoulos, V 291, 292
Bandini, L 152
Bandura, A 334, 335, 336
Bannister, Roger 335
Baron, RA 310
Bartlett, R 231, 258, 275, 283
Bartonietz, KE 242
basal metabolic rate (BMR) 132–3, 137–8
base molecules (DNA) 6
basement membrane 30
BASES (British Association of Sports Sciences) 282
basketball 294
basophils 77
Basu, S 157
batting (baseball/softball) 238
Batty, D 149
Baumeister, RF 310
Beamon, Bob 335
Bell's palsy 54
Bengston, VL 339
Benson, H 305
Berger, SM 310
Bergstrom, Jonas 27, 112, 132
Bernouilli effect 272, 274
Berthoin, S 193
beta-oxidation 107
biacromial distance 39, 121

Acknowledgements

The publisher and authors would like to thank the following for their permission to reproduce the following images:

1.1 (a), The Granger Collection/Topfoto; 1.8 (b), Professors P. Motta and T. Naguro/Science Photo Library; image from Harvey's *Exercito anatomica de motus et sanguinis in animalibus* on page 22, Interfoto/Alamy; 2.1 (a), Zephyr/Science Photo Library; 2.1 (b), Dean Sewell; (a) on page 33, Dean Sewell; (b) on page 33, Dean Sewell; 2.7 (a), Dr Robert Calentine, Visuals Unlimited/Science Photo Library; 2.7 (b), Dr Gladden Willis, Visuals Unlimited/ Science Photo Library; 2.7 (c) Manfred Kage/Science Photo Library; 4.2, Churchill Livingstone/Elsevier; 4.3 (a) & (b), Dr Michael Brooke/Bailliere Tindall; 4.4, Churchill Livingstone/Elsevier; 4.5, Churchill Livingstone/Elsevier; 4.13, Professor Mathias Gautel; 5.3, Topfoto/Image Works; 6.2, © Macduff Everton/Corbis; 7.1 (d), Gerard Julien/ AFP/Getty Images; 7.2, Action Plus Sports; 8.1 (a), Andreas Solano/AFP/Getty Images; 8.1 (b), Nick Ansell/PA Archive/Press Association Images; 8.1 (c), © Gary Kufner/Corbis; 8.2 (a) & (b), Philip Watkins; 8.17 (a) – (e), Philip Watkins; 8.28 (a) – (c), Action Plus Sports; 9.5, Antonin Thuillier/AFP/Getty Images; 9.6, Adrian Dennis/AFP/Getty Images; 9.8, Action Plus Sports; 9.27 (b), Action Plus Sports; 10.8 (a) – (c), Philip Watkins: 10.14, Glyn Kirk/Action Plus; 11.23, Michael Sohn/AP/Press Association Images; 12.4, John Giles/PA Archive/Press Association Images; 14.1, Inspirestock Inc./Alamy; 14.4, Michael Chamberlin – Fotolia; 14.5, joreks – Fotolia; 14.7, Xof711 – Fotolia; 14.8, © Roberto Caucino/iStockphoto.com; 14.9, Marshall Ulrich; 15.1, Topfoto/UPP; 15.2, redsnapper/Alamy; 15.4, Topfoto/ImageWorks; 15.5 (a), © Corbis; 15.5 (b), © 1997 Suza Scalora/Photodisc/Getty Images; 15.6, Bill Dobbins/ Rex Features; 15.8 (a) & (b), Topfoto.co.uk; 16.1, Mike Hewitt/Allsport/Getty Images; 16.2, Eisenhans – Fotolia; 16.4, © biker3 – Fotolia; 16.5, © Imagestate Media (John Foxx); 16.6, Galina Barskaya – Fotolia; 16.10, National Motor Museum/HIP/Topfoto; 16.11, Soleilc1 – Fotolia.

Every effort has been made to trace copyright holders. Any omissions brought to our attention will be corrected in any future printings.

9781138128989

#0119 - 041016 - C0 - 280/208/21 [23] - CB - 9781138128989